WITHDRAWN

BIBLIOGRAPHY AND INDEX OF PALEOZOIC CRINOIDS
1942-1968

The Geological Society of America, Inc.
Memoir 137

Bibliography and Index of Paleozoic Crinoids 1942-1968

G. D. WEBSTER
Department of Geology
Washington State University
Pullman, Washington 99163

1973

Copyright 1973 by The Geological Society of America, Inc.
Library of Congress Catalog Card Number 73-76885
I.S.B.N. 0-8137-1137-1

550.62
G345m
no. 137
1973

Published by
THE GEOLOGICAL SOCIETY OF AMERICA, INC.
3300 Penrose Place
Boulder, Colorado 80301

Printed in The United States of America

*The printing of this volume
has been made possible through the bequest of
Richard Alexander Fullerton Penrose, Jr.*

Contents

Introduction ix
Bibliography 1
Part I. Crowns and parts of crowns 29
    Section 1. Identified crowns and parts of crowns 29
    Section 2. Unidentified crowns and parts of crowns 271
Part II. Columnals 277
    Section 1. Identified columnals 277
    Section 2. Unidentified columnals 325
Appendix. New genera introduced since 1942 335

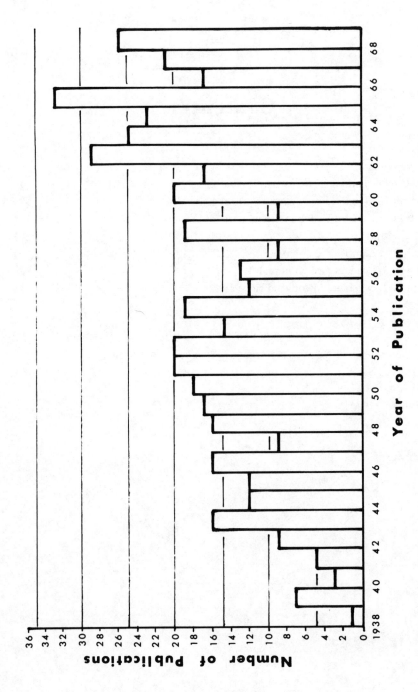

Figure 1. Graph showing the number of crinoid publications per year for the period 1938 through 1968. Note that part of the publications for 1938 through 1942 were included in Bassler and Moodey (1943) and are not repeated here.

Introduction

The monumental *Bibliographic and Faunal Index of Paleozoic Pelmatozoan Echinoderms* was initially conceived and data compilation started by Frank Springer, who died in 1927. Some of the tabulation of data was done by his daughter, Ada Springer Davis, and the bibliography was subsequently completed and published as *Bibliographic and Faunal Index of Paleozoic Pelmatozoan Echinoderms,* by Bassler and Moodey, in 1943. This tome consisted of a bibliographic and faunal index, as well as a systematic revision of the generic allocations of many species, although no supporting reasons for any of the changes were given. The uncounted hours of tedious literature searching which have been saved by investigators of Paleozoic echinoderms who referred to Bassler and Moodey's volume will remain known only to those investigators.

The bibliography for this compilation includes 488 references spanning the period 1938 through 1968 and one reference for 1930. Much of the literature for 1938 through 1942 was included in Bassler and Moodey's work (1943). Most of the papers for 1938 through 1942 not indexed by Bassler and Moodey were papers published outside North America. Due to the political conditions preceding and at the beginning of World War II they probably were not received by the authors in time to be included.

A study of the number of references for each year shows the expected over-all increase in number of papers published each year. From 1943 through 1949 there is an average of 14 papers per year (Fig. 1), for 1950 through 1959 the average is 15.4 papers per year, and for 1960 through 1968 the average is 23.4 papers per year. Thus there is a 64.3 percent increase in the average number of papers per year, comparing 1943 through 1949 to 1960 through 1968. This increase is not expected to reverse or level off during the next ten years; rather, an equal or even greater increase is anticipated.

Papers concerning crinoid morphology, paleoecology, optical properties of plates, stem orientation, parasites, and injured specimens are included in the bibliography. Although most of these papers do not include illustrated or described species, they are listed because they add significantly to information about Paleozoic crinoids. Monographs and papers of special importance, or considered by me to be major contributions to the knowledge of Paleozoic crinoids, are indicated by an asterisk (*) in the bibliography.

Papers overlooked in this volume can be included in up-dated bibliographies published in the future. Post-1968 papers are now being indexed by use of computer data retrieval cards. This information will be made available to crinoid investigators who contact me.

Only crinoids are indexed in this volume although an attempt is made to index all figured columnals, thecae, and crowns, as well as all systematically or morphologically treated species, except those in standard textbooks or popular amateur fossil guides. Section one of Part I of this index lists the species represented by thecae and crowns. Rarely, columnals are included in this section, but only where there is no question of the identification and thecae or crowns are also reported for the genus in the same reference. Section two of Part I lists thecae, crowns or plates unidentified or assigned to a family or higher taxon (often somewhat questionably). Part II is the section on columnals. Because so much study of columnals is yet to be done, and coordinating columnals to the theca will long be a problem, along with the valid species collated in section one of Part II, I have listed a number of invalid genera such as *Cyclocyclicus, Pentogonopentagonalis*, and so forth, noting that they are *nomen vetitum*. These names, although invalid according to the rules of zoological nomenclature, as pointed out by Moore and Jeffords (1968), are descriptive and make excellent repositories for many species until lineage studies and other systematic investigations provide appropriate names. Unidentified columnals are placed in section two of Part II. In both Parts I and II, type species are preceded by an asterisk (*).

Species included in synonymies differing from the allocation given in Bassler and Moodey (1943) are listed under the latest worker's assignment; reference to the original combination is cross-indexed. When genera have been put in synonymy, I have made the new combinations if species are referred to in the literature included in this compilation. Thus, this volume must be used in conjunction with Bassler and Moodey for a complete synonymy of many species.

The reported geologic epoch, or range to epoch, for each species follows European or other local terms if the species is reported from outside North America, and United States terms for those recorded from North America. If the species is reported from both Europe and North America either the original report for epoch names or the latest author's terminology is followed, usually the latter. If no epoch name was given but a formation is reported in the paper cited, I have followed the correlation charts of The Geological Society of America for epoch names.

Precise locality data for each species are not included in this volume. The locality data thus must be obtained by reference to the papers cited. Therefore, locality data are given only to the state for the United States and to the province for Canada, Australia, and China. Russian localities are listed by geologic area, province, or local geographic feature. Otherwise only the country is given or a provincial name, if in common usage. All localities are reported as given by the author cited, unless the author did not list a locality (usually when figuring a specimen from an earlier publication), in which case the locality data are taken from the original reference or from Bassler and Moodey. Van Sant (1964) cast doubt on the locality data given by Bassler and Moodey for several species from Indiana. N. Gary Lane (1967, personal commun.) stated that some of the accessory localities reported by Bassler and

Moodey are questionable because some of their data are derived from misidentified specimens in the Springer collection of the U.S. National Museum. Thus, I have tried to avoid using Bassler and Moodey's locality data.

Review of the literature for this compilation has revealed numerous instances where I question or do not agree with the generic assignment or validity of listed species. Generally I have refrained from making any changes here, but have included a comment where I felt compelled to do so.

A faunal list of new genera, introduced since Bassler and Moodey, is included as an Appendix. Genera and species known only from columnals are preceded in it by the symbol ϕ.

Although most authors are to be congratulated on the fine work they have done, a few common omissions, oversights, and mistakes should be called to the attention of future workers in the hope they will not repeat them. Foremost is the lack of original dates of proposal for the species. Second is the omission of the original author's name after the species. Incorrect statement of the ending for many older generic names (that is, *crinus* for *crinites*, or vice versa) is equally common. Failure to designate a new combination can also be frustrating. Although these mistakes and others, which I will not detail, seem trite, they add confusion to systematic studies and time-devouring hours of literature checking.

My sincere appreciation is expressed to Yu. A. Arendt for considerable help on the Russian literature, and for placing me in contact with the Russian crinoid workers. N. Gary Lane offered several suggestions on additional literature which illustrated some specimens. R. C. Moore and Russell Jeffords made available copies of several foreign papers which were unavailable by interlibrary loan. Numerous authors sent copies and bibliographies of their works, many of which were not readily available to me. Lastly, I have deep appreciation for the encouragement offered by fellow workers when the boring tedium of such a compilation set in.

Bibliography

Ahlfeld, F., and Braniša, L., 1960, Geología de Bolivia: La Paz, Institut Boliviano del Petróleo, 245 p., 60 figs.

Almela, A., and Revilla, J., 1950, Especies fosiles nuevas del Devoniano de Leon: España Inst. Geol. y Minero, Notas y Comun., no. 20, p. 45-60, 3 pls.

Almela, A., and Sanz, R., 1958, Resumen de la historia geológica de la Tierra: España Inst. Geol. y Minero, Mem., v. 59, 174 p.

Amsden, T. W., 1949, Stratigraphy and paleontology of the Brownsport formation (Silurian) of western Tennessee: Peabody Mus. Nat. Hist. Bull. 5, 138 p., 34 pls.

Antropov, I. A., 1954, Ob ostatkakj *Cupressocrinus* i ikj stratigraficheskom znachenii v devone Volgo-Uralskoi neftenoshoy oblasti (On fossil *Cupressocrinus* and its stratigraphical significance for the Devonian of the Volga-Ural oil bearing region): Akad. Nauk SSSR Izv. Kazan. Filiala, Ser. Geol. Nauk, no. 2, p. 12-16, 1 pl.

Arendt, Yu. A., 1961, O povrezhdeniiakh morekikh lilii, vyzvannykh *Schizoproboscina* (On the injuries in crinoids made by *Schizoproboscina*): Paleont. Zhur., no. 2, p. 101-106, 1 pl.

— 1962, *Rhabdocrinus vatagini* sp. nov. iz podmoskovnogo nizhnego karbona (*Rhabdocrinus vatagini* sp. nov. from the lower Carboniferous of the Moscow region): Paleont. Zhur., no. 2, p. 117-121, 2 figs.

— 1963a, Krona morskoi lilii iz srednego ordovika r. Podkamennoi Tunguski (The crown of a crinoid from the Middle Ordovician of the Podkamennaya Tunguska River): Paleont. Zhur., no. 2, p. 131-135, 2 figs.

— 1963b, *Rhabdocrinus vatagini* sp. nov., a new species from the lower Carboniferous deposits of the Moscow Basin: Internat. Geology Rev., v. 5, no. 12, p. 1674-1677, 2 figs. (English translation of Arendt, 1962).

— 1964, 5000 chashechek iskopaemykh morskikh lilii (5000 calyces of fossilized crinoids): Priroda [U.S.S.R.], no. 7, p. 111-113, 2 figs.

— 1965a, K poznaniyu morskikh lilii kaltseokrinid (Contribution to the knowledge of crinoids from the family Calceocrinidae): Paleont. Zhur., no. 1, p. 89-96, pl. 10, 2 figs.

— 1965b, Crown of crinoid from Middle Ordovician of Podkamennaya Tunguska River: Internat. Geology Rev., v. 7, no. 6, p. 1116-1119, 2 figs. (English translation of Arendt, 1963a).

— 1968a, Regressivnye i neotenicheskie iavleniya u krinoidei Hypocrinidae (Regressive and neothenic phenomena in Hypocrinidae crinoids): Trudy XII sessii Vses. Paleont. Obshchestva VPO Nauka, Leningrad, Izdatel'stvo Nauka, p. 97-107, 8 figs.

— 1968b, Pirazokrinidi iz Krasnoufimska (Pirasocrinids from Krasnoufimsk): Paleont. Zhur., no. 4, p. 99-101, 1 fig.

Arendt, Yu. A., and Hecker, R. T., 1964, Klass Crinoidea. Morskie lilii. Sistematicheskaia chasti (Class Crinoidea. Crinoids. Systematic part) *in* Orlov, Yu. A., ed., Osnovy paleontologii, Iglokozhi, Gemikhordovye,

Pogonofory, i Shchetinkochelyustnye (Fundamentals of paleontology, Echinodermata, Hemichordata, Pogonophora, and Chaetognatha): Moscow, Izdatel'stvo Nedra, p. 76, 214-231, pls. 8-16, figs. 115-140.

Author unknown, 1962, Sección Quinta Laboratorios: España, Inst. Geol. y Minero, Mem. Gen., 1961, p. 89-104, 1 fig.

Avrov, D. P., and Stukalina, G. A., 1964, Novie dannye o siluriiskikh otlozheniakh yuga Gornogo Altaya. Sbornik statei. Materialy po geologii i poleznym iskopaemym Altaia i Kazakhstana (New information about Silurian deposits of the southern Altay Mountains, *in* Collected articles of materials on geology and useful fossils of Altay and Kazakhstan): Vses. Nauchno-Issled. Geol. Inst. Trudy, nov. seria, v. III, p. 25-30, 7 figs.

Bassler, R. S., and Moodey, M. W., 1943, Bibliographic and faunal index of Paleozoic pelmatozoan echinoderms: Geol. Soc. America Spec. Paper 45, 734 p.

Bates, D.E.B., 1965, A new Ordovician crinoid from Dolgellau, north Wales: Palaeontology, v. 8, p. 355-357, pl. 45.

— 1968, on *"Dendrocrinus" cambriensis* Hicks, the earliest known crinoid: Palaeontology, v. 11, p. 406-409, pl. 76.

Bergounioux, F. M., 1938, *Cupressocrinus elongatus* Goldf. du Givetien des Asturies: Bull. Soc. Hist. Nat. Toulouse, tome 72, 1[er] trimestre, p. 63-68, 1 fig.

Bolton, T. E., 1957, Silurian stratigraphy and paleontology of the Niagara escarpment in Ontario: Canada Geol. Survey Mem. 289, 153 p., 12 pls.

Boucot, A. J., Macdonald, G.J.F., Milton, C., and Thompson, J. B., Jr., 1958, Metamorphosed middle Paleozoic fossils from central Massachusetts, eastern Vermont, and western New Hampshire: Geol. Soc. America Bull., v. 69, p. 855, 870, 6 pls., 1 fig.

Bouška, Josef, 1943, Krinoidengattung *Pycnosaccus* Angelin im bohmischen Silur: Česká Akad. Bull. Internat. Ann. 43 (1942), p. 254-256, 1 pl.

— 1944, Die Vertreter der Gattung *Gissocrinus* Angelin im bohmischen Silur: Česká Akad. Bull. Internat. Ann. 44 (1943), p. 584-592, 2 pls.

— 1946, *Pygmaeocrinus*, new crinoid from the Devonian of Bohemia: Věstnik Královské české společnosti nauk, Třida matematicko-přirodovědecká, p. 1-4, 1 pl.

— 1948, *Holynocrinus*, new crinoid genus from the Middle Devonian of Bohemia: Jour. Paleontology, v. 22, p. 520-524, 8 figs.

— 1950, On Crotalocrinidae (Angelin) from the Silurian and Devonian of Bohemia: Česká Akad. Bull. Internat. Ann. 47, p. 11-29, 4 pls.

*Bouška, Josef, 1956a, Pisocrinidae Angelin from the Silurian and Devonian of Bohemia (Crinoidea): Rozpravy Ústředního Ústavu Geologického, Svazek 20, 138 p., 6 pls. (Text in Czechoslovakian, p. 1-54; stratigraphical summary and systematics in Russian, p. 55-96, and English, p. 97-138.)

— 1956b, O rodu *Protaxocrinus* Springer (Crinoidea) ze středočeského siluru [On the occurrence of the genus *Protaxocrinus* Springer (Crinoidea) in the Silurian of Bohemia]: Česká Ustředniho Ustavu Geol., Sbornik Paleont., v. 22 (1955) p. 323-333. (English summary, p. 332-333.)

Bowsher, A. L., 1953, A new Devonian crinoid from western Maryland: Smithsonian Misc. Colln., v. 121, no. 9, 8 p., 1 pl.

— 1954, The stratigraphic significance of a crinoid from the Redwall limestone of Arizona: Jour. Paleontology, v. 28, p. 113-116, 3 figs.

— 1955a, Origin and adaptation of platyceratid gastropods: Kansas Univ. Paleont. Contr., Mollusca Art. 5, 11 p., 2 pls.

*— 1955b, New genera of Mississippian camerate crinoids: Kansas Univ. Paleont. Contr., Echinodermata Art. 1, 23 p., 6 pls.

Bramlette, W. A., 1943, *Triceracrinus*, a new Upper Pennsylvanian and lower Permian crinoid: Jour. Paleontology, v. 17, p. 550-553, pl. 96.

Branson, C. C., 1962, Additional illustrations of some Oklahoma crinoids: Oklahoma Geology Notes, v. 22, p. 163-164, 4 figs.

Branson, E. B., 1944, The geology of Missouri: Missouri Univ. Studies, v. 19, no. 3, 535 p., 49 pls., 51 figs.

Breimer, A., 1960, On the structure and systematic position of the genus *Rhipidocrinus* Beyrich, 1879: Leidse Geol. Meded., v. 25, p. 247-260.

*— 1962a, A monograph on Spanish Paleozoic Crinoidea: Overdr. Leidse Geol. Meded., Deel 27, 189 p., 16 pls.

— 1962b, Application to suppress under the plenary powers three specific names of Spanish Paleozoic Crinoidea. A. N. (S.) 1513: London, Bull. Zool. Nomenclature, v. 19, p. 325-327.

Breitbach, J., 1966, Einlagerungen im Lederschifer von Selbitz/Frankenwald—ein Beitrag zur Frage der Lederschiefergerölle: Geol. Blätter Nordost-Bayern, H. 16, no. 2-3, p. 114-123, 2 pls.

*Brower, J. C., 1965, The genus *Steganocrinus*: Jour. Paleontology, v. 39, p. 773-793, pls. 91-94, 7 figs.

*— 1966, Functional morphology of Calceocrinidae with description of some new species: Jour. Paleontology, v. 40, p. 613-634, pl. 75, 2 figs.

*— 1967, The actinocrinitid genera *Abactinocrinus*, *Aacocrinus* and *Blairocrinus*: Jour. Paleontology, v. 41, p. 675-705, pls. 75-78.

Burke, J. J., 1966a, *Endelocrinus kieri*, a new crinoid from the Ames Limestone: Ohio Jour. Sci., v. 66, p. 459-464, 1 pl.

— 1966b, On the occurrence of *Oklahomacrinus* in Ohio and Timor: Ohio Jour. Sci., v. 66, p. 464-468, 6 figs.

— 1967a, A new *Endelocrinus* from the Brush Creek Limestone (Pennsylvanian) of Pennsylvania: Carnegie Mus. Annals, v. 39, p. 75-83, 2 figs.

— 1967b, Tegmen roof of *Plaxocrinus mooresi* (Whitfield): Ohio Jour. Sci., v. 67, p. 298-300, 3 figs.

— 1968, Pachylocrinids from the Conemaugh Group, Pennsylvanian: Kirtlandia, no. 3, 18 p., 4 figs.

Busch, D. A., 1943, Some unusual cystoids and crinoids from the Niagaran (Silurian) of west-central Ohio: Jour. Paleontology, v. 17, p. 105-109, pl. 20.

Butts, Charles, 1940, Geology of the Appalachian Valley in Virginia: Part I, Geologic text and illustrations: Virginia Geol. Survey Bull. 52, 568 p., pls. 1-64.

— 1941, Geology of the Appalachian Valley in Virginia: Part II, Fossil plates and explanations: Virginia Geol. Survey Bull. 52, 271 p., pls. 65-135.

— 1948, Geology and mineral resources of the Paleozoic area in northwest Georgia: Georgia Geol. Survey Bull. 54, 79 p., 8 pls.

Byrne, F., and Seeberger, E., 1942, Fragmentary crinoids from the Lower Permian of the Manhattan area: Kansas Acad. Sci. Trans., v. 45, p. 221-228, 1 pl.

Cain, J.D.B., 1968, Aspects of the depositional environment and palaeoecology of crinoidal limestones: Scottish Jour. Geology, v. 4, pt. 3,
p. 191-208, 2 pls., 5 figs.

China, W. E. (Acting Secretary to ICZN), 1965, Opinion 727, Three specific
names of Spanish Paleozoic Crinoidea: Suppressed under the plenary
powers: London, Bull. Zool. Nomenclature, v. 22, p. 45-46.

Croneis, Carey, 1930, Geology of the Arkansas Paleozoic area with
especial reference to oil and gas possibilities: Arkansas Geol.
Survey Bull. 3, 457 p., 45 pls., 30 figs.

Cronoble, William R., 960, An occurrence of *Ulocrinus buttsi* Miller and
Gurley in Oklahoma: Oklahoma Geology Notes, v. 19, p. 96-99, 3 figs.

Cuénot, L., 1948, Anatomie, éthologie et systématique des Echinodermes,
in Grassé, P. P., ed., Traité de zoologie: Paris, Crinoides, v. 11,
p. 30-74.

Delpey, G., 1943, Description d'un nouveau crinoide du devonien francais,
Botryocrinus montis-guyonensis nov. sp.: Soc. Geol. France Bull.
Ser. 5, tome 12, p. 15-19, 3 figs.

— 1945, Histoire des echinodermens jusqu'au devonien inferieur: Soc.
Geol. France Bull., Ser. 5, tome 14, p. 247-278.

*Dubatolova, Yu. A., 1964, Morskie lilii devona Kuzbassa (Devonian crinoids
of the Kuznetz Basin): Akad. Nauk SSSR Sibirsk. Otdeleniye Inst.
Geologii i Geofiziki Trudy, 153 p., 14 pls.

— 1967, Devonskie krinoidei khrebta Tash-Khaiakhtakh (Severto-Vostok
SSSR). Sbornik Novye dannye po biostratigrafii devona i verkhnego
paleozoa Sibiri [Devonian crinoids of the Tash-Hayahtah mountain ridge
(northeastern USSR), *in* Collected articles of new facts in biostratigraphy of the Devonian and upper Paleozoic of Siberia]: Akad. Nauk
SSSR Sibirsk. Otdeleniye Inst. Geologii i Geofiziki Trudy, p. 32-41,
pl. 6, 6 figs.

— 1968, Stebli morskikh lilii tom-chumshskogo gorizonta, *in* Sokolov,
B. S., and Ivanovsky, A. B., eds., Biostratigrafiya pogranichnkh
otlozhenii silura i devona (Crinoid stems from the Tom-Chumshsk Horizon,
in Biostratigraphy of Silurian-Devonian boundary deposits): Akad. Nauk
SSSR Sibirsk. Otdeleniye Inst. Geologii i Geofiziki Trudy, p. 141-157,
pls. 15, 16, 13 figs.

Dubatolova, Yu. A., and Shao, C., 1959, Stebli morskikh lilii
kamennougolnkh, permskikh i triasovykh otlozhenii Yuzhnogo Kitaia
(Crinoid stems of the Carboniferous, Permian, and Triassic deposits of
southern China): Acta Paläeont. Sinica, v. 7, no. 1, p. 41-83,
pls. 1, 2. (Chinese, p. 41-56; Russian, p. 57-76.)

Dubatolova, Yu. A., and Yeltyschewa, R. S., 1967, Stratigraphic importance
of the Devonian crinoids of Siberia, *in* Oswald, D. H., ed., International symposium on the Devonian system: Alberta Soc. Petroleum
Geologists, p. 537-542.

Easton, W. H., 1962, Carboniferous formations and faunas of central
Montana: U.S. Geol. Survey Prof. Paper 348, 126 p., 13 pls.

Ehlers, G. M., and Kesling, R. V., 1963, Two new crinoids from Lower
Mississippian rocks in southeastern Kentucky: Jour. Paleontology,
v. 37, p. 1028-1041, pls. 133-134, 3 figs.

Ehrenberg, Kurt, 1954, Biohistorisches über die Echinodermen in
allgeneinen und über die Nebenformen der Crinoiden: Neues Jahrb.
Geologie u. Paläontologie Monatsh., H. 11, p. 491-508.

Elias, M. K., 1957, Late Mississippian fauna from the Redoak Hollow

Formation of southern Oklahoma, part I: Jour. Paleontology, v. 31, p. 370-427, pls. 39-50.

Fay, R. O., 1960, The "pores" of *Stephanocrinus* Conrad: Oklahoma Geology Notes, v. 20, p. 256-259, 2 pls.

— 1961, The type species of *Stephanocrinus* Conrad: Oklahoma Geology Notes, v. 21, p. 236-238, 1 pl.

— 1962a, Ventral structures of *Stephanocrinus angulatus* Conrad: Jour. Paleontology, v. 36, p. 206-210, pl. 35, 1 fig.

— 1962b, *Mespilocystites*, an Ordovician Coronate crinoid from Czechoslovakia: Oklahoma Geology Notes, v. 22, p. 156-161, 2 pls., 3 figs.

Fraunfelter, G. H., 1965, *Gennaeocrinus* sp. from the Glen Park Formation of northeastern Missouri: Illinois Acad. Sci. Trans, v. 58, p. 204-205, 2 figs.

Frederickson, E. A., and Waddell, D. E., 1960, An unusual crinoid from the Pennsylvanian of Oklahoma: Oklahoma Geology Notes, v. 20, p. 172-174, 1 pl.

Genser, H., 1965, Fossilfunde aus dem Schiefern von Weiber bei Weissenburg/Elsass: Uber. Mitt. Oberrhein. Geol., Ver. no. 47, p. 153-157, 3 figs.

Goldring, W., 1943, Geology of the Coxsackie quadrangle, New York: New York State Mus. and Sci. Service Bull. 332, 374 p., 71 figs.

— 1945, Notes on *Thamnocrinus springeri* Goldring and other Hamilton crinoids: Am. Jour. Sci., v. 243, p. 57-65, 1 pl.

— 1946, A new lower Chemung crinoid: Bull. Amer. Paleontology, v. 31, no. 119, p. 37-39, pl. 4.

— 1948a, Occurrence of *Gennaeocrinus kentuckiensis* (Shumard) in Pennsylvania: Bull. Wagner Free Inst. Sci., v. 23, no. 1, p. 1-3, 1 pl.

— 1948b, Status of *"Homocrinus" cylindricus* Hall: Bull. Wagner Free Inst. Sci., v. 23, p. 25-32, 1 pl.

— 1950, Devonian crinoids: New and old: Bull. Wagner Free Inst. Sci., v. 25, no. 4, p. 29-35.

— 1951, A new species of the genus *Craterocrinus* Goldring: New York State Mus. Circ. 27, 7 p., 1 pl.

*— 1954, Devonian crinoids: New and old, II: New York State Mus. Circ. 37, 51 p., 6 pls.

Gross, Karl, 1948, Ein neuer Fund von *Lodanella mira* Kayser in Unterdevon des Siegerlandes: Neues Jahrb. Geologie u. Paläontologie Monatsh., Abt. B., Jg. 1945-1948, H. 1-4, p. 40-42.

Gutschick, R. C., 1965, *Pterotocrinus* from the Kinkaid Limestone (Chester, Mississippian) of Illinois and Kentucky: Jour. Paleontology, v. 39, p. 636-646, pls. 79-80, 6 figs.

— 1968, Late Mississippian (Chester) *Allagecrinus* (Crinoidea) from Illinois and Kentucky: Jour. Paleontology, v. 42, p. 987-999, 11 figs.

Gwynne, C. S., 1962, Fossil starfish and crinoid slabs: Geotimes, v. 7, no. 1, p. 17, 1 fig.

Haaf, E. T., 1950, La couverture anale de *Hypocrinus:* Koninkl. Nederlandse Akad. Wetensch. Proc., v. 53, p. 891-893, 4 figs.

Hattin, D. E., 1958, Regeneration in a Pennsylvanian crinoid spine: Jour. Paleontology, v. 32, p. 701-702, pl. 98.

Hemming, F. (Secretary to ICZN), 1953, Case No. 36, Article 25: Proposed insertion of a provision that certain classes of generic names published for fossil species and bearing the termination -*ites*, -*ytes* or -*ithes* possess no status in zoological nomenclature: Bull. Zool. Nomenclature, v. 10, p. 360-366.

1958, Opinion 527, Interpretation under the plenary powers of the nominal species *Actinocrinus gilbertsoni* Phillips, 1836 (Class Crinoidea): Opinions and Declarations, Internat. Comm. Zool. Nomen., v. 19, p. 16, p. 301-314.

Hendriks, E.M.L., 1951, Geological succession and structure in western south Devonshire, England: Royal Geol. Soc. Cornwall Trans., v. 18, pt. 3, p. 255-308.

Hernández-Sampelayo, P., 1954, Fosiles do la zona carbonifera de Viñon y Torazo (Asturias): Estudios Geol., Inst. Mallada, v. 10, p. 7-48, 12 pls., 1 fig.

Heuer, Edward, 1958, Comments on the nomenclature revision of the Strawn and Canyon megafossil plates, *in* Hamilton, W. C., Jr., ed., A guide to the Strawn and Canyon series of the Pennsylvanian System in Palo Pinto County, Texas: Wichita Falls, Texas, North Texas Geol. Soc., p. 36-47, pls. 1-5.

Hill, D., and Woods, J. T., 1964, Permian index fossils of Queensland: Queensland Palaeontographical Soc., 32 p., 15 pls.

Horný, R., 1964, Impressions muscalaires de Platydératides (Gastropodes de Bohême): Časopis Národni Muz., Oddill prirod., v. 133, no. 2, p. 88-92, 1 pl., 3 figs.

*Horowitz, A. S., 1965, Crinoids from the Glen Dean Limestone (middle Chester) of southern Indiana and Kentucky: Indiana Geol. Survey Bull. 34, 52 p., 5 pls.

Hudson, R.G.S., Clarke, M. J., and Sevastopulo, G. D., 1966, The palaeoecology of a lower Visean crinoid fauna from Feltrim, Co. Dublin: Royal Dublin Soc. Sci Proc., Ser. A, v. 2, p. 273-286, 5 figs.

Ivanova, E. A., 1958, Razvitie fauny sredne- i Verkhnekamennougolnogo morya zapadnoi chasti moskovskoi sindklizy v svyazi s ego istoriey kniga 3. Razvitie fauny v. svyazi s uslovi yami sushchestvovaniya (Development of the fauna of the middle and upper Carboniferous sea of the western part of the Moscow syncline in connection with its history, v. 3. Development of the fauna in connection with conditions of existence): Akad. Nauk SSSR Inst., Trudy Paleontologicheskogo, v. 69, 303 p., 21 pls.

*Jeffords, R. M., and Miller, T. H., 1968, Ontogenetic development in Late Pennsylvanian crinoid columnals and pluricolumnals: Kansas Univ. Paleont. Contr., Echinodermata Art. 10, p. 1-14, pls. 1-4, 5 figs.

Jillson, W. R., 1958, The geology of Barren County, Kentucky: Frankfort, Ky., Perry Publishing Co., 101 p., illus.

— 1959, Geology of the Pitman oil pool in Green County, Kentucky: Frankfort, Ky., Perry Publishing Co., 94 p., illus.

— 1960, Geology of the Goose Creek dome in Casey and Russell Counties, Kentucky: Frankfort, Ky., Perry Publishing Co., 88 p., illus.

— 1963, Geology of the Winchester disturbance: Frankfort, Ky., Perry Publishing Co., 24 p., illus.

— 1965, The geology of Casey County, Kentucky: Frankfort, Ky., Roberts Printing Co., 108 p., illus.

Kaljo, D., 1958, Ka kivistised Vajavad Looduskaitset, *in* Eluta Looduse Kaitse (The need for preservation of fossils): Esti NSV Teaduste Akadeemia Geoloogia Instituut, Tallinn, p. 39-46, figs. 17-24.

Kesling, R. V., 1963, Occurrence and variations of *Botryocrinus thomasi* Laudon in the Thunder Bay limestone of Michigan: Michigan Univ. Mus. Paleontology Contr., v. 18, p. 231-244, 3 pls., 2 figs.

*— 1964a, A new species of *Melocrinites* from the Middle Devonian Bell Shale of Michigan: Michigan Univ. Mus. Paleontology Contr., v. 19, p. 89-103, 2 pls., 2 figs.

— 1964b, *Decadocrinus hughwingi*, a new Middle Devonian crinoid from the Silica Formation in northwestern Ohio: Michigan Univ. Mus. Paleontology Contr., v. 19, p. 135-142, 1 pl.

— 1964c, Two new crinoids of the family Periechocrinitidae from the Middle Devonian Thunder Bay limestone of Michigan: Michigan Univ. Mus. Paleontology Contr., v. 19, p. 143-155, 2 pls.

— 1965a, Primibrachials and arms of *Alloprosallocrinus conicus* Casseday and Lyon, a Lower Mississippian camerate crinoid: Michigan Univ. Mus. Paleontology Contr., v. 19, p. 257-263, 2 pls.

— 1965b, Nature and occurrence of *Gennaeocrinus goldringae* Ehlers: Michigan Univ. Mus. Paleontology Contr., v. 19, p. 265-280, 5 pls.

— 1965c, *Proctothylacocrinus esseri*, a new crinoid from the Middle Devonian Silica Formation of northwestern Ohio: Michigan Univ. Mus. Paleontology Contr., v. 20, p. 75-87, 5 pls.

— 1966, *Botryocrinus niemani*, a new crinoid from the Middle Devonian Silica Formation of Ohio: Michigan Univ. Mus. Paleontology Contr., v. 20, p. 271-276, 1 pl.

— 1968a, *Gennaeocrinus chilmanae*, a new crinoid from the Middle Devonian Silica Formation in southeastern Michigan: Michigan Univ. Mus. Paleontology Contr., v. 22, p. 127-131, pl. 1, 1 fig.

— 1968b, Note on ontogeny of the Middle Devonian crinoid *Proctothylacocrinus esseri* Kesling: Michigan Univ. Mus. Paleontology Contr., v. 22, p. 133-138, 2 pls., 4 figs.

— 1968c, *Ameliacrinus benderi*, a new dicyclic camerate crinoid from the Middle Devonian Silica Formation in northwestern Ohio: Michigan Univ. Mus. Paleontology Contr., v. 22, p. 155-162, 3 pls., 2 figs.

— 1968d, *Logocrinus conicus*, a simple new Middle Devonian inadunate crinoid from Michigan: Michigan Univ. Mus. Paleontology Contr., v. 22, p. 163-167, 1 pl., 1 fig.

Kesling, R. V., and Mintz, L. W., 1963a, Species of the crinoid *Dolatocrinus* from the Middle Devonian Dock Street Clay of Michigan: Michigan Univ. Mus. Paleontology Contr., v. 18, p. 67-100, 7 pls., 2 figs.

*— 1963b, *Dolatocrinus* and *Stereocrinus*, its junior synonym: Michigan Univ. Mus. Paleontology Contr., v. 18, p. 229-237, 2 pls.

*Kesling, R. V., and Paul, C.R.C., 1968, New species of Porocrinidae and brief remarks upon these unusual crinoids: Michigan Univ. Mus. Paleontology Contr., v. 22, p. 1-32, 8 pls. 14 figs.

Kesling, R. V., and Smith, R. N., 1962, *Gennaeocrinus variabilis*, a new species of crinoid from the Middle Devonian Bell Shale of Michigan: Michigan Univ. Mus. Paleontology Contr., v. 17, p. 173-194, 9 pls., 2 figs.

— 1963, The crinoid *Synbathocrinus* in the Middle Devonian Traverse Group of Michigan: Michigan Univ. Mus. Paleontology Contr., v. 18, p. 185-196, 1 pl.

Kettner, R., and Prantl, F., 1942, O novém nalezišti zkamenělin v břidlicích moravského devonu u Vratíkova sv. od Boskovic: Vest. Ceske Spolec. Nauk, Praha, v. 18, p. 1-18, 1 pl., 3 figs.

Kier, P. M., 1952, Echinoderms of the Middle Devonian Silica Formation of Ohio: Michigan Univ. Mus. Paleontology Contr., v. 10, p. 59-81, 4 pls.

— 1958, Infrabasals in the crinoid *Opsiocrinus* Kier: Michigan Univ. Mus. Paleontology Contr., v. 14, p. 201-206, 1 pl., 2 figs.

Kirk, Edwin, 1943a, A revision of the genus *Steganocrinus*: Washington Acad. Sci. Jour., v. 33, p. 259-265, 5 figs.

— 1943b, *Zygotocrinus*, a new fossil inadunate crinoid genus: Am. Jour. Sci., v. 241, p. 640-646, 1 pl.

— 1943c, Identification of *Actinocrinus chloris* Hall: Washington Acad. Sci. Jour., v. 33, p. 346-347.

— 1944a, *Cribanocrinus*, a new rhodocrinoid genus: Washington Acad. Sci. Jour., v. 34, p. 13-16.

— 1944b, *Cytidocrinus*, new name for *Cyrtocrinus* Kirk: Washington Acad. Sci. Jour., v. 34, p. 85.

— 1944c, *Aphelecrinus*, a new inadunate crinoid genus from the Upper Mississippian: Am. Jour. Sci., v. 242, p. 190-203, 1 pl.

— 1944d, *Cymbiocrinus*, a new inadunate crinoid genus from the Upper Mississippian: Am. Jour. Sci., v. 242, p. 233-245, 1 pl.

— 1944e, *Thyridocrinus*, a new inadunate crinoid genus from the Silurian: Washington Acad. Sci. Jour., v. 34, p. 388-390.

— 1945a, Four new genera of camerate crinoids from the Devonian: Am. Jour. Sci., v. 243, p. 341-355, 1 pl.

— 1945b, *Gaulocrinus*, a new inadunate crinoid genus from the Mississippian: Washington Acad. Sci. Jour., v. 35, p. 180-182.

— 1945c, *Holcocrinus*, a new inadunate crinoid genus from the Lower Mississippian: Am. Jour. Sci., v. 243, p. 517-521.

— 1946a, *Stiptocrinus*, a new camerate crinoid genus from the Silurian: Washington Acad Sci. Jour., v. 36, p. 33-36.

— 1946b, A new species of *Dolatocrinus* from the Travers (Middle Devonian) of Michigan: Jour. Paleontology, v. 20, p. 267-268.

— 1946c, *Corythocrinus*, a new inadunate crinoid genus from the Lower Mississippian: Jour. Paleontology, v. 20, p. 269-274, pl. 40.

— 1946d, *Plemnocrinus*, a new crinoid genus from the Lower Mississippian: Jour. Paleontology, v. 20, p. 435-441, pls. 65-66.

— 1946e, Redefinition of *Actinocrinus parvus* Shumard: Am. Jour. Sci., v. 244, p. 658-660.

— 1947, Three new genera of inadunate crinoids from the Lower Mississipian: Am. Jour. Sci., v. 245, p. 287-303, 1 pl.

— 1948, Two new inadunate crinoid genera from the Middle Devonian: Am. Jour. Sci., v. 246, p. 701-710, 1 pl.

Kleinschmidt, G., 1966, Krinoiden aus dem epizonalen Kristallin der Saualpe, Karnten: Neues Jahrb. Geologie u. Paläontologie Monatsh., no. 12, p. 707-716, 9 figs.

*Koenig, J. W., 1965, Ontogeny of two Devonian crinoids: Jour. Paleontology, v. 39, p. 398-413, 6 figs.

Koenig, J. W., and Meyer, D. L., 1965, Two new crinoids from the Devonian of New York: Jour. Paleontology, v. 39, p. 391-397, 4 figs.

Koenig, J. W., and Niewoehner, W., 1959, *Pentececrinus*, a new microcrinoid from the Louisiana formation of Missouri: Jour. Paleontology, v. 33, p. 462-470, 3 figs.

Kongiel, Roman, 1958, Nowy gatunek *Ammonicrinus* i jego wystepowanie w Polsce: Muz. Ziemi, Warsaw Inst. Geol., Pr. no. 2, p. 31-40, 6 figs.

Korejwo, K., and Teller, L., 1964, Upper Silurian non-graptolite fauna from the Chelm borehole (eastern Poland): Acta Geol. Polonica, v. 14, no. 2, p. 233-300, 26 pls., 37 figs.

Kostic-Podgorska, V., 1958, Fauna i biostratigrafija paleozojskih slojeva Prače: Geol. Glasnik [Sarajevo], v. 4, 222 p., 44 pls.

Kovalevsky, O. P., Nilova, N. V., and Stukalina, G. A., 1967, O pogranichnykh otlozheniyakh silura i devona v vorskikh fatsiyakh Tarbagataia (On Silurian-Devonian boundary beds in marine facies of Tarbagatai, Kazakhstan): Bull. Nauchnotekhicheskoy informatsii Tematicheskiy vypusk Moskva, Ser. Regional Geol., no. 7, p. 49-52.

Lakeman, Rienk, 1950, On the crinoid nature of *Timorocidaris sphaeracantha* Wanner: Amsterdam, Nederl. Akad. Wetens., Pr. v. 53, no. 1, p. 100-108.

Lane, B. O., 1962, The fauna of the Ely Group in the Illipah area of Nevada: Jour. Paleontology, v. 36, p. 888-911, pls. 125-128.

Lane, N. Gary, 1963a, Two new Mississippian camerate (Batocrinidae) crinoid genera: Jour. Paleontology, v. 37, p. 691-702, 3 figs.

*— 1963b, Meristic variation in the dorsal cup of monobathrid camerate crinoids: Jour. Paleontology, v. 37, p. 917-930.

— 1963c, The Berkeley crinoid collection from Crawfordsville, Indiana: Jour. Paleontology, v. 37, p. 1001-1008, pl. 128, 2 figs.

— 1964a, Inadunate crinoids from the Pennsylvanian of Brazil: Jour. Paleontology, v. 38, p. 362-366, pl. 57.

— 1964b, New Pennsylvanian crinoids from Clark County, Nevada: Jour. Paleontology, v. 38, p. 677-684, pl. 112, 1 fig.

— 1967a, *Cyathocrinites* Miller, 1821 (Crinoidea): Proposed designation of a type-species under the plenary powers. Z. N. (S) 1795: London, Bull. Zool. Nomenclature, v. 24, p. 237-238.

— 1967b, Revision of suborder Cyathocrinina (Class Crinoidea): Kansas Univ. Paleont. Contr. Paper 24, p. 1-3, 8 figs.

Lane, N. Gary, and Webster, G. D., 1966, New Permian crinoid fauna from southern Nevada: California Univ. Pubs. Geol. Sci., v. 63, 87 p., 13 pls.

— 1967, Symmetry planes of Paleozoic crinoids: Kansas Univ. Paleont. Contr. Paper 25, p. 14-16, 8 figs.

La Rocque, J.A.A., and Marple, M. F., 1955, Ohio fossils: Ohio Div. Geol. Survey Bull., v. 54, 152 p., 413 figs.

Larrauri y Mercadilla, L. A. de, 1944, El Devoniano en Smara (Sahara Español): España Inst. Geol. y Minero, Notas y Comun., v. 13, p. 49-51, 9 pls.

Laudon, L. R., 1948, Osage-Meramec contact: Jour. Geology, v. 56, p. 288-302, 3 pls.

— 1967, Ontogeny of the Mississippian crinoid *Platycrinites bozemanensis* (Miller and Gurley), 1897: Jour. Paleontology, v. 41, p. 1492-1497, pls. 193, 194.

Laudon, L. R., Parks, J. M., and Spreng, A. C., 1952, Mississippian crinoid fauna from the Banff formation, Sunwapta Pass, Alberta: Jour. Paleontology, v. 26, p. 544-575, pls. 65-69.

*Laudon, L. R., and Severson, J. L., 1953, New crinoid fauna, Mississippian, Lodgepole formation, Montana: Jour. Paleontology, v. 27, p. 505-536, pls. 51-55.

Laursen, Dan, 1940, Cyrtograptusskifrene paa Bornholm 1. Øleaa: Danmarks Geologiske Undersøgelse, Raekke 2, no. 64, 39 p., 5 pls.

— 1943, Cyrtograptusskifrene paa Bornholm 2. Laesaa: Danmarks Geologiske Undersøgelse, Raekke 2, no. 70, 20 p., 3 pls.

Leary, R. L., 1967, Crinoids: Living Mus., Illinois State Mus., v. 29, no. 2, p. 106-107, 2 figs.

Lehmann, W. M., 1939, Neue Beobachtungen an Versteinerungen aus dem Hunsrückschiefer: Abh. Preuss. Akad. Wiss., Math-Naturw. Kl., Jg. 1939, no. 13, p. 1-13, pls. 1-7.

— 1955, Beobachtungen und Rontgenuntersuchungen an einigen Crinoiden aus dem rheinischen Unterdevon: Neues Jahrb. Geologie u. Paläontologie Abh., v. 101, p. 135-140, pls. 14-17.

Le Maître, D., 1952, Le faune du Devonien inferieur et moyen de la Saoura et des abords de l'Erg el Djemel: Mat. Carte Géol. Algérie, Ser. Paleo., no. 12, 171 p., 22 pls.

— 1954, Echinodermes nouveaux du Devonien sudoranais: Acad. Sci. Comptes Rendus, v. 238, no. 21, p. 2107-2108.

— 1958a, Crinoides du Devonien d'Afrique du Nord: Soc. Géol. France Compte Rendu, no. 14, p. 344-346.

— 1958b, Contribution à l'étude des faunes devoniennes d'Afrique du Nord, I Echinodermes: Publ. Serv. Carte Géol. Algérie, Bull. 20, p. 114-151, 3 pls.

— 1958c, Sur le genre *Tiaracrinus* Schultze: Acad. Sci. Comptes Rendus, v. 246, no. 7, p. 1068-1071.

Le Maître, D., and Heddebaut, C., 1962, Decouverte d'un gisement à *Gastrocrinus* dans le Devonien inférieur des Aldudes (Basses-Pyrénées): Acad. Sci. Comptes Rendus, v. 254, no. 13, p. 2399-2400.

— 1963, Présence de gisements à *Scyphocrinites* dans les Pyrénées basques: Soc. Géol. France Compte Rendu, no. 8, p. 273-274.

Levitskiy, E. S., Stukalina, G. A., Polozhikhina, A. I., and Ushatinskaya, G. T., 1968, Karaespinskii gorizont Severnogo Pribalkhashya, *in* Probleme granitsy silura i devona (Karaespin unit of the North Pribalkhash, *in* Solving the Siluro-Devonian boundary problem): Vestnik MGU, seria geologii, no. 2, p. 62-74, 3 pls., 2 figs.

Lochman, C., and Hu, C. H., 1960, Upper Cambrian faunas from the northwest Wind River Mountains, Wyoming, Part I: Jour. Paleontology, v. 34, p. 793-834, pls. 95-100, 4 figs.

Lue, Gui-De, and Hy, Bo, 1965, On the occurrence of Antedoniidae from Lungtan coal series, Nanking: Acta Palaeont. Sinica, v. 13, no. 2, p. 373-375, 1 pl. (Chinese, p. 373, 375; English, p. 374.)

Luttig, Gerd. 1951, Neue Melocrinitiden aus dem rheinischen Devon: Paläont. Zeitschr., v. 24, nos. 3-4, p. 120-125.

*Macurda, D. B., Jr., 1968, Ontogeny of the crinoid *Eucalyptocrinites*: Jour. Paleontology, v. 42, no. 5, pt. II, p. 99-118, 10 figs.

Melendez, B., 1947, Tratado de Paleontologia, tomo 2: Madrid, Inst. Lucas Mallada, 710 p., figs. I-VI; 414 figs.

Mellado, T. R., 1949, Crinoides del Devónico de Santa Lucia (León): Bol. Real Soc. Española Hist. Nat., v. 47, p. 657-662, pls. 29-30.

Meyer, D. L., 1965, Plate growth in some platycrinid crinoids: Jour. Paleontology, v. 39, p. 1207-1209, 1 fig.

Mianil, R. M., 1959, Novye predstaviteli roda *Hoplocrinus* iz srednego ordovika Estonii (New representatives of the genus *Hoplocrinus* from the Middle Ordovician of Estonia): Uchenye zapiski Tartusskogo Gosudarstvennogo Univ. vypusk 75, p. 82-97, 2 pls.

Minato, Masao, 1951, On the lower Carboniferous fossils of the Kitakami Massif, northeast Honsyu, Japan: Hokkaido Univ. Fac. Sci. Jour., Ser. IV, v. 7, no. 4, p. 355-382, 5 pls.

Modzalevskaya, E. A., 1967, Biostratigraphic subdivision of the Devonian in the Far East and Transbaikal region, USSR, *in* Oswald, D. H., ed., International symposium on the Devonian System: Alberta Soc. Petroleum Geologists, p. 543-556, 6 pls.

Moore, R. C., 1948, Paleontological features of Mississippian rocks in North America and Europe: Jour. Geology, v. 56, p. 373-402, 17 figs.

*— 1950, Evolution of the Crinoidea in relation to major paleogeographic changes in earth history: Internat. Geol. Cong. Rept., 18th Sess. London, 1948, pt. 12, p. 27-53, 18 figs.

— 1952, Evolution rates among crinoids: Jour. Paleontology, v. 26, p. 338-352, 13 figs.

— 1953, Document 36/6: London, Bull. Zool. Nomenclature, v. 10, p. 373.

— 1954, Status of invertebrate paleontology, 1953, IV Echinodermata; Pelmatozoa: Harvard Univ. Mus. Comp. Zoology Bull., v. 112, no. 3, p. 125-149.

*— 1962a, Revision of Calceocrinidae: Kansas Univ. Paleont. Contr., Echinodermata Art. 4, 40 p., 3 pls.

*— 1962b, Ray structures of some inadunate crinoids: Kansas Univ. Paleont. Contr., Echinodermata Art. 5, 47 p., 4 pls.

Moore, R. C., and Ewers, J. D., 1942, A new species of *Synbathocrinus* from Mississippian rocks of Texas, with description of ontogeny. Denison Univ. Sci. Lab. Jour., v. 37, p. 92-106, 28 figs.

*Moore, R. C., and Jeffords, R. M., 1968, Classification and nomenclature of fossil crinoids based on studies of dissociated parts of their columns: Kansas Univ. Paleont. Contr., Echinodermata Art. 9, 86 p., 28 pls., 6 figs.

*Moore, R. C., Jeffords, R. M., and Miller, T. H., 1968a, Morphological features of crinoid columns: Kansas Univ. Paleont. Contr., Echinodermata Art. 8, 30 p. 4 pls., 5 figs.

— 1968b, Supplement to echinodermata articles 8-10: Kansas Univ. Paleont. Contr., 18 p.

Moore, R. C., and Laudon, L. R., 1942, *Megaliocrinus*, a new camerate crinoid genus from the Morrow series of northeastern Oklahoma: Denison Univ. Sci. Lab. Jour., v. 37, p. 67-79, 5 figs.

— 1943a, *Trichinocrinus*, a new camerate crinoid from Lower Ordovician (Canadian?) rocks of Newfoundland: Am. Jour. Sci., v. 241, p. 262-268, 2 pls.

*— 1943b, Evolution and classification of Paleozoic crinoids: Geol. Soc. America Spec. Paper 46, 167 p., 14 pls.

*— 1944, Class Crinoidea, *in* Shimer, H. W., and Shrock, R. R., Index fossils of North America: New York, John Wiley & Sons, Inc., p. 137-209, pls. 52-79.

Moore, R. C., and Strimple, 1942, *Metacatillocrinus*, a new inadunate crinoid genus from Pennsylvanian rocks of Oklahoma: Denison Univ. Sci. Lab. Jour., v. 37, p. 77-84, 6 figs.

Morzadec, P., 1967, Sur la présence du genre *Ancyrocrinus* Hall 1862 (crinoide) dans le dévonien moyen du massif Armoricain: Soc. Géol. et Minéral Bretagne Bull., n. ser. (1966), p. 26-32, 1 pl.

Mu, A. T., 1950, *Petalocrinus* from the Shihniulan Limestone of Wuchuan: Geol. Soc. China Bull., v. 29, p. 93-96, 3 figs.

*— 1954, On the occurrence of *Pisocrinus* in China: Acta Paläeont. Sinica, v. 2, no. 3, p. 323-332, 1 pl., 3 figs. (Chinese, p. 323-328, English résumé, p. 329-332.)

Nelson, S. J., 1965, Field methods in paleontology: Bull. Canadian Petroleum Geology, v. 13, 138 p., 42 pls., 9 figs.

Newell, N. D., Chronic, J., and Roberts, T. G., 1953, *Upper Paleozoic of Peru:* Geol. Soc. America Mem. 58, 276 p., 44 pls., 43 figs.

Newell, N. D., Rigby, J. R., Fischer, A. G., Whiteman, A. J., Hickox, J. E., and Bradley, J. S., 1953, The Permian reef complex of the Guadalupe Mountains region, Texas and New Mexico: San Francisco, W. H. Freeman and Co., 236 p., 32 pls., 85 figs.

Nicosia, M. L., 1954, Nuovo franmento de crinoide rinvenuto nella Pietra di Salomone (Sicilia): Italia Servizio Geol. Boll., v. 76, f. 1, p. 85-91.

Nissen, H. U., 1964, Dynamic and kinematic analysis of deformed crinoid stems in a quartz graywacke: Jour. Geology, v. 72, p. 346-360.

Norford, B. S., 1962, The Silurian fauna of the Sandpile Group of northern British Columbia: Canada Geol. Survey Bull., no. 78, 51 p., 16 pls.

Peck, R. E., 1956, Support for the proposal by J. Wright on *Actinocrinus gilbertsoni* Phillips, 1836 (Class Crinoidea): London, Bull. Zool. Nomenclature, v. 12, p. 319.

Peck, R. E., and Connelly, J. L., Jr., 1951, *Octocrinus* Peck and *Tytthocrinus* Weller, synonyms of *Amphipsalidocrinus* Weller: Jour. Paleontology, v. 25, p. 414-415, 8 figs.

Philip, G. M., 1961, Lower Devonian crinoids from Toongabbie, Victoria, Australia: Geol. Mag., v. 98, p. 143-160, pl. 8.

— 1963, Australian fossil crinoids. I. Introduction and terminology for the anal plates of crinoids: Linnean Soc. N. S. Wales Proc., v. 88, p. 259-272, 5 figs.

— 1964, Australian fossil crinoids. II. *Tribrachiocrinus clarkei* McCoy: Linnean Soc. N. S. Wales Proc., v. 89, p. 199-202, pl. 3.

— 1965a, Plate homologies in inadunate crinoids: Jour. Paleontology, v. 39, p. 146-149, 2 figs.

— 1965b, Ancestry of sea-stars: Nature, v. 208, p. 766-768, 2 figs.

Pickett, J. W., 1960, Note on a Carboniferous crinoid from Swain's Bully, Babbinboon, N.S.W.: Australian Jour. Sci., v. 23, p. 88, 2 figs.

Pinna, G., 1963, Escursione Sui Terreni Paleozoici (Silurico Superiore) Fra Ledbury E Le Malvern Hills (Herefordshire, Inghilterra): Natura, Milana, v. 54, p. 103-110, 2 figs.

Plummer, F. B., 1950, The Carboniferous rocks of the Llano region of central Texas: Texas Univ. Pub. 4329, p. 1-170, pls. 1-22, 14 figs.

Plummer, F. B., and Moore, R. C., 1921, Stratigraphy of the Pennsylvanian formations of north-central Texas: Texas Univ. Bull. 2132, 237 p., 27 pls., 19 figs.

Prokop, R. J., 1967, Některé způsoby zakotvení krinoidů ze spodnodevonských vrstev u Koněprus v Čechách (Some manners of attachment of crinoids in the lower Devonian beds at Koneprusy in Bohemia): Vestník Ustredniho Ustavu Geologisckého, v. 42, p. 367-368, 2 pls.

Ramsbottom, W.H.C., 1950, A new species of *Lyriocrinus* from the Wenlock Limestone: Ann. and Mag. Nat. Hist., Ser. 12, v. 3, no. 32, p. 651-656.

— 1951a, Two species of *Gissocrinus* from the Wenlock Limestone: Ann. and Mag. Nat. Hist., Ser. 12, v. 4, no. 41, p. 490-497.

— 1951b, The type species of *Periechocrinites* Austin and Austin: Ann. and Mag. Nat. Hist., Ser. 12, v. 4, no. 46, p. 1040-1043.

— 1952, Calceocrinidae from the Wenlock Limestone of Dudley: Great Britain Geol. Survey Bull. 4, p. 33-46.

— 1954, *Periechocrinus* versus *Periechocrinites:* Ann. and Mag. Nat. Hist., Ser. 12, v. 7, no. 81, p. 687-688.

— 1958, British Upper Silurian crinoids from the Ludlovian: Palaeontology, v. 1, p. 106-115, pls. 20, 21.

— 1960a, Synopsis of families and genera, *in* Wright, J., The British Carboniferous Crinoidea: Palaeont. Soc. Mon., v. 2, pt. 6, p. 335-341.

— 1960b, Stratigraphical distribution, *in* Wright, J., The British Carboniferous Crinoidea: Palaeont. Soc. Mon., v. 2, pt. 7, p. 342-343.

*— 1961, A monograph on British Ordovician Crinoidea: Palaeont. Soc., v. 114, 37 p., 8 pls.

*Regnéll, G., 1948, Swedish hybocrinida (Crinoidea inadunata disparata; Ordovician-Lower Silurian): Arkiv. Zoology, v. 40, no. 9, p. 1-27, 4 pls.

— 1960, Donnees concernant le developpement ontogenetique des Pelmatozoaires du Paleozoique (Echinodermes): Soc. Géol. France Bull., Ser. 7, v. I, p. 773-783, 6 figs.

Roy, S. K., 1941, The Upper Ordovician fauna of Frobisher Bay, Baffin Land: Field Mus. Nat. Hist. Geology Mem. 2, 212 p., 146 figs.

Rusconi, C., 1951, Fosiles cámbricos de Salagasta, Mendoza: Anal. Soc. Cient. Argentina, v. 152, p. 255-264.

— 1952, Los fosiles cámbricos de Salagasta, Mendoza: Rev. Mus. Nat. Hist. Mendoza, v. 6, p. 19-62, pls. 1-4, 17 figs.

Sahni, M. R., and Gupta, V. J., 1965a, An *Agassizocrinus* from the *Syringothyris* Limestone of Kashmir: Current Sci., v. 34, no. 7, p. 216-217, 1 fig.

— 1965b, Silurian crinoids from the Kashmir Himalayas: Panjab Univ. Res. Bull. (N. Ser.), v. 16, pt. 3, p. 247-248, 1 fig.

Sampelayo, P. H., 1946, Estudios Acerca del Carbonifero en España: España Inst. Geol. y Minero Bol., v. 59, p. 1-19, 4 pls.

Sangil, M. M., 1955, Sobre el valor estratigráfico y la clasificación de las placas columnares de los Crinoides: Estudios Geol., v. 11, no. 26, p. 249-257, 3 figs.

Schewtschenko, T. V., 1959, Stebli *Cupressocrinus* iz srednedevonskikh otlozhenii Zeravshano-Gissarskoi gornoi oblasti (The stems of *Cupressocrinus* from the Middle Devonian deposits of Zeravshan-Gissar mountain region): Akad. Nauk Tadzhik SSR Doklady, v. II, no. 4, p. 7-10, 1 pl.

*— 1964, Nizhnesiluriiskie krinoidei Tsentralnogo Tadzhikistana. V. sbornike Paleontologiia Tadzhikistana (Lower Silurian Crinoidea of central Tajikistan, *in* Collected works on the paleontology of Tajikistan): Akad. Nauk Tadzhik SSR Doklady, p. 8-20, 2 pls.

* 1966, Morskie lilii iz verkhnesiluriiskikh i nizhnedevonskikh otlozhenii yugo-zapadnogo Tian-Shanya i ikh stratigraficheskoe zhachenie (Crinoids from the Upper Silurian and Lower Devonian deposits of southwestern Tien Shan and their stratigraphical significance): Upravleniia Geologii Soveta Ministorov Tadzh. SSR. Trudy, Paleontologiia i stratigrafiia, vyp. 2, p. 123-188, 8 pls.

— 1967, Rannedevonskie morskie lilii semeistva Parahexacrinidae fam. nov. Zeravshanskogo khrebta (Early Devonian crinoids of the Parahexacrinidae fam. nov. of the Zeravshan mountain): Paleont. Zhur., no. 3, p. 76-88, 2 pls.

— 1968a, Klass Crinoidea, *in* Novye vidy drevnikh rasteny i bespozvonochnykh SSSR Vypusk II, chast 2 (Class Crinoidea, *in* New species of plants and invertebrates from the USSR, Issue 2, part 2): Vses. Nauchno-Issled. Geol. Inst. Trudy Gosgeoltekhizdat, p. 278-285, pl. 66, figs. 29-36.

— 1968b, Early Devonian sea lilies of the family Parahexacrinidae fam. nov. from the Zeravshan Stage: Paleont. Zhur., 1967, no. 3, p. 76-88, pls. 9, 10; 9 figs. (English translation of Schewtschenko, 1967.)

*Schmidt, W. E., 1941, Die Crinoideen des Rheinischen Devons II Teil: Abh. Reichs. Bodenfors., n. f., H. 182, 253 p., 26 pls., 62 figs.

— 1952, Crinoides y Blastoideos del Devónico inferior de Asturias: Pub. estranjeras sobre Geol. España Inst. L. Mallada, p. 119-170, pls. 1-10; figs. 1-11. (Spanish translation of Schmidt, 1931.)

Seilacher, Adolf, 1961, Ein Füllhorn aus dem Hunsrückschiefer: Natur u. Volk, v. 91, H. 1, p. 15-19, 4 figs.

Sieverts-Doreck, H., 1942, Crinoiden aus dem Perm Tasmanicus: Zentrbl. Miner., Abt. B, no. 7, p. 222-231, 7 figs.

— 1950, Über *Hexacrinus* und *Bactrocrinus*: Neues Jahrb. Geologie u. Paläontologie Monatsh., H. 3, p. 80-87, 4 figs.

— 1951a, Crinoiden aus dem Unterkarbondes Oberharzes: Neues Jahrb. Geologie u. Paläontologie Abh., bd. 93, H. 1, p. 117-143, pls. 8, 9.

— 1951b, Echinodermen aus dem spanischen Ober-Karbon: Paläont. Zeitschr., v. 24, no. 3-4, p. 104-119. (Spanish translation, 1953, Pub. extranjeras Geol. España, t. 7, no. 1, p. 23-46; unseen.)

— 1952, Über die sogenannten "Deckplattchen" gotlandischer Cyathocrinidae: Neues Jahrb. Geologie u. Paläontologie Monatsh., H. 9, p. 420-430.

— 1953, Über einige inadunate Crinoiden aus dem rheinischen Devon: Notizbl. Hess. Landesamt Bodenforsch, bd. 81, p. 75-87, pl. 4.

— 1954, Weitere Mitteilungen über *Myelodactylus* aus dem Mittle Devon der Eifel: Notizbl. Hess. Landesamt Bodenforsch, bd. 82, p. 46-49.

— 1957a, Neufunde von *Diamenocrinus* und *Ctenocrinus* aus der Siegen-Stufe des Siegerlandes: Notizbl. Hess. Landesamt Bodenforsch, bd. 85, p. 62-66, 1 fig.

— 1957b, Bemerkungen über altpaläozoische Crinoiden aus Argentinien: Neues Jahrb. Geologie u. Paläontologie Monatsh, H. 4, p. 151-156.

— 1962, Über eine neue Form von Tubusplatten devonischer Crinoiden: Hess. Landesamt Bodenforsch Wiesbaden, bd. 90, p. 106-116, 1 pl.

— 1963, Über Missbildungen bei *Cupressocrinus elongatus* aus dem Mittledevon der Eifel: Decheniana, bd. 115, p. 239-244, 1 pl.

— 1964, Crinoiden aus dem Paläozoikum des Kantabrischen Gebirges (Nordspanien): Mitt. Bayer. Staatssamml. Paläont. Hist. Geol., bd. 4, p. 1-12, 3 pls.

Simon, Wilhelm, 1950, Der Genotypus *Acanthocrinus longispina* F. A. Roemer (Rhodocrinitidae, Camerata, Crinoidea): Neues Jahrb. Geologie u. Paläontologie Monatsh., H. 12, p. 363-370, 3 figs.

Sinclair, G. W., 1945, Some Ordovician echinoderms from Oklahoma: Am. Midland Naturalist, v. 34, p. 707-716, pls. 1, 2; fig. 1.

Sizova, E. N., 1960, Znachenie iskoraemikh steblei morskikh lilii dlia stratigrafii devona i karbona Tsentralnogo Kazakhstana. Sbornik statei Materialy po geologii i poleznym iskopaemym Altaia i Kazakhstana (Significance of crinoid stems for the Devonian and Carboniferous stratigraphy of central Kazakhstan, *in* Collected articles on materials on the geology and mineral resources of Altay and Kazakhstan): Vses. Nauchno-Issled. Geol. Inst. Trudy, vyp. 33, p. 51-65.

Strimple, H. L., 1947, Three new crinoid species from the Virgil Series of southeastern Kansas: Bull. Am. Paleontology, v. 31, no. 124, 12 p., 2 pls.

— 1948a, *Peremistocrinus* [sic] from the Dewey limestone formation, Oklahoma: Geol. Mag., v. 85, p. 113-116, pl. 10.

— 1948b, Notes on *Phanocrinus* from the Fayetteville Formation of northeastern Oklahoma: Jour. Paleontology, v. 22, p. 490-493, pl. 77.

— 1948c, Crinoid studies: Part I, Two new species of *Allagecrinus* from the Pennsylvanian of Kansas and Texas; Part II, *Apographiocrinus* from the Altamont Limestone of Oklahoma: Bull. Am. Paleontology, v. 32, no. 130, 16 p., 2 pls.

— 1949a, *Mooreocrinus bowsheri*, new species from the Chester Series of northeastern Oklahoma: Am. Jour. Sci., v. 247, p. 128-131, 1 pl.

— 1949b, Evolution of *Delocrinus* to *Paradelocrinus*, and description of *Stuartwellercrinus argentinei*, sp. nov.: Geol. Mag., v. 86, p. 123-127, pl. 4.

*— 1949c, Crinoid studies: Part III, *Apographiocrinus arcuatus*, new species from the Missouri series of Oklahoma; Part IV, *Exocrinus*, new genus from the Pennsylvanian of Oklahoma; Part V, *Allosocrinus*, a new crinoid genus from the Pennsylvanian of Oklahoma; Part VI, *Allagecrinus copani*, new species from the Pennsylvanian of Oklahoma; Part VII, New species of crinoids from southeastern Kansas: Bull. Am. Paleontology, v. 32, no. 133, 42 p., 7 pls.

*— 1949d, Studies of Carboniferous crinoids: Part I, A group of Pennsylvanian crinoids from the Ardmore basin; Part II, Delocrinids of the Brownville formation of Oklahoma; Part III, Description of two new cromyocrinids from the Pennsylvanian of Nebraska; Part IV, On new species of *Alcimocrinus* and *Ulrichicrinus* from the Fayetteville Formation of Oklahoma: Palaeontographica Americana, v. 3, no. 23, 30 p., 5 pls.

— 1949e, Two new species of *Acrocrinus* from the Pennsylvanian of Oklahoma: Am. Jour. Sci., v. 247, p. 900-904, 1 pl.

— 1950a, Emendation of *Endelocrinus tumidus* (Strimple): Jour. Paleontology, v. 24, p. 112-113.

— 1950b, New species of *Utharocrinus* and *Lasanocrinus:* Jour. Paleontology, v. 24, p. 571-574, pl. 77.

— 1951a, Pennsylvanian crinoids from Lake Bridgeport, Texas: Jour. Paleontology, v. 25, p. 200-207, pls. 36-39.

— 1951b, New species of crinoids from the Pennsylvanian of Kansas: Jour. Paleontology, v. 25, p. 372-376, pls. 56, 57.

— 1951c, Some new species of Carboniferous crinoids: Bull. Am. Paleontology, v. 33, no. 137, 40 p., 5 pls.

1951d, New Desmoinesian crinoids: Washington Acad. Sci. Jour., v. 41, p. 191-194, 20 figs.

— 1951e, New crinoids from the Pitkin of Oklahoma: Washington Acad. Sci. Jour., v. 41, p. 260-263, 13 figs.

— 1951f, New Carboniferous crinoids: Jour. Paleontology, v. 25, p. 669-676, pls. 98, 99.

— 1951g, Notes on *Phanocrinus cylindricus* and description of new species of Chester crinoids: Washington Acad. Sci. Jour., v. 41, p. 291-294, 12 figs.

— 1952a, The arms of *Polusocrinus*: Washington Acad. Sci. Jour., v. 42, p. 12-14, 8 figs.

— 1952b, Some new species of crinoids from the Henryhouse Formation of Oklahoma: Washington Acad. Sci. Jour., v. 42, p. 75-79, 13 figs.

— 1952c, Notes on *Texacrinus*: Washington Acad. Sci. Jour., v. 42, p. 216-220, 16 figs.

— 1952d, The arms of *Haerteocrinus*: Washington Acad. Sci. Jour., v. 42, p. 245-248, 7 figs.

— 1952e, The arms of *Perimestocrinus*: Jour. Paleontology, v. 26, p. 784-788, pls. 112, 113.

— 1952f, New species of *Lecanocrinus*: Washington Acad. Sci. Jour., v. 42, p. 318-323, 17 figs.

— 1953a, A new species of *Archaeocrinus* from Oklahoma: Jour. Paleontology, v. 27, p. 604-606, 7 figs.

— 1953b, A new species of *Carinocrinus* from Oklahoma: Washington Acad. Sci. Jour., v. 43, p. 201-203, 2 figs.

— 1954a, New species of *Plummericrinus*: Jour. Paleontology, v. 28, p. 204-207, pl. 23.

— 1954b, Two new crinoid species from the Henryhouse of Oklahoma: Washington Acad. Sci. Jour., v. 44, p. 280-283, 10 figs.

— 1955, A new species of *Cymbiocrinus* from the Pitkin: Washington Acad. Sci. Jour., v. 45, p. 14, 1 fig.

— 1957, Two aberrant crinoid specimens: Washington Acad. Sci. Jour., v. 47, p. 369, 1 fig.

— 1959a, Crinoids from the Missourian near Bartlesville, Oklahoma: Oklahoma Geol. Notes, v. 19, p. 115-127, 2 pls.

— 1959b, The occurrence of *Galateacrinus allisoni* in Oklahoma: Oklahoma Geol. Notes, v. 19, p. 195-196.

— 1960a, The genus *Paragassizocrinus* in Oklahoma: Oklahoma Geol. Survey Circ. 55, p. 24, 3 pls.

— 1960b, Regressive evolution among Erisocrinids: Oklahoma Geol. Notes, v. 20, p. 151-155, 2 figs.

— 1960c, A new cromyocrinid from Brazil: Soc. Brasileira Geologia Bol., v. 9, p. 75-77, 3 figs.

— 1960d, The posterior interradius of Carboniferous inadunate crinoids of Oklahoma: Oklahoma Geol. Notes, v. 20, no. 10, p. 247-253.

— 1961a, Notes on two Chester crinoids: Oklahoma Geol. Notes, v. 21, p. 23-25, 3 figs.

— 1961b, An unusual *Agassizocrinus*: Oklahoma Geol. Notes, v. 21, p. 158-161, 5 figs.

— 1961c, New species of *Bronaughocrinus* and *Stuartwellercrinus* from the Carboniferous of Oklahoma: Oklahoma Geol. Notes, v. 21, p. 186-189, 1 pl.

— 1961d, New *Paradelocrinus* from Oklahoma: Oklahoma Geol. Notes, v. 21, p. 225-229, 1 pl.

— 1961e, Additional notes concerning *Paragassizocrinus:* Oklahoma Geol. Notes, v. 21, p. 294-298, 1 pl.

— 1961f, Morrowan *Hydriocrinus:* Oklahoma Geol. Notes, v. 21, p. 306-307.

*— 1961g, Late Desmoinesian crinoids: Oklahoma Geol. Survey Bull. 93, 189 p., 19 pls.

— 1962a, Platycrinitid columnals from the Pumpkin Creek Limestone: Oklahoma Geol. Notes, v. 22, p. 3-5, 7 figs.

— 1962b, *Endelocrinus bransoni*, a new species from the Lenapah Limestone: Oklahoma Geol. Notes, v. 22, p. 28-29, 3 figs.

— 1962c, Crinoids of the Brownville Formation: Oklahoma Geol. Notes, v. 22, no. 4, p. 109-110.

— 1962d, *Tarachiocrinus* and *Tholiacrinus:* Oklahoma Geol. Notes, v. 22, p. 135-136.

*— 1962e, Crinoids from the Oologah Formation: Oklahoma Geol. Survey Circ. 60, 75 p., 9 pls.

— 1962f, *Graphiocrinus stantonensis* in Oklahoma Geol. Notes, v. 22, p. 137-140, 3 figs.

— 1962g, Associated *Cryphiocrinus* and *Agassizocrinus:* Oklahoma Geol. Notes, v. 22, p. 183-188, 13 figs.

— 1962h, Microcrinoids of the St. Joe group: Oklahoma Geol. Notes, v. 22, no. 9, p. 245-246.

— 1962i, Suppression of *Ethelocrinus texasensis:* Oklahoma Geol. Notes, v. 22, p. 270-272.

— 1962j, *Zeacrinites* in Oklahoma: Oklahoma Geol. Notes, v. 22, p. 307-316, 1 pl., 7 figs.

— 1963a, Class Crinoidea, *in* Mudge, M. R., and Yochelson, E. L., Stratigraphy and paleontology of the uppermost Pennsylvanian and lowermost Permian rocks in Kansas: U.S. Geol. Survey Prof. Paper 323, p. 67-73, pl. 12, figs. 29, 30.

— 1963b, *Dasciocrinus* in Oklahoma: Oklahoma Geol. Notes, v. 23, p. 101-107, 1 pl.

— 1963c, A new species of *Graphiocrinus:* Oklahoma Geol. Notes, v. 23, p. 191-194, 3 figs.

*— 1963d, Crinoids of the Hunton Group: Oklahoma Geol. Survey Bull. 100, 169 p., 12 pls.

— 1966a, New species of cromyocrinids from Oklahoma and Arkansas: Oklahoma Geol. Notes, v. 26, p. 3-12, 2 pls.

— 1966b, A unique crinoid from the Upper Permian: Oklahoma Geol. Notes, v. 26, p. 80-84, 4 figs.

*— 1966c, Some notes concerning the Allagecrinidae: Oklahoma Geol. Notes, v. 26, p. 99-111, 1 pl., 2 figs.

— 1967, Aphelecrinidae, a new family of inadunate crinoids: Oklahoma Geol. Notes, v. 27, p. 81-85, 1 pl.

Strimple, H. L., and Beane, B. H., 1966, Reproduction of lost arms on a crinoid from Le Grand, Iowa: Oklahoma Geol. Notes, v. 26, p. 35-37, 1 fig.

Strimple, H. L., and Blythe, J. G., 1960, *Paragassizocrinus* in the Atoka of northeastern Oklahoma: Oklahoma Geol. Survey Circ. 55, p. 25-30, pl. 3.

Strimple, H. L., and Boyt, R., 1965, *Rhodocrinites beanei*, new species from the Hampton Formation (Mississippian) of Iowa: Oklahoma Geol. Notes, v. 25, p. 222-226, 2 figs.

Strimple, H. L., and Knapp, W. D., 1966, Lower Pennsylvanian fauna from eastern Kentucky; Part 2, Crinoids: Jour. Paleontology, v. 40, p. 309-314, pl. 36.

*Strimple, H. L., and Koenig, J. W., 1956, Mississippian microcrinoids from Oklahoma and New Mexico: Jour. Paleontology, v. 30, p. 1225-1247, 4 figs.

Strimple, H. L., and Watkins, W. T., 1949, *Hybocrinus crinerensis*, new species from the Ordovician of Oklahoma: Am. Jour. Sci., v. 247, p. 131-133, 1 pl.

— 1955, I. Three new genera, *in* New Ordovician echinoderms: Washington Acad. Sci. Jour., v. 45, no. 11, p. 347-353, figs. 1, 2, 4-6, 8-10.

— 1961, On *Synbathocrinus? antiquus*: Oklahoma Geol. Notes, v. 21, p. 48-49, 2 figs.

Struve, Wolfgang, 1957, Ein Massengrab kreidezeitlicher Seelilien: Natur u. Volk, v. 87, p. 361-373, 14 figs.

Stukalina, G. A., 1960, Kompleks llandoveriiskikh steblei morskikh lilii khrebta Chingiz (Complex Llandoverian crinoid stems from the Chingiz Mountain): Informatsionnyi Sbornik, Vses. Nauchno-Issled. Geol. Inst. Trudy, no. 35, p. 95-110, fig. 7.

— 1961, Stebli krinoidei iz otlozhenii verkhnego silura gor Aksarly (Tsentralnyi Kazakhstan) (Crinoid stems from Upper Silurian deposits of the Aksarli Mountains [central Kazakhstan]): Informatsionnyi Sbornik, Vses. Nauchno-Issled. Geol. Inst. Trudy, no. 42, p. 31-42, 2 pls., fig. 1.

— 1964, K metodike izucheniia i sboram iskopaemykh ostatkov steblei morskikh lilii. Sbornik statei. Materialy po geologii i poleznym iskopaemym Altaia i Kazakhstana (On the methods of research and collecting of fossil remains of crinoid stems, *in* Collected articles on materials of the geology and mineral resources of Altay and Kazakhstan): Vses. Nauchno-Issled. Geol. Inst. Trudy, nov. ser., t. III, p. 31-35, fig. 1.

— 1965a, O taksonomicheskom znachenii steblei drevnikh morskikh lilii (On taxonomical significance of ancient crinoid stems): Biostratigraficheskiy Sbornik, Vses. Nauchno-Issled. Geol. Inst. Trudy, nov. ser., t. 115, vyp. I, p. 210-217, 2 figs.

— 1965b, Morskie lilii karaespinskogo gorizonta. Sbornik statei. Stratigrafiia nizhnepaleozoiskikh i siluriiskikh otlozhenii Tsentralnogo Kazakhstana (Crinoids of the Karaespin horizon, *in* Collected articles on stratigraphy of the lower Paleozoic and Silurian deposits of central Kazakhstan): Vses. Nauchno-Issled. Geol. Inst. Trudy, p. 134-145, 2 pls., 7 figs.

— 1965c, Novye vidy *Hexacrinites* Tsentralnogo Kazakhstana (New species of *Hexacrinites* from central Kazakhstan): Vses. Ezhegodnik Paleontologicheskogo Obshch. t. XVII, p. 188-195, 1 pl.

*— 1966, O printisipakh klassifikatsii steblei drevnikh morskikh lilii

(On the principles of classification of Paleozoic crinoid stems):
Paleont. Zhur., no. 3, p. 94-102, 2 figs.

— 1967, O taksonomicheskikh priznakakh segmehtirovannykh steblei morskikh
lilii (On taxonomic features of segmented stems of crinoids): Biostratigraficheskiy Sbornik: Vses. Nauchno-Issled. Geol. Inst. Trudy,
nov. ser., t. 192, vyp. 3, p. 200-206, 6 figs.

— 1968a, Stratigraphic significance of the stems of crinoids, in Oswald,
D. H., ed., Solving the Siluro-Devonian boundary problem: Internat.
Symposium Devonian System Proc., Calgary, Canada, v. 2, p. 893-896.

*— 1968b, K sistematike semeistva Decacrinidae (On the systematics of the
family Decacrinidae): Vses. Ezhegodnik Paleontologicheskogo Obshch.,
t. XVIII, p. 250-267, 1 pl., 2 figs.

— 1968c, K sistematike gruppy Pentamerata (Crinoidea) (On the systematics
of the group Pentamerata [Crinoidea]): Paleont. Zhur., no. 1, p. 81-91,
2 figs.

Stukalina, G. A., and Tuyutyan, Yu A., 1967, Novye Ordovikskie morskie
lilii Kazakhstana (New Ordovician crinoids from Kazakhstan): Akad.
Nauk Kazakh. SSR Izv. Ser. Geol., no. 4, p. 72-76, 12 figs.

Stürmer, W., 1965, Rontgehaufnahmen von einigen Fossilien aus dem
Geologischen Institut der Universitat Erlangen-Nurnberg: Geol. Blätter
Nordost-Bayern, v. 15, no. 4, p. 217-223, pl. 6, 7 figs.

Svoboda, Josef, and others, 1968, Regional geology of Czechoslovakia:
Part I, The Bohemian Massif: Geol. Survey Czechoslovakia, 668 p.

Talent, J. A., 1963, The Devonian of the Mitchell and Wentworth Rivers:
Geol. Survey Victoria Mem., no. 24, p. 1-122, pls. 1-78, 33 figs.

*Teichert, Curt, 1949, Permian crinoid *Calceolispongia:* Geol. Soc. America
Mem. 34, 132 p., 26 pls.

— 1954, A new Permian crinoid from western Australia: Jour. Paleontology,
v. 28, p. 70-75, pls. 13, 14.

Termier, G., and Termier, H., 1947, Paléontologie Marocaine I. Généralités
sur les invertébres fossiles: Notes et Mém. no. 69, Serv. Géol.
Morocco, 391 p., 22 pls.

*— 1950, Paléontologie Marocaine II. Invertébres de l'Ere Primaire: Notes
et Mém. no. 79, fasc. IV, Serv. Géol. Morocco, 279 p., pls. 184-241.

Termier, H., and Termier, G., 1949, Hiérarchie et corrélations des
caractères chez les crinoides fossiles: Algeria, Serv. Cart. Géol.
Bull., Ser. 1, no. 10, 91 p., 8 pls.

— 1955, Rectification de nomenclature; *Becharocrinus* nov. gen. pro
Triacrinus H. and G. Termier, 1950: Soc. Géol. France Compte Rendu
no. 15-16, p. 316-317.

— 1958, Les Echinoderms Permiens du Djebel Tebaga (Extrême Sud Tunisien):
Soc. Géol. France Bull., Ser. 6, v. 8, p. 51-64, 7 pls.

Tesmer, I. H., 1964, Geology of Chautauqua Co., N. Y. Part 1, Stratigraphy
and paleontology (Upper Devonian): New York State Mus. and Sci. Service
Bull., no. 391, 65 p., 5 pls., 17 figs.

Tischler, Herbert, 1963, Fossils, faunal zonation, and depositional environment of the Madera Formation, Huerfano Park, Colorado: Jour. Paleontology, v. 37, p. 1054-1068, pls. 139-142, 6 figs.

Trunko, Laszlo, 1966, Ein seltener Crinoiden-Kelch aus dem rheinischen
Mitteldevon: Beitr. Naturk. Forsch. SW Deutschl., bd. 25, hf. 1,
p. 93-95, pls. 3-6.

*Ubaghs, G., 1943, Note sur la morphologie, la biologie et la systématique du genre *Mespilocrinus* de Koninck et Le Hon: Belgique Mus. Royal Hist. Nat. Bull., v. 19, no. 15, 16 p., 11 figs.

* —— 1945a, Contribution à la Connaissance des Crinoides de l'Éodévonien de la Belgique. I. Revision systematique des Melocrinitidae: Belgique Mus. Royal Hist. Nat. Bull., v. 21, no. 15, 24 p., 2 pls., 3 figs.

*—— 1945b, Contribution à la Connaissance des Crinoides de l'Éodévonien de la Belgique. II. La morphologie des bras chez *Ctenocrinus* Bronn: Belgique Mus. Royal Hist. Nat. Bull., v. 21, no. 16, 24 p., 1 pl.

*—— 1947, Contribution à la Connaissance des Crinoides de l'Éodévonien de la Belgique. III. L'appareil brachial d'*Acanthocrinus* Roemer et de *Diamenocrinus* Oehlert: Belgique Mus. Royal Hist. Nat. Bull., v. 23, no. 4, 31 p., 3 pls., 2 figs.

*—— 1950, Le genre *Spyridiocrinus* Oehlert: Annales Paléont., v. 36, p. 107-122, pl. 7.

*—— 1952, *Ammonicrinus* Springer, Crinoidea Flexibilia du Dévonien moyen d'Allemagne: Senckenbergiana Lethaea, bd. 33, no. 4-6, p. 203-226.

—— 1953, Crinoides, *in* Piveteau, J., director, Traité de Paléontologie, T. III: Paris, Masson, p. 658-756.

*—— 1956a, Recherches sur les Crinoides Camerata du Silurien de Gotland (Suède), Introduction generale et Partie I: Morphologie et Paleonbiologie de *Barrandeocrinus sceptrum* Angelin: Arkiv Zoologi, ser. 2, bd. 9, no. 26, p. 515-550, 7 pls., 15 figs.

*—— 1956b, Recherches sur les Crinoides Camerata du Silurien de Gotland (Suède), Partie II: Morphologie et position systematique de *Polypeltes granulatus* Angelin: Arkiv Zoologi, ser. 2, v. 9, no. 27, p. 551-572, 4 pls.

—— 1958a, Morphologie et position systematique de *Methabocrinus erraticus* Jaekel (Crinoidea Camerata): Paläont. Zeitschr., bd. 32, no. 1-2, p. 52-62.

*—— 1958b, Recherches sur les Crinoides Camerata du Silurien de Gotland (Suède), Partie III: Melocrinicae: Arkiv Zoologi, ser. 2, v. 11, no. 16, p. 259-306, 5 pls.

Ubaghs, G., and Bouček, B., 1962, Sur la présence du genre *Tiaracrinus* Schultze (Crinoidea) dans le Dévonien inférieur de Moravie *(Tiaracrinus moravicus* n. sp.): [Czechoslovakia] Ustřed. Ústav. Geol. Sborník, Svazek 27, p. 41-52, 2 pls.

Van Sant, J. F., 1965, *Actinocrinites grandissimus*, a new name for a camerate crinoid from Borden (lower-Middle Mississippian) rocks of Indiana: Jour. Paleontology, v. 39, p. 290-292, 2 figs.

*Van Sant, J. F., and Lane, N. G., 1964, Crawfordsville (Indiana) crinoid studies: Kansas Univ. Paleont. Contr., Echinodermata Art. 7, 136 p., 8 pls.

Vialov, O. P., 1953a, K voprosu o klassifikatsii stebelkov morskikh lilii (On the question of classification of crinoid stems): Akad. Nauk SSSR Doklady, t. 89, no. 6, p. 1087-1090.

—— 1953b, O klassifikatsii stebelkov morskikh lilii (On the classification of crinoid stems): Trudy Lvovskogo geologicheskogo obshchestva, Paleontologicheskaia seriia, vyp. 2, p. 30-45.

Walter, J. C., Jr., 1953, Paleontology of Rustler Formation, Culberson County, Texas: Jour. Paleontology, v. 27, p. 679-702, pls. 70-73, 3 figs.

Wanner, J., 1942a, Einige neue Krinoiden aus dem Mittel Devon der Eifel: Decheniana, bd. 101 A B, Festschrift, p. 25-38, 1 pl.

— 1942b, Beiträge zur Paläontologie des Ostindischen Archipels; XIX, Die Crinoidengattung *Paradoxocrinus* aus dem Perm von Timor: Zentralb. Mineralogie, u. Palaeontologie, Abt. B, no. 7, p. 201-214, 6 figs.

— 1948, Revue of numerous echinoderm articles: Zentralb. Mineralogie, Geologie u. Paläeontologie, Teil, III, pt. 2, p. 319-336.

— 1949a, Neue Beiträge zur Kenntnis der permischen Echinodermen von Timor, XVI. Poteriocrinidae, Teil 4: Palaeontographica Supp., bd. IV, 56 p., 3 pls.

— 1949b, Ungleiche Primaxillaria und andere Abweichungen von der normalen symmetrie bei palaozoischen Crinoiden: Neues Jahrb. Mineralogie, Geologie u. Paläontologie, Abh., Abt. B, bd. 91, II. 1, p. 81-100.

— 1950, Über die Crinoidengattung *Timorocidaris:* Neues Jahrb. Geologie u. Paläontologie Monatsh., H. 12, p. 360-370, 3 figs.

— 1953, Document 36/3, -*crinites* or -*crinus?:* London, Bull. Zool. Nomenclature, v. 10, p. 367-371.

— 1954, Die Analstruktur von *Ammonicrinus* Springer nebst Bemerkungen über Aberrauzen und Anomalien bei Krinoiden: Neues Jahrb. Geologie u. Paläontologie Monatsh., H. 5, p. 231-236, 3 figs.

Washburn, A. T., 1968, Early Pennsylvanian crinoids from the south-central Wasatch Mountains of central Utah: Brigham Young Univ. Geology Studies, v. 15, pt. 1, p. 115-132, 3 pls., 1 fig.

Webby, B. D., 1961, A Middle Devonian inadunate crinoid from West Somerset, England: Palaeontology, v. 4, p. 538-541, pl. 67.

— 1965, *Quantoxocrinus*, a new Devonian inadunate crinoid from West Somerset, England: Palaeontology, v. 8, p. 11-15, pl. 4.

*Webster, G. D., and Lane, N. G., 1967, Additional Permian crinoids from southern Nevada: Kansas Univ. Paleont. Contr. Paper 27, 32 p., 8 pls.

Welby, C. W., 1962, Paleontology of the Champlain basin in Vermont: Vermont Geol. Survey Spec. Pub., v. 1, 99 p., 16 pls., 19 figs.

Weller, J. Marvin, 1947, Tegmen structure of *Pterotocrinus capitalis* (Lyon): Jour. Paleontology, v. 21, p. 581-584, 4 figs.

Wells, J. W., 1963, Early investigations of the Devonian System in New York, 1656-1836: Geol. Soc. America Spec. Paper 74, 74 p., 11 pls.

Weyer, Dieter, 1965, *Triacrinus* Münster, 1839 (Crinoidea) aus der *Wocklumeria*- Stufe des thuringischen Oberdevons: Deutschenia, Geol. Jg. 14, H. 8, p. 969-981, 2 pls.

Whitcomb, Lawrence, 1963, *Dolatocrinus* from the Centerfield, Monroe County, Pennsylvania: Pennsylvania Acad. Sci. Proc., v. 36, p. 38-41, 1 fig.

Williams, J. S., 1943, Stratigraphy and fauna of the Louisiana limestone of Missouri: U.S. Geol. Survey Prof. Paper 203, 133 p., 9 pls., 9 figs.

*Wilson, A. E., 1946, Echinodermata of the Ottawa Formation of the Ottawa-St. Lawrence Lowland: Canada Geol. Survey Bull. 4, 61 p., 6 pls., 2 figs.

Wilson, C. W., Jr., 1949, Pre-Chattanooga stratigraphy in central Tennessee: Tennessee Div. Geology Bull. 56, 407 p., 28 pls., 89 figs.

Wright, James, 1942, New British Carboniferous crinoids: Geol. Mag., v. 79, p. 269-283, pls. 9-12.

— 1943a, *Pimlicocrinus* gen. nov. and two new species of *Amphoracrinus* from the Carboniferous limestone: Geol. Mag., v. 80, p. 81-94, pls. 3-6.

1943b, Notes on *Actinocrinites elephantinus* Austin and *Amphoracrinus brevicalix* Rofe: Geol. Mag., v. 80, p. 231-236, pl. 8.

— 1944, *Rhabdocrinus* n. g. from the Scottish Carboniferous limestone: Geol. Mag., v. 81, p. 266-271, pls. 10, 11.

— 1945, *Tyrieocrinus* (gen. nov.) and *Scotiacrinus* (gen. nov.) and seven new species of inadunate crinoids from the Carboniferous limestone of Scotland and Yorkshire: Geol. Mag., v. 82, p. 114-125, pls. 2, 3.

— 1946a, *Caldenocrinus* gen. nov. and related Scottish crinoids: Geol. Mag., v. 83, p. 33-38, pls. 3, 4.

— 1946b, New species of *Taxocrinus* and *Synbathocrinus* and other rare crinoids from the Carboniferous limestone of Coplow Knoll, Clitheroe: Geol. Mag., v. 83, p. 121-128, pls. 7-9.

— 1947, *Steganocrinus westheadi* n. sp. and note on a rare crinoid and a blastoid from the Carboniferous limestone of Coplow Knoll, Clitheroe: Geol. Mag., v. 84, p. 101-105, pl. 3.

— 1948, *Scytalocrinus seafieldensis* sp. nov. and a rare *Ureocrinus* from the Carboniferous limestones of Fife, with notes on a blastoid and two crinoids from the Carboniferous limestone of the Clitheroe area: Geol. Mag., v. 85, p. 48-52, pl. 5.

*— 1950-1960, The British Carboniferous Crinoidea: Palaeont. Soc. Mon., v. 1, pt. 1, 24 p., pls. 1-7, 1950; v. 1, pt. 2, p. 24-46, pls. 8-12, 1951a; v. 1, pt. 3, p. 47-102, pls. 13-31, 1951b; v. 1, pt. 4, p. 103-148, pls. 32-40, 1952; v. 1, pt. 5, p. 149-190, pls. 41-47, 1954; v. 2, pt. 1, p. 191-254, pls. 48-63, 1955a; v. 2, pt. 2, p. 255-272, pls. 64-67, 1955b; v. 2, pt. 3, p. 273-306, pls. 68-75, 1956; v. 2, pt. 4, p. 307-328, 1958; v. 2, pt. 5, p. 329-334, pls. A, B, 1960.

— 1952b, Crinoids from Thornton Burn, East Lothian: Geol. Mag., v. 89, p. 320-327, pl. 13.

— 1952c, Anomalous excavations in the radials of *Hydreionocrinus parkinsoni* Wright: Edinburgh Geol. Soc. Trans., v. 15, p. 406-409.

— 1954b, *Idosocrinus* gen. nov. and other crinoids from Thornton Burn, East Lothian: Geol. Mag., v. 91, p. 167-170.

— 1956a, Proposed determination under the plenary powers of the interpretation of the nominal species *Actinocrinus gilbertsoni* Phillips, 1836 (Class Crinoidea): London, Bull. Zool. Nomenclature, v. 12, p. 156-158.

— 1957, Addendum to application relating to the proposed determination under the plenary powers of the interpretation of the nominal species *Actinocrinus gilbertsoni* Phillips, 1836 (Class Crinoidea): London, Bull. Zool. Nomenclature, v. 13, p. 25.

Wright, James, and Ramsbottom, W.H.C., 1960, The British Carboniferous Crinoidea: Palaeont. Soc. Mon., v. 2, pt. 5, p. 329-347, pls. A, B.

Wright, James, and Strimple, H. L., 1945, *Mooreocrinus* and *Ureocrinus* gen. nov., with notes on the family Cromyocrinidae: Geol. Mag., v. 82, p. 221-229, pl. 9.

Xu, I-Wen, 1962, *Caelocrinus*--nouveau genre of crinoid of Middle Silurian age from the province of Szuchuan: Acta Paläeont. Sinica, v. 10, no. 1, p. 45-54, 1 pl., 2 figs. (Chinese, p. 45-49; Russian, p. 50-53.)

— 1963, Some "hexagonal lilies" (*Hexacrinites*) from the Middle Devonian section in the Syansyan district of Kwangsi Province: Acta Paläeont. Sinica, v. 11, no. 1, p. 108-118. (Chinese, p. 108-113; Russian, p. 113-118.)

Yakovlev, N. N., 1939a, Klass Morskie Lilii—Crinoidea. V kn Atlas rukovodyashchikh form iskopaemkh faun SSSR T. 5 Srednii i Verkhnii Otdel Kamennougolnoi Sistem (Class Marine Lilies—Crinoidea, *in* Gorsky, I., ed., Atlas of the leading forms of the fossil faunas of USSR, v. 5, middle and upper Carboniferous): Leningrad Central Geol. and Prospecting Inst., p. 64-67, pls. 11, 12.

— 1939b, Tip Echinodermata Iglokozhie V kn Atlas rukovodyashchikh form iskopaemkh faun SSSR T. 6 Permskaya Sistem (Phylum Echinodermata, *in* Licharew, B., ed., Atlas of the leading forms of the fossil fauna of USSR, v. 6, Permian System): Leningrad Central Geol. and Prospecting Inst., p. 58-63, pl. 10.

— 1939c, Ob otkrtii originalnogo parazita kamennougolnikh morskikh lilii (On the discovery of an original parasite on a marine crinoid from the Carboniferous): Akad. Nauk SSSR Doklady, v. 22, no. 3, p. 146-148, 7 figs.

— 1939d, Sur la decouverte d'un parasite original des crinoides marins du carbonifère: Acad. Sci. URSS Comptes Rendus (Doklady), v. 22, no. 3, p. 146-148, 7 figs.

— 1939e, Zametki o permskikh Pelmatozoa (Notes on Permian Pelmatozoa): Akad. Nauk SSSR Doklady, v. 24, no. 8, p. 832-833.

— 1939f, Notes sur quelques Pelmatozoa Permiens: Acad. Sci. URSS Comptes Rendus (Doklady), v. 24, no. 8, p. 832-833.

— 1940a, O nakhodke *Eucalyptocrinus* v nizhnem devone Urala (On an *Eucalyptocrinus* from the Lower Devonian of the Urals): Akad. Nauk SSSR Doklady, v. 27, no. 2, p. 192, fig.

— 1940b, Sur une trouvaille d'*Eucalyptocrinus* dans le Dévonien inférieur de l'Oural: Acad. Sci. URSS Comptes Rendus (Doklady), v. 27, no. 2, p. 193, fig.

— 1941a, Morskie lilii glavnogo devonskogo polia V kn Fauna glavnogo devonskogo polia. M.-L. (Crinoids of the main Devonian field, *in* Fauna of the main Devonian field, Middle and Lower): Akad. Nauk SSSR, p. 323-331, 2 pls. (English resume, p. 329-331.)

— 1941b, Deux nouveaux échinodermes des dépôts permiens du Timan: Akad Sci. URSS Comptes Rendu (Doklady), v. 32, no. 1, p. 102-104, illus.

— 1944a, Novoe o stroenii i dykhatelnoi funktsii analonogo meshke morskikh lilii (News on the structure and respiratory function of the anal sac of crinoids): Akad. Nauk SSSR Doklady, v. 44, no. 3, p. 116-117.

— 1944b, On the structure and respiratory function of the anal sac of the crinoids: Acad. Sci. URSS Comptes Rendus (Doklady), v. 44, no. 3, p. 116-117.

— 1945a, O mshankovykh i krinoidiykh rifakh permskogo perioda na Urale (On bryozoan and crinoidal reefs of the Permian period in the Urals): Akad. Nauk SSSR Doklady, v. 48, no. 5, p. 374-376.

— 1945b, On bryozoan and crinoidal reefs of the Permian period in the Urals: Acad. Sci. URSS Comptes Rendus (Doklady), v. 48, no. 5, p. 352-354.

— 1946a, Un hexacrinide du silurien superieur: Acad. Sci. URSS Comptes Rendus (Doklady), v. 51, no. 2, p. 153-154.

— 1946b, Ob atavisticheskikh iavleniiakh neotenii u morskikh lilii (On atavistic phenomena of neotenia in Crinoidea): Akad. Nauk SSSR Doklady, v. 51, no. 3, p. 225-227.

— 1946c, On atavistic phenomena of neotenia in Crinoidea: Acad. Sci. URSS Comptes Rendus (Doklady), v. 51, no. 3, p. 229-231.

— 1946d, Onakhodke roda *Wachsmuthicrinus* v Rossii i ego proiskhozhdenii (On the finding of the genus *Wachsmuthicrinus* in Russia and on its origin): Akad. Nauk SSSR Doklady, v. 54, no. 3, p. 263-265, 2 figs.

— 1946e, On the finding of the genus *Wachsmuthicrinus* in Russia and on its origin: Acad. Sci. URSS Comptes Rendus (Doklady), v. 54, no. 3, p. 263-265, 2 figs.

— 1947a, Tip Echinodermata. Iglokozhie. Class Crinoidea. Morskie lilii V kn Nalivkin, D., ed., Atlas rukovodiashikikh form iskopaemykh fauna SSSR, T. 3, Devonskaia sistema M.-L. (Type Echinodermata, class Crinoidea, crinoids, *in* Atlas of guide forms of fossil faunas of the SSSR, T. 3, Lower and Middle Devonian): Gosgeolizdat, p. 55-57, pl. 10.

— 1947b, Izmeneniia skeletnykh chastei morskikh lilii vsledstvie mekhanicheskikh faktorov (Changes in the skeletal parts of crinoids on account of mechanical factors): Akad. Nauk SSSR Doklady, v. 56, no. 7, p. 747-749, 2 figs.

— 1947c, *Jaekelicrinus* gen. nov. nedostavav shii chlen fileticheskogo riada Pisocrinidae (*Jaekelicrinus* gen. nov. member of the family Pisocrinidae): Akad. Nauk SSSR Doklady, v. 57, no. 6, p. 609-611, illus.

— 1947d, Vliianie mekhanicheskikh uslovii na stroenie morskikh lilii (Influence of mechanical conditions on the structure of crinoids): Priroda [U.S.S.R.], no. 11, p. 41-47, 14 figs.

— 1948, Novye permskie morskiy lilii iz Severnogo Timana (New Permian crinoids from northern Timan): Akad. Nauk SSSR Izv., Ser. Biol., no. 1, p. 119-122, illus.

— 1949a, Eshche odna kategoriya vliyaniya mekhanicheskikh usloviy no stroenie morskikh lilii (Another category of influence of mechanical conditions on the structure of crinoids): Akad. Nauk SSSR Doklady, v. 66, no. 2, p. 265, 267, 1 fig.

— 1949b, Proiskhozhdenie roda *Indocrinus* ot *Ulocrinus* i faktory evolyutsii (Origin of the genus *Indocrinus* from *Ulocrinus* and factors of evolution): Akad. Nauk SSSR Doklady, v. 67, no. 5, p. 897-900, 2 figs.

— 1949c, O sushchestvovanii v verkhnem silure i nizhnem devone SSSR morskikh lilii sem Crotalocrinitidae (On the existence of crinoids of the family Crotalocrinitidae in the Upper Silurian and Lower Devonian of Russia): Vses. Paleont. Obshch. Ezhegodnik ob-va, v. 13, p. 14-19, 2 pls.

— 1949d, *Jaekelicrinus bashkiricus* Yakovlev, n. gen., n. sp.: Jour. Paleontology, v. 23, p. 435, 1 fig.

— 1950, O tipakh skulptury chashechek morskikh lilii, ikh proiskhozhdenii i naznachenii (On sculpture patterns of crinoid calyces, their origin and setting): Akad. Nauk SSSR Doklady, v. 70, no. 1, p. 93-96, 1 pl., 1 fig.

— 1951, Vozniknovenie odnorukosti u morskikh lilii (Origin of one-armed crinoids): Akad. Nauk SSSR Doklady, v. 78, no. 3, p. 577-579, 1 fig.

— 1952a, Samoregulirovanie i formoobrazovanie u morskikh lilii (Self-regulation and production of form in crinoids): Akad. Nauk SSSR Doklady, v. 86, no. 4, p. 827-828.

— 1952b, Organizm i sreda (no Paleontologicheskom materiale) (Organisms and environment on paleontological material): Zhur. Obshch. Biologii, v. 13, no. 2, p. 143-152, 1 pl., 1 fig.

— 1953, O nakhodke lobolitov v SSSR i o biologicheskom znachenii ikh (On the discoveries of loboliths in the USSR and thie biological significance): Vses. Paleont. Obshch. Ezhegodnik, v. 14, p. 18-31, pls. 1-3.

— 1954a, Gigantskaia morskaia liliia iz kamennougolnykh otlozhenii Kazakhstana *Blothrocrinus litvinovitschae* sp. n. (A gigantic crinoid from the Carboniferous deposits of Kazakhstan, *Blothrocrinus litvinovitschae* sp. n.): Akad. Nauk SSSR Izv., Ser. Geol., v. 99, no. 4, p. 113-115, 3 figs.

— 1954b, O napravlenii izmeneniia bazisa chashechki morskikh lilii i o prichinakh etogo izmeneniia (About the order of change in the base of the calyx of crinoids and reasons for this change): Akad. Nauk SSSR Doklady, v. 99, no. 6, p. 1087-1090, 1 fig.

*— 1956a, Organizm i sreda. Stati po paleoekologii besnozvonochnykh 1913-1956 gg (Organisms and environment. Writings on paleoecology of invertebrates, 1913-1956): Akad. Nauk SSSR, 140 p., 2 pls.

— 1956b, Morskie lilii ovronezhskogo devona (Crinoids from the Devonian of the Voronezh district): Akad. Nauk SSSR Izv., Ser. Biol., no. 2, p. 91-93, 1 pl.

— 1956c, K peresmotru kharakteristiki roda *Ristnacrinus* Öpik (A revision of the characteristics of the genus *Ristnacrinus* Opik): Vses. Paleont. Obshch. Ezhegodnik, ob-va, v. 15, p. 155-157, 2 figs.

— 1957, Dve linii razvitiia morskikh lilii kromiokrinid v sviazi s ikh geograficheskim rasprostraneniem. V kn Voprosy paleobiogeografii i biostratigrafii (Two lines of development of cromyocrinid crinoids in connection with their geographical dissemination, *in* Questions of paleobiogeography and biostratigraphy): Trudy I sessii Vsesoiuznogo Paleontologicheskogo obshchestva 24-28 ianvaria 1955 g. M., Gosgeoltekhizdat, p. 11-14, 1 pl.

— 1958, Organizm i sreda: Peking, Kesyue chubanshe, 122 p., illus.

— 1961, Morskie lilii iz nizhnego karbona Donbassa (Crinoids from the Lower Carboniferous of the Donetz Basin): Geol. Sbornik (Lvov. Geol. Obshch.), no. 7-8, p. 417-420, 3 pls. (English résumé).

— 1964a, Organizm i sreda. Stati po paleoekologii Bespozvonochnykh 1913-1960 gg (Organisms and environment. Writings of palaeoecology of invertebrates 1913-1960): Akad. Nauk SSSR Paleont. Inst. Izdatel'stvo 2, 148 p., 1 pl.

*— 1964b, Klass Krinoidea. Morskie lilii. Obshchaya chast. V kn Orlov, Yu. A., ed., Osnovy paleontologii, Iglokozhi, Gemikhordovye, Pogonofory, i Shchetinkocheltustny (Class Crinoidea, crinoids, general part, characteristics, and morphology, *in* Fundamentals of paleontology, Echinodermata, Hemichordata, Pogonophora, and Chaetognatha): Moscow, Izdatel'stvo Nedra, p. 54-74, figs. 51-113.

*Yakovlev, N. N., and Ivanov, A. P., 1956, Morskie lilii i blastoidei kamennougolnykh i permskikh otlozhenii SSSR (Crinoids and blastoids of the Carboniferous and Permian deposits of Russia): Vses. Nauchno-Issled. Geol. Inst. Trudy, 142 p., 21 pls., 23 figs.

Yelkin, E. A., Gratsianova, R. T., and Dubatolova, Yu. A., 1967, O stratigrafii terrigenno-karbonatnykh otlozhenii srednego devona r. Chumysha (Salair) i ikh korreliatsii. Sbornik Novye dannye po biostratigrafii devona i verkhnego paleozoia Sibiri (On the stratigraphy of the Middle Devonian terrigeno-carbonaceous deposits of the Chumysh river (Salair) and on their correlation, *in* Collected articles on new facts in the biostratigraphy of the Devonian and upper Paleozoic of Siberia): Moscow, Izdatel'stvo Nauka, p. 1-9, pl. 6, 6 figs.

Yeltyschewa, R. S., 1955a, Morskie lilii (stebli morskikh lilii. V kn Polevoi atlas ordovikskoi i siluriiskoi fauny Sibirskoi platformy (Crinoids [crinoid stems], *in* Field atlas of the Ordovician and Silurian fauna of the Siberian Platform): Leningrad. Vses. Geol. Inst. (Gosgeoltekhizdat), p. 40-47, 3 pls.

— 1955b, Morskie lilii (stebli morskikh lilii V Kn Polevoi atlas kharaklernikh kompleksov fauny i flory devonskikh otlozenii Minusinskoi kotloviny (Crinoids [crinoid stems], *in* Field atlas of characteristic complex fauna and flora from the Devonian deposits of the Minusinsk Trough): Leningrad, Vses. Geol. Inst. Gosgeoltekhizdat, p. 36-37, pl. 15.

*— 1956, Stebli morskikh lilii i ikh klassifikatsiia (Crinoid stems and their classification): Leningrad. Univ. Vestnik, Ser. Geologiya i Geografii, vyp. 2, p. 40-46, 3 figs.

— 1957, O novom semeistve paleozoiskikh morskikh lilii (On a new family of Paleozoic crinoids): Ezheg. Vses. Paleont. Soc., v. 14, p. 218-234, 3 pls.

*— 1959, Printsipy klassifikatsii, metodika izudeniia i stratigraficheskoe znachenie steblei morskikh lilii (Principles of classification, techniques of study, and stratigraphical significance of crinoid stems): Trudy II sessii Vses. Paleont. Soc., p. 230-235, 2 figs.

*— 1960, Ordovikskie i siluriiskie krinoidei Sibirskoi platformy (The Ordovician and Silurian Crinoidea of the Siberian Platform): Vses. Nauchno-Issled. Geol. Inst. Trudy, vyp. 3, p. 1-39, 6 pls.

*— 1964a, Stebli ordovikskikh morskikh lilii Pribaltiki (nizhnii ordovik) (The Ordovician crinoid stems of the Baltic Sea area [Lower Ordovician]). Voprosy Paleontologii, v. 4: Leningrad, Univ. Izdatel'stvo, p. 59-84, 4 pls.

— 1964b, O klassifikatsii steblei morskikh lilii. V kn Orlov, Yu. A., ed., Osnovy Paleontologii, Iglokozhi, Gemikhordovye, Pogonofory, i Shchetinkochelyustnye (On the classification of crinoid stems, *in* Fundamentals of paleontology, Echinodermata, Hemichordata, Pogonophora, and Chaetognatha): Moscow, Izdatel'stvo Nedra, p. 74-75, 80, 228, pl. 115, figs. 9-23; fig. 114.

— 1965, Morskie lilii silura SSSR. V kn Stratigrafiia SSSR Siluriiskaia sistema (The Silurian Crinoidea of the USSR, *in* Stratigraphy of the USSR, Silurian System): Moscow, Izdatel'stvo Nauka, p. 450-453.

*— 1966, Stebli ordovikskikh morskikh lilii Pribaltiki (srednii ordovik) (Stems of the Ordovician Crinoidea of the Baltic Sea area [Middle Ordovician]): Voprosy Paleontologii: Leningrad Univ. Izdatel'stvo, v. 5, p. 53-70, 4 pls.

*— 1968, Krinoidei Skalskogo i Borschovskogo gorizontov Podolii V kn Siluriiskodevonskaia fauna Podolii (Crinoid from the Skalski and Borschovski horizons of Podolsky, *in* Silurian-Devonian fauna from Podoli): Leningrad. Univ. Izdatel'stvo, p. 30-50, pls. 1-5.

Yeltyschewa, R. S., and Dubatolova, Yu. A., 1960, Novye vidy devonskikh krinoidei Verkhnego Amura. V kn Sbornik Novye vidy drevnikh rastenii i bespozvonochnykh SSSR, chast II (New species of Devonian crinoids of the upper Amur, *in* Collected articles on new species of the ancient plants and invertebrates of the USSR, pt. II): Moscow, Vses. Geol. Inst. Gosgeoltekhizdat, p. 367-372, pl. 70.

— 1961, Morskie lilii. V kn Biostratigrafiia paleozoia Sayano-Alraiskoi gornoi oblasti, Srednii paleozoi, t. II (Crinoidea, *in* Biostratigraphy of the Paleozoic of the Sayan-Altay Mountain region, middle Paleozoic, v. II): Moscow, Izdatel'stvo Nauka, p. 294-296, 552-560, pls. D86-D87.

*— 1967, Opisanie steblei morskikh lilii V kn Dubatolova, Yu. A., Yeltyschewa, R. S., and Modzalevskaya, E. A., Morskie lilii devona i nizhnego karbona Dalnego Vostoka (Description of stems of crinoids, *in* Devonian and Lower Carboniferous crinoids of the Far East): Akad. Nauk SSSR Sibirsk. Otdeleniye Inst. Geologii i Geofiziki Trudy, 72 p. 7 pls.

Yeltyschewa, R. S., and Schewtschenko, T. V., 1960, Stebli morskikh lilii iz kamennongolnykh otlozhenii Tian-Shania i Darvaza (Crinoid stems from the Carboniferous deposits of Tien-Shan and Darvas): Akad. Nauk Tadzh. SSR Izv., otdel. geologii, geokhimii i tekhnicheskikh nauk, vyp. I(2), p. 119-125, 2 pls.

*Yeltyschewa, R. S., and Stukalina, G. A., 1963, Stebli ordovikskikh i nizhnesiluriiskikh krinoidei Tsentralnogo Taimyra, Novoi Zemli i Vaigacha (Stems of the Ordovician and Lower Silurian Crinoidea of the Central Taymyr, Nova Zemlya, and Vaigach): Uchenya zapiski Sci. Invest. Inst. Geol. Arctic Paleont. i Biostratigrafiya, Izdatel'stvo vyp. 2, p. 23-62, 4 pls., 22 figs.

Zankl, H., 1965, Zur microfaunistischen Charakteristik des Dachsteinkalkes (Nor/Rat) mit Hilfe einer Losungstechnik: Zeit Deutsch. Geol. Gesellsch., H. 116, no. 2, p. 549-567, 3 pls.

Part I. Crowns and Parts of Crowns

SECTION 1. IDENTIFIED CROWNS AND PARTS OF CROWNS

AACOCRINUS Bowsher, 1955. **A. nododorsatus* Bowsher, 1955. Bowsher, 1955b, p. 3, Pl. 1, figs. 2, 3; Figs. 1, 2. Brower, 1967, p. 681.
Mississippian, Kinderhookian

Aacocrinus boonensis (Peck and Keyte), 1938. Bowsher, 1955b, p. 5. Brower, 1967, p. 686, Pl. 75, figs. 4, 5, 7-9; Fig. 3B.
Actinocrinus boonensis Peck and Keyte, 1938. Bowsher, 1955b, p. 5.
Actinocrinites boonensis (Peck and Keyte), 1938. Moore and Laudon, 1944, p. 193, Pl. 77, fig. 12. Laudon, 1948, Pl. 3.
Mississippian, Kinderhookian, Chouteau Limestone, Compton Limestone: United States: Missouri

Aacocrinus chouteauensis (Miller), 1892. Bowsher, 1955b, p. 5. Brower, 1967, p. 693, Pl. 75, figs. 2, 3, 6; Figs. 5C-E.
Actinocrinus? chouteauensis Miller, 1892. Bowsher, 1955b, p. 5.
Actinocrinites chouteauensis (Miller), 1892. Branson, 1944, p. 210, Pl. 38, figs. 16, 17.
Mississippian, Kinderhookian, Chouteau Limestone, Compton Limestone: United States: Missouri

Aacocrinus milleri (Peck and Keyte), 1938. Bowsher, 1955b, p. 5.
Actinocrinus milleri Peck and Keyte, 1938. Bowsher, 1955b, p. 5.
Actinocrinites milleri (Peck and Keyte), 1938. Branson, 1944, p. 210, Pl. 38, figs. 14, 15.
Mississippian, Kinderhookian, Chouteau Limestone: United States: Missouri

*Aacocrinus nododorsatus Bowsher, 1955b, p. 5, Pl. 1, figs. 2, 3; Figs. 1, 2. Brower, 1967, p. 685, Pl. 76, figs. 12, 15, 16; Figs. 1A, D; 3A.
Mississippian, Kinderhookian, Chouteau Limestone, Compton Limestone: United States: Missouri

Aacocrinus protuberoarmatus Brower, 1967, p. 689, Pl. 75, fig. 1; Fig. 3E.
Mississippian, Kinderhookian, upper Sedalia Limestone: United States: Missouri

Aacocrinus sampsoni (Miller and Gurley), 1895. Bowsher, 1955b, p. 7. Brower, 1967, p. 685, Pl. 76, fig. 10; Pl. 78, figs. 12-16; Fig. 3D.
Melocrinus sampsoni Miller and Gurley, 1895. Bowsher, 1955b, p. 7.
Mississippian, Kinderhookian, Chouteau Limestone: United States: Missouri

Aacocrinus senectus (Miller and Gurley), 1897. Bowsher, 1955b, p. 7. Brower, 1967, p. 687, Pl. 76, figs. 1-8; Figs. 3F-L.
Actinocrinus senectus Miller and Gurley, 1897. Bowsher, 1955b, p. 7.
Actinocrinites senectus (Miller and Gurley), 1897. Branson, 1944, p. 210, Pl. 38, figs. 12-13.
Mississippian, Kinderhookian, Compton Limestone, Compton-Sedalia transition beds: United States: Missouri

Aacocrinus spinosulus (Miller and Gurley), 1893. Bowsher, 1955b, p. 7. Brower, 1967, p. 692.
Blairocrinus spinosulus Miller and Gurley, 1893. Bowsher, 1955b, p. 7.

Mississippian, Kinderhookian, Compton Limestone: United States:
Missouri

Aacocrinus tetradactylus Brower, 1967, p. 695, Pl. 75, fig. 11; Pl. 77,
figs. 15, 17, 19; Pl. 78, fig. 18; Figs. 1B, E; 5A, B.
Mississippian, Kinderhookian, Sedalia Limestone: United States:
Missouri

Aacocrinus triarmatus Brower, 1967, p. 690, Pl. 78, figs. 17, 19; Fig. 4.
Mississippian, Kinderhookian, McCraney Limestone: United States:
Missouri

Aacocrinus sp. Brower, 1967, p. 689, Pl. 76, figs. 11, 13, 14, 17;
Fig. 3C.
Mississippian, Osagean, St. Joe Formation: United States: Oklahoma

AATOCRINUS Moore and Plummer, 1940.*Zeacrinus? robustus* Beede, 1900.
Moore and Laudon, 1943b, p. 58; 1944, p. 163.
Pennsylvanian-Permian

Aatocrinus mooresi (Whitfield), 1882. Moore and Laudon, 1944, p. 163,
Pl. 63, fig. 10.
Zeacrinus mooresi Whitfield, 1882.
Pennsylvanian, Atokan, and Desmoinesian, Mercer Formation, Cherokee
Group, Savanna Formation: United States: Ohio, Kansas, Oklahoma

Aatocrinus mucrospinus (McChesney), 1860. Moore and Laudon, 1944, p. 163,
Pl. 63, fig. 3.
Pennsylvanian, Desmoinesian, Des Moines Formation: United States:
Illinois

Aatocrinus patulus (Girty), 1911. Moore and Laudon, 1944, p. 163, Pl. 63,
fig. 4.
Pennsylvanian, Desmoinesian, Wewoka Formation: United States: Oklahoma

Aatocrinus permicus Webster and Lane, 1967, p. 27, Pl. 5, figs. 7, 8.
Permian, Wolfcampian, Bird Spring Formation: United States: Nevada

**Aatocrinus robustus* (Beede), 1900. Moore and Laudon, 1944, p. 163,
Pl. 62, fig. 21; Pl. 63, fig. 11.
Pennsylvanian, Missourian: United States: Missouri

ABACOCRINUS Angelin, 1878. *Actinocrinites tesseracontadactylus* Goldfuss,
1831. Moore and Laudon, 1943b, p. 92. Ubaghs, 1953, p. 738, Fig. 56b.
Silurian

ABACTINOCRINUS Laudon and Severson, 1953. *A. rossei* Laudon and Severson,
1953. Laudon and Severson, 1953, p. 528. Brower, 1967, p. 679.
Mississippian, Kinderhookian

**Abactinocrinus rossei* Laudon and Severson, 1953, p. 529, Pl. 53, figs. 8-10;
Pl. 55, figs. 22, 23. Brower, 1967, p. 680, Fig. 2C.
Mississippian, Kinderhookian, Lodgepole Limestone: United States:
Montana

ABATHOCRINUS Strimple, 1963. *Mariacrinus? rotundus* Springer, 1926.
Strimple, 1963d, p. 109.
Silurian, Niagaran

**Abathocrinus rotundus* (Springer), 1926. Strimple, 1963d, p. 110.
Silurian, Niagaran, Decatur Limestone: United States: Tennessee

ABATOCRINUS Lane, 1963. *Actinocrinus turbinatus* Hall, 1858. Lane, 1963a,
p. 696; 1963b, p. 925.
Mississippian, Kinderhookian-Osagean

Abatocrinus aequalis (Hall), 1858. Lane, 1963a, p. 696.
Actinocrinus aequalis Hall, 1858. Lane, 1963a, p. 696.
Actinocrinus formosus Hall, 1860. Lane, 1963a, p. 696.

Actinocrinus subaequalis McChesney, 1860. Lane, 1963a, p. 696.
Mississippian, Osagean, lower Burlington Limestone: United States: Illinois, Iowa, Missouri

Abatocrinus calvini (Rowley), 1890. Lane, 1963a, p. 696.
Batocrinus calvini Rowley, 1890. Lane, 1963a, p. 696.
Mississippian, Osagean, lower Burlington Limestone: United States: Missouri

Abatocrinus clavigerus (Hall), 1860. Lane, 1963a, p. 697.
Actinocrinus clavigerus Hall, 1860. Lane, 1963a, p. 697.
Mississippian, Osagean, Keokuk Limestone: United States: Illinois

Abatocrinus clypeatus (Hall), 1860. Lane, 1963a, p. 697.
Actinocrinus clypeatus Hall, 1860. Lane, 1963a, p. 697.
Batocrinus quasillus Meek and Worthen, 1868. Lane, 1963a, p. 697.
Mississippian, Osagean, lower Burlington Limestone: United States: Illinois, Iowa, Missouri

Abatocrinus curiosus (Rowley, *in* Greene), 1908 (*non* Miller and Gurley, 1895). Lane, 1963a, p. 697.
Batocrinus curiosus Rowley, *in* Greene, 1908. Lane, 1963a, p. 697.
Mississippian, Osagean, lower Burlington Limestone: United States: Missouri

Abatocrinus grandis (Lyon and Casseday), 1859. Lane, 1963a, p. 697. Van Sant, *in* Van Sant and Lane, 1964, p. 108, Pl. 7, figs. 5-8; Figs. 17, no. 1; 19, no. 4; 36.
Actinocrinus sp. *nobis* (*grandis*) Lyon and Casseday, 1859. Lane, 1963a, p. 697.
Batocrinus grandis (Lyon), 1859. Moore and Laudon, 1943b, p. 140, Pl. 11, fig. 9.
Batocrinus wachsmuthi (White), 1880. Moore and Laudon, 1944, p. 195, Pl. 75, fig. 6. Laudon, 1948, Pl. 2. Lane, 1963a, p. 697.
Mississippian, Osagean, Keokuk Formation, Edwardsville Formation, Borden Group, Fort Payne Chert: United States: Indiana, Kentucky, Tennessee

Abatocrinus laura (Hall), 1861. Lane, 1963a, p. 697.
Actinocrinus laura Hall, 1861. Lane, 1963a, p. 697.
Actinocrinus sinuosus Hall, 1860. Lane, 1963a, p. 697.
Batocrinus laura var. *sinuosus* Wachsmuth and Springer, 1897. Lane, 1963a, p. 697.
Batocrinus remotus Miller and Gurley, 1896. Lane, 1963a, p. 697.
Batocrinus repositus Miller and Gurley, 1896. Lane, 1963a, p. 697.
Batocrinus selectus Miller and Gurley, 1896. Lane, 1963a, p. 697.
Mississippian, Osagean, upper Burlington Limestone: United States: Illinois, Iowa, Missouri

Abatocrinus lepidus (Hall), 1860. Lane, 1963a, p. 697.
Actinocrinus lepidus Hall, 1860. Lane, 1963a, p. 697.
Mississippian, Osagean, lower Burlington Limestone: United States: Illinois, Iowa, Missouri

Abatocrinus macbridei (Wachsmuth and Springer, *in* Miller), 1889. Lane, 1963a, p. 697.
Batocrinus macbridei Wachsmuth and Springer, *in* Miller, 1889. Moore and Laudon, 1944, p. 195, Pl. 75, fig. 7. Lane, 1963a, p. 697.
Mississippian, Kinderhookian, Hampton Formation: United States: Iowa

Abatocrinus pistillus (Meek and Worthen), 1865. Lane, 1963a, p. 697.
Actinocrinus pistillus Meek and Worthen, 1865. Lane, 1963a, p. 697.
Mississippian, Osagean, upper Burlington Limestone: United States: Iowa

Abatocrinus poculum (Miller and Gurley), 1890. Lane, 1963a, p. 697.
  Batocrinus poculum Miller and Gurley, 1890. Lane, 1963a, p. 697.
  Mississippian, Kinderhookian, Hampton Formation: United States:
    Iowa

Abatocrinus rotadentatus (Rowley and Hare), 1891. Lane, 1963a, p. 697.
  Batocrinus rotadentatus Rowley and Hare, 1891. Lane, 1963a, p. 697.
  Mississippian, Osagean, lower Burlington Limestone: United States:
    Missouri

Abatocrinus steropes (Hall), 1860. Lane, 1963a, p. 697.
  Actinocrinus steropes Hall, 1860. Lane, 1963a, p. 697.
  Actinocrinites gibbosus Troost, 1849. Lane, 1963a, p. 697.
  Batocrinus sayi Wood, 1909. Lane, 1963a, p. 697.
  Mississippian, Osagean, Keokuk Limestone, Fort Payne Chert: United
    States: Illinois, Kentucky

Abatocrinus tuberculatus (Wachsmuth and Springer), 1897. Lane, 1963a,
    p. 697.
  Batocrinus tuberculatus Wachsmuth and Springer, 1897. Lane, 1963a,
    p. 697.
  Mississippian, Osagean, lower Burlington Limestone: United States:
    Iowa, Missouri

**Abatocrinus turbinatus* (Hall), 1858. Lane, 1963a, p. 696, fig. 3C.
  Actinocrinus turbinatus Hall, 1858. Lane, 1963a, p. 696.
  Actinocrinus turbinatus var. *elegans* Hall, 1858. Lane, 1963a, p. 696.
  Mississippian, Osagean, lower Burlington Limestone: United States:
    Illinois, Iowa, Missouri

ABRACHIOCRINUS Wanner, 1920. **Sycocrinites clausus* Austin and Austin,
    1843. Moore and Laudon, 1943b, p. 50. Arendt and Hecker, 1964,
    p. 85. Lane, 1967b, p. 11, fig. 4.
  Mississippian-Permian

ABROTOCRINUS Miller and Gurley, 1890. **A. cymosus* Miller and Gurley,
    1890. Moore and Laudon, 1943b, p. 56, fig. 6; 1944, p. 159.
    Moore, 1950, p. 34, fig. 4i. Arendt and Hecker, 1964, p. 88.
  Lower Carboniferous

Abrotocrinus angustatus Wright, 1934. Wright, 1951b, p. 69, Pl. 31,
    figs. 3, 6-9; Figs. 34-36.
  Carboniferous, Visean, shales above Ardross Limestone, calciferous
    sandstone: Scotland

Abrotocrinus coreyi (Meek and Worthen), 1869. Van Sant, *in* Van Sant and
    Lane, 1964, p. 85, Pl. 3, figs. 2, 9.
  Poteriocrinites (Graphiocrinus; Scaphiocrinus) coreyi Meek and Worthen,
    1869.
  Mississippian, Osagean, Edwardsville Limestone: United States:
    Indiana

**Abrotocrinus cymosus* Miller and Gurley, 1890. Moore and Laudon, 1943b,
    p. 132, Pl. 3, fig. 5; 1944, p. 159, Pl. 60, fig. 1. Van Sant,
    in Van Sant and Lane, 1964, p. 85.
  Mississippian, Osagean, Keokuk Limestone: United States: Indiana

Abrotocrinus ornatus Termier, G. and H., 1950, p. 91, 100, Pl. 221,
    figs. 9-14.
  Carboniferous, upper Visean: Morocco

Abrotocrinus unicus (Hall), 1861. Moore and Laudon, 1944, p. 159, Pl. 59,
    fig. 15. Van Sant, *in* Van Sant and Lane, 1964, p. 85, Pl. 3,
    figs. 1, 6-8, 12, 13; Figs. 23, 32.
  Mississippian, Osagean, Borden Group: United States: Indiana

Abrotocrinus spp.
  Termier and Termier, 1950, p. 91, Pl. 218, figs. 27, 28 (typographical

error, given as Pl. 217, p. 91).
Carboniferous, Westphalian: Morocco
Laudon and others, 1952, p. 554, Pl. 65, figs. 29-31.
Mississippian, Kinderhookian, Banff formation: Canada: Alberta
Wright, 1960, p. 331, Appendix, Pl. B, fig. 8.
Carboniferous, Tournaisian: England

ABYSSOCRINUS Strimple, 1963. *Synbathocrinus antiquus Strimple, 1952.
Strimple, 1963d, p. 29.
Silurian, Niagaran

*Abyssocrinus antiquus (Strimple), 1952. Strimple, 1963d, p. 30, figs. 4a, 5d, 6a-d.
Synbathocrinus antiquus Strimple, 1952b, p. 76, figs. 1-3. Kesling and Smith, 1963, p. 190.
Synbathocrinus? antiquus Strimple, 1952. Strimple and Watkins, 1961, p. 48, figs. 1, 2.
Silurian, Niagaran, Henryhouse Formation: United States: Oklahoma

ACACOCRINUS Wachsmuth and Springer, 1897. *A. elrodi Wachsmuth and Springer, 1897. Moore and Laudon, 1943b, p. 93. Ubaghs, 1953, p. 740.
Silurian-Devonian

ACANTHOCRINUS Roemer, 1850. *A. longispina Roemer, 1850. Moore and Laudon, 1943b, p. 83; 1944, p. 185. Ubaghs, 1953, p. 737, figs. 78, 122. Arendt and Hecker, 1964, p. 95.
Devonian

Acanthocrinus gracilior Jaekel, 1895. Schmidt, 1941, p. 209, Pl. 14, fig. 1; Fig. 58b. Ubaghs, 1947, p. 5, Pl. 3, figs. 11, 12.
Acanthocrinus gregarius Wirtgen and Zeiler, 1854. Ubaghs, 1947, p. 2.
Acanthocrinus longispina Wirtgen and Zeiler, 1854 (non Roemer, 1850). Ubaghs, 1947, p. 2.
Devonian, Coblentzian: Belgium, Germany

Acanthocrinus gregarius Wirtgen and Zeiler, 1854 = A. gracilior

Acanthocrinus gregarius Schmidt, 1941 = A. jaekeli

Acanthocrinus jaekeli Schmidt, 1941, p. 207, Pl. 14, fig. 2; Fig. 57. Ubaghs, 1947, p. 4, Pl. 2, figs. 5, 6; Pl. 3, fig. 13.
Acanthocrinus gregarius Schmidt, 1941 (non Wirtgen and Zeiler, 1854), p. 222. Ubaghs, 1947, p. 2.
Devonian, Coblentzian: Belgium, Germany

Acanthocrinus aff. jaekeli Schmidt, 1941. Le Maitre, 1958b, p. 125, Pl. 2, figs. 1-3.
Devonian, Coblentzian: northern Africa

Acanthocrinus cf. jaekeli Schmidt, 1941. Ubaghs, 1947, p. 4, Pl. 2, fig. 8.
Devonian, Coblentzian: Belgium

Acanthocrinus lingenbachensis Lehmann, 1939, p. 6, Pls. 2-4; Fig. 1. Schmidt, 1941, p. 32.
Devonian, Coblentzian, Hunsrück Schiefer: Germany

*Acanthocrinus longispina Roemer, 1850. Simon, 1950, p. 364, figs. 1-3.
Devonian, ?Wissenbacher beds: Germany

Acanthocrinus rex Jaekel, 1895. Ubaghs, 1953, p. 737, fig. 122. Struve, 1957, p. 362, fig. 1.
Devonian, Coblentzian, Hunsrück Schiefer: Germany

Acanthocrinus spinosus (Hall), 1862. Moore and Laudon, 1943b, p. 138, Pl. 9, fig. 6; 1944, p. 185, Pl. 72, fig. 13.
Devonian, Erian, Hamilton Group: United States: New York

Acanthocrinus sp. Ubaghs, 1947, p. 4, Pl. 1, figs. 1-4; Pl. 2, figs. 1-4, 7, 9; Pl. 3, figs. 1-10; Fig. 2; 1953, p. 737, Fig. 78.
Devonian, Emsian: Belgium

Acanthocrinus sp. indet. Schmidt, 1941, p. 212, fig. 59.
Devonian, Coblentzian: Germany

ACARIAIOCRINUS Wanner, 1924. *A. clavulus* Wanner, 1924. Moore and Laudon, 1943b, p. 50. Arendt and Hecker, 1964, p. 84. Lane, 1967b, p. 13, figs. 3, 7 (*non* fig. 8 as listed).
Permian

Acariaiocrinus caryophylloides (Yakovlev), 1927. Yakovlev, *in* Yakovlev and Ivanov, 1956, p. 57, Pl. 10, figs. 7-13; Fig. 15; 1964b, p. 58, fig. 68. Arendt and Hecker, 1964, p. 84, Pl. 9, figs. 7-9. Arendt, 1968a, p. 100, Fig. 1v.
Permian, Artinskian: Russia: Ural Mountains

*Acariaiocrinus clavulus Wanner, 1924. Arendt and Hecker, 1964, p. 84, Pl. 9, fig. 10.
Pormian: Timor

ACHRADOCRINUS Schultze, 1866. *A. ventrosus* Schultze, 1866. Moore and Laudon, 1943b, p. 50. Ubaghs, 1953, p. 750.
Devonian

Achradocrinus patulus Slocum, 1908 = *Thyridocrinus patulus*

ACROCRINUS Yandell, 1855. *A. shumardi* Yandell, 1855. Moore and Laudon, 1943b, p. 96, fig. 16; 1944, p. 199. Cuénot, 1948, p. 69. Ubaghs, 1953, p. 741, fig. 2. Arendt and Hecker, 1964, p. 97.
Carboniferous

Acrocrinus alvestonensis Wright, 1958, p. 308, Pl. 79, figs. 1-4.
Carboniferous, Visean (S_2): England: Greenhill, Gloucestershire

Acrocrinus amphora Wachsmuth and Springer, 1897. Moore and Laudon, 1943b, p. 142, Pl. 13, fig. 3; 1944, p. 199, Pl. 78, fig. 15. Cuénot, 1948, p. 69, fig. 91. Ubaghs, 1953, p. 741, fig. 2.
Mississippian, Meramecian, St. Louis Limestone: United States: Alabama

Acrocrinus elegans Strimple, 1949e, p. 903, Pl. 1, figs. 4-6; Fig. 2.
Pennsylvanian, Missourian, Ochelata Group: United States: Oklahoma

Acrocrinus expansus Strimple, 1951d, p. 192, figs. 1, 17-20; 1962e, p. 48.
Pennsylvanian, Desmoinesian, Oologah Limestone: United States: Oklahoma

Acrocrinus mjatschkowensis Yakovlev, 1926. Yakovlev and Ivanov, 1956, p. 34, Pl. 8, fig. 7. Arendt and Hecker, 1964, p. 97, Pl. 15, fig. 6.
Carboniferous (C_2m): Russia: Moscow Basin

Acrocrinus pirum Moore and Plummer, 1938. Moore and Laudon, 1944, p. 199, Pl. 78, fig. 14.
Pennsylvanian, Morrowan: United States: Oklahoma

Acrocrinus pumpkensis Strimple, 1949e, p. 900, Pl. 1, figs. 1-3; Fig. 1.
Pennsylvanian, Atokan, Pumpkin Creek Limestone: United States: Oklahoma

ACTINOCRINITES Miller, 1821. *A. triacontadactylus* Miller, 1821. Moore and Laudon, 1943b, p. 93, Fig. 15; 1944, p. 193, Pl. 71, fig. 10. Cuénot, 1948, p. 69. Ubaghs, 1953, p. 739, figs. 60, 61, 139e. Philip, 1963, p. 270, fig. 5b. Arendt and Hecker, 1964, p. 96.
Actinocrinus Termier and Termier, 1947, p. 255, Pl. 19, fig. 11.
Carboniferous

Actinocrinites aculeatus Austin and Austin, 1842 = *Dialutocrinus aculeatus*

Actinocrinites aculeatus Austin and Austin, 1843, or *A. longispinosus* Austin and Austin, 1843. Wright. 1943b, p. 233, Pl. 8, fig. 2.
Carboniferous: Ireland

Actinocrinites alatus Wright, 1955a, p. 224, Pl. 56, figs. 12-15.
Carboniferous, Tournaisian: Scotland

Actinocrinites albersi (Miller and Gurley), 1897 = *Actinocrinites multiradiatus*

Actinocrinites benedicti (Miller), 1892. Brower, 1965, p. 791.
Mississippian, Osagean, Keokuk Limestone: United States: Indiana

Actinocrinites blairi (Miller and Gurley), 1897. Brower, 1965, p. 791.
Mississippian, Osagean, Burlington Limestone: United States: Missouri

Actinocrinites boonensis (Peck and Keyte), 1938 = *Aacocrinus boonensis*

Actinocrinites cataphractus Austin and Austin, 1842 = *Dialutocrinus cataphractus*

Actinocrinites chouteauensis (Miller), 1892 = *Aacocrinus chouteauensis*

Actinocrinites comptus Wright, 1955a, p. 226, Pl. 58, figs. 7, 10.
Carboniferous, Tournaisian: Scotland

Actinocrinites coplowensis Wright, 1955a, p. 218, Pl. 55, figs. 12, 13; Pl. 56, figs. 2, 3; Pl. 57, figs. 5-10.
Carboniferous, Tournaisian: Scotland

Actinocrinites costus (McCoy), 1844. Moore, 1948, p. 390, fig. 11.
Actinocrinus costus McCoy, 1844.
Carboniferous, Tournaisian: England

Actinocrinites depressus Wright, 1955a, p. 222, Pl. 55, figs. 1-3.
Carboniferous, Tournaisian: Scotland

Actinocrinites elephantinus Austin and Austin, 1842 = *Dialutocrinus elephantinus*

Actinocrinites elongatus Wright, 1955a, p. 219, Pl. 53, figs. 10-12.
Carboniferous, Tournaisian: Scotland

Actinocrinites eximius (Kirk), 1943, nom. correct
Actinocrinus griffithi Wachsmuth and Springer, 1897. Kirk, 1943a, p. 264.
Actinocrinus eximius Kirk, 1943a, p. 264.
Mississippian, Osagean, upper Burlington Limestone: United States: Illinois, Missouri

Actinocrinites gibbosus Troost, 1849 = *Abatocrinus steropes*

Actinocrinites gibsoni (Miller and Gurley), 1894. Van Sant, *in* Van Sant and Lane, 1964, p. 106.
Actinocrinus multiramosus Wachsmuth and Springer, 1897. Van Sant, *in* Van Sant and Lane, 1964, p. 106 (misspelled as *Actionocrinus multiramosus*).
Mississippian, Osagean, Borden Group: United States: Indiana

Actinocrinites grandissimus Van Sant, 1965, p. 290, figs. 1, 2.
Actinocrinus grandis Miller and Gurley, 1890 (*non* Lyon and Casseday, 1859), Van Sant, 1965, p. 290.
Mississippian, Osagean, Borden Group: United States: Indiana

Actinocrinites hemisphericus (Miller and Gurley), 1895. Branson, 1944, p. 210, Pl. 38, figs. 1-4 (listed on plate explanation as *Sampsonocrinus hemisphericus* Miller and Gurley).

Mississippian, Kinderhookian, Chouteau Group: United States: Missouri

Actinocrinites higuchisawensis (Minato), 1951, nom. correct
   Actinocrinus higuchisawensis Minato, 1951, p. 358, Pl. 1, fig. 16.
   Lower Carboniferous, Hikoroiti Series: Japan

Actinocrinites icosidactylus Portlock, 1843 = *Dialutocrinus icosidactylus*

Actinocrinites intermedius Wright, 1955a, p. 223, Pl. 56, figs. 10,17.
   Carboniferous, Tournaisian: Scotland

Actinocrinites jessieae (Miller and Gurley), 1896. Branson, 1944, p. 210, Pl. 38, fig. 5.
   Mississippian, Kinderhookian, Chouteau Group: United States: Missouri

Actinocrinites laevissimus Austin and Austin, 1842 = *Dialutocrinus laevissimus*

Actinocrinites lobatus (Hall), 1860. Van Sant, in Van Sant and Lane, 1964, p. 107.
   Mississippian, Osagean: United States: (non Crawfordsville), Indiana

Actinocrinites longispinosus Austin and Austin, 1842 = *Dialutocrinus longispinosus*

Actinocrinites lowei (Hall), 1858. Moore and Laudon, 1944, p. 193, Pl. 76, fig. 4. Laudon, 1948, Pl. 3.
   Mississippian, Osagean, Keokuk Limestone: United States: Iowa, Missouri

Actinocrinites milleri (Peck and Keyte), 1938 = *Aacocrinus milleri*

Actinocrinites moderatus Wright, 1955a, p. 225, Pl. 57, figs. 11, 14-16.
   Carboniferous, Tournaisian: Scotland

Actinocrinites multiradiatus (Shumard), 1857. Moore and Laudon, 1943b, p. 139, Pl. 10, fig. 10; 1944, p. 193, Pl. 76, fig. 3. Laudon, 1948, Pl. 3. Brower, 1965, p. 791.
   Actinocrinites albersi (Miller and Gurley), 1897. Brower, 1965, p. 792.
   Actinocrinus multiradiatus Shumard, 1857. Kirk, 1943a, p. 264.
   Steganocrinus albersi Miller and Gurley, 1897. Brower, 1965, p. 792.
   Mississippian, Osagean, upper Burlington Limestone: United States: Illinois, Iowa, Missouri

Actinocrinites nodosus Wright, 1955a, p. 221, Pl. 55, figs. 7, 8. Pickett, 1960, p. 88, 2 figs.
   Carboniferous, Tournaisian: Scotland; Australia: New South Wales

Actinocrinites ohmoriensis (Minato), 1951, nom. correct
   Actinocrinus ohmoriensis Minato, 1951, p. 359, Pl. 4, fig. 5 (non Pl. 5 as given in text).
   Lower Carboniferous, Kikoroiti Series: Japan

Actinocrinites parkinsoni Wright, 1955a, p. 221, Pl. 54, figs. 1-4; Pl. 57, fig. 12.
   Carboniferous, Tournaisian: Scotland

Actinocrinites penicillus Meek and Worthen, 1869 = *Cusacrinus penicillus*

Actinocrinites pernodosus (Hall), 1858. Laudon, 1948, Pl. 3.
   Mississippian, Osagean, Keokuk Limestone: United States: Iowa

Actinocrinites polydactylus J. S. Miller, 1821 = *Dialutocrinus polydactylus*

Actinocrinites polydactylus Bonny, 1835 (non Miller, 1821) = *Melocrinites pachydactylus*

Actinocrinites rotundatus Wright, 1955a, p. 219, Pl. 55, figs. 11, 14; Pl. 56, figs. 6-9.
   Carboniferous, Tournaisian: Scotland

Actinocrinites rubra (Weller), 1909 = *Nunnacrinus rubra*
Actinocrinites scitulus (Meek and Worthen), 1860. Brower, 1965, p. 791.
 Mississippian, Osagean, Burlington Limestone: United States: Iowa
Actinocrinites senectus (Miller and Gurley), 1897 = *Aacocrinus senectus*
Actinocrinites sharonensis (Miller and Gurley), 1897. Brower, 1965, p. 791.
 Actinocrinus scitulus Meek and Worthen, 1860. Kirk, 1943a, p. 264.
 Steganocrinus sharonensis Miller and Gurley, 1897. Kirk, 1943a, p. 264.
 Mississippian, Osagean, Burlington Limestone: United States: Missouri
Actinocrinites skourensis (Termier and Termier), 1950, nom. correct
 Actinocrinus skourensis Termier and Termier, 1950, p. 84, Pl. 208,
   fig. 36.
 Carboniferous, upper Visean: Morocco
Actinocrinites spergenensis (Miller and Gurley), 1896. Brower, 1965,
   p. 791.
 Mississippian, Meramecian, Harrodsburg Limestone: United States:
   Indiana
Actinocrinites spinulosus (Miller and Gurley), 1894. Moore and Laudon,
   1944, p. 193, Pl. 77, fig. 13.
 Mississippian, Kinderhookian, Chouteau Group: United States: Missouri
Actinocrinites stellaris (de Koninck and Le Hon), 1854. Wright, 1955a,
   p. 217, Pl. 53, figs. 1-3; Pl. 54, fig. 17; Pl. 55, fig. 6; Pl. 56,
   fig. 16.
 Actinocrinus stellaris de Koninck and Le Hon. Termier and Termier,
   1950, p. 84, Pl. 208, fig. 32; Pl. 212, figs. 7, 8; Pl. 214,
   figs. 1-8, 14-16.
 Carboniferous, Tournaisian, Westphalian: Belgium, Scotland, Morocco
*Actinocrinites triacontadactylus Miller, 1821. Wright, 1955a, p. 215,
   Pl. 54, figs. 5-8, 14, 16; Pl. 55, figs. 9, 10; Pl. 56, fig. 11;
   Pl. 57, figs. 1-4, 13; Pl. 80, fig. 1; Fig. 112, figs. 1A, 2A.
 Carboniferous, Tournaisian: Scotland
Actinocrinites triacontadactylus Miller, 1821 (part) = *Dialutocrinus
   milleri*
Actinocrinites triplus Wright, 1955a, p. 223, Pl. 53, figs. 9, 13, 14.
 Carboniferous, Tournaisian: Scotland
Actinocrinites sp., ?*triplus* Wright, 1955a, p. 223, Pl. 54, fig. 21.
 Carboniferous, Tournaisian: England
Actinocrinites tripus Ehlers and Kesling, 1963, p. 1038, Pl. 134, figs. 1-6;
   Fig. 3.
 Mississippian, Osagean, Fort Payne Formation: United States: Kentucky
Actinocrinites vermiculatus Wright, 1955a, p. 225, Pl. 54, fig. 15.
 Carboniferous, Tournaisian: Scotland
Actinocrinites verrucosus (Hall), 1858. Moore and Laudon, 1944, p. 193,
   Pl. 76, fig. 1.
 Mississippian, Osagean, Burlington Formation: United States: Iowa,
   Missouri
Actinocrinites spp.
 Laudon and Severson, 1953, p. 528, Pl. 53, fig. 7.
   Mississippian, Kinderhookian, Lodgepole Group: United States:
   Montana
 Ubaghs, 1953, p. 739, figs. 60, 61.
   Carboniferous
 Wright, 1960, p. 334, Appendix, Pl. A, fig. 10.
   Carboniferous, Tournaisian: England

Actinocrinus aequalis Hall, 1858 = *Abatocrinus aequalis*

Actinocrinus (Alloprosallocrinus) euconus Meek and Worthen, 1865 = *Batocrinus euconus*

Actinocrinus (Amphoracrinus) brevicalix Rofe, 1865 = *Pimlicocrinus brevicalix*

Actinocrinus (Amphoracrinus?) olla McCoy, 1849 = *Ectocrinus olla*

Actinocrinus andrewsianus McChesney, 1860 = *Azygocrinus andrewsianus*

Actinocrinus araneolus Meek and Worthen, 1860 = *Steganocrinus araneolus*

Actinocrinus armatus de Koninck and Le Hon, 1854 = *Nunnacrinus armatus*

Actinocrinus arnoldi Wachsmuth and Springer, *in* Miller, 1889 = *Cusacrinus arnoldi*

Actinocrinus bischoffi Miller and Gurley, 1896 = *Cusacrinus bischoffi*

Actinocrinus caroli Hall, 1860 = *Dizygocrinus caroli*

Actinocrinus chloris Hall, 1861 = *Cusacrinus chloris*

Actinocrinus clavigerus Hall, 1860 = *Abatocrinus clavigerus*

Actinocrinus clypeatus Hall, 1860 = *Abatocrinus clypeatus*

Actinocrinus coelatus Hall, 1859 = *Cusacrinus coelatus*

Actinocrinus costus McCoy, 1844 = *Actinocrinites costus*

Actinocrinus dalayanus Miller, 1881 = *Nunnacrinus dalayanus*

Actinocrinus daphne Hall, 1863 = *Cusacrinus daphne*

Actinocrinus deornatus de Koninck and Le Hon, 1854 = *Nunnacrinus deornatus*

Actinocrinus dodecadactylus Meek and Worthen, 1861 = *Azygocrinus dodecadactylus*

Actinocrinus dorsatus de Koninck and Le Hon, 1854 = *Nunnacrinus dorsatus*

Actinocrinus formosus Hall, 1860 = *Abatocrinus aequalis*

Actinocrinus foveatus Miller and Gurley, 1895 = *Nunnacrinus foveatus*

Actinocrinus gilbertsoni Form I Miller, *in* Phillips, 1836 = *Amphoracrinus gilbertsoni*

Actinocrinus gracilis Wachsmuth and Springer, 1897 = *Cusacrinus gracilis*

Actinocrinus grandis Miller and Gurley, 1890 (*non* Lyon and Casseday, 1859) = *Actinocrinites grandissimus*

Actinocrinus griffithi (Wachsmuth and Springer), 1897 (*non* Miller and Gurley, 1897) = *A. eximius*

Actinocrinus higuchisawensis Minato, 1951 = *Actinocrinites higuchisawensis*

Actinocrinus hurdianus McChesney, 1860 = *Cusacrinus hurdianus*

Actinocrinus jessieae Miller and Gurley, 1897 = *Nunnacrinus jessieae*

Actinocrinus lagunculus Hall, 1860 = *Macrocrinus mundulus*

Actinocrinus laura Hall, 1861 = *Abatocrinus laura*

Actinocrinus lepidus Hall, 1860 = *Abatocrinus lepidus*

Actinocrinus limabrachiatus Hall, 1861 = *Cusacrinus limabrachiatus*

Actinocrinus locellus Hall, 1861 = *Nunnacrinus locellus*

Actinocrinus longus Meek and Worthen, 1869 = *Cusacrinus longus*

Actinocrinus (Megistocrinus) whitei Hall, 1861 = *Aryballocrinus whitei*

Actinocrinus multiradiatus Shumard, 1857 = *Actinocrinites multiradiatus*
Actinocrinus multiramosus Wachsmuth and Springer, 1897 = *Actinocrinites gibsoni*
Actinocrinus mundulus Hall, 1860 = *Macrocrinus mundulus*
Actinocrinus sp. *nobis* (*grandis*) Lyon and Casseday, 1859 = *Abatocrinus grandis*
Actinocrinus nodobrachiatus Wachsmuth and Springer, 1889 = *Cusacrinus nodobrachiatus*
Actinocrinus ohmoriensis Minato, 1951 = *Actinocrinites ohmoriensis*
Actinocrinus ornatissimus Wachsmuth and Springer, 1889 = *Cusacrinus ornatissimus*
Actinocrinus pallubrum Miller and Gurley, 1896 = *Nunnacrinus pallubrum*
Actinocrinus parvus Shumard, 1855 = *Dizygocrinus parvus*
Actinocrinus pettisensis Miller and Gurley, 1896 = *Nunnacrinus pettisensis*
Actinocrinus pistillus Meek and Worthen, 1865 = *Abatocrinus pistillus*
Actinocrinus puteatus Rowley and Hare, 1891 = *Nunnacrinus puteatus*
Actinocrinus rotundus Shumard, 1855 = *Azygocrinus rotundus*
Actinocrinus rubra Weller, 1909 = *Nunnacrinus rubra*
Actinocrinus sampsoni Miller and Gurley, 1896 = *Nunnacrinus sampsoni*
Actinocrinus scitulus Meek and Worthen, Kirk, 1943 = *Actinocrinites sharonensis*
Actinocrinus sinuosus Hall, 1860 = *Abatocrinus laura*
Actinocrinus spino-tentaculus Hall, 1860 = *Cusacrinus spinotentaculus*
Actinocrinus stellaris de Koninck and Le Hon, 1854 = *Nunnacrinus stellaris*
Actinocrinus steropes Hall, 1860 = *Abatocrinus steropes*
Actinocrinus subaequalis McChesney, 1860 = *Abatocrinus aequalis*
Actinocrinus symmetricus Hall, 1858 = *Aorocrinus symmetricus*
Actinocrinus? tenuidiscus Hall, 1861 = *Aryballocrinus tenuidiscus*
Actinocrinus tenuisculptus McChesney, 1860 = *Cusacrinus tenuisculptus*
Actinocrinus tessellatus Phillips, 1836 = *Dialutocrinus tessellatus*
Actinocrinus thetis Hall, 1861 = *Cusacrinus thetis*
Actinocrinus tuberculosus Wachsmuth and Springer, 1897 = *Cusacrinus tuberculosus*
Actinocrinus turbinatus Hall, 1858 = *Abatocrinus turbinatus*
Actinocrinus turbinatus var. *elegans* Hall, 1858 *Abatocrinus turbinatus*
Actinocrinus viaticus White, 1874 = *Cusacrinus viaticus*
Actinocrinus wachsmuthi White, 1880 = *Abatocrinus grandis*

ACYLOCRINUS Kirk, 1947. *A. tumidus* Kirk, 1947. Kirk, 1947, p. 293.
  Mississippian, Kinderhookian-Osagean

Acylocrinus depressus Kirk, 1947, p. 294, Pl. 1, fig. 9.
  Mississippian, Osagean, upper Burlington Limestone: United States: Iowa

Acylocrinus sampsoni (Weller), 1909. Kirk, 1947, p. 295.
  Graphiocrinus sampsoni Weller, 1909. Kirk, 1947, p. 295.
  Mississippian, Osagean, Fern Glen Limestone: United States: Missouri

Acylocrinus striatus (Meek and Worthen), 1869. Kirk, 1947, p. 296.
  Poteriocrinites (Scaphiocrinus) striatus Meek and Worthen, 1869.
    Kirk, 1947, p. 296.
  Mississippian, Kinderhookian, Burlington Limestone: United States:
    Iowa

Acylocrinus tortuosus (Hall), 1861. Kirk, 1947, p. 296.
  Scaphiocrinus tortuosus Hall, 1861. Kirk, 1947, p. 296.
  Mississippian, Kinderhookian, Burlington Limestone: United States:
    Iowa

*Acylocrinus tumidus Kirk, 1947, p. 297, Pl. 1, figs. 10-12.
  Mississippian, Kinderhookian, Burlington Limestone: United States:
    Iowa

ADINOCRINUS Kirk, 1938. **Zeacrinus nodosus* Wachsmuth and Springer, 1885.
  Moore and Laudon, 1943b, p. 58; 1944, p. 161.
  Carboniferous

Adinocrinus compactilis (Worthen), 1873. Moore and Laudon, 1943b, p. 134,
  Pl. 5, fig. 4; 1944, p. 161, Pl. 60, fig. 23.
  Mississippian, Osagean: United States: Kentucky

*Adinocrinus nodosus (Wachsmuth and Springer), 1885. Moore and Laudon,
  1944, p. 161, Pl. 60, fig. 25; Pl. 62, fig. 4.
  Mississippian, Osagean, Keokuk Limestone: United States: Tennessee

Adinocrinus pentagonus (Miller and Gurley), 1890. Moore and Laudon, 1944,
  p. 161, Pl. 60, fig. 24.
  Pennsylvanian, Missourian: United States: Missouri

AESIOCRINUS Miller and Gurley, 1890. **A. magnificus* Miller and Gurley,
  1890. Moore and Laudon, 1943b, p. 59, fig. 8; 1944, p. 167.
  Arendt and Hecker, 1964, p. 89.
  Carboniferous-Permian

Aesiocrinus amplus Lane and Webster, 1966 = *Polusocrinus amplus*

Aesiocrinus basilicus Miller and Gurley, 1890. Strimple, 1962e, p. 47,
  Pl. 3, figs. 8-11.
  Pennsylvanian, Desmoinesian, Oologah Limestone: United States:
    Oklahoma

Aesiocrinus detrusus Strimple, 1951c, p. 21, Pl. 4, figs. 1-3.
  Pennsylvanian, Missourian, Wann Formation: United States: Oklahoma

Aesiocrinus delicatulus Lane and Webster, 1966, p. 53, Pl. 7, fig. 6;
  Fig. 19.
  Permian, Wolfcampian, Bird Spring Formation: United States: Nevada

Aesiocrinus erectus Strimple, 1951 = *Anobasicrinus erectus*

Aesiocrinus cf. *harei* Miller and Gurley, Wright, 1937 = *Ampelocrinus
  plumosus*

Aesiocrinus inornatus Lane and Webster, 1966, P. 53, Pl. 11, figs. 3, 6;
  Fig. 19.
  Permian, Wolfcampian, Bird Spring Formation: United States: Nevada

Aesiocrinus ivanovi Yakovlev, *in* Yakovlev and Ivanov, 1956, p. 29, Pl. 7,
  fig. 6.
  Carboniferous (C_2m): Russia: Moscow Basin

Aesiocrinus luxuris Strimple, 1951c, p. 20, Pl. 5, figs. 7-10.
  Pennsylvanian, Missourian, Wann Formation: United States: Oklahoma

*Aesiocrinus magnificus Miller and Gurley, 1890. Moore and Laudon, 1943b,
  p. 135, Pl. 6, fig. 4; 1944, p. 167, Pl. 56, fig. 32; Pl. 62,
  fig. 32.
  Pennsylvanian, Missourian, Lane Shale: United States: Missouri

Aesiocrinus nodosus Webster and Lane, 1967, p. 31, Pl. 6, fig. 5 (*non* fig. 6 as listed in plate explanation).
Permian, Wolfcampian, Bird Spring Formation: United States: Nevada

Aesiocrinus patens (Trautschold), 1879. Yakovlev and Ivanov, 1956, p. 28, Pl. 7, figs. 3-5. Arendt and Hecker, 1964, p. 89, Pl. 11, figs. 7, 8.
Phialocrinus patens Trautschold, 1879. Yakovlev, 1939a, p. 67, Pl. 12, fig. 6.
Middle Carboniferous (C_2m): Russia: Moscow Basin

Aesiocrinus paucus Strimple, 1951c, p. 22, Pl. 5, figs. 1-3.
Pennsylvanian, Missourian, Wann Formation: United States: Oklahoma

Aesiocrinus prudentia Strimple, 1963a, p. 72, fig. 30.
Pennsylvanian, Virgilian, Dover Limestone: United States: Kansas

Aesiocrinus secundus Washburn, 1968, p. 119, Pl. 1, figs. 1-3.
Pennsylvanian, Morrowan, Bridal Veil Falls Member: United States: Utah

Aesiocrinus tactilis Strimple, 1951 = *Polusocrinus tactilis*

AGARICOCRINUS Hall, 1858. **Amphoracrinus americanus* Roemer, 1854.
Cuénot, 1948, p. 69. Ubaghs, 1953, p. 740, fig. 22d. Lane, 1963b, p. 927.
Agaricocrinites Troost, 1858. Moore and Laudon, 1943b, p. 93, fig. 16; 1944, p. 195, Pl. 71, fig. 8.
Lower Mississippian

Agaricocrinus adamsensis Miller and Gurley, 1896 = *A. hodgsoni*

*Agaricocrinus americanus (Roemer), 1854. Ehlers and Kesling, 1963, p. 1030. Van Sant, *in* Van Sant and Lane, 1964, p. 118, Pl. 8, fig. 1; Fig. 17, no. 3.
Lower Mississippian, Burlington Limestone; Keokuk Limestone; Fort Payne Chert, Edwardsville Formation: United States: Iowa, Illinois, Indiana, Kentucky, Tennessee

Agaricocrinus arculus Miller and Gurley, 1895. Ehlers and Kesling, 1963, p. 1030.
Mississippian, Osagean, Fort Payne Chert: United States

Agaricocrinus attenuatus Wood, 1909. Ehlers and Kesling, 1963, p. 1030.
Mississippian, Osagean, Fort Payne Chert: United States

Agaricocrinus blairi Miller, 1894. Ehlers and Kesling, 1963, p. 1030.
Mississippian, Kinderhookian, Chouteau Limestone: United States

Agaricocrinus brevis (Hall), 1858. Ehlers and Kesling, 1963, p. 1031.
Agaricocrinites brevis (Hall), 1858. Moore and Laudon, 1944, p. 195, Pl. 76, fig. 15.
Mississippian, Kinderhookian, Burlington Limestone: United States: Missouri, Iowa

Agaricocrinus bullatus Hall, 1858. Ehlers and Kesling, 1963, p. 1030.
Agaricocrinites bullatus (Hall), 1858. Moore and Laudon, 1942, p. 74, fig. 4c; 1943b, p. 141, Pl. 12, fig. 4; 1944, p. 195, Pl. 76, fig. 14.
Mississippian, Kinderhookian, Burlington Limestone: United States: Iowa, Missouri

Agaricocrinus conicus Wachsmuth and Springer, 1897. Ehlers and Kesling, 1963, p. 1031.
Mississippian, Osagean, Keokuk Limestone: United States

Agaricocrinus convexus (Hall), 1860. Ehlers and Kesling, 1963, p. 1030.
Mississippian, Osagean, Keokuk Limestone: United States

Agaricocrinus coreyi (Lyon and Casseday), 1860. Ehlers and Kesling, 1963, p. 1031.
Mississippian, Osagean, Keokuk Limestone, Fort Payne Chert: United States

Agaricocrinus crassus Wetherby, 1881. Ehlers and Kesling, 1963, p. 1030.
Mississippian, Osagean, Keokuk Limestone, Fort Payne Chert: United States

Agaricocrinus elegans Wetherby, 1881. Ehlers and Kesling, 1963, p. 1031.
Mississippian, Osagean, Keokuk Limestone, Fort Payne Chert: United States

Agaricocrinus excavatus (Hall), 1861. Ehlers and Kesling, 1963, p. 1031.
Mississippian, Kinderhookian, Burlington Limestone: United States

Agaricocrinus fiscellus (Hall), 1861. Ehlers and Kesling, 1963, p. 1031.
Mississippian, Kinderhookian, Burlington Limestone: United States

Agaricocrinus geometricus (Hall), 1859. Ehlers and Kesling, 1963, p. 1031.
Actinocrinus (*Agaricocrinus*)? *geometricus* Hall, 1859.
Mississippian, Kinderhookian, Burlington Limestone: United States

Agaricocrinus gracilis Meek and Worthen, 1861. Ehlers and Kesling, 1963, p. 1030.
Mississippian, Kinderhookian, Burlington Limestone: United States

Agaricocrinus hodgsoni Miller and Gurley, 1896. Ehlers and Kesling, 1963, p. 1031.
Agaricocrinus adamsensis Miller and Gurley, 1896. Ehlers and Kesling, 1963, p. 1031.
Mississippian, Kinderhookian, Burlington Limestone: United States

Agaricocrinus illinoisensis Miller and Gurley, 1896. Ehlers and Kesling, 1963, p. 1030.
Mississippian, Kinderhookian, Burlington Limestone: United States

Agaricocrinus inflatus (Hall), 1861. Ehlers and Kesling, 1963, p. 1030.
Mississippian, Kinderhookian, Burlington Limestone: United States

Agaricocrinus iowaensis Miller and Gurley, 1897. Bowsher, 1955a, p. 3, Pl. 2, fig. 6. Ehlers and Kesling, 1963, p. 1031.
Mississippian, Osagean, Edwardsville Formation, Keokuk Limestone: United States: Indiana

Agaricocrinus keokukensis Miller and Gurley, 1897 = *A. nodulosus* Worthen, 1889.

Agaricocrinus louisianensis Rowley, 1900. Ehlers and Kesling, 1963, p. 1031.
Mississippian, Kinderhookian, Burlington Limestone: United States

Agaricocrinus montgomeryensis Peck and Keyte, 1938. Ehlers and Kesling, 1963, p. 1030.
Mississippian, Kinderhookian, Chouteau Limestone: United States

Agaricocrinus nodosus Meek and Worthen, 1860. Ehlers and Kesling, 1963, p. 1030.
Mississippian, Kinderhookian, Burlington Limestone: United States

Agaricocrinus nodulosus Worthen, 1889. Ehlers and Kesling, 1963, p. 1031.
Agaricocrinus keokukensis Miller and Gurley, 1897. Ehlers and Kesling, 1963, p. 1031.
Mississippian, Osagean, Keokuk Limestone, Fort Payne Chert: United States

Agaricocrinus nodulosus macadamsi Worthen, 1889. Van Sant, *in* Van Sant and Lane, 1964, p. 119.
Mississippian, Osagean: United States: Indiana

Agaricocrinus planoconvexus Hall, 1861. Ehlers and Kesling, 1963, p. 1030.
  Agaricocrinites planoconvexus (Hall), 1861. Moore and Laudon, 1943b,
    p. 141, Pl. 12, fig. 3; 1944, p. 195, Pl. 76, fig. 13. Laudon,
    1948, Pl. 2.
  Mississippian, Kinderhookian, Burlington Limestone, Chouteau Limestone:
    United States: Iowa

Agaricocrinus podagricus Ehlers and Kesling, 1963, p. 1035, Pl. 133,
    figs. 1-6; Fig. 2.
  Mississippian, Osagean, Fort Payne Chert: United States: Kentucky

Agaricocrinus ponderosus Wood, 1909. Ehlers and Kesling, 1963, p. 1030.
  Mississippian, Osagean, Fort Payne Chert: United States

Agaricocrinus praecursor Rowley, 1902. Ehlers and Kesling, 1963, p. 1031.
  Mississippian, Osagean, Fern Glen Limestone: United States

Agaricocrinus profundus Miller and Gurley, 1895. Ehlers and Kesling, 1963,
    p. 1030.
  Mississippian, Osagean, Fort Payne Chert: United States

Agaricocrinus pyramidatus (Hall), 1858. Ehlers and Kesling, 1963, p. 1031.
  Mississippian, Kinderhookian, Burlington Limestone: United States

Agaricocrinus sampsoni Miller, 1894. Ehlers and Kesling, 1963, p. 1030.
  Agaricocrinites sampsoni (Miller), 1894. Laudon, 1948, Pl. 2.
  Mississippian, Kinderhookian, Chouteau Limestone: United States

Agaricocrinus splendens Miller and Gurley, 1890. Ehlers and Kesling, 1963,
    p. 1031. Van Sant, *in* Van Sant and Lane, 1964, p. 119, Pl. 8,
    figs. 2, 4; Figs. 19, no. 3; 20, nos. 2, 6.
  Mississippian, Osagean, Borden Group, Keokuk Limestone: United States:
    Indiana

Agaricocrinus stellatus (Hall), 1858. Ehlers and Kesling, 1963, p. 1030.
  Mississippian, Kinderhookian, Burlington Limestone: United States

Agaricocrinus tuberosus Troost, 1849. Ehlers and Kesling, 1963, p. 1030.
  Mississippian, Osagean, Keokuk Limestone, Fort Payne Chert: United
    States

Agaricocrinus tugurium Miller and Gurley, 1895. Ehlers and Kesling, 1963,
    p. 1030 (incorrectly given as *A. tuguriinus*).
  Mississippian, Osagean, Fort Payne Chert: United States

Agaricocrinus whitfieldi Hall, 1858. Ehlers and Kesling, 1963, p. 1031.
  Mississippian, Osagean, Keokuk Limestone, Fort Payne Chert: United
    States

Agaricocrinus wortheni Hall, 1858. Ubaghs, 1953, p. 740, fig. 22d. Ehlers
    and Kesling, 1963, p. 1031.
  Agaricocrinites wortheni (Hall), 1958. Moore and Laudon, 1942, p. 74,
    figs. 4A-B; 1943b, p. 141, Pl. 12, fig. 2; 1944, p. 195, Pl. 76,
    fig. 22. Laudon, 1948, Pl. 2.
  Mississippian, Osagean, Keokuk Limestone, Fort Payne Chert: United
    States: Illinois, Indiana, Iowa, Missouri

AGASSIZOCRINUS Owen and Shumard, 1852. *A. conicus* Owen and Shumard, 1852.
    Yakovlev, 1939e, p. 832; 1939f, p. 832. Moore and Laudon, 1943b,
    p. 62, fig. 8; 1944, p. 175. Yakovlev, 1947d, p. 42, fig. 4.
    Moore, 1950, fig. 4m. Yakovlev, 1956a, p. 78, fig. 24; 1958, p. 63,
    fig. 24; 1964a, p. 73, fig. 26. Arendt and Hecker, 1964, p. 91.
  Carboniferous, Chesterian

Agassizocrinus caliculus Moore and Plummer, 1938 = *Paragassizocrinus
    caliculus*

*Agassizocrinus conicus Owen and Shumard, 1852. Moore and Laudon, 1944,
    p. 177, Pl. 61, fig. 7.
  Mississippian, Chesterian: United States: Illinois

Agassizocrinus cf. *A. conicus* Owen and Shumard, 1852. Horowitz, 1965, p. 38, Pl. 4, figs. 1, 2.
Mississippian, Chesterian, Glen Dean Formation: United States: Kentucky

Agassizocrinus dactyliformis Shumard, 1854. Horowitz, 1965, p. 38, Pl. 4, figs. 3, 4.
Mississippian, Chesterian, Glen Dean Formation: United States: Kentucky

Agassizocrinus gibbosus Hall, 1858. Moore and Laudon, 1944, p. 177, Pl. 61, fig. 5.
Mississippian, Chesterian: United States: Illinois

Agassizocrinus globosus Worthen, 1873. Moore and Laudon, 1944, p. 177, Pl. 61, fig. 6. Strimple, 1962g, p. 186, figs. 8-13.
Mississippian, Chesterian, Gasper Formation, Hindsville Limestone: United States: Kentucky, Oklahoma

Agassizocrinus hemisphericus Worthen, 1882 = *Mooreocrinus hemisphericus*

Agassizocrinus laevis (Roemer), 1852. Moore and Laudon, 1943b, p. 135, Pl. 6, fig. 10; 1944, p. 177, Pl. 61, fig. 8.
Mississippian, Chesterian, Okaw Formation: United States: Kentucky

Agassizocrinus magnus Moore and Plummer, 1938 = *Paragassizocrinus magnus*

Agassizocrinus ovalis Miller and Gurley, 1896. Moore and Laudon, 1944, p. 177, Pl. 61, fig. 9.
Mississippian, Chesterian, Okaw Formation: United States: Illinois

Agassizocrinus cf. *A. ovalis* Miller and Gurley, 1896. Butts, 1941, p. 251, Pl. 132, figs. 43, 44; 1948, p. 47, Pl. 6, figs. 19, 20.
Mississippian, Chesterian, Gasper Formation: United States: Georgia, Virginia

Agassizocrinus papillatus Worthen, 1882 = *Mooreocrinus papillatus*

Agassizocrinus patulus Strimple, 1951f, p. 672, Pl. 98, figs. 9, 10; Pl. 99, fig. 12.
Mississippian, Chesterian, Pitkin Limestone: United States: Oklahoma

Agassizocrinus spp.
Strimple, 1961b, p. 158, figs. 1-5.
Mississippian, Chesterian, Goddard Formation: United States: Oklahoma
Sahni and Gupta, 1965a, p. 216, fig. 1.
Carboniferous, *Syringothyris* Limestone: India: Kashmir
Sahni and Gupta, 1965b, p. 247 (*not* an *Agassizocrinus*).
Silurian, Naubug Limestone: India: Kashmir

AGATHOCRINUS Schewtschenko, 1967. *A. globosus* Schewtschenko, 1967. Schewtschenko, 1967, p. 84; 1968b, p. 80.
Lower Devonian

Agathocrinus globosus Schewtschenko, 1967, p. 84, Pl. 10, figs. 1-6; Figs. 7, 8; 1968b, p. 81, Pl. 10, figs. 1-6; Figs. 7, 8.
Lower Devonian, Kshtutskian horizon: Russia: Zeravshan Range

AGENERACRINUS Sutton and Winkler, 1940. Moore and Laudon, 1943b, p. 62.
Mississippian, Chesterian

Ageneracrinus subtumidus (Meek and Worthen), 1873. Moore and Laudon, 1944, p. 209, Pl. 79, fig. 29.
Mississippian, Chesterian, Chester Group: United States: Illinois

AGLAOCRINUS Strimple, 1961 = PARETHELOCRINUS Strimple, 1961.

Aglaocrinus compactus (Moore and Plummer), 1940, Strimple, 1961 = *Paracromyocrinus compactus*

Aglaocrinus expansus (Strimple), 1938, Strimple, 1961 = *Parethelocrinus expansus*

Aglaocrinus iatani (Strimple), 1949, Strimple, 1961 = *Parethelocrinus iatani*

Aglaocrinus magnus (Strimple), 1949, Strimple, 1961 = *Parethelocrinus magnus*

Aglaocrinus pustulosus (Moore and Plummer), 1940, Strimple, 1961 = *Paracromyocrinus pustulosus*

Aglaocrinus? rectilatus Lane and Webster, 1966 = *Parethelocrinus rectilatus*

AGNOSTOCRINUS Webster and Lane, 1967. *A. typus* Webster and Lane, 1967. Webster and Lane, 1967, p. 15.
Permian, Wolfcampian

Agnostocrinus typus Webster and Lane, 1967, p. 16, Pl. 3, figs. 1, 2.
Permian, Wolfcampian, Bird Spring Formation: United States: Nevada

AINACRINUS Wright, 1939. *Synerocrinus? smithi* Wright, 1934. Moore and Laudon, 1943b, p. 74. Ubaghs, 1953, p. 755. Arendt and Hecker, 1964, p. 94.
Carboniferous, Visean

Ainacrinus smithi (Wright), 1934. Wright, 1954a, p. 151, Pl. 41, figs. 2, 4, 5, 7, 9; Figs. 82-84.
Carboniferous, Visean, Lower Limestone Group: Scotland

ALCIMOCRINUS Kirk, 1938. *Zeacrinus girtyi* Springer, 1926. Moore and Laudon, 1943b, p. 58; 1944, p. 167. Arendt and Hecker, 1964, p. 89.
Mississippian-Pennsylvanian, Chesterian-Morrowan

Alcimocrinus girtyi (Springer), 1926. Moore and Laudon, 1943b, p. 134, Pl. 5, fig. 10; 1944, p. 167, Pl. 61, fig. 15. Moore, 1962b, p. 24, fig. 4.
Pennsylvanian, Morrowan, Morrow Formation: United States: Oklahoma

Alcimocrinus ornatus Strimple, 1949d, p. 28, Pl. 5, figs. 1-7.
Mississippian, Chesterian, Fayetteville Formation: United States: Oklahoma

ALISOCRINUS Kirk, 1929. *Mariacrinus warreni* Ringueberg, 1888. Moore and Laudon, 1943b, p. 96; 1944, p. 199. Ubaghs, 1953, p. 741, fig. 82a; 1958b, p. 294, fig. 16.
Silurian, Niagaran

Alisocrinus carleyi (Hall), 1863. Moore and Laudon, 1943b, p. 142, Pl. 13, fig. 5.
Mariacrinus carleyi (Hall), 1877 [sic]. Wilson, 1949, p. 251, Pl. 25, figs. 13, 14.
Silurian, Niagaran, Laurel Limestone, Waldron Shale: United States: Indiana, Tennessee

Alisocrinus laevis Kirk, 1929. Moore and Laudon, 1944, p. 199, Pl. 74, fig. 2.
Silurian, Niagaran, Laurel Limestone: United States: Indiana

Alisocrinus warreni (Ringueberg), 1888. Ubaghs, 1953, p. 741, fig. 82a.
Silurian, Niagaran, Rochester Shale: United States: New York

ALLAGECRINUS Carpenter and Etheridge, 1881. *A. austinii* Carpenter and Etheridge, 1881. Moore and Laudon, 1943b, p. 30; 1944, p. 149, Pl. 52, fig. 4. Ubaghs, 1953, p. 747, figs. 11f, 47-51, 140G. Moore, 1962b, p. 16, fig. 12, no. 3. Arendt and Hecker, 1964, p. 81. Strimple, 1966c, p. 103.
Carboniferous

Allagecrinus acutus Wanner, 1929 = *Metallagecrinus acutus*

Allagecrinus americanus Rowley, 1895. Williams, 1943, p. 64, Pl. 6, figs. 14, 15. Strimple and Koenig, 1956, p. 1228, figs. 4, 28-31. Strimple, 1966c, p. 103, fig. 2d. Gutschick, 1968, p. 988, fig. 6.
Mississippian, Kinderhookian, Louisiana Limestone: United States: Missouri

*Allagecrinus austinii Carpenter and Etheridge, 1881. Wright, 1952a, p. 139, Pl. 39, figs. 1, 2, 4-7, 9, 11, 12; Pl. 41, figs. 19-22, 24, 25, 30-34 (*non* Pl. 41, fig. 23 as given). Gutschick, 1968, p. 988, fig. 2.
Carboniferous, Visean, Lower Limestone Group: Scotland

Allagecrinus bassleri Strimple, 1938 = *Isoallagecrinus bassleri*

Allagecrinus bassleri status Strimple, 1951 = *Isoallagecrinus status*

Allagecrinus sp. cf. *A. bassleri* Moore, 1940 = *Isoallagecrinus* sp. cf. *I. bassleri*

Allagecrinus constellatus Moore, 1940 = *Isoallagecrinus constellatus*

Allagecrinus copani Strimple, 1949 = *Isoallagecrinus copani*

Allagecrinus coronarius Gutschick, 1968, p. 996, figs. 2, 6-9.
Mississippian, Chesterian, Renault Formation, Ridenhower Shale: United States: Illinois, Kentucky

Allagecrinus dignatus Moore, 1940 = *Isoallagecrinus dignatus*

Allagecrinus donetzensis Yakovlev, 1930 = *Isoallagecrinus? donetzensis*

Allagecrinus dux Wanner, 1930 = *Metallagecrinus dux*

Allagecrinus elegantulus Gutschick, 1968, p. 997, figs. 2, 5-7, 10, 11.
Mississippian, Chesterian, Kinkaid Limestone: United States: Illinois

Allagecrinus excavatus Wanner, 1929 = *Metallagecrinus excavatus*

Allagecrinus garpelensis Wright, 1932. Wright, 1952a, p. 142, Pl. 39, figs. 3, 8, 10, 17-19, 21; Pl. 40, figs. 27, 29. Gutschick, 1968, p. 988, fig. 2.
Carboniferous, Namurian: Scotland

Allagecrinus graffhami Strimple, 1948 = *Isoallagecrinus graffhami*

Allagecrinus indoaustralicus Wanner, 1916 = *Metallagecrinus indoaustralicus*

Allagecrinus inflatus Wanner, 1929 = *Metallagecrinus inflatus*

Allagecrinus kylensis Strimple, 1948 = *Isoallagecrinus kylensis*

Allagecrinus multibrachiatus Yakovlev, 1927 = *Metallagecrinus multibrachiatus*

Allagecrinus ornatus Wanner, 1929 = *Metallagecrinus ornatus*

Allagecrinus procerus Wanner, 1929 = *Metallagecrinus procerus*

Allagecrinus quinquebrachiatus Wanner, 1929 = *Metallagecrinus quinquebrachiatus*

Allagecrinus quinquelobus Wanner, 1929 = *Metallagecrinus quinquelobus*

Allagecrinus sculptus Strimple and Koenig, 1956, p.1229, fig.2, nos. 1-12. Gutschick, 1968, p. 988, fig. 6.
Mississippian, Osagean, Nunn Member: United States: New Mexico

Allagecrinus strimplei Kirk, 1936 = *Isoallagecrinus strimplei*

Allagecrinus uralensis Yakovlev, 1927 = *Metallagecrinus uralensis*

Allagecrinus uralensis var. *nodocarinatus* Yakovlev, 1927 = *Metallagecrinus uralensis* var. *nodocarinatus*

Allagecrinus sp. Ubaghs, 1953, p. 747, figs. 47, 48.

ALLOCATILLOCRINUS Wanner, 1937. **Allagecrinus carpenteri* Wachsmuth, 1882. Moore and Laudon, 1943b, p. 30, fig. 3; 1944, p. 149, Pl. 52, fig. 4. Ubaghs, 1953, p. 747, figs. 52, 53. Moore, 1962b, p. 16, fig. 12, no. 10. Arendt and Hecker, 1964, p. 81. Strimple, 1966c, p. 107.
Carboniferous

*Allocatillocrinus carpenteri (Wachsmuth), 1882. Moore and Laudon, 1943b, p. 130, Pl. 1, fig. 17; 1944, p. 149, Pl. 55, fig. 4.
Mississippian, Chesterian: United States: Illinois, Indiana, Alabama

Allocatillocrinus morrowensis (Strimple), 1940. Moore and Laudon, 1944, p. 149, Pl. 56, fig. 18.
Pennsylvanian, Morrowan, Wapanucka Limestone: United States: Oklahoma

Allocatillocrinus rotundus Moore, 1940. Moore and Laudon, 1943b, p. 130, Pl. 1, fig. 18; 1944, p. 149, Pl. 56, fig. 19. Ubaghs, 1953, p. 747, fig. 52.
Pennsylvanian, Morrowan, Brentwood Limestone, Hale Formation: United States: Oklahoma

Allocatillocrinus rotundus multibrachiatus Moore, 1940. Ubaghs, 1953, p. 747, fig. 53.
Pennsylvanian, Morrowan, Brentwood Limestone: United States: Oklahoma

Allocatillocrinus scoticus (Wright), 1933. Wright, 1952a, p. 147, Pl. 39, figs. 13-15; Figs. 76-81.
Carboniferous, Visean, Lower Limestone Group: Scotland

ALLOCRINUS Wachsmuth and Springer, 1889. **A. typus* Wachsmuth and Springer, 1889. Moore and Laudon, 1943b, p. 98, fig. 18; 1944, p. 200, Pl. 70, fig. 11. Ubaghs, 1953, p. 742.
Silurian, Niagaran

Allocrinus divergens Strimple, 1952b, p. 78, figs. 8, 9.
Silurian, Niagaran, Henryhouse Formation: United States: Oklahoma

Allocrinus globulus Strimple, 1963d, p. 107, Pl. 10, figs. 2-4.
Silurian, Niagaran, Henryhouse Formation: United States: Oklahoma

Allocrinus irroratus Strimple, 1963d, p. 106, Pl. 8, figs. 3-7.
Silurian, Niagaran, Henryhouse Formation: United States: Oklahoma

Allocrinus longidactylus Springer, 1926. Moore and Laudon, 1943b, p. 143, Pl. 14, fig. 4; 1944, p. 200, Pl. 73, fig. 21.
Silurian, Niagaran, Beech River Formation: United States: Tennessee

*Allocrinus typus Wachsmuth and Springer, 1889. Moore and Laudon, 1943b, p. 143, Pl. 14, fig. 5; 1944, p. 200, Pl. 73, fig. 18.
Silurian, Niagaran, Beech River Formation: United States: Tennessee

ALLOPROSALLOCRINUS Casseday and Lyon, 1862. **A. conicus* Casseday and Lyon, 1862. Moore and Laudon, 1943b, p. 95, fig. 16; 1944, p. 195. Ubaghs, 1953, p. 740. Lane, 1963b, p. 925.
Mississippian

*Alloprosallocrinus conicus Casseday and Lyon, 1862. Moore and Laudon, 1943b, p. 141, Pl. 12, fig. 10; 1944, p. 195, Pl. 75, fig. 4. Laudon, 1948, Pl. 2. Van Sant, *in* Van Sant and Lane, 1964, p. 110, Pl. 7, figs. 2-4; Figs. 17, no. 5; 18; 37 (*non* fig. 11 as given). Kesling, 1965a, p. 257, Pl. 1, figs. 3-6; Pl. 2, figs. 1-5.
Conocrinites tuberculosus Troost, 1849. Van Sant, *in* Van Sant and Lane, 1964, p. 110.

Lower Mississippian, Warsaw Group, Keokuk Limestone, Borden Group:
United States: Iowa, Kentucky, Missouri, Tennessee, Indiana

ALLOSOCRINUS Strimple, 1949. *A. bronaughi Strimple, 1949. Strimple,
1949c, p. 17. Arendt and Hecker, 1964, p. 89.
Pennsylvanian, Desmoinesian-Permian, Wolfcampian

*Allosocrinus bronaughi Strimple, 1949c, p. 18, Pl. 4, figs. 1-4.
Pennsylvanian, Missourian, Wann Formation: United States: Oklahoma

Allosocrinus libratus Strimple, 1961g, p. 105, Pl. 5, figs. 4-6; Pl. 15,
fig. 5.
Pennsylvanian, Desmoinesian, Holdenville Shale: United States:
Oklahoma

Allosocrinus porus Strimple, 1951c, p. 19, Pl. 4, fig. 7.
Pennsylvanian, Missourian, Wann Formation: United States: Oklahoma

Allosocrinus quinarius Lane and Webster, 1966, p. 56, Pl. 13, fig. 3.
Permian, Wolfcampian, Bird Spring Formation: United States: Nevada

ALLOSYCOCRINUS Wanner, 1924. *A. pusillus Wanner, 1924. Moore and Laudon,
1943b, p. 50. Arendt and Hecker, 1964, p. 85. Lane, 1967b, p. 12,
fig. 6.
Permian

ALSOPOCRINUS Tansey, 1924. *A. anna Tansey, 1924. Moore and Laudon, 1943b,
p. 55.
Devonian

AMBICOCRINUS Kirk, 1945. *Thysanocrinus arborescens Talbot, 1905. Kirk,
1945a, p. 353. Ubaghs, 1953, p. 737.
Upper Silurian?—Lower Devonian

*Ambicocrinus arborescens (Talbot), 1905. Kirk, 1945a, p. 353.
Upper Silurian? Manlius Limestone or Lower Devonian, Coeymans Limestone:
United States: New York

AMBLACRINUS d'Orbigny, 1849. *Platycrinus rosaceus Roemer, 1844. Moore
and Laudon, 1943b, p. 101. Ubaghs, 1953, p. 743.
Devonian—lower Carboniferous

AMELIACRINUS Kesling, 1968. *A. benderi Kesling, 1968. Kesling, 1968c,
p. 159.
Devonian, Erian

*Ameliacrinus benderi Kesling, 1968c, p. 159, Pls. 1-3; Figs. 1, 2.
Devonian, Erian, Silica Formation: United States: Ohio

AMMONICRINUS Springer, 1926. *A. wanneri Springer, 1926. Moore and
Laudon, 1943b, p. 29.
Devonian-Permian

Ammonicrinus doliiformis Wolburg, 1937. Ubaghs, 1952, p. 216, Pl. 3;
1953, p. 729, fig. 134. Kongiel, 1958, p. 31, fig. 5. Yakovlev,
1946b, p. 69, fig. 107.
Devonian, Givetian, Rommersheimer Schichten: Germany

Ammonicrinus(?) nordicus Yakovlev, in Yakovlev and Ivanov, 1956, p. 69,
Pl. 11, fig. 15.
Permian: Russia: Ural Mountains

Ammonicrinus sulcatus Kongiel, 1958, p. 34, fig. 6.
Devonian, lower Givetian: Poland

*Ammonicrinus wanneri Springer, 1926. Cuénot, 1948, p. 38, fig. 49.
Ubaghs, 1952, p. 206, Pls. 1, 2; Figs. 1, 2, 3F, 4, 5; 1953,
p. 729, figs. 132, 133. Wanner, 1954, p. 231, fig. 1. Kongiel,
1958, p. 31, figs. 1-4.
Devonian, Eifelian-Givetian: Germany

AMONOHEXACRINUS Schewtschenko, 1967. *A. adelius* Schewtschenko, 1967.
Schewtschenko, 1967, p. 82; 1968b, p. 78.
Lower Devonian, Kshtut horizon: Russia: Zeravshan Mountains

*Amonohexacrinus adelius Schewtschenko, 1967, p. 82, Pl. 9, fig. 4; Figs. 5, 6; 1968b, p. 79, Pl. 9, fig. 4; Figs. 5, 6.
Lower Devonian, Kshtut horizon: Russia: Zeravshan Mountains

AMPELOCRINUS Kirk, 1942. *A. bernhardinae* Kirk, 1942. Moore and Laudon, 1943b, p. 59. Arendt and Hecker, 1964, p. 89.
Mississippian

Ampelocrinus erectus Strimple, 1949c, p. 16, Pl. 4, figs. 5, 6.
Mississippian, Chesterian, Fayetteville Formation: United States: Oklahoma

Ampelocrinus plumosus Wright, 1951b, p. 93, Pl. 30, fig. 3.
Aesiocrinus cf. *harei* Wright, 1937. Wright, 1951b, p. 93.
Aesiocrinus cf. *harei* Wright, 1939. Wright, 1951b, p. 93.
?*Ampelocrinus* n. sp. Kirk, 1942. Wright, 1951b, p. 93.
Lower Carboniferous, Visean, Lower Limestone Group: Scotland

?*Ampelocrinus* n. sp. Kirk, 1942 = *Ampelocrinus plumosus*

AMPHERISTOCRINUS Hall, 1879. *A. typus* Hall, 1879. Moore and Laudon, 1943b, p. 51, fig. 4; 1944, p. 154, Pl. 54, fig. 29. Moore, 1962b, p. 15, fig. 10, no. 3. Arendt and Hecker, 1964, p. 85.
Silurian

*Ampheristocrinus typus Hall, 1879. Moore and Laudon, 1943b, p. 131, Pl. 2, fig. 9; 1944, p. 154, Pl. 54, fig. 28.
Silurian, Niagaran, Waldron Shale, Beech River Formation: United States: Indiana, Tennessee

AMPHICRINUS Springer, 1906. *A. scoticus* Springer, *in* Wright, 1914. Moore and Laudon, 1943b, p. 74, fig. 10; 1944, p. 179, Pl. 67, fig. 2. Cuénot, 1948, p. 67. Ubaghs, 1953, p. 755. Arendt and Hecker, 1964, p. 94.
Carboniferous

Amphicrinus carbonarius Springer, 1920. Moore and Laudon, 1944, p. 179, Pl. 79, fig. 32. Strimple, 1962e, p. 50, Pl. 5, figs. 3-5.
Pennsylvanian, Desmoinesian, Marmaton Formation: United States: Kansas, Oklahoma

Amphicrinus luberculatus Yakovlev, 1961, p. 417, Pl. 1, figs. 3, 4 (incorrectly spelled as *A. luberculatum*, p. 417).
Lower Carboniferous (C_1): Russia: Donetz Basin

*Amphicrinus scoticus Springer, *in* Wright, 1914. Wright, 1946a, p. 37, Pl. 3, figs. 7, 10-12; Pl. 4, figs. 1-3, 5-7, 10-12; 1954a, p. 149, Pl. 41, figs. 1, 14; Pl. 42, figs. 8, 9; Pl. 43, figs. 16-18, 20-23; Pl. 47, figs. 26, 27. Yakovlev, 1961, p. 417, Pl. 1, figs. 1, 2.
Lower Carboniferous, Visean, Lower Limestone Group: Scotland; Russia: Donetz Basin

Amphicrinus? sp. Wright, 1954a, p. 151, Pl. 42, fig. 5.
Carboniferous, Namurian, E_1, red beds above Main Limestone: England

AMPHIPSALIDOCRINUS Weller, 1930. *A. scissurus* Weller, 1930. Moore and Laudon, 1943b, p. 50. Ubaghs, 1953, p. 747. Arendt and Hecker, 1964, p. 85. Lane, 1967b, p. 11, fig. 4.
Carboniferous

Amphipsalidocrinus comptus (Weller), 1930. Peck and Connelly, 1951 = *Tytthocrinus comptus*

Amphipsalidocrinus inconsuetus (Peck), 1936. Peck and Connelly, 1951 = *Tytthocrinus inconsuetus*

*Amphipsalidocrinus scissurus Weller, 1930. Peck and Connelly, 1951, p. 414, figs. 5, 6.
Pennsylvanian, Morrowan: United States: Indiana

AMPHORACRINUS Austin, 1848. **Actinocrinus gilbertsoni* Miller, *in* Phillips, 1836. Moore and Laudon, 1943b, p. 92, fig. 15; 1944, p. 191. Termier and Termier, 1947, p. 255, Pl. 19, fig. 47. Ubaghs, 1953, p. 739. Hemming, 1958, p. 301.
Carboniferous

Amphoracrinus amphora Gilbertson, Portlock, 1843 = *Amphoracrinus portlocki*

Amphoracrinus atlas (McCoy), 1849. Wright, 1943a, p. 87, Pl. 4, figs. 6, 7; 1955a, p. 199, Pl. 50, figs. 5, 10-14.
Carboniferous, Visean: Scotland

Amphoracrinus bollandensis Wright, 1955a, p. 200, Pl. 50, figs. 19, 20; Pl. 51, figs. 26, 28-32.
Amphoracrinus gilbertsoni Form III (Miller, *in* Phillips), 1836. Wright, 1943a, p. 87, Pl. 4, fig. 11; Pl. 6, figs. 4, 8-12; 1955a, p. 200.
Carboniferous, Tournaisian-Visean: Scotland

Amphoracrinus brevicalix (Rofe), 1865. Wright, 1943b, p. 234.
Carboniferous: Scotland

Amphoracrinus clitheroensis Wright, 1942, p. 272, Pl. 10, figs. 1-4.
Lower Carboniferous, Tournaisian (C_1): Scotland

Amphoracrinus compressus Wright, 1943a, p. 88, Pl. 5, fig. 11; 1955a, p. 202, Pl. 49, fig. 21.
Carboniferous, Visean: England

Amphoracrinus crassus (Austin and Austin), 1843. Wright, 1955a, p. 199, Pl. 51, fig. 38; Fig. 110, no. 5.
Actinocrinus crassus Austin and Austin, 1843. Wright, 1955a, p. 199.
Carboniferous, Tournaisian: England

Amphoracrinus divergens (Hall), 1860. Moore and Laudon, 1943b, p. 139, Pl. 10, fig. 13; 1944, p. 191, Pl. 77, fig. 28.
Mississippian, Osagean, Burlington Limestone, Lake Valley Limestone: United States: Iowa, Missouri, New Mexico

Amphoracrinus gigas Austin, *in* Wright, 1955a, p. 202, Pl. 48, figs. 6, 7, 9-12, 14, 15; Pl. 51, fig. 5; Fig. 110, nos. 1-4.
Carboniferous, ?Tournaisian: Ireland

*Amphoracrinus gilbertsoni Form I (Miller, *in* Phillips), 1836. Wright, 1943a, p. 86, Pl. 4, figs. 2-4, 8-10, 12-14; Pl. 5, figs. 1-9, 12, 13; Pl. 6, figs. 13-15; 1955a, p. 194, Pl. 48, fig. 16; Pl. 49, figs. 2-4, 6, 16, 17, 20; Pl. 50, figs. 3, 9; Pl. 51, figs. 15-20; Fig. 110, nos. 6-8. Hemming, 1958, p. 301. Wright, 1960, p. 333.
Actinocrinus gilbertsoni Miller, *in* Phillips, 1836. Wright, 1956a, p. 156. Peck, 1956, p. 319. Hemming, 1958, p. 301.
Carboniferous, Tournaisian-Visean: Scotland

Amphoracrinus gilbertsoni Form II (Miller, *in* Phillips), 1836 = *Amphoracrinus rotundus*

Amphoracrinus gilbertsoni Form III (Miller, *in* Phillips), 1836 = *Amphoracrinus bollandensis*

Amphoracrinus gilbertsoni var. *triangularis* Wright, 1955a, p. 198, Pl. 49, figs. 9-14, 18.
Carboniferous, Tournaisian-Visean: Scotland

Amphoracrinus granulatus (Austin and Austin), 1842. Wright, 1955a, p. 198,
   Pl. 52, fig. 11; Pl. 80, figs. 8, 16; Fig. 110, nos. 9, 9b.
   Carboniferous, Tournaisian: England
Amphoracrinus portlocki Wright, 1955a, p. 203, Pl. 48, fig. 5; Pl. 50,
   figs. 15-18, 21.
   Amphoracrinus amphora Gilbertson, Portlock, 1843. Wright, 1955a,
   p. 203.
   Carboniferous, ?Tournaisian: Ireland
Amphoracrinus rochi Delpey, 1939. Termier and Termier, 1950, p. 83,
   Pl. 208, figs. 37-41.
   Carboniferous, upper Visean: Morocco
Amphoracrinus rotundus Wright, 1955a, p. 199, Pl. 48, fig. 13; Pl. 49,
   figs. 1, 5, 7, 8, 15.
   Amphoracrinus gilbertsoni Form II (Miller, *in* Phillips), 1836.
   Wright, 1943a, p. 87, Pl. 4, figs. 1, 5; Pl. 5, fig. 10; 1955a,
   p. 199.
   Carboniferous, Tournaisian: Scotland
Amphoracrinus sampsoni Miller and Gurley, 1896 = *Aryballocrinus sampsoni*
Amphoracrinus turgidus Wright, 1943a, p. 88, Pl. 6, figs. 1-3, 5-7.
   Moore, 1948, p. 390, fig. 11. Wright, 1955a, p. 201, Pl. **49**,
   fig. 19; Pl. 50, figs. 1, 2, 4, 6-8.
   Carboniferous, Visean: Scotland
Amphoracrinus sp. Minato, 1951, p. 360.
   Lower Carboniferous: Japan
ANAMESOCRINUS Goldring, 1923. *A. lutheri* Goldring, 1923. Moore and
   Laudon, 1943b, p. 30, fig. 3; 1944, p. 147, Pl. 52, fig. 1.
   Ubaghs, 1953, p. 746, fig. 140D. Moore, 1962b, p. 12, fig. 3, no. 5.
   Devonian
Anamesocrinus lutheri Goldring, 1923. Moore and Laudon, 1943b, p. 130,
   Pl. 1, fig. 14; 1944, p. 147, Pl. 56, fig. 12.
   Devonian, Chautauquan, Chemung Group, Laona Sandstone: United States:
   New York
ANARCHOCRINUS Jaekel, 1918. *A. rossicus* Jaekel, 1918. Moore and Laudon,
   1943b, p. 51. Ubaghs, 1953, p. 750. Arendt and Hecker, 1964, p. 83.
   Ordovician
Anarchocrinus rossicus Jaekel, 1918. Arendt and Hecker, 1964, p. 83,
   fig. 121.
   Middle Ordovician: Estonia
ANARTIOCRINUS Kirk, 1940. *A. lyoni* Kirk, 1940. Moore and Laudon, 1943b,
   p. 62; 1944, p. 177. Arendt and Hecker, 1964, p. 91.
   Mississippian
Anartiocrinus maxvillensis (Whitfield), 1891. Moore and Laudon, 1944,
   p. 177, Pl. 61, fig. 11.
   Mississippian, Chesterian, Glen Dean Formation: United States:
   Kentucky
ANCISTROCRINUS Wanner, 1924. *A. vermistriatus* Wanner, 1924. Moore
   and Laudon, 1943b, p. 74. Ubaghs, 1953, p. 756. Arendt and Hecker,
   1964, p. 94.
   Permian
ANCYROCRINUS Hall, 1862. *A. bulbosus* Hall, 1862. Moore and Laudon,
   1943b, p. 103. Ubaghs, 1953, p. 752, fig. 120. Arendt and Hecker,
   1964, p. 86. Moore and Jeffords, 1968, p. 11, fig.1, no. 7.
   Devonian

Ancyrocrinus armoricanus Morzadec, 1967, p. 27, Pl. 1, figs. 1-6; Figs. 1, 2.
 Devonian, Couvinian: France: Brittany

*Ancyrocrinus bulbosus Hall, 1862. Cuénot, 1948, p. 34, fig. 42. Ubaghs, 1953, p. 752, fig. 120. Yakovlev, 1964b, p. 67, fig. 96.
 Devonian: United States

ANEMETOCRINUS Wright, 1938. *A. biserialis* Wright, 1938. Moore and Laudon, 1943b, p. 59. Arendt and Hecker, 1964, p. 89.
 Lower Carboniferous

Anemetocrinus ardrossensis Wright, 1938. Wright, 1951a, p. 27, Pl. 7, figs. 3, 4, 6; Fig. 6.
 Carboniferous, Visean: Scotland

*Anemetocrinus biserialis Wright, 1938. Wright, 1950, p. 24, Pl. 7, figs. 1, 2, 5, 8, 9; Fig. 5 (*non* Figs. 5-8 as given).
 Carboniferous, Visean, Lower Limestone Group: Scotland

Anemetocrinus covensis (Wright), 1934. Wright, 1951a, p. 28, Pl. 7, fig. 7; Fig. 7.
 Carboniferous, Visean, Calciferous Sandstone series: Scotland

Anemetocrinus pentonensis (Wright), 1934. Wright, 1951a, p. 28, Pl. 7, fig. 10; Fig. 8.
 Carboniferous, Visean, Lower Limestone Group: Scotland

Anemetocrinus wilsoni Wright, 1952b, p. 323, Pl. 13, fig. 10; Figs. 3, 4; 1954b, p. 170, Pl. 3, figs. 5, 6; 1960, p. 330, Appendix, Pl. A, fig. 12; Pl. B, figs. 3, 4.
 Carboniferous, Visean, Calciferous Sandstone Series: Scotland

ANGELINOCRINUS Jaekel, 1918. **Briarocrinus angustus* Angelin, 1878. Moore and Laudon, 1943b, p. 98.
 Silurian

ANISOCRINUS Angelin, 1878. **A. interradiatus* Angelin, 1878. Moore and Laudon, 1943b, p. 74, fig. 11; 1944, p. 181, Pl. 66, fig. 4. Ubaghs, 1953, p. 756, fig. 15c. Arendt and Hecker, 1964, p. 94.
 Silurian

Anisocrinus greenei (Miller and Gurley), 1896. Moore and Laudon, 1944, p. 181, Pl. 54, fig. 8. Ubaghs, 1953, p. 756, fig. 15c.
 Silurian, Niagaran, Louisville Limestone: United States: Kentucky, Tennessee

Anisocrinus oswegoensis (Miller and Gurley), 1894. Moore and Laudon, 1944, p. 181, Pl. 54, fig. 7.
 Silurian, Niagaran: United States: Illinois

ANOBASICRINUS Strimple, 1961. **A. bulbosus* Strimple, 1961. Strimple, 1961g, p. 114.
 Pennsylvanian

Anobasicrinus braggsi (Strimple), 1951. Strimple, 1961g, p. 114. *Plummericrinus braggsi* Strimple, 1951c, p. 17, Pl. 3, figs. 9-12.
 Pennsylvanian, Morrowan, Brentwood Limestone: United States: Oklahoma

*Anobasicrinus bulbosus Strimple, 1961g, p. 115, Pl. 3, figs. 1-3; Pl. 7, figs. 1-3; Fig. 21.
 Pennsylvanian, Desmoinesian, Holdenville Shale: United States: Oklahoma

Anobasicrinus erectus (Strimple), 1951. Strimple, 1961g, p. 115. *Aesiocrinus erectus* Strimple, 1951d, p. 191, figs. 9-12.
 Pennsylvanian, Desmoinesian, Oologah Limestone: United States: Oklahoma

Anobasicrinus granulosus (Strimple), 1954. Strimple, 1961g, p. 115.
  Plummericrinus granulosus Strimple, 1954a, p. 207, Pl. 23, figs. 6, 7.
  Pennsylvanian, Missourian, Wann Formation: United States: Oklahoma

Anobasicrinus obscurus Strimple, 1961g, p. 120, Pl. 7, figs. 4-7; Fig. 20.
  Pennsylvanian, Morrowan, Wapanucka Limestone: United States: Oklahoma

Anobasicrinus perplexus Strimple, 1961g, p. 119.
  Plummericrinus erectus Strimple, 1954a, p. 206, Pl. 23, figs. 9-12.
  Pennsylvanian, Missourian, Wann Formation: United States: Oklahoma

Anobasicrinus praecursor (Moore and Plummer), 1940. Strimple, 1962e, p. 23, Pl. 9, figs. 7-10.
  Neozeacrinus praecursor Moore and Plummer, 1940. Moore and Laudon, 1944, p. 167, Pl. 63, fig. 21.
  Pennsylvanian, Desmoinesian, Oologah Limestone: United States: Texas

ANOMALOCRINUS Meek and Worthen, 1865. *Heterocrinus? incurvus* Meek and Worthen, 1865. Moore and Laudon, 1943b, p. 31, fig. 3; 1944, p. 151, Pl. 52, fig. 8. Ubaghs, 1953, p. 747. Moore, 1962b, p. 14, fig. 5, no. 7. Philip, 1963, fig. 3c.
Ordovician

Anomalocrinus incurvus (Meek and Worthen), 1865. Moore and Laudon, 1943b, p. 130, Pl. 1, fig. 12; 1944, p. 151, Pl. 53, fig. 26.
  Ordovician, Cincinnatian, Maysville Formation: United States: Ohio

ANTHEMOCRINUS Wachsmuth and Springer, 1881. *Eucrinus venustus* Angelin, 1878. Moore and Laudon, 1943b, p. 82. Ubaghs, 1953, p. 737. Arendt and Hecker, 1964, p. 95.
Silurian

ANTHRACOCRINUS Strimple and Watkins, 1955. *A. primitivus* Strimple and Watkins, 1955. Strimple and Watkins, 1955, p. 348.
Ordovician, Champlainian

Anthracocrinus primitivus Strimple and Watkins, 1955, p. 349, figs. 1a-c, 2a, 4-6.
  Ordovician, Champlainian, Bromide Formation: United States: Oklahoma

ANTIHOMOCRINUS Schmidt, 1934. *Homocrinus tenuis* Bather, 1893. Moore and Laudon, 1943b, p. 52. Ubaghs, 1953, p. 751. Arendt and Hecker, 1964, p. 86.
Silurian—Lower Devonian

Antihomocrinus gertrudianus Schmidt, 1941, p. 111, Pl. 13, fig. 1.
  Devonian, upper Coblentzian: Germany

Antihomocrinus multifidus Schmidt, 1941, p. 109, Pl. 26, figs. 2, 3.
  Devonian, Coblentzian: Germany

Antihomocrinus rarifissus (Jaekel), 1921. Schmidt, 1941, p. 110, Pl. 26, fig. 1.
  Nassoviocrinus rarifissus Jaekel, 1921. Schmidt, 1941, p. 110.
  Devonian, lower Coblentzian: Germany

Antihomocrinus zeileri (Jaekel), 1895. Schmidt, 1941, p. 108, Pl. 12, fig. 5.
  Bactrocrinus zeileri Jaekel, 1895. Schmidt, 1941, p. 108.
  Poteriocrinus rhenanus Müller, *in* Zeiler and Wirtgen, 1855, Pl. 7, fig. 2 only. Schmidt, 1941, p. 108.
  Devonian, upper Coblentzian: Germany

ANULOCRINUS Ramsbottom, 1961. *A. thraivensis* Ramsbottom, 1961. Ramsbottom, 1961, p. 8. Moore, 1962a, p. 21. Arendt and Hecker, 1964, p. 81.
Ordovician, Champlainian-Silurian, Niagaran

Anulocrinus drummuckensis Ramsbottom, 1961, p. 9, Pl. 2, figs. 9, 10.

Moore, 1962a, p. 22, Pl. 1, fig. 5; Fig. 14, nos. 1, 2.
Ordovician, Ashgillian, Upper Drummuck Group: Scotland

Anulocrinus sp. aff. *A. drummuckensis* Ramsbottom, 1960 [sic]. Brower, 1966, p. 619, Pl. 75, figs. 10-12, 25; Figs. 1A-G.
Middle Ordovician, Champlainian, Bromide Formation: United States: Oklahoma

Anulocrinus simplex (Springer), 1926. Ramsbottom, 1961, p. 8. Moore, 1962a, p. 24, Pl. 3, fig. 5; Fig. 14, nos. 3, 4.
Cremacrinus simplex Springer, 1926. Ramsbottom, 1961, p. 8.
Silurian, Niagaran, Brownsport Formation, Beech River Formation: United States: Tennessee

**Anulocrinus thraivensis* Ramsbottom, 1961, p. 8, Pl. 2, figs. 1-8; Fig. 7. Moore, 1962a, p. 22, Pl. 1, fig. 4; Figs. 12, 13.
Ordovician, Ashgillian, Upper Drummuck Group: Scotland

AOROCRINUS Wachsmuth and Springer, 1897. **Dorycrinus immaturus* Wachsmuth and Springer, 1889. Moore and Laudon, 1943b, p. 93; 1944, p. 195. Ubaghs, 1953, p. 740.
Devonian-Mississippian

Aorocrinus armatus Goldring, 1923. Moore and Laudon, 1944, p. 195, Pl. 76, fig. 10.
Devonian, Erian, Hamilton Group: United States: New York

Aorocrinus banffensis Laudon, Parks and Spreng, 1952, p. 568, Pl. 67, figs. 18, 19; Pl. 69, fig. 19.
Mississippian, Kinderhookian, Banff formation: Canada: Alberta

Aorocrinus douglassi (Miller and Gurley), 1897. Laudon and Severson, 1953, p. 527, Pl. 53, figs. 11, 12.
Batocrinus douglassi Miller and Gurley, 1897. Laudon and Severson, 1953, p. 527.
Mississippian, Kinderhookian, Lodgepole Limestone: United States: Montana

**Aorocrinus immaturus* (Wachsmuth and Springer), 1889. Moore and Laudon, 1943b, p. 141, Pl. 12, fig. 1; 1944, p. 195, Pl. 76, fig. 9. Laudon, 1948, Pl. 2. Bowsher, 1955a, p. 3, Pl. 2, fig. 5.
Mississippian, Kinderhookian, Hampton Formation: United States: Iowa

Aorocrinus iola Laudon, 1933. Laudon and others, 1952, p. 569, Pl. 67, figs. 16, 17.
Mississippian, Kinderhookian, Banff formation: Canada: Alberta

Aorocrinus nodosus Springer, 1926 = *Stiptocrinus nodosus*

Aorocrinus parvus (Shumard), 1855. Moore and Laudon, 1944, p. 195, Pl. 76, fig. 11. Laudon, 1948, Pl. 2.
Mississippian, Osagean, Burlington Limestone: United States: Iowa, Missouri

Aorocrinus symmetricus (Hall), 1858. Kirk, 1946e, p. 660.
Actinocrinus symmetricus Hall, 1858. Kirk, 1946e, p. 660.
Mississippian, Osagean, Burlington Limestone: United States: Iowa

Aorocrinus spp.
Laudon and others, 1952, p. 569, Pl. 67, fig. 20.
Mississippian, Kinderhookian, Banff formation: Canada: Alberta
Breimer, 1962a, p. 78, Pl. 8, figs. 5, 6.
Carboniferous, Namurian, Rabanal Limestone: Spain

APHELECRINUS Kirk, 1944. **A. elegans* Kirk, 1944. Kirk, 1944c, p. 190. Arendt and Hecker, 1964, p. 88.
Mississippian

Aphelecrinus bayensis (Meek and Worthen), 1865. Kirk, 1944c, p. 192.

Horowitz, 1965, p. 24, Pl. 2, figs. 3, 4.
Poteriocrinus (Scaphiocrinus) bayensis Meek and Worthen, 1865. Kirk, 1944c, p. 192.
Mississippian, Chesterian, Glen Dean Formation: United States: Indiana, Illinois

Aphelecrinus crassus Kirk, 1944c, p. 192, Pl. 1, fig. 11.
Mississippian, Chesterian: United States: Illinois

Aphelecrinus crineus (Hall), 1864. Strimple, 1967, p. 81.
Poteriocrinus crineus Hall, 1864.
Lower Mississippian, Cuyahoga Group: United States: Ohio

Aphelecrinus delicatus (Meek and Worthen), 1869. Strimple, 1967, p. 81.
Poteriocrinus (Graphiocrinus; Scaphiocrinus) delicatus Meek and Worthen, 1869.
Mississippian, Osagean, Burlington Limestone: United States: Iowa

Aphelecrinus dilatatus Wright, 1945, p. 117, Pl. 2, figs. 5-15. Moore, 1948, p. 390, fig. 11. Wright, 1951a, p. 45, Pl. 12, figs. 5, 6, 9-13, 26, 30, 35, 37.
Carboniferous, Visean, Lower Limestone Group: Scotland

Aphelecrinus dunlopi (Wright), 1936. Kirk, 1944c, p. 193. Wright, 1951a, p. 43, Pl. 11, figs. 1, 5, 9; Pl. 13, figs. 2-4; Figs. 13, 14.
Pachylocrinus dunlopi Wright, 1936. Kirk, 1944c, p. 193.
Carboniferous, Visean, Lower Limestone Group: Scotland

*Aphelecrinus elegans Kirk, 1944c, p. 194, Pl. 1, figs. 6-8.
Mississippian, Meramecian, Ste. Genevieve Limestone: United States: Alabama

Aphelecrinus elegantulus (Wachsmuth and Springer, *in* Miller), 1889. Laudon and Severson, 1953, p. 520 (misspelled as *Amphelecrinus elegantatus*). Strimple, 1967, p. 81.
Scaphiocrinus elegantulus Wachsmuth and Springer, *in* Miller, 1889.
Mississippian, Kinderhookian, Hampton Formation: United States: Iowa

Aphelecrinus exoticus Strimple, 1951g, p. 292, fig. 5.
Mississippian, Chesterian, Fayetteville Formation: United States: Oklahoma

Aphelecrinus? greenhillensis Wright, 1951b, p. 47, Pl. 30, fig. 1.
Carboniferous, Visean: Scotland

Aphelecrinus limatus Kirk, 1944c, p. 196, Pl. 1, fig. 10.
Mississippian: United States: Kentucky

Aphelecrinus madisonensis Laudon and Severson, 1953 = *Paracosmetocrinus madisonensis*

Aphelecrinus meeki (Kirk), 1941. Laudon and Severson, 1953, p. 521 (misspelled as *Amphelecrinus*). Strimple, 1967, p. 81.
Cosmetocrinus meeki Kirk, 1941. Laudon and Severson, 1953, p. 521.
Mississippian, Osagean, Burlington Limestone: United States: Iowa

Aphelecrinus mundus Kirk, 1944c, p. 197, Pl. 1, fig. 9.
Mississippian: United States: Kentucky

Aphelecrinus okawensis (Worthen), 1882. Kirk, 1944c, p. 198.
Poteriocrinus okawensis Worthen, 1882. Kirk, 1944c, p. 198.
Mississippian, Chesterian: United States: Illinois

Aphelecrinus oweni Kirk, 1944c, p. 198, Pl. 1, figs. 1-3 (figures mislabeled as *A. lyoni* Kirk, n. sp.). Horowitz, 1965, p. 25, Pl. 2, figs. 15-17.
Mississippian, Chesterian, Glen Dean Formation: United States: Indiana, Kentucky

Aphelecrinus parvus Wright, 1945, p. 118, Pl. 3, figs. 2-5; 1951a, p. 46, Pl. 12, fig. 14.
   Carboniferous, Visean, Lower Limestone Group: Scotland
Aphelecrinus peculiaris (Worthen), 1882. Kirk, 1944c, p. 202.
   Poteriocrinus peculiaris Worthen, 1882. Kirk, 1944c, p. 202.
   Mississippian, Chesterian: United States: Illinois
Aphelecrinus planus Strimple, 1951g, p. 292, figs. 9-11.
   Mississippian, Chesterian, Fayetteville Formation: United States: Oklahoma
Aphelecrinus randolphensis (Worthen), 1873. Kirk, 1944c, p. 200. Horowitz, 1965, p. 26, Pl. 2, fig. 18.
   Poteriocrinus (Scaphiocrinus) randolphensis Worthen, 1873. Kirk, 1944c, p. 200.
   Mississippian, Chesterian, Glen Dean Formation: United States: Indiana, Illinois
Aphelecrinus richfieldensis (Worthen), 1882. Laudon and Severson, 1953, p. 521 (misspelled as *Amphelecrinus*). Strimple, 1967, p. 81.
   Poteriocrinus richfieldensis Worthen, 1882.
   Lower Mississippian, Cuyahoga Group: United States: Ohio
Aphelecrinus róscobiensis Wright, 1945, p. 118, Pl. 3, figs. 16, 18, 22; 1951a, p. 46, Pl. 12, figs. 31-33.
   Carboniferous, Visean, Lower Limestone Group: Scotland
Aphelecrinus sacculatus Laudon and Severson, 1953, p. 522, Pl. 51, fig. 1; Pl. 55, fig. 7 (misspelled as *Amphelecrinus*).
   Mississippian, Kinderhookian, Lodgepole Limestone: United States: Montana
Aphelecrinus scoparius (Hall), 1858. Kirk, 1944c, p. 200.
   Scaphiocrinus scoparius Hall, 1858. Kirk, 1944c, p. 200.
   Mississippian, Meramec, St. Louis Limestone: United States: Illinois
Aphelecrinus simplex Kirk, 1944c, p. 201, Pl. 1, figs. 4, 5.
   Mississippian, Chesterian, Renault Formation: United States: Illinois
Aphelecrinus venustus (Worthen), 1882. Kirk, 1944c, p. 202.
   Poteriocrinus venustus Worthen, 1882. Kirk, 1944c, p. 202.
   Mississippian, Chesterian: United States: Illinois
Aphelecrinus sp. Wright, 1951b, p. 47, Pl. 12, fig. 4.
   Carboniferous, Visean: Scotland
APOGRAPHIOCRINUS Moore and Plummer, 1940. *A. typicalis* Moore and Plummer, 1940. Moore and Laudon, 1943b, p. 55, fig. 5; 1944, p. 158, Pl. 62, fig. 28.
   Carboniferous
Apographiocrinus angulatus Strimple, 1948c, p. 11, Pl. 2, figs. 5-8; Fig. 2; 1962e, p. 46.
   Pennsylvanian, Desmoinesian, Altamont Limestone, Oologah Limestone: United States: Oklahoma
Apographiocrinus arcuatus Strimple, 1949c, p. 6, Pl. 1, figs. 1-10; Fig. 1.
   Pennsylvanian, Missourian, Wann Formation: United States: Oklahoma
Apographiocrinus exculptus Moore and Plummer, 1940. Moore and Laudon, 1943b, p. 132, Pl. 3, fig. 6; 1944, p. 158, Pl. 58, fig. 8.
   Pennsylvanian, Missourian, Brownwood Shale: United States: Texas
Apographiocrinus obtusus Strimple, 1948c, p. 8, Pl. 2, figs. 1-4; Fig. 1; 1962e, p. 46.
   Pennsylvanian, Desmoinesian, Altamont Limestone, Oologah Limestone: United States: Oklahoma

Apographiocrinus quietus Strimple, 1948c, p. 9, Pl. 2, figs. 13-16; Fig. 4; 1962e, p. 46.
  Pennsylvanian, Desmoinesian, Altamont Limestone, Oologah Limestone: United States: Oklahoma

Apographiocrinus rotundus Strimple, 1948c, p. 8, Pl. 2, figs. 9-12; Fig. 3; 1962e, p. 46.
  Pennsylvanian, Desmoinesian, Altamont Limestone, Oologah Limestone: United States: Oklahoma

Apographiocrinus? scoticus (Wright), 1942. Wright, 1945, p. 122, Pl. 4, figs. 9, 11, 12; Fig. 5; 1952a, p. 119, Pl. 35, figs. 7-10; Fig. 66.
  Erisocrinus scoticus Wright, 1942, p. 275, Pl. 12, figs. 7-10, 12.
  Carboniferous, Visean, Lower Limestone Group: Scotland

*Apographiocrinus typicalis Moore and Plummer, 1940. Moore and Laudon, 1944, p. 158, Pl. 56, fig. 21; Pl. 62, fig. 22. Strimple, 1959a, p. 117, Pl. 1, figs. 1-11; Pl. 2, fig. 1.
  Pennsylvanian, Missourian: United States: Kansas, Oklahoma, Texas

Apographiocrinus sp. Termier and Termier, 1950, p. 89, Pl. 222, figs. 1-3.
  Carboniferous, Westphalian: Morocco

ARACHNOCRINUS Meek and Worthen, 1866. **Cyathocrinus bulbosus* Hall, 1862.
  Moore and Laudon, 1943b, p. 50; 1944, p. 153. Ubaghs, 1953, p. 750.
  Devonian

*Arachnocrinus bulbosus (Hall), 1862. Moore and Laudon, 1943b, p. 131, Pl. 2, fig. 14; 1944, p. 153, Pl. 57, figs. 10, 21.
  Devonian, Ulsterian, Onondaga Limestone: United States: New York

Arachnocrinus extensus Wachsmuth and Springer, 1879. Moore and Laudon, 1944, p. 153, Pl. 57, fig. 29.
  Devonian, Ulsterian, Onondaga Limestone: United States: New York, Kentucky

Arachnocrinus ignotus Stauffer, 1918. Moore and Laudon, 1944, p. 153, Pl. 57, fig. 14.
  Devonian, Ulsterian, Onondaga Limestone: United States: New York

ARCHAEOCRINUS Wachsmuth and Springer, 1881. **Glyptocrinus lacunosus* Billings, 1857. Moore and Laudon, 1943b, p. 82, fig. 13; 1944, p. 183, Pl. 70, fig. 6. Ubaghs, 1950, p. 120, fig. 8; 1953, p. 736. Ramsbottom, 1961, p. 23, fig. 10. Arendt and Hecker, 1964, p. 95.
  Ordovician-Silurian

Archaeocrinus desideratus Billings, 1885. Moore and Laudon, 1943b, p. 138, Pl. 9, fig. 2; 1944, p. 183, Pl. 73, fig. 14. Wilson, 1946, p. 25. Strimple, 1963d, p. 70, fig. 19c.
  Ordovician, Champlainian: United States: Ohio; Canada: Ontario

Archaeocrinus elevatus Ramsbottom, 1961, p. 23, Pl. 8, figs. 1-4; Fig. 11.
  Ordovician, Caradocian, Stinchar Limestone Group: England

*Archaeocrinus lacunosus (Billings), 1857. Wilson, 1946, p. 25. Strimple, 1963d, p. 70, fig. 19a.
  Ordovician, Champlainian, Cobourg Limestone: Canada: Ontario

Archaeocrinus marginatus (Billings), 1857.
  Glyptocrinus marginatus Billings, 1857. Wilson, 1946, p. 28.
  Ordovician, Champlainian, Cobourg Limestone: Canada: Ontario

Archaeocrinus microbasalis (Billings), 1857. Wilson, 1946, p. 26.
  Ordovician, Champlainian, Cobourg Limestone and Hull Limestone: Canada: Ontario, Quebec

Archaeocrinus ottawaensis Wilson, 1946, p. 26, Pl. 5, fig. 1; Fig. 1.
  Ordovician, Champlainian, Cobourg Limestone: Canada: Ontario

cf. *Archaeocrinus ottawaensis* Wilson, 1946, p. 27.
   Ordovician, Champlainian, Cobourg Limestone: Canada: Ontario

Archaeocrinus pyriformis (Billings), 1857. Wilson, 1946, p. 26.
   Ordovician, Champlainian, Cobourg Limestone: Canada: Ontario

Archaeocrinus subovalis Strimple, 1953a, p. 606, figs. 1-7; 1963d, p. 70, fig. 19d.
   Ordovician, Champlainian, Bromide Formation: United States: Oklahoma

Archaeocrinus sp. Philip, 1965, p. 767, fig. 1f.
   Ordovician

ARROYOCRINUS Lane and Webster, 1966. *A. popenoei* Lane and Webster, 1966. Lane and Webster, 1966, p. 40.
   Permian, Wolfcampian

**Arroyocrinus popenoei* Lane and Webster, 1966, p. 41, Pl. 9, figs. 1-6; Figs. 15-17. Webster and Lane, 1967, p. 22, Pl. 6, figs. 4, 6 (*non* fig. 5 as given); Pl. 7, fig. 2.
   Permian, Wolfcampian, Bird Spring Formation: United States: Nevada

ARTHROACANTHA Williams, 1883. **A. ithacensis* Williams, 1883. Moore and Laudon, 1943b, p. 95, fig. 16; 1944, p. 197. Cuénot, 1948, p. 69. Ubaghs, 1953, p. 740. Arendt and Hecker, 1964, p. 96.
   Devonian

Arthroacantha carpenteri (Hinde), 1885. Kier, 1952, p. 69, Pl. 3, figs. 6-9. Bowsher, 1955a, p. 2, Pl. 2, figs. 1, 2. La Rocque and Marple, 1955, p. 94, fig. 232.
   Devonian, Erian, Silica Formation, Arkona beds: United States: Ohio; Canada: Ontario

Arthroacantha ornata (Schmidt), 1913. Schmidt, 1941, p. 52, fig. 10.
   Devonian, Eifelian: Germany

Arthroacantha punctobrachiata (Hall), 1872. Moore and Laudon, 1943b, p. 141, Pl. 12, fig. 12; 1944, p. 197, Pl. 78, fig. 9. (Moore and Laudon attributed authorship to Williams in both references.)
   Devonian, Erian, Arkona shale: Canada: Ontario

Arthroacantha schwerdii (Follmann), 1911. Schmidt, 1941, p. 50, fig. 9.
   Devonian, upper Coblentzian, Laubach Beds: Germany

Arthroacantha splendens (Goldring), 1923. Moore and Laudon, 1944, p. 197, Pl. 78, fig. 10.
   Devonian, Erian, Chemung Group: United States: New York

Arthroacantha tenuispinata Schmidt, 1915. Schmidt, 1941, p. 54, Pl. 10, fig. 3.
   Devonian, upper Coblentzian: Germany

ARTICHTHYOCRINUS Wright, 1923. **A. springeri* Wright, 1923. Moore and Laudon, 1943b, p. 74. Ubaghs, 1953, p. 755. Arendt and Hecker, 1964, p. 94.
   Lower Carboniferous

**Artichthyocrinus springeri* Wright, 1923. Wright, 1954a, p. 160, Pl. 41, figs. 6, 10-12; Pl. 47, figs. 1-6, 12.
   Carboniferous, Visean, Lower Limestone Group: Scotland

ARYBALLOCRINUS Breimer, 1962. *Actinocrinus (*Megistocrinus*) *whitei* Hall, 1861. Breimer, 1962a, p. 72.
   Lower Carboniferous

Aryballocrinus awthornsensis (Wright), 1955. Breimer, 1962a, p. 73. *Periechocrinus? awthornsensis* Wright, 1955a, p. 192, Pl. 63, fig. 7; Fig. 109. Breimer, 1962a, p. 73.
   Carboniferous, Tournaisian: England

Aryballocrinus parvus (Wachsmuth and Springer, *in* Miller), 1889. Breimer, 1962a, p. 74.
  Megistocrinus parvus Wachsmuth and Springer, *in* Miller, 1889. Breimer, 1962a, p. 74.
  Mississippian, Kinderhookian, Hampton Formation: United States: Iowa
Aryballocrinus sampsoni (Miller and Gurley), 1896. Breimer, 1962a, p. 73.
  Amphoracrinus sampsoni Miller and Gurley, 1896. Breimer, 1962a, p. 73.
  Mississippian, Kinderhookian, Chouteau Limestone: United States: Missouri
Aryballocrinus tenuidiscus (Hall), 1861. Breimer, 1962a, p. 73.
  Actinocrinus? tenuidiscus Hall, 1861. Breimer, 1962a, p. 73.
  Mississippian, Osagean, Burlington Limestone: United States: Iowa
*Aryballocrinus whitei (Hall), 1861. Breimer, 1962a, p. 72.
  Actinocrinus (Megistocrinus) whitei Hall, 1861. Breimer, 1962a, p. 72.
  Mississippian, Osagean, Burlington Limestone: United States: Iowa
Aryballocrinus sp. Breimer, 1962a, p. 74.
  Mississippian, Kinderhookian, Lodgepole Limestone: United States: Montana
ASAPHOCRINUS Springer, 1920. *A. bassleri* Springer, 1920. Moore and Laudon, 1943b, p. 74, fig. 11; 1944, p. 181, Pl. 66, fig. 3. Ubaghs, 1953, p. 756. Arendt and Hecker, 1964, p. 94.
  Silurian
*Asaphocrinus bassleri Springer, 1920. Moore and Laudon, 1944, p. 181, Pl. 54, fig. 9.
  Silurian, Niagaran, Beech River Formation: United States: Tennessee
Asaphocrinus densus Strimple, 1963d, p. 111, Pl. 9, figs. 3-5.
  Silurian, Niagaran, Henryhouse Formation: United States: Oklahoma
Asaphocrinus ornatus (Hall), 1852. Moore and Laudon, 1943b, p. 136, Pl. 7, fig. 6; Pl. 8, fig. 5; 1944, p. 181, Pl. 68, fig. 2; Pl. 69, fig. 17.
  Silurian, Niagaran, Rochester Shale: United States: New York
ASCETOCRINUS Kirk, 1940. **Scaphiocrinus rusticellus* White, 1963. Moore and Laudon, 1943b, p. 56; 1944, p. 159.
  Mississippian
*Ascetocrinus rusticellus (White), 1863. Moore and Laudon, 1944, p. 159, Pl. 55, fig. 14.
  Mississippian, Osagean, Burlington Limestone: United States: Iowa
Ascetocrinus scoparius (Hall), 1861. Moore and Laudon, 1944, p. 159, Pl. 55, fig. 17.
  Zeacrinus scoparius Hall, 1861.
  Mississippian, Chesterian: United States: Kentucky
Asterocrinites Roemer, 1851 = *Haplocrinites*
Asterocrinus Munster, 1838 = *Haplocrinites*
Astrocrinites Conrad, 1841 = *Melocrinites*
Astrocrinus Bather, 1900 = *Melocrinites*
ASUTURAECRINUS Yakovlev, *in* Yakovlev and Ivanov, 1956. *A. dorofeievi* Yakovlev, *in* Yakovlev and Ivanov, 1956. Yakovlev, *in* Yakovlev and Ivanov, 1956, p. 71. Arendt and Hecker, 1964, p. 94.
  Permian, Artinskian
*Asuturaecrinus dorofeievi Yakovlev, *in* Yakovlev and Ivanov, 1956, p. 71, Pl. 11, fig. 17. Arendt and Hecker, 1964, p. 94, Pl. 14, fig. 10.
  Permian, Artinskian: Russia: Central Urals

ASYMMETROCRINUS Wanner, 1937. *A. poteriocrinoides Wanner, 1937. Moore
and Laudon, 1943b, p. 50. Arendt and Hecker, 1964, p. 85. Lane,
1967b, p. 11, fig. 5.
Permian

ATACTOCRINUS Weller, 1916. *A. wilmingtonensis Weller, 1916. Moore and
Laudon, 1943b, p. 82. Ubaghs, 1953, p. 737. Arendt and Hecker,
1964, p. 95.
Silurian

Ataxiacrinus Strimple, 1961 = Tarachiocrinus

Ataxiacrinus meadowensis (Strimple), 1949, Strimple, 1961 = Tarachiocrinus
meadowensis

Ataxiacrinus multiramus Strimple, 1961 = Tarachiocrinus multiramus

Ataxiacrinus optimus (Strimple), 1951, Strimple, 1961 = Tarachiocrinus
optimus

Ataxiacrinus periodus (Strimple), 1951, Strimple, 1961 = Tarachiocrinus
periodus

ATELESTOCRINUS Wachsmuth and Springer, 1886. *A. delicatus Wachsmuth and
Springer, 1886. Moore and Laudon, 1943b, p. 54. Ubaghs, 1953,
p. 751. Arendt and Hecker, 1964, p. 86.
Mississippian

ATHABASCACRINUS Laudon, Parks and Spreng, 1952. *A. colemanensis Laudon,
Parks and Spreng, 1952. Laudon and others, 1952, p. 566.
Mississippian, Kinderhookian

*Athabascacrinus colemanensis Laudon, Parks and Spreng, 1952, p. 567,
Pl. 68, figs. 1-3; Pl. 69, figs. 16-18.
Mississippian, Kinderhookian, Banff formation: Canada: Alberta

ATHLOCRINUS Moore and Plummer, 1940. *A. placidus Moore and Plummer, 1940.
Moore and Laudon, 1943b, p. 58; 1944, p. 163.
Pennsylvanian

Athlocrinus clarus Strimple, 1962e, p. 33, Pl. 1, figs. 1-4.
Pennsylvanian, Desmoinesian, Oologah Formation: United States:
Oklahoma

Athlocrinus clypeiformis Moore and Plummer, 1940. Moore and Laudon, 1944,
p. 163, Pl. 63, fig. 12.
Pennsylvanian, Missourian, Merriman Limestone: United States: Texas

*Athlocrinus placidus Moore and Plummer, 1940. Moore and Laudon, 1944,
p. 163, Pl. 62, fig. 1; Pl. 63, fig. 13.
Pennsylvanian, Missourian, Plattsburg Limestone: United States: Kansas

ATRACTOCRINUS Kirk, 1948. *A. consinnus Kirk, 1948. Kirk, 1948, p. 701.
Devonian, Erian-Senecan

Atractocrinus campanulatus Kirk, 1948, p. 703, Pl. 1, figs. 2, 3.
Devonian, Erian, Thunder Bay Limestone: United States: Michigan

*Atractocrinus consinnus Kirk, 1948, p. 704, Pl. 1, figs. 7-11.
Devonian, Erian, Arkona shale: Canada: Ontario

Atractocrinus curtus Kirk, 1948, p. 705, Pl. 1, figs. 4, 5.
Devonian, Erian, Norway Point Formation: United States: Michigan

Atractocrinus tenuis Kirk, 1948, p. 707, Pl. 1, fig. 1.
Devonian, Erian, Alpena Limestone: United States: Michigan

Atractocrinus westoni (Belanski), 1928. Kirk, 1948, p. 708.
Bactrocrinus westoni Belanski, 1928. Kirk, 1948, p. 708.
Devonian, Senecan, Shellrock Limestone: United States: Iowa

ATREMACRINUS Wanner, 1929. **A. calyculus* Wanner, 1929. Moore and Laudon, 1943b, p. 50. Arendt and Hecker, 1964, p. 85. Lane, 1967b, p. 13, fig. 3.
Permian

ATYPHOCRINUS Ulrich, 1924. **A. corryvillensis* Ulrich, 1924. Moore and Laudon, 1943b, p. 31. Ubaghs, 1953, p. 747. Moore, 1962b, p. 14, fig. 5, no. 6.
Ordovician

AULOCRINUS Wachsmuth and Springer, 1897. **A. agassizi* Wachsmuth and Springer, 1897. Moore and Laudon, 1943b, p. 59; 1944, p. 171. Cuénot, 1948, p. 62. Arendt and Hecker, 1964, p. 89.
Mississippian

**Aulocrinus agassizi* Wachsmuth and Springer, 1897. Moore and Laudon, 1944, p. 171, Pl. 61, fig. 14. Ubaghs, 1953, p. 686, fig. 25.
Mississippian, Osagean, Keokuk Limestone: United States: Indiana

Aulocrinus? sp. Ahlfeld and Braniša, 1960, p. 104, Pl. 7, fig. 20 (probably not an *Aulocrinus*).
Permian, Zudañez Group: Bolivia

AULODESOCRINUS Wright, 1942. **A. parvus* Wright, 1942. Wright, 1942, p. 278. Arendt and Hecker, 1964, p. 88.
Lower Carboniferous

**Aulodesocrinus parvus* Wright, 1942, p. 279, Pl. 12, figs. 1, 2; Fig. 1; 1951b, p. 73, Pl. 35, figs. 11, 12; Fig. 37.
Carboniferous, Visean, Lower Limestone Group: Scotland

Aviadocrinus Almela and Revilla, 1950 = *Cupressocrinites*

Aviadocrinus sampelayoi Almela and Revilla, 1950 = *Cupressocrinites sampelayoi*

AZYGOCRINUS Lane, 1963. **Actinocrinus dodecadactylus* Meek and Worthen, 1861. Lane, 1963a, p. 698; 1963b, p. 923.
Mississippian, Osagean

Azygocrinus andrewsianus (McChesney), 1860. Lane, 1963a, p. 698.
Actinocrinus andrewsianus McChesney, 1860. Lane, 1963a, p. 698.
Mississippian, Osagean, Burlington Limestone: United States: Iowa

**Azygocrinus dodecadactylus* (Meek and Worthen), 1861. Lane, 1963a, p. 698, fig. 3A.
Actinocrinus dodecadactylus Meek and Worthen, 1861. Lane, 1963a, p. 698.
Mississippian, Osagean, Burlington Limestone: United States: Illinois, Iowa

Azygocrinus rotundus (Shumard), 1855. Lane, 1963a, p. 698; 1963b, p. 923, fig. 4.
Actinocrinus rotundus Shumard, 1855. Lane, 1963a, p. 698.
Batocrinus complanatus Miller and Gurley, 1896. Lane, 1963a, p. 698.
Batocrinus enodis Miller and Gurley, 1896. Lane, 1963a, p. 698.
Batocrinus glaber Miller and Gurley, 1896. Lane, 1963a, p. 698.
Batocrinus levigatus Miller and Gurley, 1896. Lane, 1963a, p. 698.
Batocrinus subovatus Miller and Gurley, 1896. Lane, 1963a, p. 698.
Batocrinus subrotundus Miller and Gurley, 1896. Lane, 1963a, p. 698.
Mississippian, Osagean, Burlington Limestone: United States: Illinois, Iowa, Missouri

BACTROCRINITES Schnur, 1849. **Poteriocrinus fusiformis* Roemer, 1844 (part). Moore and Laudon, 1943b, p. 52. Arendt and Hecker, 1964, p. 86. Ubaghs, 1953, p. 751.
Bactrocrinus Sieverts-Doreck, 1953, p. 79.
Silurian-Devonian

Bactrocrinites cyathus Schmidt, 1941, p. 100, Pl. 12, fig. 9.
Devonian, upper Coblentzian: Germany

*Bactrocrinites fusiformis (Roemer), 1844.
Bactrocrinus fusiformis Steininger, 1849. Sieverts-Doreck, 1950, p. 83, figs. 2-4.
Devonian, Eifelian: Germany

Bactrocrinites oklahomaensis Strimple, 1952b, p. 78, figs. 12, 13; 1963d, p. 62.
Silurian, Niagaran, Henryhouse Formation: United States: Oklahoma

Bactrocrinites onondagensis Goldring, 1954, p. 13, Pl. 1, figs. 12, 13.
Devonian, Erian, Onondaga Limestone: United States: New York

Bactrocrinites reimanni Goldring, 1954, p. 15, Pl. 1, figs. 8-11.
Devonian, Erian, Ludlowville Shale: United States: New York

Bactrocrinites sp. Breimer, 1962a, p. 168, Pl. 15, figs. 7, 8.
Devonian, Emsian, La Vid Formation: Spain

Bactrocrinus auct. = *Bactrocrinites*

Bactrocrinus fusiformis Sieverts-Doreck, 1950 - *Bactrocrinites fusiformis*

?*Bactrocrinus nanus* Opitz, 1932 = *Parisangulocrinus cucumis*

Bactrocrinus westoni Belanski, 1928 = *Atractocrinus westoni*

Bactrocrinus zeileri Jaekel, 1895 = *Antihomocrinus zeileri*

BAEROCRINUS Volborth, 1864. *B. ungerni* Volborth, 1864. Moore and Laudon, 1943b, p. 32. Ubaghs, 1953, p. 749, fig. 80b. Moore, 1962b, p. 14, fig. 8, no. 2. Arendt and Hecker, 1964, p. 82. Philip, 1965, fig. 1f.
Ordovician

Baerocrinus parvus Jaekel, 1902. Regnell, 1948, p. 9, fig. 3A. Ubaghs, 1953, p. 749, fig. 80b.
Ordovician: Russia

*Baerocrinus ungerni Volborth, 1864. Arendt and Hecker, 1964, p. 82, figs. 118, 119.
Ordovician: Russia: Estonia

BALACRINUS Ramsbottom, 1961. *Glyptocrinus basalis* McCoy, 1850. Ramsbottom, 1961, p. 24. Arendt and Hecker, 1964, p. 95.
Ordovician, Caradocian

*Balacrinus basalis (McCoy), 1850. Ramsbottom, 1961, p. 25, Pl. 7, figs. 1-7.
Glyptocrinus basalis McCoy, 1850. Ramsbottom, 1961, p. 25.
Ordovician, Caradocian: England

BARRANDEOCRINUS Angelin, 1878. *B. sceptrum* Angelin, 1878. Moore and Laudon, 1943b, p. 90. Cuénot, 1948, p. 69. Ubaghs, 1953, p. 740.
Silurian

*Barrandeocrinus sceptrum Angelin, 1878. Ubaghs, 1953, p. 740, fig. 81; 1956a, p. 522, Pls. 1-7; Figs. 1-14.
Silurian, Wenlockian: Sweden

BARYCRINUS Wachsmuth, *in* Meek and Worthen, 1868. *Cyathocrinus spurius* Hall, 1858. Moore and Laudon, 1943b, p. 51, figs. 1, 4; 1944, p. 154. Cuénot, 1948, p. 33, fig. 41, no. 1. Ubaghs, 1953, p. 750.
Mississippian

Barycrinus asteriscus Van Sant, *in* Van Sant and Lane, 1964, p. 73, Pl. 1, figs. 3-5, 9, 13; Fig. 29.
Mississippian, Osagean, Edwardsville Formation: United States: Indiana

Barycrinus herculeus Wachsmuth and Springer, 1879 = *Barycrinus hoveyi*
Barycrinus hoveyi (Hall), 1861. Van Sant, *in* Van Sant and Lane, 1964, p. 74,Pl. 1, fig. 16; Pl. 2, figs. 5, 7-9; Figs. 21, no. 2; 22; 28; 30.
 Barycrinus herculeus Wachsmuth and Springer, 1879. Van Sant, *in* Van Sant and Lane, 1964, p. 74.
 Mississippian, Osagean, Edwardsville Formation, Indian Creek crinoid beds: United States: Indiana, *non* Iowa
Barycrinus? neglectus Miller and Gurley, 1896. Van Sant, *in* Van Sant and Lane, 1964, p. 76, Pl. 1, figs. 10, 12.
 Mississippian, Osagean, Edwardsville Formation: United States: Indiana
Barycrinus princeps Miller and Gurley, 1890. Moore and Laudon, 1943b, p. 131, Pl. 2, fig. 15; 1944, p. 154, Pl. 57, fig. 26. Van Sant, *in* Van Sant and Lane, 1964, p. 77, Pl. 1, fig. 14.
 Mississippian, Osagean, Edwardsville Formation: United States: Indiana
Barycrinus protuberans (Hall), 1858. Moore and Laudon, 1944, p. 154, Pl. 57, fig. 15.
 Cyathocrinus protuberans Hall, 1858.
 Mississippian, Osagean, Keokuk Limestone: United States: Ilinois
Barycrinus ribblesdalensis (Wright), 1942. Wright, 1950, p. 1, Pl. 1, figs. 4-10.
 Poteriocrinus [sic] *ribblesdalensis* Wright, 1942, p. 273, Pl. 9, figs. 7-9; 1950, p. 1.
 Carboniferous, ?Tournaisian: Scotland
Barycrinus stellifer Miller, 1892. Van Sant, *in* Van Sant and Lane, 1964, p. 77.
 Mississippian: United States: probably *not* Crawfordsville, Indiana
Barycrinus tumidus (Hall), 1858. Van Sant, *in* Van Sant and Lane, 1964, p. 77.
 Mississippian: United States: probably *not* Crawfordsville, Indiana
Barycrinus wachsmuthi (Meek and Worthen), 1868. Moore and Laudon, 1944, p. 154, Pl. 57, fig. 18.
 Mississippian, Osagean, Burlington Limestone: United States: Iowa
Barycrinus sp. Ubaghs, 1953, p. 750, fig. 108. Yakovlev, 1946b, p. 66, fig. 93.
 Mississippian

BASLEOCRINUS Wanner, 1916. **B. pocillum* Wanner, 1916. Moore and Laudon, 1943b, p. 55. Arendt and Hecker, 1964, p. 92.
 Permian

Basleocrinus striategranulatus Wanner, 1937. Wanner, 1949a, p. 40.
 Permian: Timor

BATHERICRINUS Jaekel, 1918. **Botryocrinus ramosus* Bather, 1891. Moore and Laudon, 1943b, p. 54.
 Silurian-Devonian

Bathronocrinus Strimple, 1962. **B. turioformis* Strimple, 1962. Strimple, 1962e, p. 37.
 Pennsylvanian

Bathronocrinus deweyensis (Strimple), 1939. Strimple, 1962e, p. 37.
 Hydreionocrinus deweyensis Strimple, 1939. Strimple, 1962e, p. 37.
 Pennsylvanian, Missourian, Dewey Limestone: United States: Oklahoma

*Bathronocrinus turioformis Strimple, 1962e, p. 38, Pl. 1, figs. 15-18.
 Pennsylvanian, Desmoinesian, Oologah Limestone: United States: Oklahoma

BATOCRINUS Casseday, 1854. **B. icosidactylus* Casseday, 1854. Moore and

Laudon, 1943b, p. 95, fig. 16; 1944, p. 195, Pl. 71, fig. 6. Cuénot, 1948, p. 69. Ubaghs, 1953, p. 740, fig. 139d. Lane, 1963b, p. 921.
Mississippian

Batocrinus abscissus Rowley and Hare, 1891 = *Dizygocrinus biturbinatus*

Batocrinus aequalis (Hall), 1858 = *Abatocrinus aequalis*

Batocrinus arcula Miller and Gurley, 1895 = *Batocrinus irregularis*

Batocrinus boonvillensis Miller, 1891 = *Dizygocrinus mediocris*

Batocrinus broadheadi Miller and Gurley, 1895 = *Dizygocrinus mediocris*

Batocrinus burketi Miller and Gurley, 1895 = *Dizygocrinus biturbinatus*

Batocrinus calvini Rowley, 1890 = *Abatocrinus calvini*

Batocrinus calyculus (Hall), 1860. Lane, 1963a, p. 695.
   Batocrinus irregularis Rowley, 1904, *non* Casseday, 1854. Lane, 1963a, p. 695.
   Batocrinus salemensis Miller and Gurley, 1896. Lane, 1963a, p. 695.
Mississippian, Meramecian, Salem Limestone: United States: Illinois

Batocrinus complanatus Miller and Gurley, 1896 = *Azygocrinus rotundus*

Batocrinus crassitestus Rowley, *in* Greene, 1904 = *Batocrinus irregularis*

Batocrinus curiosus Rowley, 1908, *non* Miller and Gurley, 1895 = *Abatocrinus curiosus*

Batocrinus davisi var. *lanesvillensis* Rowley, *in* Greene, 1904 = *Batocrinus irregularis*

Batocrinus davisi var. *sculptus* Rowley, *in* Greene, 1904 = *Batocrinus unionensis*

Batocrinus decoris Miller, 1891. Lane, 1963a, p. 695.
Mississippian, Meramecian, Salem Limestone: United States: Indiana

Batocrinus decrepitus Miller, 1892 = *Dizygocrinus crawfordsvillensis*

Batocrinus douglassi Miller and Gurley, 1897 = *Aorocrinus douglassi*

Batocrinus enodis Miller and Gurley, 1896 = *Azygocrinus rotundus*

Batocrinus euconus (Meek and Worthen), 1865. Lane, 1963a, p. 695.
   Actinocrinus (*Alloprosallocrinus*) *euconus* Meek and Worthen, 1865. Lane, 1963a, p. 695.
Mississippian, Meramecian, St. Louis or Salem Limestone: United States: Illinois

Batocrinus facetus Miller and Gurley, 1890 = *Dizygocrinus whitei*

Batocrinus gallatinensis Laudon and Severson, 1953, p. 526, Pl. 52, figs. 23-27; Pl. 55, figs. 19-21.
Mississippian, Kinderhookian, Lodgepole Limestone: United States: Montana

Batocrinus glaber Miller and Gurley, 1896 = *Azygocrinus rotundus*

Batocrinus gorbyi Miller, 1891 = *Dizygocrinus whitei*

Batocrinus grandis (Lyon and Casseday), 1859 = *Abatocrinus grandis*

Batocrinus gurleyi Rowley and Hare, 1891 = *Dizygocrinus biturbinatus*

Batocrinus heteroclitus Miller and Gurley, 1895 = *Dizygocrinus venustus*

Batocrinus icosidactylus Casseday, 1854. Moore and Laudon, 1943b, p. 140, Pl. 11, figs. 7, 8; 1944, p. 195, Pl. 75, fig. 1. Lane, 1963a, p. 695, figs. 2, 3B.
   Batocrinus pileus Miller and Gurley, 1895. Lane, 1963a, p. 695.

Mississippian, Osagean-Meramecian, Warsaw Limestone, Salem Limestone:
United States: Kentucky, Indiana

Batocrinus ignotus Miller and Gurley, 1895 = *Dizygocrinus mediocris*

Batocrinus inconsuetus Miller and Gurley, 1895 = *Dizygocrinus venustus*

Batocrinus inopinatus Miller and Gurley, 1895 = *Dizygocrinus venustus*

Batocrinus irregularis Casseday, 1854. Jillson, 1958, p. 64, fig. p. 64; 1959, p. 36, fig. p. 5. Lane, 1963a, p. 695.
Batocrinus arcula Miller and Gurley, 1895. Lane, 1963a, p. 695.
Batocrinus crassitestus Rowley, *in* Greene, 1904. Lane, 1963a, p. 695.
Batocrinus davisi var. *lanesvillensis* Rowley, *in* Greene, 1904. Lane, 1963a, p. 695.
Batocrinus magnirostris Rowley, *in* Greene, 1904. Lane, 1963a, p. 695.
Mississippian, Osagean-Meramecian, Warsaw Formation, Salem Limestone:
United States: Kentucky, Indiana

Batocrinus irregularis Rowley, *in* Greene, 1904, *non* Casseday, 1854 = *Batocrinus calyculus*

Batocrinus jucundus Miller and Gurley, 1890 = *Macrocrinus mundulus*

Batocrinus laura var. *sinuosus* Wachsmuth and Springer, 1897 = *Abatocrinus laura*

Batocrinus levigatus Miller and Gurley, 1896 = *Azygocrinus rotundus*

Batocrinus macbridei Wachsmuth and Springer, *in* Miller, 1889 = *Abatocrinus macbridei*

Batocrinus magnirostris Rowley, *in* Greene, 1904 = *Batocrinus irregularis*

Batocrinus marinus Miller and Gurley, 1890 = *Uperocrinus marinus*

Batocrinus mediocris Miller, 1891 = *Dizygocrinus mediocris*

Batocrinus mitidulus Miller and Gurley, 1895 = *Dizygocrinus intermedius*

Batocrinus modestus Miller and Gurley, 1895 = *Dizygocrinus mediocris*

Batocrinus nanus Miller and Gurley, 1896 = *Stinocrinus nanus*

Batocrinus pileus Miller and Gurley, 1895 = *Batocrinus icosidactylus*

Batocrinus poculum Miller and Gurley, 1890 = *Abatocrinus poculum*

Batocrinus polydactylus Miller and Gurley, 1895 = *Dizygocrinus venustus*

Batocrinus pulchellus Miller, 1891 = *Dizygocrinus whitei*

Batocrinus quasillus Meek and Worthen, 1868 = *Abatocrinus clypeatus*

Batocrinus remotus Miller and Gurley, 1896 = *Abatocrinus laura*

Batocrinus repositus Miller and Gurley, 1896 = *Abatocrinus laura*

Batocrinus rotadentatus Rowley and Hare, 1891 = *Abatocrinus rotadentatus*

Batocrinus sacculus Miller and Gurley, 1894 = *Batocrinus spergenensis*

Batocrinus salemensis Miller and Gurley, 1896 = *Batocrinus calyculus*

Batocrinus sampsoni Miller and Gurley, 1895 = *Dizygocrinus mediocris*

Batocrinus sayi Wood, 1909 = *Abatocrinus steropes*

Batocrinus selectus Miller and Gurley, 1896 = *Abatocrinus laura*

Batocrinus serratus Miller and Gurley, 1895 = *Dizygocrinus venustus*

Batocrinus spergenensis Miller, 1891. Lane, 1963a, p. 695.
Batocrinus sacculus Miller and Gurley, 1894. Lane, 1963a, p. 695.
Mississippian, Meramecian, Salem Limestone: United States: Indiana

Batocrinus stelliformis Miller and Gurley, 1896 = *Dizygocrinus venustus*

Batocrinus subaequalis (McChesney), 1860. Moore and Laudon, 1943b, p. 140, Pl. 11, fig. 10; 1944, p. 195, Pl. 75, fig. 5. Laudon, 1948, Pl. 2, Mississippian, Osagean, Burlington Limestone: United States: Iowa, Missouri

Batocrinus subconicus Worthen, 1884 = *Dizygocrinus montgomeryensis*

Batocrinus subovatus Miller and Gurley, 1896 = *Azygocrinus rotundus*

Batocrinus subrotundus Miller and Gurley, 1896 = *Azygocrinus rotundus*

Batocrinus sweeti Rowley and Hare, 1891 = *Dizygocrinus biturbinatus*

Batocrinus tuberculatus Wachsmuth and Springer, 1897 = *Abatocrinus tuberculatus*

Batocrinus unionensis Worthen, 1890. Lane, 1963a, p. 696.
Batocrinus davisi var. *sculptus* Rowley, in Greene, 1904. Lane, 1963a, p. 696.
Dizygocrinus persculptus Ulrich, 1917. Lane, 1963a, p. 696.
Dizygocrinus superctes Ulrich, 1917. Lane, 1963a, p. 696.
Mississippian, Meramecian, Ste. Genevieve Limestone, Salem Limestone, Warsaw Limestone: United States: Illinois, Indiana, Kentucky, Missouri, Tennessee, Virginia

Batocrinus venustulus Miller and Gurley, 1895 = *Dizygocrinus caroli*

Batocrinus venustus Miller, 1891 = *Dizygocrinus venustus*

Batocrinus veterator Miller and Gurley, 1895 = *Dizygocrinus caroli*

Batocrinus wachsmuthi (White), 1880, Moore and Laudon, 1944 = *Abatocrinus grandis*

BECHAROCRINUS Termier and Termier, 1955. *Triacrinus paradoxus* Termier and Termier, 1950. Termier and Termier, 1955, p. 316.
Triacrinus Termier and Termier, 1950 (*non* Munster, 1839), p. 84, 100.
Carboniferous, Westphalian

**Becharocrinus paradoxus* (Termier and Termier), 1950. Termier and Termier, 1955, p. 316.
Triacrinus paradoxus Termier and Termier, 1950, p. 84, 100, Pl. 214, figs. 10-13.
Carboniferous, lower Westphalian: Morocco

BELEMNOCRINUS White, 1862. *B. typus* White, 1862. Moore and Laudon, 1943b, p. 31. Ubaghs, 1953, p. 747.
Mississippian

BENTHOCRINUS Wanner, 1937. *B. cryptobasalis* Wanner, 1937. Moore and Laudon, 1943b, p. 59.
Permian

BEYRICHOCRINUS Waagen and Jahn, 1899. *B. humilis* Waagen and Jahn, 1899. Moore and Laudon, 1943b, p. 92. Ubaghs, 1953, p. 739,
Lower Devonian

BLAIROCRINUS Miller, 1891. *B. trijugis* Miller, 1891. Brower, 1967, p. 696.
Lower Mississippian

Blairocrinus arrosus Miller, 1892. Brower, 1967, p. 701, Pl. 77, figs. 1-4, 6-14, 16, 18; Fig. 8.
Mississippian, Kinderhookian-Osagean, Compton Limestone, Burlington Limestone: United States: Missouri

Blairocrinus spinosulus Miller and Gurley, 1893 = *Aacocrinus spinosulus*

**Blairocrinus trijugis* Miller, 1891. Brower, 1967, p. 698, Pl. 77, fig. 5; Pl. 78, figs. 1-12; Figs. 1C, 1F, 6.

Mississippian, Kinderhookian, Compton Limestone, Compton-Sedalia transition beds: United States: Missouri

BLOTHROCRINUS Kirk, 1940. *Poteriocrinus jesupi Whitfield, 1881. Moore and Laudon, 1943b, p. 55, fig. 6; 1944, p. 158. Moore, 1962b, p. 15, fig. 9, no. 6. Arendt and Hecker, 1964, p. 87.
Mississippian

Blothrocrinus balconi Termier and Termier, 1950, p. 90, 101, Pl. 218, figs. 18-26.
Carboniferous, Westphalian: Morocco

Blothrocrinus brevidactylus (Austin and Austin), 1843. Wright, 1951a, p. 38, Pl. 10, fig. 6.
Cladocrinites brevidactylus Austin and Austin, 1843. Wright, 1951a, p. 38.
Carboniferous, Tournaisian: England

Blothrocrinus impressus (Phillips), 1836. Wright, 1951a, p. 39, Pl. 13, fig. 6.
Poteriocrinus impressus Phillips, 1836. Wright, 1951a, p. 39.
Carboniferous, Tournaisian: England

*Blothrocrinus jesupi (Whitfield), 1881. Moore and Laudon, 1943b, p. 133, Pl. 4, fig. 9; 1944, p. 158, Pl. 58, fig. 20.
Mississippian, Osagean, Burlington Limestone: United States: Iowa

Blothrocrinus litvinovitschae Yakovlev, 1954a, p. 113, figs. 1-3. Arendt and Hecker, 1964, p. 88, Pl. 10, fig. 3; Fig. 128.
Carboniferous, Visean: Russia: Central Kazakhstan

Blothrocrinus longidactylus (Austin and Austin), 1847. Wright, 1951a, p. 36, Pl. 8, figs. 5, 6; Pl. 10, figs. 1, 3.
Poteriocrinus longidactylus Austin and Austin, 1847. Wright, 1951a, p. 36.
Carboniferous, Tournaisian: England

Blothrocrinus rugosus (Grenfell), 1876. Wright, 1951a, p. 39, Pl. 10, fig. 5.
Poteriocrinus rugosus Grenfell, 1876. Wright, 1951a, p. 39.
Carboniferous, Tournaisian, Lower Limestone Shales: England

Blothrocrinus spartarius (Miller and Gurley), 1890. Moore and Laudon, 1944, p. 158, Pl. 58, fig. 4.
Mississippian, Kinderhookian, Hampton Formation: United States: Iowa

Blothrocrinus swallovi (Meek and Worthen), 1860. Moore and Laudon, 1944, p. 158, Pl. 58, fig. 11. Ubaghs, 1953, p. 692, fig. 33j.
Mississippian, Osagean, Burlington Limestone: United States: Iowa

Blothrocrinus thorntonensis Wright, 1952b, p. 322, Pl. 13, fig. 9. Wright, 1960, p. 331, Appendix, Pl. B, fig. 2.
Carboniferous, Visean, Calciferous Sandstone Series: Scotland

Blothrocrinus spp.
Laudon and others, 1952, p. 553, Pl. 65, fig. 27.
Mississippian, Kinderhookian, Banff formation: Canada: Alberta
Laudon and others, 1952, p. 553, Pl. 65, figs. 23, 33.
Mississippian, Kinderhookian, Banff formation: Canada: Alberta
Laudon and others, 1952, p. 553, Pl. 65, fig. 26.
Mississippian, Kinderhookian, Banff formation: Canada: Alberta

BOGOTACRINUS Schmidt, 1937. *B. scheibei Schmidt, 1937. Moore and Laudon, 1943b, p. 101. Ubaghs, 1953, p. 743.
Silurian

BOHEMICOCRINUS Waagen and Jahn, in Jahn, 1899. *B. pulverens Waagen and Jahn, in Jahn, 1899. Ubaghs, 1953, p. 740.
Silurian

BOLBOCRINUS Wanner, 1916. **B. hieroglyphicus* Wanner, 1916. Moore and Laudon, 1943b, p. 50. Arendt and Hecker, 1964, p. 84. Lane, 1967b, p. 11, fig. 5.
Permian

Bolbocrinus eudoxiae Yakovlev, 1927. Yakovlev, *in* Yakovlev and Ivanov, 1956, p. 55, Pl. 9, figs. 12-14. Arendt and Hecker, 1964, p. 84, Pl. 9, fig. 1.
Permian, Artinskian: Russia: Ural Mountains

*Bolbocrinus hieroglyphicus Wanner, 1916. Arendt and Hecker, 1964, p. 84, Pl. 9, fig. 2; Fig. 122.
Permian: Timor

BOLLANDOCRINUS Wright, 1951. **Poteriocrinus conicus* Phillips, 1836 (part). Wright, 1951b, p. 62. Arendt and Hecker, 1964, p. 89.
Lower Carboniferous

*Bollandocrinus conicus (Phillips), 1836. Wright, 1951b, p. 63, Pl. 14, figs. 1-10, 14-16. Cain, 1968, p. 196, fig. 1-viii.
Poteriocrinus conicus Phillips, 1836. Wright, 1951b, p. 63.
Carboniferous, Tournaisian-?Visean, Limestone Series: England

Bollandocrinus cravenensis (Wright), 1942. Wright, 1951b, p. 64, Pl. 17, fig. 18.
Pachylocrinus cravenensis Wright, 1942, p. 274, Pl. 11, fig. 17.
Carboniferous, Visean: England

Bollandocrinus erectus Wright, 1951b, p. 64, Pl. 14, figs. 11-13.
Poteriocrinus conicus Phillips, 1843 (part, Pl. 16, fig. 12 only). Wright, 1951b, p. 64.
Carboniferous, Tournaisian-?Visean: England

BOTRYOCRINUS Angelin, 1878. **B. ramosissimus* Angelin, 1878. Moore and Laudon, 1943b, p. 54, figs. 1, 5; 1944, p. 155, Pl. 52, fig. 9. Cuénot, 1948, p. 61. Moore, 1950, p. 34, fig. 4d. Ubaghs, 1953, p. 751, fig. 14d. Arendt and Hecker, 1964, p. 86.
Silurian-Devonian

Botryocrinus americanus Rowley, *in* Greene, 1904. Moore and Laudon, 1944, p. 155, Pl. 57, fig. 11.
Devonian, Erian, Hamilton Group: United States: Indiana

Botryocrinus angularis Goldring, 1954, p. 18, Pl. 2, figs. 5-7.
Devonian, Erian, Moscow Shale: United States: New York

Botryocrinus arkonensis Goldring, 1950, p. 31, Pl. 1, figs. 14, 15; Pl. 2, figs. 1-8.
Devonian, Erian, Arkona shale: Canada: Ontario

Botryocrinus (Bathericrinus) ericius Schmidt, 1934 = *Dictenocrinus ericius*

Botryocrinus (Bathericrinus) hystrix Schmidt, 1934 = *Dictenocrinus hystrix*

Botryocrinus (Bathericrinus) semipinnulatus Schmidt, 1934 = *Dictenocrinus semipinnulatus*

Botryocrinus (Bathericrinus) speciosus Schmidt, 1934 = *Dictenocrinus speciosus*

Botryocrinus bethaniensis Goldring, 1954, p. 20, Pl. 2, figs. 8-10 (given as figs. 8, 9).
Devonian, Erian, Moscow Shale: United States: New York

Botryocrinus consolidatus Schmidt, 1941, p. 125, Pl. 19, fig. 5; Pl. 25, fig. 8; Fig. 35.
Devonian, lower Coblentzian: Germany

Botryocrinus costatus Goldring, 1954, p. 22, Pl. 3, figs. 1-7.
Devonian, Erian, Moscow Shale: United States: New York

Botryocrinus crassus (Whiteaves), 1887. Moore and Laudon, 1943b, p. 131, Pl. 2, fig. 4; 1944, p. 155, Pl. 57, fig. 23. Goldring, 1945, p. 63 (misspelled as *B. cressus*); 1954, p. 24, Pl. 3, figs. 8-10.
Devonian, Erian, Moscow Shale: United States: New York

Botryocrinus cucurbitaceus (Angelin), 1878. Ubaghs, 1953, p. 751, fig. 14d.
Silurian: Gotland

Botryocrinus cyathiformis Haarmann, 1920 = *Dictenocrinus cyathiformis*

Botryocrinus(?) *cylindricus* (Hall), 1852. Goldring, 1948b, p. 27, Pl. 1, figs. 1-6.
Homocrinus cylindricus Hall, 1852. Goldring, 1948b, p. 27.
Silurian, Niagaran, Rochester Shale: United States: New York; Canada: Ontario

Botryocrinus s. str. *dactylus* Schmidt, 1941, p. 130, Pl. 26, fig. 6.
Lower Devonian, Herdorf Schichten: Germany

Botryocrinus cf. *dactylus* Schmidt, 1941, p. 130, Pl. 26, fig. 7.
Lower Devonian, Herdorf Schichten: Germany

Botryocrinus decadactylus Bather, 1891 = *Dictenocrinus decadactylus*

Botryocrinus (*Dictenocrinus*) *ericius* Schmidt, 1934 = *Dictenocrinus ericius*

Botryocrinus (*Dictenocrinus*) *hystrix* Schmidt, 1934 = *Dictenocrinus hystrix*

Botryocrinus (*Dictenocrinus*) *semipinnulatus* Schmidt, 1934 = *Dictenocrinus semipinnulatus*

Botryocrinus (*Dictenocrinus*) *speciosus* Schmidt, 1934 = *Dictenocrinus speciosus*

Botryocrinus helenae Schmidt, 1941, p. 128, Pl. 25, fig. 7.
Devonian, Coblentzian, Rauhflaser Schichten: Germany

Botryocrinus irregularis Haarmann, 1920. Schmidt, 1941, p. 114, Pl. 6, fig. 1; Pl. 26, fig. 5; Fig. 30.
Devonian, Coblentzian, Herdorf Schichten: Germany

Botryocrinus(?) *jambus* Schmidt, 1941, p. 131, Pl. 26, fig. 4. Sieverts-Doreck, 1957, p. 66.
Devonian, Coblentzian, Herdorf Schichten: Germany

Botryocrinus mirandus Yeltyschewa, 1968, p. 33, Pl. 2, fig. 6; Pl. 3, figs. 1-4, 26.
Silurian, Borschovskian horizon: Russia: Podolsky

Botryocrinus montisguyonensis Delpey, 1943, p. 15, figs. 1-5 (incorrectly given as *B. montis-guyonensis*).
Devonian, Coblentzian, graywackes schisteux: France

Botryocrinus niemani Kesling, 1966, p. 273, Pl. 1, figs. 1-5; Fig. 1.
Devonian, Erian, Silica Formation: United States: Ohio

Botryocrinus nycteus (Hall), 1862. Moore and Laudon, 1944, p. 155, Pl. 57, figs. 9, 22.
Devonian, Erian, Hamilton Group: United States: New York

Botryocrinus ornatus Goldring, 1954, p. 26, Pl. 3, figs. 11, 12.
Devonian, Erian, Moscow Shale: United States: New York

Botryocrinus pachydactylus (Sandberger and Sandberger), 1855. Schmidt, 1941, p. 124, Pl. 6, fig. 3; Pl. 18, fig. 2.
Devonian, Coblentzian, Herdorf Schichten: Germany

Botryocrinus? sp. cf. *pachydactylus* (Sandberger and Sandberger), 1855. Sieverts-Doreck, 1957a, p. 66.
Devonian, Coblentzian, Herdorf Schichten: Germany

Botryocrinus parcus Strimple, 1963d, p. 63, figs. 17a, b.
   Silurian, Niagaran, Henryhouse Formation: United States: Oklahoma

Botryocrinus patellaris Haarmann, 1920. Schmidt, 1941, p. 114, Pl. 6, fig. 2.
   Devonian, Coblentzian, Herdorf Schichten: Germany

Botryocrinus polyxo (Hall), 1863. Moore and Laudon, 1944, p. 155, Pl. 54, fig. 22.
   Silurian, Niagaran, Waldron Shale: United States: Indiana

Botryocrinus procerus Haarmann, 1920 = *Dictenocrinus procerus*

Botryocrinus reimanni Goldring, 1934. Goldring, 1950, p. 29, Pl. 1, figs. 1-11.
   Devonian, Erian, Ludlowville Group: United States: New York

Botryocrinus reimanni var. *granilineatus* Goldring, 1950, p. 30, Pl. 1, figs. 12, 13.
   Devonian, Erian, Dock Street Clay: United States: Michigan

Botryocrinus schultzei Haarmann, 1920. Schmidt, 1941, p. 116, Pl. 6, figs. 4-6; Pl. 12, figs. 6-8; Pl. 24, fig. 1; Figs. 31, 32A.
   Devonian, Coblentzian: Germany

Botryocrinus schultzei goldenbergi Schmidt, 1941, p. 119, Pl. 24, fig. 2; Pl. 25, fig. 9; Figs. 31, 32B, 33.
   Devonian, Coblentzian: Germany

Botryocrinus n. sp. aff. *schultzei* Haarmann, 1920. Schmidt, 1941, p. 123, fig. 34.
   Devonian, Coblentzian, Herdorf Schichten: Germany

Botryocrinus thomasi Laudon, 1936. Kesling, 1963, p. 233, Pls. 1-3; Figs. 1, 2.
   Devonian, Erian, Thunder Bay Limestone: United States: Michigan

BRACHIOCRINUS Hall, 1859. *B. nodosarius* Hall, 1859. Moore, 1962b, p. 43. Moore and Jeffords, 1968, p. 12, fig. 4, no. 1.
   Devonian

Brachiocrinus nodosarius Hall, 1859. Moore, 1962b, p. 43, Pl. 3, fig. 5.
   Devonian, Ulsterian, New Scotland Limestone: United States: New York

BRACHYPUS Wanner, 1929. *B. adhaerens* Wanner, 1929. Moore and Laudon, 1943b, p. 76. Ubaghs, 1953, p. 756. Arendt and Hecker, 1964, p. 94.
   Permian

Brachypus adhaerens Wanner, 1929. Ubaghs, 1953, p. 756, fig. 137. Yakovlev, 1964b, p. 69, fig. 101.
   Permian: Timor

BRAHMACRINUS Sollas, 1900. *B. ponderosus* Sollas, 1900. Moore and Laudon, 1943b, p. 98; 1944, p. 200. Ubaghs, 1953, p. 743. Arendt and Hecker, 1964, p. 98 (misspelled as *Brachmacrinus*).
   Silurian-lower Carboniferous

Brahmacrinus elongatus Springer, 1926. Moore and Laudon, 1943b, p. 143, Pl. 14, fig. 2; 1944, p. 200, Pl. 77, fig. 18.
   Silurian, Niagaran, Laurel Limestone: United States: Indiana

Brahmacrinus ponderosus Sollas, 1900. Wright, 1955b, p. 260, Pl. 63, figs. 5, 6, 13-16.
   Carboniferous, Tournaisian: Scotland

BRIAROCRINUS Angelin, 1878. *B. inflatus* Angelin, 1878. Moore and Laudon, 1943b, p. 98. Ubaghs, 1953, p. 742.
   Silurian

BRIDGEROCRINUS Laudon and Severson, 1953. *B. fairyensis* Laudon and Sever-

son, 1953. Laudon and Severson, 1953, p. 517.
Mississippian, Kinderhookian

Bridgerocrinus fairyensis Laudon and Severson, 1953, p. 518, Pl. 51, figs. 6-8; Pl. 55, fig. 5.
Mississippian, Kinderhookian, Lodgepole Limestone: United States: Montana

BRONAUGHOCRINUS Strimple, 1951. *B. figuratus* Strimple, 1951. Strimple, 1951f, p. 671.
Mississippian, Chesterian

Bronaughocrinus cherokeensis Strimple, 1961c, p. 186, Pl. 1, figs. 1-3.
Mississippian, Chesterian, Hindsville Limestone: United States: Oklahoma

Bronaughocrinus figuratus Strimple, 1951f, p. 671, Pl. 99, figs. 9-11; 1961a, p. 24, figs. 2, 3; 1961c, p. 186, Pl. 1, figs. 4-6.
Mississippian, Chesterian, Pitkin Limestone: United States: Oklahoma

BURSACRINUS Meek and Worthen, 1861. *B. wachsmuthi* Meek and Worthen, 1861. Moore and Laudon, 1943b, p. 55. Arendt and Hecker, 1964, p. 88.
Mississippian

Cacabocrinus glyptus Hall, 1862 = *Dolatocrinus liratus*

Cacabocrinus glyptus var. *intermedius* Hall, 1862 = *Dolatocrinus liratus*

CACTOCRINUS Wachsmuth and Springer, 1897. *Actinocrinus proboscidialis* Hall, 1858. Moore and Laudon, 1943b, p. 93, fig. 15; 1944, p. 191, Pl. 71, fig. 11. Termier and Termier, 1947, Pl. 19, figs. 22, 23. Cuénot, 1948, p. 47, fig. 60. Ubaghs, 1953, p. 739. Yakovlev, 1964b, p. 60, fig. 71. Arendt and Hecker, 1964, p. 96.
Lower Carboniferous

Cactocrinus arnoldi (Wachsmuth and Springer, *in* Miller), 1889. Moore and Laudon, 1944, p. 191, Pl. 77, fig. 4. Laudon, 1948, Pl. 3. Laudon and others, 1952, p. 571, Pl. 68, fig. 5. Laudon and Severson, 1953, p. 530, Pl. 53, figs. 1-3.
Mississippian, Kinderhookian, Hampton Formation, Lodgepole Limestone, Banff formation: United States: Iowa, Montana; Canada: Alberta

Cactocrinus coelatus (Hall), 1858. Laudon, 1948, Pl. 3.
Mississippian, Osagean, Burlington Limestone

Cactocrinus denticulatus Wachsmuth and Springer, 1897 = *Cusacrinus denticulatus*

Cactocrinus glans (Hall), 1860. Moore and Laudon, 1943b, p. 140, Pl. 11, fig. 2; 1944, p. 193, Pl. 77, fig. 10. Laudon, 1948, Pl. 3.
Mississippian, Osagean, Burlington Limestone: United States: Iowa

Cactocrinus imperator Laudon, 1933 = *Cusacrinus imperator*

Cactocrinus kuenzii Laudon, Parks and Spreng, 1952, p. 572, Pl. 68, figs. 7-10; Pl. 69, fig. 22.
Mississippian, Kinderhookian, Banff formation: Canada: Alberta

Cactocrinus magnidactylus Laudon and Severson, 1953, p. 531, Pl. 53, fig. 5.
Mississippian, Kinderhookian, Lodgepole Limestone: United States: Montana

Cactocrinus multibrachiatus (Hall), 1858. Moore and Laudon, 1943b, p. 140, Pl. 11, fig. 1; 1944, p. 191, Pl. 77, fig. 2.
Mississippian, Osagean, Burlington Limestone, Lake Valley Limestone: United States: Iowa, Missouri, New Mexico

Cactocrinus nodobrachiatus Wachsmuth and Springer, 1890. Laudon and Severson, 1953, p. 530, Pl. 53, fig. 6.

Mississippian, Kinderhookian, Lodgepole Limestone: United States: Montana

Cactocrinus ornatissimus Wachsmuth and Springer, 1890. Laudon and Severson, 1953, p. 530, Pl. 53, fig. 4.
Mississippian, Kinderhookian, Lodgepole Limestone: United States: Montana

*Cactocrinus proboscidialis (Hall), 1858. Moore and Laudon, 1944, p. 191, Pl. 77, fig. 3. Sieverts-Doreck, 1952, p. 424, fig. 2 (misspelled as *C. proboscidalis*). Ubaghs, 1953, p. 739, fig. 22e (misspelled as *C. proboscidalis*).
Mississippian, Osagean, Burlington Limestone: United States: Iowa, Illinois, Missouri

Cactocrinus tenuisculptus (McChesney), 1860 = *Cusacrinus tenuisculptus*

Cactocrinus spp.
Wright, 1943b, p. 233, Pl. 8, fig. 4.
Lower Carboniferous: Ireland
Laudon and others, 1952, p. 571, Pl. 68, fig. 11.
Mississippian, Kinderhookian, Banff formation: Canada: Alberta
Laudon and others, 1952, p. 571, Pl. 68, fig. 6.
Mississippian, Kinderhookian, Banff formation: Canada: Alberta

CADISCOCRINUS Kirk, 1945. **C. southworthi* Kirk, 1945. Kirk, 1945a, p. 346. Ubaghs, 1953, p. 737.
Devonian, Erian

*Cadisocrinus southworthi Kirk, 1945a, p. 348, Pl. 1, figs. 1-3.
Devonian, Erian, Arkona shale: Canada: Ontario

CADOCRINUS Wanner, 1924. **Hydreionocrinus variabilis* Wanner, 1916. Moore and Laudon, 1943b, p. 62. Teichert, 1949, p. 27, figs. 9, 10. Arendt and Hecker, 1964, p. 91.
Permian

Cadocrinus saltanajewi Yakovlev, *in* Yakovlev and Ivanov, 1956, p. 65, Pl. 13, fig. 5.
Permian: Russia: Ural Mountains

Cadocrinus timanicus Yakovlev, 1948, p. 119, figs. 1-4; 1950, p. 95, fig. 1, no. VI; Yakovlev, *in* Yakovlev and Ivanov, 1956, p. 73, Pl. 13, figs. 1, 2. Arendt and Hecker, 1964, p. 91, Pl. 13, fig. 7.
Lower Permian: Russia: Timan

*Cadocrinus variabilis (Wanner), 1916. Wanner, 1949a, p. 5, Pl. 1, figs. 1, 2; Figs. 1-7. Moore, 1962b, p. 24, fig. 13, no. 1.
Permian: Timor

CAELOCRINUS Xu, 1962. **C. stellifer* Xu, 1962. Xu, 1962, p. 46, 51.
Middle Silurian

*Caelocrinus stellifer Xu, 1962, p. 46, 51, Pl. 1, figs. 1-6; Figs. 1, 2.
Middle Silurian: China: Szuchuan Province

CALCEOCRINUS Hall, 1852. **Cheirocrinus chrysalis* Hall, 1860. Moore and Laudon, 1943b, p. 30, fig. 3; 1944, p. 145. Cuenot, 1948, p. 60. Ramsbottom, 1952, p. 34, fig. 1. Ubaghs, 1953, p. 746. Moore, 1962a, p. 18, figs. 3, no. 2; 5, no. 3; 1962b, p. 12, fig. 4, no. 2. Strimple, 1963d, p. 51, fig. 13 (part). Arendt and Hecker, 1964, p. 81.
Eucheirocrinus Meek and Worthen, 1869. Moore and Laudon, 1943b, p. 29, fig. 3; 1944, p. 145. Ubaghs, 1953, p. 746, fig. 140J. Moore, 1962a, p. 18. Arendt and Hecker, 1964, p. 81.
Middle Ordovician–Silurian

Calceocrinus alleni Rowley, *in* Greene, 1904 = *Deltacrinus alleni*

Calceocrinus anglicus (Springer), 1926. Moore, 1962a, p. 20, figs. 7F, 11A.
  Eucheirocrinus anglicus Springer, 1926. Ramsbottom, 1952, p. 43, Pl. 5, figs. 8, 9; Fig. 3.
  Silurian, Wenlockian, Wenlock Limestone: England

Calceocrinus barrandei Walcott, 1883. Moore, 1962a, p. 21, figs. 7D, 11C.
  Ordovician, Champlainian, Trenton Limestone: United States

Calceocrinus barrisi Worthen, 1875 = *Synchirocrinus barrisi*

Calceocrinus bassleri Springer, 1926 = *Synchirocrinus bassleri*

Calceocrinus bifurcatus Springer, 1926 = *Synchirocrinus bifurcatus*

Calceocrinus centervillensis Foerste, 1936 = *Synchirocrinus centervillensis*

*Calceocrinus chrysalis (Hall), 1860. Moore, 1962a, p. 20, Pl. 1, fig. 2; Figs. 7E, 11D.
  Eucheirocrinus chrysalis (Hall), 1860. Moore and Laudon, 1943b, p. 130, Pl. 1, fig. 3; 1944, p. 145, Pl. 55, fig. 6.
  Silurian, Niagaran, Rochester Shale: United States: New York

Calceocrinus constrictus Brower, 1966, p. 623, Pl. 75, figs. 21, 26, 27; Figs. 2A, E.
  Ordovician, Cincinnatian, Girardeau Limestone: United States: Illinois

Calceocrinus fletcheri (Salter), 1873 = *Chirocrinus fletcheri*

Calceocrinus foerstei Springer, 1926 = *Synchirocrinus foerstei*

Calceocrinus gotlandicus (Angelin), 1878 = *Chirocrinus gotlandicus*

Calceocrinus gradatus (Salter), 1873. Ramsbottom, 1952, p. 42, Pl. 5, fig. 1; Fig. 2B.
  Cheirocrinus gradatus Salter, 1873. Ramsbottom, 1952, p. 42.
  Silurian, Wenlockian, Wenlock Limestone: England

Calceocrinus? granuliferus Rowley, *in* Green, 1903 = *Halysiocrinus granuliferus*

Calceocrinus halli Ringueberg, 1889 = *Synchirocrinus? halli*

Calceocrinus humilis Strimple, 1963d, p.58, Pl.1, figs. 12-14; Figs. 14a-d, 15d-f.
  Silurian, Niagaran, Henryhouse Formation: United States: Oklahoma

Calceocrinus? incertus Foerste, 1936 = *Chirocrinus incertus*

Calceocrinus inclinus Ramsbottom, 1952 = *Synchirocrinus inclinus*

Calceocrinus indianensis Miller, 1891. Moore, 1962a, p. 20.
  Silurian, Niagaran, Laurel Limestone: United States: Indiana

Calceocrinis interpres Bather, 1893 = *Synchirocrinus interpres*

Calceocrinus minor (Springer), 1926. Moore, 1962a, p. 20, figs. 7C, 11B.
  Eucheirocrinus minor Springer, 1926. Moore, 1962a, p. 20.
  Silurian, Niagaran, Beech River Formation: United States: Tennessee

Calceocrinus multibifurcatus Brower, 1966, p. 624, Pl. 75, figs. 17, 18; Figs. 2D, H.
  Ordovician, Champlainian, Kirkfield Limestone: Canada: Ontario

Calceocrinus nitidus Bather, 1893 = *Synchirocrinus nitidus*

Calceocrinus ontario (Springer), 1919. Moore, 1962a, p. 20, fig. 7G.
  Eucheirocrinus ontario Springer, 1919. Moore, 1962a, p. 20.
  Silurian, Albionian, Cataract Formation: Canada: Ontario

Calceocrinus pinnulatus Bather, 1893 = *Chiropinna pinnulatus*

Calceocrinus pugil Bather, 1893 = *Synchirocrinus pugil*

Calceocrinus pustulosus Johnson, *in* Brower, 1966, p. 627, Pl. 75, fig. 13; Fig. 2C.
Silurian, Albionian, Cataract Formation: Canada: Ontario

Calceocrinus serialis (Austin, *in* Salter), 1859. Ramsbottom, 1952, p. 40, Pl. 5, figs. 2-7; Fig. 2D.
Cheirocrinus serialis (Austin,*in* Salter, 1859. Ramsbottom, 1952, p. 40.
Silurian, Wenlockian, Wenlock Limestone: England

Calceocrinus tenax Bather, 1893 = *Synchirocrinus tenax*

Calceocrinus tucanus Bather, 1893 = *Synchirocrinus tucanus*

Calceocrinus typus Ringueberg, 1889 = *Synchirocrinus typus*

Calceocrinus? n. sp. Brower, 1966, p. 620, Pl. 75, fig. 24; Figs. 1J, K.
Ordovician, Champlainian, Bromide Formation, Poolville Member: United States: Oklahoma

CALCEOLISPONGIA Etheridge, 1914. *C. hindei* Etheridge, 1914. Moore and Laudon, 1943b, p. 62. Teichert, 1949, p. 54, figs. 2-5, 7. Arendt and Hecker, 1964, p. 91.
Dinocrinus Wanner, 1916. Teichert, 1949, p. 54.
Permian

Calceolispongia abundans Teichert, 1949, p. 54, Pl. 1, figs. 2, 7, 12, 21, 22; Pl. 6, figs. 21, 22; Pl. 8, fig. 2; Pl. 9; Pl. 10, figs. 1-10, 13-38; Pls. 11, 12.
Permian, Artinskian, Wandagee Series: Western Australia

Calceolispongia aculeata (Wanner), 1924. Teichert, 1949, p. 41, Pl. 25, figs. 23, 24.
Dinocrinus aculeatus Wanner, 1924. Teichert, 1949, p. 41.
Permian: Timor

Calceolispongia acuminata Teichert, 1949, p. 59, Pl. 3, figs. 8-35.
Permian, Artinskian, Bulgadoo Series, Barrabiddy Shale Series, Byro Series: Western Australia

Calceolispongia barrabiddiensis Teichert, 1949, p. 60, Pl. 3, figs. 1-7.
Permian, Artinskian, Barrabiddy Shale Series: Western Australia

Calceolispongia bifurca (Wanner), 1937. Teichert, 1949, p. 41, Pl. 25, figs. 25-27.
Dinocrinus bifurcatus Wanner, 1937. Teichert, 1949, p. 41.
Permian: Timor

Calceolispongia cornuta (Wanner), 1916. Teichert, 1949, p. 41, Pl. 25, figs. 1-6.
Dinocrinus cornutus Wanner, 1916. Teichert, 1949, p. 41.
Permian: Timor

Calceolispongia digitata Teichert, 1949, p. 61, Pl. 2, figs. 1-45.
Permian, Artinskian, Callytharra Series: Western Australia

Calceolispongia elegans (Oyens), 1940. Teichert, 1949, p. 41, Pl. 25, figs. 28-30.
Dinocrinus elegans Oyens, 1940. Teichert, 1949, p. 41.
Permian: Timor

Calceolispongia elegantula Teichert, 1949, p. 63, Pl. 1, figs. 1, 8, 13, 25; Pl. 5, figs. 14-16; Pl. 6, figs. 1-20, 23-80; Pl. 7; Pl. 8, fig. 1.
Permian, Artinskian, Wandagee Series: Western Australia

Calceolispongia hindei Etheridge, 1914. Teichert, 1949, p. 71, Pls. 21, 22; Pl. 23, figs. 1-12, 22-31.
Permian, Artinskian, Nooncanbah Series, Wandagee Series: Western Australia

Calceolispongia mammeata (Wanner), 1916. Teichert, 1949, p. 41, Pl. 25,

figs. 7-11.
Dinocrinus mammeatus Wanner, 1916. Teichert, 1949, p. 41.
   Permian: Timor

Calceolispongia multiformis Teichert, 1949, p. 75, Pl. 1, figs. 3, 5, 6, 10, 11, 14, 15, 17, 18, 23, 24; Pls. 13-16.
   Permian, Artinskian, Wandagee Series: Western Australia

Calceolispongia noetlingi (Sieverts-Doreck), 1942. Teichert, 1949, p. 7.
Dinocrinus noetlingi Sieverts-Dorcck, 1942, p. 228, figs. 5, 6.
   Teichert, 1949, p. 7.
   Permian, Artinskian: Australia; Tasmania

Calceolispongia robusta Teichert, 1949, p. 80, Pl. 24.
   Permian, Artinskian, Wandagee Series: Western Australia

Calceolispongia rotundata Teichert, 1949, p. 83, Pl. 1, fig. 16; Pl. 4, figs. 25-42; Pl. 5, figs. 1-13.
   Permian, Artinskian, Wandagee Series: Western Australia

Calceolispongia rubra Teichert, 1949, p. 85, Pl. 10, figs. 11, 12; Pl. 17; Pl. 18, figs. 1-36, 44-47.
   Permian, Artinskian, Wandagee Series: Western Australia

Calceolispongia spectabilis Teichert, 1949, p. 89, Pl. 1, figs. 4, 9; Pl. 18, figs. 37-43, 48; Pls. 19, 20.
   Permian, Artinskian, Wandagee Series: Western Australia

Calceolispongia spinosa Teichert, 1949, p. 92, Pl. 2, figs. 46-72.
   Permian, Artinskian, Callytharra Series: Western Australia

Calceolispongia truncata Teichert, 1949, p. 93, Pl. 3, figs. 36-56; Pl. 4, figs. 1-21.
   Permian, Artinskian, Cundlego Series: Western Australia

Calceolispongia sp. ind. A Teichert, 1949, p. 95, Pl. 1, figs. 19, 20; Pl. 23, figs. 13-16.
   Permian, Artinskian, Wandagee Series: Western Australia

Calceolispongia sp. ind. B Teichert, 1949, p. 96, Pl. 23, figs. 17, 18.
   Permian, Artinskian, Wandagee Series: Western Australia

Calceolispongia sp. ind. C Teichert, 1949, p. 96, Pl. 23, figs. 19-21.
   Permian, Artinskian, Wandagee Series: Western Australia

Calceolispongia sp. ind. D Teichert, 1949, p. 96.
   Permian, Artinskian, Wandagee Series: Western Australia

Calceolispongia sp. ind. E Teichert, 1949, p. 97, Pl. 4, figs. 22-24.
   Permian, Artinskian, Wandagee Series: Western Australia

Calceolispongia sp. ind. F Teichert, 1949, p. 97, Pl. 2, figs. 73-76.
   Permian, Artinskian, Barrabiddy Shale Series: Western Australia

Calceolispongia sp. nov.
   Teichert, 1949, p. 41, Pl. 25, figs. 12-14.
      Permian, Basleo Beds: Timor
   Teichert, 1949, p. 41, Pl. 25, figs. 15, 16.
      Permian, Basleo Beds: Timor
   Teichert, 1949, p. 41, Pl. 25, figs. 17, 18.
      Permian, Basleo Beds: Timor
   Hill and Woods, 1964, p. 24, Pl. P 12, figs. 21a-d.
      Permian: Australia: Queensland

Calceolispongia? sp. nov.
   Teichert, 1949, p. 41, Pl. 25, figs. 19-22.
      Permian, Umaria Beds: India

CALDENOCRINUS Wright, 1946. *C. curtus* Wright, 1946. Wright, 1946a, p. 33.

Ubaghs, 1953, p. 755.
Carboniferous, Visean

Caldenocrinus curtus Wright, 1946a, p. 34, Pl. 3, figs. 1-5, 14, 17;
Pl. 4, figs. 4, 8, 9;   1954a, p. 158, Pl. 41, fig. 13; Pl. 43,
figs. 1-4, 8, 10-13, 19.
Carboniferous, Visean, Lower Limestone Group: Scotland

CALEIDOCRINUS Waagen and Jahn, *in* Jahn, 1892. *C. multiramus* Waagen and
Jahn, *in* Jahn, 1892. Moore and Laudon, 1943b, p. 31.
Ordovician

cf. *Caleidocrinus* Whittard, 1931 = *Iocrinus shelvensis*

Caleidocrinus turgidulus Ramsbottom, 1961, p. 6, Pl. 1, figs. 1, 2.
Cyathocrinus sp. Huxley and Etheridge, 1878. Ramsbottom, 1961, p. 6.
Ordovician, Llanvirnian, red agglomerate and ash: England

CALLIOCRINUS d'Orbigny, 1849. *Eugeniacrinites? costatus* Hisinger, 1837.
Moore and Laudon, 1943b, p. 99; 1944, p. 201. Cuénot, 1948,
p. 69. Ubaghs, 1953, p. 742. Arendt and Hecker, 1964, p. 97.
Silurian-Devonian?

Calliocrinus beachleri Wachsmuth and Springer, 1892. Moore and Laudon,
1944, p. 201, Pl. 75, fig. 18.
Silurian, Niagaran, Laurel Limestone: United States: Indiana

Calliocrinus primibrachialis Busch, 1943, p. 107, Pl. 20, figs. 6-10;
Fig. 1 (misspelled as *Callicrinus primibrachialis*).
Silurian, Niagaran, Cedarville Dolomite: United States: Ohio

CALPIOCRINUS Angelin, 1878. *C. fimbriatus* Angelin, 1878. Moore and Laudon,
1943b, p. 74. Ubaghs, 1953, p. 756. Arendt and Hecker, 1964, p. 94.
Silurian

CALYCANTHOCRINUS Follmann, 1887. *C. decadactylus* Follmann, 1887. Moore
and Laudon, 1943b, p. 30. Ubaghs, 1953, p. 746, fig. 11e. Moore,
1962b, p. 14, fig. 6, no. 2. Arendt and Hecker, 1964, p. 81.
Devonian

Calycanthocrinus decadactylus Follmann, 1887. Strimple, 1963d, figs. 9b,
10h.
Devonian

CALYCOCRINUS Wanner, 1916. *C. curvatus* Wanner, 1916. Moore and Laudon,
1943b, p. 76. Yakovlev, 1947d, p. 47, fig. 14. Ubaghs, 1953, p. 756,
fig. 136. Yakovlev, 1956a, p. 83, fig. 34; 1958, p. 68, fig. 34;
1964a, p. 78, fig. 36; 1964b, p. 68, fig. 99. Arendt and Hecker,
1964, p. 94.
Permian

Calycocrinus curvatus Wanner, 1916. Arendt and Hecker, 1964, p. 94,
Pl. 14, fig. 9.
Permian: Timor

Calycocrinus curvatus depressus Wanner, 1924. Ubaghs, 1952, p. 213,
figs. 3A-C.
Permian: Timor

Calycocrinus curvatus subturbinatus Wanner, 1916. Yakovlev, 1949a, p. 266,
fig. 1a; 1956a, p. 86, fig. 35, no. 1; 1958, p. 71, fig. 35, no. 1;
1964a, p. 79, fig. 37, no. 1; 1964b, p. 58, fig. 63.
Permian: Timor

Calycocrinus major Wanner, 1916. Yakovlev, 1949a, p. 265, fig. 1b; 1956a,
p. 86, fig. 35, no. 2; 1958, p. 71, fig. 35, no. 2; 1964a, p. 79,
fig. 37, no. 2; 1964b, p. 58, fig. 64.
Permian: Timor

Calycocrinus perplexus Wanner, 1929. Ubaghs, 1953, p. 756, fig. 136.
Permian: Timor

Calycocrinus rossicus Yakovlev, *in* Yakovlev and Ivanov, 1956, p. 70, Pl. 9, figs.5, 6, 8, 10, 15-18. Arendt and Hecker, 1964, p. 94, Pl. 14, figs. 5-8.
Permian, Artinskian: Russia: Ural Mountains

Calycocrinus sp. Yakovlev, *in* Yakovlev and Ivanov, 1956, p. 71, Pl. 9, figs. 7, 9.
Permian, Artinskian: Russia: Ural Mountains

Camarocrinus Hall, 1879 = *Scyphocrinites*

Camarocrinus mariannae Yakovlev, 1953 = *Scyphocrinites mariannae*

Camarocrinus saffordi Hall, 1879 = *Scyphocrinites saffordi*

Camarocrinus ulrichi Schuchert, 1903 = *Scyphocrinites ulrichi*

CAMPTOCRINUS Wachsmuth and Springer, 1897. *Dichocrinus (Camptocrinus) myelodactylus* Wachsmuth and Springer, 1897. Moore and Laudon, 1943b, p. 96; 1944, p. 199. Cuénot, 1948, p. 69. Ubaghs, 1953, p. 741. Arendt and Hecker, 1964, p. 97.
Mississippian-Permian

Camptocrinus compressus Wright, 1937. Ubaghs, 1953, p. 741, fig. 131. Wright, 1958, p. 307, Pl. 75, figs. 6-9.
Carboniferous, Visean, Lower Limestone Group: Scotland

Camptocrinus crawfordsvillensis Springer, 1926. Van Sant, *in* Van Sant and Lane, 1964, p. 123, fig. 41, no. 1.
Mississippian, Osagean, Borden Group: United States: Indiana

Camptocrinus multicirrus Springer, 1926. Moore and Laudon, 1944, p. 199, Pl. 78, fig. 13.
Mississippian, Chesterian: United States: Alabama

**Camptocrinus myelodactylus* (Wachsmuth and Springer), 1897. Ubaghs, 1953, p. 741, figs. 129, 130.
Mississippian

Camptocrinus plenicirrus Springer, 1926. Van Sant, *in* Van Sant and Lane, 1964, p. 124, fig. 41, no. 2.
Mississippian, Osagean, Borden Group: United States: Indiana

CANISTOCRINUS Wachsmuth and Springer, 1885. *Glyptocrinus richardsoni* Wetherby, 1880. Ubaghs, 1953, p. 738.
Early Silurian

CANTHAROCRINUS Breimer, 1962. *C. minor* Breimer, 1962. Breimer, 1962a, p. 117.
Lower-Middle Devonian

**Cantharocrinus minor* Breimer, 1962a, p. 117, Pl. 11, figs. 5-7; Figs. 27, 32.
Devonian, Emsian, La Vid Shale: Spain

Cantharocrinus simplex Breimer, 1962a, p. 120, Pl. 11, figs. 8-11; Fig. 28.
Devonian, Couvinian, Santa Lucia Formation: Spain

CARABOCRINUS Billings, 1857. *C. radiatus* Billings, 1857. Moore and Laudon, 1943b, p. 49, figs. 1, 4; 1944, p. 151, Pl. 52, fig. 19. Moore, 1950, p. 34, fig. 4a. Ubaghs, 1953, p. 750, fig. 14b. Moore, 1962b, p. 15, fig. 9, no. 2. Philip, 1963, p. 266, fig. 4a. Arendt and Hecker, 1964, p. 83.
Ordovician-Early Silurian

Carabocrinus esthonus Jaekel, 1918. Arendt and Hecker, 1964, p. 83,

fig. 120.
Middle Ordovician: Estonia

Carabocrinus huronensis Foerste, 1924. Sinclair, 1945, p. 716, Pl. 2, fig. 17.
Ordovician, Champlainian, Lower Trenton Group: Canada: Ontario

Carabocrinus radiatus Billings, 1857. Moore and Laudon, 1943b, p. 131, Pl. 2, fig. 5; 1944, p. 153, Pl. 53, fig. 5. Wilson, 1946, p. 38, Pl. 5, fig. 5.
Ordovician, Champlainian, Trenton Limestone, Cobourg Limestone, Hull Limestone: United States: New York; Canada: Ontario

Carabocrinus treadwelli Sinclair, 1945, p. 715, Pl. 2, figs. 14-16.
Ordovician, Champlainian, Bromide Formation: United States: Oklahoma

Carabocrinus vancortlandti Billings, 1859. Moore and Laudon, 1943b, p. 131, Pl. 2, fig. 6; 1944, p. 153, Pl. 53, fig. 6. Wilson, 1946, p. 39 (misspelled as *C. vancortlandi*).
Ordovician, Champlainian, Trenton Limestone, Cobourg Limestone?: Canada: Ontario

Carabocrinus sp. Ubaghs, 1953, p. 750, fig. 14a.
Ordovician

CARCINOCRINUS Laudon, 1941. *C. stevensi* Laudon, 1941. Moore and Laudon, 1943b, p. 55. Arendt and Hecker, 1964, p. 88.
Mississippian

Carinocrinus [sic] *eventus* Strimple, 1953 = *Cosmetocrinus eventus*

CARLOPSOCRINUS Wright, 1933. *C. bullatus* Wright, 1933. Moore and Laudon, 1943b, p. 76. Ubaghs, 1953, p. 756. Arendt and Hecker, 1964, p. 94.
Lower Carboniferous

Carlopsocrinus bullatus Wright, 1933. Wright, 1954a, p. 181, Pl. 39, figs. 20, 22; Figs. 106-108.
Carboniferous, Visean, Lower Limestone Group: Scotland

CAROLICRINUS Waagen and Jahn, 1899. *C. barrandei* Waagen and Jahn, 1899. Ubaghs, 1953, p. 741. Arendt and Hecker, 1964, p. 97.
Silurian

CARPOCRINUS Müller, 1840. *Actinocrinites simplex* Phillips, *in* Murchison, 1839. Moore and Laudon, 1943b, p. 93, fig. 16; 1944, p. 193, Pl. 71, fig. 4. Ubaghs, 1953, p. 740.
Silurian

Carpocrinus laurelianus (Springer), 1926. Moore and Laudon, 1943b, p. 139, Pl. 10, fig. 14; 1944, p. 193, Pl. 75, fig. 13.
Desmidocrinus laurelianus Springer, 1926. Moore and Laudon, 1943b, p. 139.
Silurian, Niagaran, Laurel Limestone: United States: Indiana

CATACTOCRINUS Goldring, 1923. *C. leptodactylus* Goldring, 1923. Moore and Laudon, 1943b, p. 54.
Devonian

CATILLOCRINUS Shumard, 1866. *C. tennesseeae* Shumard, 1866. Moore and Laudon, 1943b, p. 30, fig. 3; 1944, p. 149, Pl. 52, fig. 4. Ubaghs, 1953, p. 747, fig. 11d. Moore, 1962b, p. 16, fig. 12, no. 11. Arendt and Hecker, 1964, p. 81.
Mississippian

Catillocrinus bradleyi Meek and Worthen, 1868 = *Eucatillocrinus bradleyi*

Catillocrinus tennesseeae Shumard, 1866. Moore and Laudon, 1943b, p. 130, Pl. 1, fig. 15; 1944, p. 149, Pl. 55, fig. 2. Ubaghs, 1953, p. 747, figs. 44-46.

Mississippian, Osagean, Keokuk Limestone, New Providence Formation:
United States: Indiana, Kentucky

Catillocrinus turbinatus Springer, 1923. Ubaghs, 1953, p. 747, fig. 43.
Mississippian

Catillocrinus wachsmuthi (Meek and Worthen), 1866. Moore and Laudon, 1944,
p. 149, Pl. 55, fig. 1. Van Sant, *in* Van Sant and Lane, 1964, p. 68,
fig. 24, no. 2.
Mississippian, Osagean, Burlington Limestone: United States: Iowa

CELONOCRINUS Lane and Webster, 1966. *C. expansus* Lane and Webster, 1966.
Lane and Webster, 1966, p. 47.
Permian

**Celonocrinus expansus* Lane and Webster, 1966, p. 48, Pl. 11, figs. 1, 2,
4, 5; Fig. 18. Webster and Lane, 1967, p. 14.
Permian, Wolfcampian, Bird Spring Formation: United States: Nevada

CENTRIOCRINUS Bather, 1899. *Actinocrinus pentaspinus* Lyon, 1869. Moore
and Laudon, 1943b, p. 99. Ubaghs, 1953, p. 742.
Devonian

CERATOCRINUS Wanner, 1937. *C. exornatus* Wanner, 1937. Moore and Laudon,
1943b, p. 51. Ubaghs, 1953, p. 750. Arendt and Hecker, 1964, p. 83.
Permian

CERCIDOCRINUS Kirk, 1938. *Poteriocrinus bursaeformis* White, 1862. Moore
and Laudon, 1943b, p. 56, fig. 6; 1944, p. 159, Pl. 52, fig. 20.
Mississippian

**Cercidocrinus bursaeformis* (White), 1862. Moore and Laudon, 1943b, p. 133,
Pl. 4, fig. 4; 1944, p. 159, Pl. 59, fig. 10.
Mississippian, Osagean, Burlington Limestone: United States: Iowa

Cercidocrinus infrequens (Laudon and Beane), 1937. Moore and Laudon, 1944,
p. 159, Pl. 60, fig. 20.
Mississippian, Kinderhookian, Hampton Formation: United States: Iowa

Ceriocrinus crassus Wanner, 1924 = *Graphiocrinus*? *crassus*

Ceriocrinus serratomarginatus Yakovlev, 1930 = *Delocrinus serratomarginatus*

CESTOCRINUS Kirk, 1940. *C. striatus* Kirk, 1940. Moore and Laudon, 1943b,
p. 51, fig. 4. Ubaghs, 1953, p. 750. Moore, 1962b, p. 15, fig. 10,
no. 2.
Mississippian

CHARIENTOCRINUS Goldring, 1923. *C. ithacensis* Goldring, 1923. Moore and
Laudon, 1943b, p. 54.
Devonian

Cheirocrinus chrysalis Hall, 1860 = *Calceocrinus chrysalis*

Cheirocrinus fletcheri Salter, 1873 = *Chirocrinus fletcheri*

Cheirocrinus gradatus Salter, 1873 = *Synchirocrinus gradatus*

Cheirocrinus lamellosus Hall, 1860 = *Halysiocrinus? lamellosus*

Cheirocrinus serialis Salter, 1873 = *Synchirocrinus serialis*

Cheirocrinus stigmatus Hall, 1863 = *Deltacrinus stigmatus*

Cheirocrinus? sp. Butts, 1941, p. 54, Pl. 76, figs. 24-27.
Ordovician, Champlainian, Lenoir Limestone: United States: Virginia

CHICAGOCRINUS Weller, 1900. *C. ornatus* Weller, 1900. Moore and Laudon,
1943b, p. 99.
Silurian

CHIROCRINUS Angelin, 1878. *C. gotlandicus* Angelin, 1878. Moore, 1962a,

p. 24, figs. 5, nos. 5a, b.
Silurian

Chirocrinus fletcheri (Salter), 1873. Moore, 1962a, p. 26, Pl. 3, fig. 7; Fig. 5, no. 5c (*non* Fig. 5, no. 3c as given in text).
Calceocrinus fletcheri (Salter), 1873. Ramsbottom, 1952, p. 36, Pl. 4, fig. 1; Fig. 2F.
Silurian, Wenlockian, Wenlock Limestone: England

*Chirocrinus gotlandicus Angelin, 1878. Moore, 1962a, p. 26, fig. 9.
Silurian, Gotlandian: Sweden: Isle of Gotland

Chirocrinus incertus (Foerste), 1936. Moore, 1962a, p. 26, fig. 7B.
Calceocrinus incertus Foerste, 1936. Moore, 1962a, p. 26.
Silurian, Medinian, Brassfield Limestone: United States: Ohio

CHIROPINNA Moore, 1962. **Calceocrinus pinnulatus* Bather, 1893. Moore, 1962a, p. 28.
Silurian

*Chiropinna pinnulatus (Bather), 1893. Moore, 1962a, p. 28, Pl. 2, fig. 2; Fig. 5, no. 2; Fig. 18 (*non* Fig. 5, no. 5 as given in text).
Calceocrinus pinnulatus Bather, 1893. Moore, 1962a, p. 28.
Silurian, Gotlandian: Sweden

CHOLOCRINUS Springer, 1906. **Forbesiocrinus obesus* Angelin, 1878. Moore and Laudon, 1943b, p. 74. Ubaghs, 1953, p. 756. Arendt and Hecker, 1964, p. 94.
Silurian

CHYTROCRINUS Jaekel, 1918. **Stelidiocrinus laevis* Angelin, 1878 (part).
Moore and Laudon, 1943b, p. 98.
Silurian

CIBOLOCRINUS Weller, 1909. **C. typus* Weller, 1909. Moore and Laudon, 1943b, p. 76, fig. 11; 1944, p. 181, Pl. 66, fig. 8. Moore, 1950, p. 37, fig. 40. Ubaghs, 1953, p. 756. Arendt and Hecker, 1964, p. 94.
Pennsylvanian, Morrowan-Permian, Wolfcampian

Cibolocrinus abyssus Strimple, 1951c, p. 11, Pl. 1, figs. 5, 6.
Pennsylvanian, Missourian, Wann Formation: United States: Oklahoma

Cibolocrinus banioni Moore, 1939. Moore and Laudon, 1944, p. 183, Pl. 56, fig. 29; Pl. 62, fig. 16.
Pennsylvanian, Virgilian, Brownville Limestone: United States: Oklahoma

Cibolocrinus conicus Strimple, 1951c, p. 10, Pl. 2, figs. 1-3.
Pennsylvanian, Missourian Series, Wann Formation: United States: Oklahoma

Cibolocrinus detectus Strimple, 1951c, p. 12, Pl. 1, figs. 7, 8.
Pennsylvanian, Missourian, Ochelata Group: United States: Oklahoma

Cibolocrinus detrusus Strimple, 1951c, p. 12, Pl. 1, figs. 1-4; 1962e, p. 49.
Pennsylvanian, Desmoinesian, Altamont Limestone, Oologah Limestone: United States: Oklahoma

Cibolocrinus erectus Strimple, 1951a, p. 203, Pl. 36, figs. 5-7.
Pennsylvanian, Missourian, Lake Bridgeport Shale: United States: Texas

Cibolocrinus punctatus Moore and Plummer, 1940. Moore and Laudon, 1943b, p. 137, Pl. 8, fig. 6; 1944, p. 183, Pl. 56, fig. 13. Plummer, 1950, p. 53, Pl. 11, figs. 9, 10.
Pennsylvanian, Morrowan, Marble Falls Limestone: United States: Texas

Cibolocrinus pusillus Wanner, 1937 = *Stuartwellercrinus pusillus*

Cibolocrinus robustus Strimple, 1939. Strimple, 1951c, p. 7, Pl. 2,

figs. 6-8.
Pennsylvanian, Missourian, Stanton Limestone: United States: Oklahoma

Cibolocrinus tumidus Moore and Plummer, 1938. Moore and Laudon, 1944, p. 183, Pl. 56, fig. 15.
Pennsylvanian, Morrowan, Brentwood Limestone: United States: Oklahoma

*Cibolocrinus typus Weller, 1909. Moore and Laudon, 1944, p. 181, Pl. 56, fig. 16. Lane and Webster, 1966, p. 22, Pl. 1, figs. 5, 6, 8, 9.
Permian, Wolfcampian, Cibolo Formation, Bird Spring Formation: United States: Texas, Nevada

Cibolocrinus sp. Strimple, 1951c, p. 8, Pl. 2, figs. 4, 5.
Pennsylvanian, Missourian, Birch Creek Limestone: United States: Oklahoma

CICEROCRINUS Sollas, 1900. *C. elegans* Sollas, 1900. Moore and Laudon, 1943b, p. 30. Ubaghs, 1953, p. 746. Moore, 1962b, p. 14, fig. 6, no. 3. Arendt and Hecker, 1964, p. 81.
Silurian

Cicerocrinus anglicus (Jaekel), 1900. Ramsbottom, 1958, p. 112.
Silurian, Ludlovian, Ludlow Beds: England

*Cicerocrinus elegans Sollas, 1900. Ramsbottom, 1958, p. 111, Pl. 21, figs. 2-5.
Legarocrinus tenuis Jaekel, 1900. Ramsbottom, 1958, p. 111, Pl. 21, fig. 3.
Silurian, Ludlovian, Ludlow Beds: England

Cladocrinites brevidactylus Austin and Austin, 1843 = *Blothrocrinus brevidactylus*

Cladocrinites pentagonus Austin and Austin, 1843 = *Phacelocrinus?* *pentagonus*

CLARKEOCRINUS Goldring, 1923. **Cacabocrinus troosti* Hall, 1862. Moore and Laudon, 1943b, p. 99; 1944, p. 201, Pl. 70, fig. 7. Ubaghs, 1953, p. 742, fig. 107.
Devonian

Clarkeocrinus spriestersbachi Schmidt, 1941, p. 92, Pl. 11, fig. 1;Fig. 26.
Devonian, Eifelian: Germany

*Clarkeocrinus troosti (Hall),1862. Moore and Laudon, 1944, p. 201, Pl. 74, fig. 1. Goldring, 1950, p. 34, Pl. 2, fig. 10. Ubaghs, 1953, p. 742, fig. 107.
Devonian, Erian, Kashong Shale: United States: New York

CLEIOCRINUS Billings, 1857. **C. regius* Billings, 1857. Moore and Laudon, 1943b, p. 101. Ubaghs, 1953, p. 735, fig. 139a.
Ordovician

*Cleiocrinus regius Billings, 1857. Wilson, 1946, p. 23, Pl. 5, fig. 2. Ubaghs, 1950, p. 119, fig. 7; 1953, p. 735, figs. 33d-f.
Ordovician, Champlainian, Cobourg Limestone, Hull Limestone: Canada: Ontario

Cleiocrinus sculptus Springer, 1911. Ubaghs, 1953, p. 735, fig. 33g.
Ordovician

CLEISTOCRINUS Springer, 1920. **Calpiocrinus humilis* Angelin, 1878. Moore and Laudon, 1943b, p. 76. Ubaghs, 1953, p. 756.
Silurian

CLEMATOCRINUS Jaekel, 1897. **Actinocrinites retiarius* Phillips, *in* Murchison, 1839. Ubaghs, 1953, p. 743.
Silurian

CLIDOCHIRUS Angelin, 1878. **C. pyrum* Angelin, 1878. Moore and Laudon,

1943b, p. 76, fig. 11; 1944, p. 183, Pl. 66, fig. 9. Ubaghs, 1953, p. 756.
Silurian-Mississippian

Clidochirus americanus Springer, 1920. Moore and Laudon, 1944, p. 183, Pl. 54, fig. 3.
Silurian, Niagaran, Clinton Formation: United States: Ohio

Clidochirus gratiosus Strimple, 1963d, p. 113, Pl. 8, figs. 9, 10; Pl. 9, fig. 7.
Devonian, Helderbergian, Haragan Formation: United States: Oklahoma

Clidochirus keyserensis Springer, 1920. Moore and Laudon, 1943b, p. 136, Pl. 7, fig. 4; 1944, p. 183, Pl. 69, fig. 18.
Silurian, Cayugan, Keyser Limestone: United States: West Virginia

**Clidochirus pyrum* Angelin, 1878. Ubaghs, 1953, p. 756, fig. 15a.
Silurian

CLISTOCRINUS Kirk, 1937. **C. pyriformis* Kirk, 1937. Lane, 1967b, p. 13, figs. 2, 7 (*non* fig. 8 as given).
Clithrocrinus Kirk, 1937. Moore and Laudon, 1943b, p. 50. Arendt and Hecker, 1964, p. 85.
Mississippian

CLONOCRINUS Quenstedt, 1876. **Eucalyptocrinus polydactylus* McCoy, 1849. Moore and Laudon, 1943b, p. 98, fig. 17; 1944, p. 200. Ubaghs, 1953, p. 742.
Silurian-?Devonian

Clonocrinus occidentalis Springer, 1926. Moore and Laudon, 1943b, p. 142, Pl. 13, fig. 12; 1944, p. 201, Pl. 73, figs. 13, 24.
Silurian, Niagaran, Decatur Limestone: United States: Tennessee

CLOSTEROCRINUS Hall, 1852. **C. elongatus* Hall, 1852. Moore and Laudon, 1943b, p. 51. Arendt and Hecker, 1964, p. 85.
Silurian

CODIACRINUS Schultze, 1866. **C. granulatus* Schultze, 1866. Moore and Laudon, 1943b, p. 50, fig. 4. Ubaghs, 1953, p. 751. Yakovlev, 1954b, p. 1088, fig. 1v; 1956a, p. 90, fig. 36, no. 3; 1958, p. 75, fig. 36, no. 3; 1964a, p. 83, fig. 38, no. 3; 1964b, p. 57, fig. 59v. Arendt and Hecker, 1964, p. 85. Lane, 1967b, p. 11, fig. 4.
Devonian

Codiacrinus sp. Breimer, 1962a, p. 159, Pl. 15, fig. 15.
Devonian, Emsian: Spain

COELIOCRINUS White, 1863. **Poteriocrinus dilatatus* Hall, 1861. Moore and Laudon, 1943b, p. 56, fig. 6; 1944, p. 159.
Mississippian

**Coeliocrinus dilatatus* (Hall), 1861. Moore and Laudon, 1943b, p. 133, Pl. 4, fig. 7; 1944, p. 159, Pl. 55, fig. 18.
Mississippian, Osagean, Burlington Limestone: United States: Iowa

Coeliocrinus lyra (Meek and Worthen), 1869. Moore and Laudon, 1944, p. 159, Pl. 57, fig. 1.
Mississippian, Osagean, Burlington Limestone: United States: Iowa

Coeliocrinus subspinosus White, 1863. Moore and Laudon, 1944, p. 159, Pl. 59, fig. 1.
Mississippian, Osagean, Burlington Limestone: United States: Iowa

Coeliocrinus ventricosus (Hall), 1861. Moore and Laudon, 1944, p. 159, Pl. 55, fig. 20.
Mississippian, Osagean, Burlington Limestone: United States: Iowa

COELOCRINUS Meek and Worthen, 1865. *Actinocrinus concavus Meek and
    Worthen, 1861. Ubaghs, 1953, p. 740.
    Mississippian
COENOCYSTIS Girty, 1908. *C. richardsoni Girty, 1908. Moore and Laudon,
    1943b, p. 50. Cuénot, 1948, p. 64. Arendt and Hecker, 1964, p. 85.
    Lane, 1967b, p. 13, figs. 3, 7 (fig. 7 given as fig. 8).
    Permian
Coencystis moreyi Peck, 1936 [sic] = Desmacriocrinus moreyi
COLUMBICRINUS Ulrich, 1924. *C. crassus Ulrich, 1924. Moore and Laudon,
    1943b, p. 31. Ubaghs, 1953, p. 747. Moore, 1962b, p. 14, fig. 5,
    no. 5. Philip, 1965a, p. 148, fig. 1c.
    Ordovician
COMANTHOCRINUS Springer, 1921. *Stereocrinus indianensis Miller and Gurley,
    1897. Moore and Laudon, 1943b, p. 99; 1944, p. 201. Ubaghs, 1953,
    p. 742.
    Devonian
*Comanthocrinus indianensis (Miller and Gurley), 1897. Moore and Laudon,
    1944, p. 201, Pl. 75, fig. 21.
    Devonian, Erian, Hamilton Group: United States: New York, Indiana,
    Kentucky
COMPSOCRINUS Miller, 1883. *Glyptocrinus harrisi Miller, 1881. Moore and
    Laudon, 1943b, p. 90, fig. 13; 1944, p. 189, Pl. 71, fig. 5.
    Ubaghs, 1953, p. 738.
    Ordovician-Silurian
*Compsocrinus harrisi (Miller), 1881. Moore and Laudon, 1943b, p. 139,
    Pl. 10, fig. 2; 1944, p. 189, Pl. 75, fig. 20. Ubaghs, 1953, p. 738,
    fig. 20c.
    Ordovician, Cincinnatian: United States: Ohio
Compsocrinus miamiensis (Miller), 1883. Moore and Laudon, 1944, p. 189,
    Pl. 75, fig. 19.
    Ordovician, Cincinnatian: United States: Ohio
CONDYLOCRINUS Eichwald, 1860. *C. verrucosus Eichwald, 1860. Moore and
    Laudon, 1943b, p. 83. Ubaghs, 1953, p. 737.
    ?Devonian
Conocrinites tuberculosus Troost, 1849 = Alloprosallocrinus conicus
CORDYLOCRINUS Angelin, 1878. *C. comtus Angelin, 1878. Moore and Laudon,
    1943b, p. 101; 1944, p. 203. Ubaghs, 1953, p. 743.
    Silurian-Devonian
Cordylocrinus plumosus (Hall), 1859. Moore and Laudon, 1943b, p. 143,
    Pl. 14, fig. 3; 1944, p. 203, Pl. 77, fig. 27.
    Devonian, Helderbergian, Coeymans Limestone: United States: New
    York
COREMATOCRINUS Goldring, 1923. *C. plumosus Goldring, 1923. Moore and
    Laudon, 1943b, p. 59; 1944, p. 167. Arendt and Hecker, 1964, p. 89.
    Devonian
*Corematocrinus plumosus Goldring, 1923. Moore and Laudon, 1944, p. 169,
    Pl. 58, fig. 7.
    Devonian, Senecan, Portage Shale: United States: New York
CORNUCRINUS Regnéll, 1948. *C. mirus Regnéll, 1948. Regnéll, 1948, p. 11.
    Ubaghs, 1953, p. 749. Moore, 1962b, p. 14, fig. 8, no. 4.
    Ordovician-Silurian
Cornucrinus longicornis Regnéll, 1948, p. 21, Pl. 1, figs. 4, 5.
    Silurian, Dalmanitina Beds: Sweden

Cornucrinus mirus Regnéll, 1948, p. 14, Pl. 1, fig. 3; Pls. 2-4;Figs. 3C, 4. Ubaghs, 1953, p. 749, figs. 80d, 145-147.
  Ordovician, Caradocian, *Macrourus* Beds: Sweden

COROCRINUS Goldring, 1923. *C. ornatus* Goldring, 1923. Moore and Laudon, 1943b, p. 92; 1944, p. 191. Ubaghs, 1953, p. 739.
  Devonian

Corocrinus? grandosensis Breimer, 1962a, p. 67, Pl. 7, figs. 8-12.
  Devonian, Emsian, Santa Lucia Limestone: Spain

Corocrinus imbecillus Schmidt, 1941, p. 97, Pl. 14, figs. 3, 4.
  Devonian, Coblentzian: Germany

Corocrinus nodosus Kier, 1952, p. 67, Pl. 4, figs. 2, 3.
  Devonian, Erian, Silica Formation: United States: Ohio

Corocrinus ornatus Goldring, 1923. Moore and Laudon, 1943b, p. 139, Pl. 10, fig. 7; 1944, p. 191, Pl. 74, fig. 3.
  Devonian, Erian, Hamilton Group: United States: New York

Corocrinus pettyesi Kesling, 1964c, p. 150, Pl. 2, figs. 1-7; Fig. 2.
  Devonian, Erian, Thunder Bay Limestone: United States: Michigan

Coronocrinus Hall, 1859 = *Himerocrinus?*

CORYNECRINUS Kirk, 1934. *C. romingeri* Kirk, 1934. Moore and Laudon, 1943b, p. 51, fig. 4. Ubaghs, 1953, p. 750. Moore, 1962b, p. 15, fig. 10, no. 4.
  Devonian

CORYTHOCRINUS Kirk, 1946. *C. insculptus* Kirk, 1946. Kirk, 1946c, p. 269, Pl. 40, fig. 1.
  Mississippian, Osagean

Corythocrinus Strimple (*non* Kirk), 1961 = *Endelocrinus*

Corythocrinus bifidus (Moore and Plummer), Strimple 1961g = *Endelocrinus bifidus*

Corythocrinus fragilis Kirk, 1946c, p. 270, Pl. 40, figs. 5-7. Van Sant, *in* Van Sant and Lane, 1964, p. 99.
  Mississippian, Osagean, upper part of Borden Group: United States: Indiana

Corythocrinus gracilis Kirk, 1946c, p. 271, Pl. 40, fig. 4. Van Sant, *in* Van Sant and Lane, 1964, p. 99.
  Mississippian, Osagean, upper part of Borden Group: United States: Indiana

Corythocrinus insculptus Kirk, 1946c, p. 272, Pl. 40, figs. 2, 3.
  Mississippian, Osagean, upper part of Borden Group: United States: Indiana

Corythocrinus parinodosarius (Strimple), 1940. Strimple, 1961g = *Endelocrinus parinodosarius*

Corythocrinus rectus (Moore and Plummer), Strimple, 1961g = *Endelocrinus rectus*

Corythocrinus tenuis Kirk, 1946c, p. 273, Pl. 40, fig. 8.
  Mississippian, Osagean, upper part of Burlington Limestone: United States: Iowa

Corythocrinus undulatus Strimple, 1961 = *Endelocrinus undulatus*

COSMETOCRINUS Kirk, 1941. *C. gracilis* Kirk, 1941. Moore and Laudon, 1943b, p. 55; 1944, p. 159. Arendt and Hecker, 1964, p. 88.
  Mississippian

Cosmetocrinus crawfordsvillensis (Miller and Gurley), 1890. Moore and

Laudon, 1944, p. 159, Pl. 58, fig. 14.
Mississippian, Osagean, Borden Group: United States: Indiana

Cosmetocrinus elegantulus (Wachsmuth and Springer), 1889. Moore and Laudon, 1944, p. 159, Pl. 58, figs. 6, 21.
Mississippian, Kinderhookian, Hampton Formation: United States: Iowa

Cosmetocrinus eventus (Strimple), 1953. Strimple, 1967, p. 81.
Carinocrinus eventus Strimple, 1953b, p. 201, figs. 1, 2 (misspelling of *Carcinocrinus*).
Mississippian, Chesterian, Pitkin Limestone: United States: Oklahoma

Cosmetocrinus extentus Wright, 1951a, p. 40, Pl. 10, fig. 8.
Carboniferous, Tournaisian: England

**Cosmetocrinus gracilis* Kirk, 1941. Moore and Laudon, 1944, p. 159, Pl. 58, fig. 16.
Mississippian, Osagean, Borden Group: United States: Indiana

Cosmetocrinus meeki Kirk, 1941 = *Aphelecrinus meeki*

Cosmetocrinus nanus (Meek and Worthen), 1869. Moore and Laudon, 1944, p. 159, Pl. 58, fig. 15.
Mississippian, Osagean, lower Burlington Limestone: United States: Iowa

Cosmetocrinus richfieldensis (Worthen), 1882 = *Amphelecrinus richfieldensis*

Cosmetocrinus spp.
Wright, 1951a, p. 41, Pl. 10, fig. 7.
Carboniferous, Tournaisian: Ireland
Laudon and others, 1952, p. 553, Pl. 65, fig. 28.
Mississippian, Kinderhookian, Banff formation: Canada: Alberta

COSTALOCRINUS Jaekel, 1918. **Poteriocrinus dilatatus* Schultze, 1866. Moore and Laudon, 1943b, p. 54.
Devonian

CRADEOCRINUS Goldring, 1923. **C. elongatus* Goldring, 1923. Moore and Laudon, 1943b, p. 55; 1944, p. 158. Ubaghs, 1953, p. 752. Arendt and Hecker, 1964, p. 87.
Devonian-Mississippian

Cradeocrinus dendratus Laudon and Severson, 1953, p. 514, Pl. 51, figs. 9-12; Pl. 55, figs. 3, 4.
Mississippian, Kinderhookian, Lodgepole Limestone: United States: Montana

**Cradeocrinus elongatus* Goldring, 1923. Moore and Laudon, 1944, p. 158, Pl. 53, fig. 18.
Upper Devonian, Chemung Formation: United States: New York

Cradeocrinus warreni Laudon, Parks and Spreng, 1952, p. 551, Pl. 65, figs. 18-21; Pl. 69, fig. 3.
Mississippian, Kinderhookian, Banff formation: Canada: Alberta

Cradeocrinus sp. Laudon and others, 1952, p. 551, Pl. 65, fig. 25.
Mississippian, Kinderhookian, Banff formation: Canada: Alberta

CRANOCRINUS Wanner, 1929. **C. timoricus* Wanner, 1929. Moore and Laudon, 1943b, p. 50. Arendt and Hecker, 1964, p. 85. Lane, 1967b, p. 11, fig. 5. Arendt, 1968a, p. 101, fig. 6.
Permian

CRATEROCRINUS Goldring, 1923. **C. ruedemanni* Goldring, 1923. Moore and Laudon, 1943b, p. 99. Ubaghs, 1953, p. 742.
Devonian

**Craterocrinus ruedemanni* Goldring,1923.Goldring,1951,p. 5,Pl. 1,figs. 4,5.
Devonian, Ulsterian, Onondaga Limestone: United States: New York

Craterocrinus schoharie Goldring, 1923. Goldring, 1951, p. 5, Pl. 1, fig. 3.
Devonian, Ulsterian, Onondaga Limestone: United States: New York

Craterocrinus skinneri Goldring, 1951, p. 3, Pl. 1, figs. 1, 2.
Devonian, Ulsterian, Onondaga Limestone: United States: New York

CREMACRINUS Ulrich, 1886. *C. punctatus* Ulrich, 1886. Moore and Laudon, 1943b, p. 29, fig. 3; 1944, p. 145. Ubaghs, 1953, p. 746, fig. 140I. Moore, 1962a, p. 20, figs. 3, no. 1; 5, no. 1; 6; 1962b, p. 12, fig. 4, no. 1. Strimple, 1963d, p. 50, fig. 13 (part). Arendt and Hecker, 1964, p. 81.
Ordovician-Silurian

Cremacrinus arctus Sardeson, 1928. Moore, 1962a, p. 21. Strimple, 1963d, fig. 12d.
Ordovician, Champlainian, Platteville Limestone: United States: Minnesota

Cremacrinus articulosus (Billings), 1859. Moore and Laudon, 1944, p. 145, Pl. 55, fig. 8. Wilson, 1946, p. 35, Pl. 6, fig. 5. Moore, 1962a, p. 21.
Ordovician, Champlainian, Curdsville Limestone, Hull Limestone, Cobourg Limestone: United States: Kentucky; Canada: Ontario

Cremacrinus billingsianus (Ringueberg), 1889. Wilson, 1946, p. 35 (misspelled as *C. billingsiana*). Moore, 1962a, p. 21.
Ordovician, Champlainian, Cobourg Limestone, Hull Limestone: Canada: Ontario

Cremacrinus decatur Springer, 1926. Moore, 1962a, p. 21.
Silurian, Niagaran, Decatur Limestone: United States: Tennessee

Cremacrinus furcillatus (Billings), 1887. Wilson, 1946, p. 35. Moore, 1962a, p. 21.
Ordovician, Champlainian, Cobourg Limestone: Canada: Ontario; United States: Michigan

Cremacrinus inaequalis (Billings), 1859. Wilson, 1946, p. 35, Pl. 5, fig. 4. Moore, 1962a, p. 21.
Ordovician, Champlainian, Hull Limestone, Cobourg Limestone, Prosser Limestone: Canada: Ontario; United States: Wisconsin

Cremacrinus kentuckiensis (Miller and Gurley), 1894. Moore, 1962a, p. 21, fig. 7A.
Ordovician, Champlainian, Curdsville Limestone: United States: Kentucky

Cremacrinus punctatus Ulrich, 1886. Moore, 1962a, p. 21, Pl. 1, fig. 3 (*non* fig. 4 as given on p. 21).
Ordovician, Champlainian, Decorah Shale: United States: Minnesota, Michigan

Cremacrinus simplex Springer, 1926 = *Anulocrinus simplex*

Cremacrinus tubuliferus Springer, 1926. Moore and Laudon, 1943b, p. 130, Pl. 1, fig. 2; 1944, p. 145, Pl. 55, fig. 7. Moore, 1962a, p. 21, Pl. 1, fig. 1; Fig. 10. Brower, 1966, p. 617, fig. 1H.
Silurian, Niagaran, Beech River Formation: United States: Tennessee

Cremacrinus ulrichi Springer, 1926. Moore, 1962a, p. 21.
Silurian, Niagaran, Beech River Formation: United States: Tennessee

CRIBANOCRINUS Kirk, 1944. *Rhodocrinus wortheni* Hall, 1858. Kirk, 1944a, p. 13. Ubaghs, 1953, p. 737.
Mississippian, Kinderhookian—Osagean

Cribanocrinus benedicti (Miller), 1892. Kirk, 1944a, p. 15.
Rhodocrinus benedicti Miller, 1892. Kirk, 1944a, p. 15.
Mississippian, Osagean, Borden Group: United States: Indiana

Cribanocrinus bridgerensis (Miller and Gurley), 1897. Kirk, 1944a, p. 15.
  Rhodocrinus bridgerensis Miller and Gurley, 1897. Kirk, 1944a, p. 15.
  Mississippian, Madison Limestone: United States: Montana

Cribanocrinus coxanus (Worthen), 1882. Kirk, 1944a, p. 15.
  Rhodocrinus coxanus Worthen, 1882. Kirk, 1944a, p. 15.
  Mississippian, Osagean, Keokuk Limestone: United States: Iowa

Cribanocrinus parvus (Miller), 1891. Kirk, 1944a, p. 15.
  Rhodocrinus parvus Miller, 1891. Kirk, 1944a, p. 15.
  Mississippian, Meramecian, Warsaw Formation: United States: Missouri

Cribanocrinus punctatus (Weller), 1909. Kirk, 1944a, p. 15.
  Rhodocrinus punctatus Weller, 1909. Kirk, 1944a, p. 15.
  Mississippian, Osagean, Fern Glen Limestone: United States: Missouri

Cribanocrinus urceolatus (Wachsmuth and Springer), 1897. Kirk, 1944a, p. 15.
  Rhodocrinus wortheni var. *urceolatus* Wachsmuth and Springer, 1897. Kirk, 1944a, p. 15.
  Mississippian, Osagean, Lake Valley Limestone: United States: New Mexico

Cribanocrinus wachsmuthi (Hall), 1861. Kirk, 1944a, p. 15.
  Rhodocrinus wachsmuthi Hall, 1861. Kirk, 1944a, p. 15.
  Mississippian, Osagean, lower Burlington Limestone: United States: Iowa

Cribanocrinus watersianus (Wachsmuth and Springer, *in* Miller), 1889. Kirk, 1944a, p. 15. Laudon and Severson, 1953, p. 526, Pl. 52, figs. 20-22.
  Rhodocrinites watersianus (Wachsmuth and Springer, *in* Miller), 1889. Moore and Laudon, 1944, p. 185, Pl. 73, fig. 8. Kirk, 1944a, p. 15. Strimple and Beane, 1966, p. 35, fig. 1.
  Mississippian, Kinderhookian, Hampton Formation, Gilmore City Limestone, Lodgepole Limestone: United States: Iowa, Montana

Cribanocrinus whitei (Hall), 1861. Kirk, 1944a, p. 15.
  Rhodocrinus whitei Hall, 1861. Branson, 1944, p. 212, Pl. 38, figs. 10, 11. Kirk, 1944a, p. 15.
  Mississippian, Kinderhookian, Chouteau Group, Burlington Limestone: United States: Missouri, Iowa

Cribanocrinus wilsoni Laudon, Parks and Spreng, 1952, p. 563, Pl. 67, figs. 10, 11; Pl. 69, fig. 15.
  Mississippian, Kinderhookian, Banff formation: Canada: Alberta

Cribanocrinus wortheni (Hall), 1858. Kirk, 1944a, p. 15.
  Rhodocrinus wortheni Hall, 1858. Kirk, 1944a, p. 15.
  Rhodocrinites wortheni (Hall), 1858. Moore and Laudon, 1943b, p. 138, Pl. 9, fig. 7; 1944, p. 185, Pl. 73, fig. 11.
  Mississippian, Osagean, Burlington Limestone: United States: Iowa

CRINOBRACHIATUS Moore, 1962. *Myelodactylus brachiatus* Hall, 1852. Moore, 1962b, p. 43.
  Silurian, Niagaran

Crinobrachiatus brachiatus (Hall), 1852. Moore, 1962b, p. 43, Pl. 3, fig. 2.
  Myelodactylus brachiatus Hall, 1852. Cuénot, 1948, p. 60, fig. 76. Ubaghs, 1953, p. 744, fig. 123. Moore, 1962b, p. 43. Yakovlev, 1964b, p. 69, fig. 104.
  Silurian, Niagaran, Rochester Shale: United States: New York

CROMYOCRINUS Trautschold, 1867. *C. simplex* Trautschold, 1867. Moore and Laudon, 1943b, p. 60, fig. 7. Moore, 1962b, p. 15, fig. 10, no. 6.

Arendt and Hecker, 1964, p. 90.
Carboniferous

Cromyocrinus cupoliformis Yakovlev and Ivanov, 1956, p. 18, Pl. 8, fig. 6.
Carboniferous: Russia: Moscow Basin

Cromyocrinus geminatus Trautschold, 1867 = *Mooreocrinus geminatus*

Cromyocrinus geniculatus Yakovlev and Ivanov, 1956, p. 18, Pl. 3, fig. 2.
Arendt and Hecker, 1964, p. 90, Pl. 12, fig. 6.
Carboniferous (C_2pd?): Russia: Moscow Basin

Cromyocrinus nuciformis (McCoy), Strimple, 1961g = *Ureocrinus bockschii*

Cromyocrinus ornatus Trautschold, 1879 = *Dicromyocrinus ornatus*

Cromyocrinus belonging to *C. ornatus* group. Yakovlev, 1944b, p. 116, fig. 1.
Lower Carboniferous: Russia: Talas Alatau

**Cromyocrinus simplex* Trautschold, 1867. Yakovlev, 1939a, p. 65, Pl. 12, figs. 1, 8, 9; 1939c, p. 146, figs. 1-7; 1939d, p. 146, figs. 1-7. Wright, 1945, p. 116, Pl. 3, fig. 20. Melendez, 1947, p. 335, fig. 177. Yakovlev, 1956a, p. 114, figs. 43, nos. 1-3; 45, nos. 1-7. Yakovlev and Ivanov, 1956, p. 15, Pl. 3, figs. 3-5. Yakovlev, 1957, p. 12, figs. 3a, b; 1958, p. 101, figs. 43, nos. 1-3; 45, nos. 1-7. Ivanova, 1958, p. 132, Pl. 11, fig. 4; Pl. 12, fig. 5; Figs. 2, 56b. Arendt, 1961, p. 102, fig. 1. Yakovlev, 1964a, p. 103, figs. 45, 47; 1964b, p. 70, fig. 110. Arendt and Hecker, 1964, p. 90, Pl. 12, figs. 7, 8.
Middle Carboniferous: Russia: Moscow Basin

Cromyocrinus sp. aff.*C. simplex* Trautschold, 1867. Breimer, 1962a, p. 170, Pl. 15, figs. 13-15.
Carboniferous, Moscovian, Cotarasso Limestone: Spain

Cromyocrinus sp. Yakovlev, 1964b, p. 70, fig. 111.
Carboniferous: Russia: Moscow Basin

CROTALOCRINITES Austin and Austin, 1842. **Cyathocrinites rugosus* Miller, 1821. Moore and Laudon, 1943b, p. 50, fig. 4; 1944, p. 153. Moore, 1950, p. 37, fig. 4n. Bouška, 1950, p. 12, figs. 3, 4. Ubaghs, 1953, p. 750. Arendt and Hecker, 1964, p. 83. Moore and Jeffords, 1968, p. 16, fig. 5, no. 4.
Crotalocrinus Cuénot, 1948, p. 64.
Silurian

Crotalocrinites baschkiricus Yakovlev, 1949c, p. 15, Pl. 1, figs. 1-3. Arendt and Hecker, 1964, p. 83, Pl. 8, figs. 6, 7.
Silurian, Wenlockian-Ludlovian transition: Russia: northern Baschkir

Crotalocrinites coniforme Bouška, 1950, p. 13, Pl. 1, figs. 6-9 (*non* figs. 6, 6a, typographical error).
Silurian, Budňany limestone: Bohemia

Crotalocrinites cora (Hall), 1865. Moore and Laudon, 1944, p. 153, Pl. 54, fig. 30.
Silurian, Niagaran, Racine Dolomite: United States: Wisconsin, Illinois

Crotalocrinites pulcher (Hisinger), 1840. Cuénot, 1948, p. 64, fig. 81. Bouška, 1950, p. 22, fig. 7. Ubaghs, 1953, p. 750, fig. 83c.
Silurian, Gotlandian: Sweden

**Crotalocrinites rugosus* (Miller), 1821. Bouska, 1950, p. 12, Pl. 1, figs. 1-5. Ubaghs, 1953, p. 750, figs. 83a, b. Arendt and Hecker, 1964, p. 83, Pl. 8, figs. 8, 9.
Silurian, Wenlockian–Ludlovian transition, Wenlock Limestone: Bohemia; Sweden

Crotalocrinites spp.
  Ubaghs, 1953, p. 750, fig. 83d.
    Silurian
  Ubaghs, 1953, p. 750, fig. 111.
    Silurian

CRYPHIOCRINUS Kirk, 1929. **C. girtyi* Kirk, 1929. Moore and Laudon, 1943b, p. 62.
  Mississippian-Pennsylvanian

Cryphiocrinus bowsheri (Strimple), 1949. Strimple, 1961g, p. 68; 1962g, p. 184, figs. 1-7.
  Mooreocrinus bowsheri Strimple, 1949a, p. 130, Pl. 1, figs. 1-3; 1961g, p. 68.
    Mississippian-Pennsylvanian, Mayes Formation, Hindsville Limestone: United States: Oklahoma

Ctenocrinites Steininger, 1849 = *Melocrinites* Goldfuss, 1826

Ctenocrinus Bronn, 1840 = *Melocrinites* Goldfuss, 1826

Ctenocrinus acicularis Follmann, 1887 = *Melocrinites acicularis*

Ctenocrinus acidepressus Schmidt, 1941 = *Melocrinites acidepressus*

Ctenocrinus aculeatus Ubaghs, 1945 = *Melocrinites aculeatus*

Ctenocrinus acutior Schmidt, 1941 = *Melocrinites acutior*

Ctenocrinus clathratus Schmidt, 1941 = *Melocrinites clathratus*

Ctenocrinus decadactylus (Goldfuss), Schmidt, 1941 = *Melocrinites decadactylus*

Ctenocrinus decadactylus densalternans Schmidt, 1941 = *Melocrinites decadactylus densalternans*

Ctenocrinus decadactylus elegans Jaekel, 1895. Schmidt, 1941, p. 220, nomen nudum

Ctenocrinus cf. *decadactylus* (Goldfuss), Ubaghs, 1945a = *Melocrinites* cf. *decadactylus*

Ctenocrinus gothlandicus (Angelin), Ubaghs, 1953 [sic] = *Melocrinites gottlandicus*

Ctenocrinus aff. *gracilis* Jaekel, 1895 = *Melocrinites* aff. *gracilis*

Ctenocrinus loricatus Schmidt, 1941 = *Melocrinites loricatus*

Ctenocrinus nobilissimus (Hall), 1859 = *Melocrinites nobilissimus*

Ctenocrinus nodiferus Follmann, 1882 = *Melocrinites nodiferus*

Ctenocrinus pachydactylus (Conrad), 1841 = *Melocrinites pachydactylus*

Ctenocrinus paucidactylus (Hall), Moore and Laudon, 1943 = *Melocrinites paucidactylus*

?*Ctenocrinus portezuelensis* Rusconi, 1952 = ?*Melocrinites portezuelensis*

Ctenocrinus pulvinatus Luttig, 1951 = *Melocrinites pulvinatus*

Ctenocrinus pyramidalis Ubaghs, 1945 = *Melocrinites pyramidalis*

Ctenocrinus rhenanoides Schmidt, 1941 = *Melocrinites rhenanoides*

Ctenocrinus rhenanoides demissa Schmidt, 1941 = *Melocrinites rhenanoides demissa*

Ctenocrinus rhenanus Follmann, 1887 = *Melocrinites rhenanus*

Ctenocrinus rhenanus multiplex Schmidt, 1941 = *Melocrinites rhenanus multiplex*

Ctenocrinus rhenanus simplex Schmidt, 1941 = *Melocrinites rhenanus simplex*
Ctenocrinus rhenanus ulcifer Schmidt, 1941 = *Melocrinites rhenanus ulcifer*
Ctenocrinus rugosus Schmidt, 1941 = *Melocrinites rugosus*
Ctenocrinus sculptus Jaekel, 1895, Schmidt, 1941 = *Melocrinites acicularis*
Ctenocrinus signatus Schmidt, 1941 = *Melocrinites signatus*
Ctenocrinus spectabilis (Angelin), Ubaghs, 1958 = *Melocrinites spectabilis*
Ctenocrinus stellifer Follmann, 1887 = *Melocrinites stellifer*
Ctenocrinus typus Bronn, 1840 = *Melocrinites typus*
Ctenocrinus umbraculum Luttig, 1951 = *Melocrinites umbraculum*
Ctenocrinus unnenbergensis Luttig, 1951 = *Melocrinites unnenbergensis*
Ctenocrinus sp. Ubaghs, 1958 = *Melocrinites* sp.
Ctenocrinus sp. indet. Ubaghs, 1945 = *Melocrinites* sp. indet.
Ctenocrinus n. sp. Sieverts-Doreck, 1957 = *Melocrinites* sp. nov.
Ctenocrinus sp. nov. Ubaghs, 1945 = *Melocrinites* sp. nov.
CULICOCRINUS Müller, 1854. *Platycrinus nodosus* Wirtgen and Zeiler, 1854. Moore and Laudon, 1943b, p. 98; 1944, p. 203. Ubaghs, 1953, p. 743. Silurian-Devonian
Culicocrinus inventriosus Schmidt, 1941, p. 43, Pl. 7, fig. 4.
    Devonian, lower Coblentzian: Germany
Culicocrinus inventriosus var. *intemperans* Schmidt, 1941, p. 44, figs. 6A, B.
    Lower Devonian: Germany
Culicocrinus nodosus (Wirtgen and Zeiler), 1854. Schmidt, 1941, p. 37, Pl. 7, figs. 1, 2; fig. 5A (Schmidt attributes authorship of species to Müller).
    Devonian, upper Coblentzian: Germany
Culicocrinus nodosus var. *confluentina* (Müller), 1854. Schmidt, 1941, p. 42, fig. 5G.
    Devonian, upper Coblentzian: Germany
Culicocrinus nodosus var. *inermis* Jaekel, 1895. Schmidt, 1941, p. 41, Pl. 7, fig. 3; Figs. 5B, D.
    Devonian, upper Coblentzian: Germany
Culicocrinus nodosus var. *inermis/virgo* Schmidt, 1941, p. 42, Pl. 7, fig. 8; Fig. 5H.
    Devonian, upper Coblentzian: Germany
Culicocrinus nodosus var. *virgo* Schmidt, 1941, p. 41, figs. 5C, E, F.
    Devonian, upper Coblentzian: Germany
Culicocrinus rhenanus Follmann, 1891. Schmidt, 1941, p. 220, *nomen nudum*.
Culicocrinus rugosus de Koninck, *in* Wirtgen and Zeiler, 1855. Schmidt, 1941, p. 220, *nomen nudum*.
Culicocrinus spinosus Springer, 1926. Moore and Laudon, 1944, p. 203, Pl. 77, fig. 20. Breimer, 1962a, p. 138, fig. 32 (mislabeled as *Pleurocrinus spinosus* in fig. 32).
    Silurian, Niagaran, Beech River Formation: United States: Tennessee
CULMICRINUS Jaekel, 1918. *Poteriocrinus regularis* von Meyer, 1858. Moore and Laudon, 1943b, p. 55; 1944, p. 159, Pl. 52, fig. 21. Cuénot, 1948, p. 62. Arendt and Hecker, 1964, p. 88.
    Lower Carboniferous

Culmicrinus elegans (Wachsmuth and Springer), 1897. Moore and Laudon,
1943b, p. 133, Pl. 4, fig. 8; 1944, p. 159, Pl. 59, fig. 9.
Mississippian, Chesterian: United States: Kentucky

Culmicrinus jeffersonensis Laudon and Severson, 1953, p. 515, Pl. 51,
figs. 29, 30; Pl. 55, fig. 1.
Mississippian, Kinderhookian, Lodgepole Limestone: United States:
Montana

Culmicrinus missouriensis (Shumard), 1857. Moore and Laudon, 1944, p. 159,
Pl. 59, fig. 6. Ubaghs, 1953, p. 692, fig. 33k.
Mississippian, Meramecian, St. Louis Limestone: United States: Missouri

CUPRESSOCRINITES Goldfuss, 1831. *C. crassus* Goldfuss, 1831. Moore and
Laudon, 1943b, p. 64, fig. 8. Ubaghs, 1953, p. 751.
Aviadocrinus Almela and Revilla, 1950, p. 55. Breimer, 1962a, p. 160.
Cupressocrinus Cuénot, 1948, p. 60. Sieverts-Doreck, 1953, p. 82,
fig. 2. Moore and Jeffords, 1968, fig. 4, no. 3.
Devonian

Cupressocrinites abbreviatus Goldfuss, 1839.
Cupressocrinus abbreviatus Sieverts-Doreck, 1953, p. 82, Pl. 4, fig. 10;
Figs. 6, 7. Struve, 1957, p. 369, fig. 13.
Devonian, Eifelian: Germany

Cupressocrinites abbreviatus granulosus Schultze, 1866 = *Cupressocrinites schlotheimi granulosa*

Cupressocrinites assimilis Dubatolova, 1964, p. 26, Pl. 2, figs. 6-8.
Devonian, Eifelian: Russia: Kuznetz Basin

Cupressocrinites crassus Goldfuss, 1831.
Cupressocrinus crassus Melendez, 1947, p. 22, fig. IV. Sieverts-Doreck,
1952, p. 424, fig. 6; 1953, p. 82, Pl. 4, figs. 8, 9; Figs. 1, 3, 4.
Devonian: Germany; Spain

Cupressocrinites elongatus Goldfuss, 1839.
Cupressocrinus elongatus Bergounioux, 1938, p. 63, fig. Sieverts-
Doreck, 1963, p. 239, Pl. 1.
Devonian, Eifelian: Germany; France

Cupressocrinites gracilis Goldfuss, 1831.
Cupressocrinus gracilis Sieverts-Doreck, 1953, p. 82, Pl. 4, fig. 7.
Devonian, Eifelian: Germany

Cupressocrinites inflatus (Schultze), 1867. Breimer, 1962a, p. 163,
Pl. 16, fig. 1; Fig. 39.
Devonian, Portilla Limestone: Spain

Cupressocrinites sampelayoi (Almela and Revilla), 1950. Breimer, 1962a,
p. 160, Pl. 16, figs. 4, 5; Fig. 38.
Aviadocrinus sampelayoi Almela and Revilla, 1950, p. 57, Pl. 3,
figs. 1-3. Sieverts-Doreck, 1953, p. 85, Pl. 4, fig. 11.
Devonian, Eifelian, Portilla Limestone: Spain

Cupressocrinites schlotheimi schlotheimi (Steininger), 1831. Schmidt,
1941, p. 104.
Cupressocrinus schlotheimi Cuénot, 1948, p. 60, fig. 74.
Middle Devonian, Eifelian

Cupressocrinites schlotheimi alta (Schultze), 1866. Schmidt, 1941, p. 104.
Cupressocrinus abbreviatus alta Schultze, 1866. Schmidt, 1941, p. 104.
Devonian: Germany

Cupressocrinites schlotheimi granulosa (Schultze), 1886. Schmidt, 1941,
p. 105.
Cupressocrinus abbreviatus granulosus Schultze, 1866. Schmidt, 1941,
p. 105.
Devonian: Germany

Cupressocrinites schlotheimi var. *minor* (Schultze), 1866. Schmidt, 1941, p. 104.
    Cupressocrinus abbreviatus minor Schultze, 1866. Schmidt, 1941, p. 104.
    Devonian: Germany
Cupressocrinites schlotheimi rectangularis Schmidt, 1941, p. 103, fig. 27.
    Devonian, lower Eifelian: Germany
Cupressocrinites schlotheimi spinosa (Sieverts), 1934. Schmidt, 1941, p. 105.
    Cupressocrinus abbreviatus spinosa Sieverts, 1934. Schmidt, 1941, p. 105.
    Devonian, Eifelian: Germany
Cupressocrinites schlotheimi urogali (Roemer), 1850. Schmidt, 1941, p. 104 (misspelled as *C. schlotheimi urogalli*).
    Devonian, Eifelian: Germany
Cupressocrinites sp. cf. *C. schlotheimi* (Steininger), 1831. Breimer, 1962a, p. 166, Pl. 16, figs. 2, 3.
    Middle Devonian: Spain
Cupressocrinites sp. aff. *C. townsendi* (Konig), 1825. Breimer, 1962a, p. 164, Pl. 16, figs. 6, 7.
    Devonian, Givetian or unidentified strata, Portilla Limestone, Caldas Limestone, unidentified strata: Spain
Cupressocrinites sp. Breimer, 1962a, p. 168.
    Devonian, Givetian, Portilla Limestone: Spain
Cupressocrinus abbreviatus alta Schultze, 1866 = *Cupressocrinites schlotheimi alta*
Cupressocrinus abbreviatus granulosus Schultze, 1866 = *Cupressocrinites schlotheimi granulosa*
Cupressocrinus abbreviatus var. *minor* Schultze, 1866 = *Cupressocrinites schlotheimi* var. *minor*
Cupressocrinus abbreviatus spinosa Sieverts, 1934 = *Cupressocrinites schlotheimi spinosa*
Cupressocrinus impressus McCoy, 1851 = *Zeacrinites impressus*
Cupressocrinus urogali Roemer, 1850 = *Cupressocrinites schlotheimi urogali*
CUPULOCRINUS d'Orbigny, 1849. *Scyphocrinus heterocostalis* Hall, 1847.
    Moore and Laudon, 1943b, p. 52, fig. 5; 1944, p. 155, Pl. 52, fig. 16. Moore, 1950, p. 34, fig. 4g. Ubaghs, 1953, p. 751. Ramsbottom, 1961, p. 12, fig. 2. Moore, 1962b, p. 15, fig. 9, no. 1. Yakovlev, 1964b, p. 65, fig. 90a.
    Ordovician-Silurian
Cupulocrinus conjugans (Billings), 1857. Wilson, 1946, p. 36, Pl. 6, fig. 6.
    Ordovician, Champlainian, Cobourg Limestone: Canada: Ontario
Cupulocrinus gracilis Ramsbottom, 1961, p. 13, Pl. 5, figs. 6, 7.
    Ordovician, Ashgillian, Upper Drummuck Group: England
Cupulocrinus heterobrachialis Ramsbottom, 1961, p. 12, Pl. 5, figs. 1-5.
    Ordovician, Ashgillian, Upper Drummuck Group: England
*Cupulocrinus heterocostalis (Hall), 1847. Moore and Laudon, 1944, p. 155, Pl. 53, fig. 9.
    Ordovician, Mohawkian, Trenton Limestone: United States: New York
Cupulocrinus humilis (Billings), 1857. Moore and Laudon, 1943b, p. 132, Pl. 3, fig. 8; 1944, p. 155, Pl. 53, figs. 7, 25. Wilson, 1946,

p. 36. Ubaghs, 1953, p. 751, figs. 14b, 148, 149.
Ordovician, Mohawkian, Hull Limestone: Canada: Ontario

Cupulocrinus jewetti (Billings), 1859. Moore and Laudon, 1944, p. 155, Pl. 53, fig. 12.
Ordovician, Champlainian, Trenton Limestone: Canada: Ontario

Cupulocrinus kentuckiensis Springer, 1911. Moore, 1962b, p. 24, fig. 13, no. 2.
Ordovician, Champlainian, Curdsville Limestone: United States: Kentucky

Cupulocrinus sepulchrum Ramsbottom, 1961, p. 14, Pl. 4, figs. 8, 9.
Heterocrinus? Huxley and Etheridge, 1878. Ramsbottom, 1961, p. 14.
Ordovician, Caradocian, Gaerfawr Group: England

CUSACRINUS Bowsher, 1955. *Actinocrinus nodobrachiatus* Wachsmuth and Springer, 1890. Bowsher, 1955b, p. 7.
Mississippian, Kinderhookian-Osagean

Cusacrinus arnoldi (Wachsmuth and Springer,*in* Miller), 1889. Bowsher, 1955b, p. 11.
Actinocrinus arnoldi Wachsmuth and Springer, *in* Miller, 1889. Bowsher, 1955b, p. 11.
Mississippian, Kinderhookian, Hampton Formation: United States: Iowa

Cusacrinus asperrimus (Meek and Worthen), 1869. Bowsher, 1955b, p. 11.
Strotocrinus? asperrimus Meek and Worthen, 1869. Bowsher, 1955b, p. 11.
Mississippian, Osagean, Burlington Limestone: United States: Illinois

Cusacrinus bischoffi (Miller and Gurley), 1896. Bowsher, 1955b, p. 11.
Actinocrinus bischoffi Miller and Gurley, 1896. Bowsher, 1955b, p. 11.
Mississippian, Osagean, Burlington Limestone: United States: Iowa

Cusacrinus chloris (Hall), 1861. Bowsher, 1955b, p. 11, Pl. 3, fig. 1.
Actinocrinus chloris Hall, 1861. Kirk, 1943c, p. 346. Bowsher, 1955b, p. 11.
Actinocrinites tenuisculptus Wachsmuth and Springer, 1897 (*non* McChesney, 1841). Moore and Laudon, 1943b, p. 139, Pl. 10, fig. 11. Kirk, 1943b, p. 346. Moore and Laudon, 1944, p. 193, Pl. 77, fig. 14.
Mississippian, Osagean, Burlington Limestone: United States: Iowa

Cusacrinus coelatus (Hall), 1859. Bowsher, 1955b, p. 11.
Actinocrinus coelatus Hall, 1859. Bowsher, 1955b, p. 11.
Mississippian, Osagean, Burlington Limestone: United States: Iowa

Cusacrinus daphne (Hall), 1863. Bowsher, 1955b, p. 12.
Actinocrinus daphne Hall, 1863. Bowsher, 1955b, p. 12.
Lower Mississippian, Waverly Sandstone Group: United States: Ohio

Cusacrinus denticulatus (Wachsmuth and Springer), 1897. Bowsher, 1955b, p. 12.
Cactocrinus denticulatus Wachsmuth and Springer, 1897. Bowsher, 1955b, p. 12.
Mississippian, Osagean, Burlington Limestone: United States: Iowa

Cusacrinus ectypus (Meek and Worthen), 1869. Bowsher, 1955b, p. 12.
Strotocrinus ectypus Meek and Worthen, 1869. Bowsher, 1955b, p. 12.
Mississippian, Osagean, Burlington Limestone: United States: Iowa

Cusacrinus elephantinus (Austin and Austin), 1842 = *Dialutocrinus elephantinus*

Cusacrinus gracilis (Wachsmuth and Springer), 1897. Bowsher, 1955b, p. 12.
Actinocrinus gracilis Wachsmuth and Springer, 1897. Bowsher, 1955b, p. 12.
Mississippian, Osagean, Burlington Limestone: United States: Iowa

Cusacrinus hurdianus (McChesney), 1860. Bowsher, 1955b, p. 12.
   Actinocrinus hurdianus McChesney, 1860. Bowsher, 1955b, p. 12.
   Mississippian, Osagean, Burlington Limestone: United States: Iowa

Cusacrinus imperator Laudon, 1933. Bowsher, 1955b, p. 13.
   Cactocrinus imperator Laudon, 1933. Moore and Laudon, 1944, p. 191, Pl. 76, fig. 8. Laudon and others, 1952, p. 571, Pl. 68, fig. 4. Bowsher, 1955b, p. 13.
   Mississippian, Kinderhookian, Gilmore City Limestone, Banff formation: United States: Iowa; Canada: Alberta

Cusacrinus limabrachiatus (Hall), 1861. Bowsher, 1955b, p. 13, Pl. 4, figs. 2-4.
   Actinocrinus limabrachiatus Hall, 1861. Bowsher, 1955b, p. 13.
   Mississippian, Osagean, Burlington Limestone: United States: Iowa

Cusacrinus longus (Meek and Worthen), 1869. Bowsher, 1955b, p. 14.
   Actinocrinus longus Meek and Worthen, 1869. Bowsher, 1955b, p. 14.
   Mississippian, Osagean, Burlington Limestone: United States: Iowa

*Cusacrinus nodobrachiatus (Wachsmuth and Springer),1890. Bowsher, 1955b, p. 14, Pl. 2, figs. 1, 2; Figs. 3A-C, 4.
   Actinocrinus nodobrachiatus Wachsmuth and Springer, 1890. Bowsher, 1955b, p. 14.
   Mississippian, Kinderhookian, Hampton Formation: United States: Iowa

Cusacrinus ornatissimus (Wachsmuth and Springer), 1890. Bowsher, 1955b, p. 16.
   Actinocrinus ornatissimus Wachsmuth and Springer, 1890. Bowsher, 1955b, p. 16.
   Mississippian, Kinderhookian, Hampton Formation: United States: Iowa

Cusacrinus penicillus (Meek and Worthen), 1869. Bowsher, 1955b, p. 16.
   Actinocrinites penicillus Meek and Worthen, 1860. Bowsher, 1955b, p.16.
   Mississippian, Osagean, Burlington Limestone: United States: Iowa

Cusacrinus spinotentaculus (Hall), 1860. Bowsher, 1955b, p. 16.
   Actinocrinus spino-tentaculus Hall, 1860. Bowsher, 1955b, p. 16.
   Mississippian, Osagean, Burlington Limestone: United States: Iowa

Cusacrinus tenuisculptus (McChesney), 1860. Bowsher, 1955b, p. 16.
   Actinocrinus tenuisculptus McChesney, 1860. Bowsher, 1955b, p. 16.
   Cactocrinus tenuisculptus (McChesney), 1860. Kirk, 1943c, p. 346.
   Mississippian, Osagean, Burlington Limestone: United States: Missouri

Cusacrinus thetis (Hall), 1861. Bowsher, 1955b, p. 16, Pl. 3, figs. 2, 3.
   Actinocrinus thetis Hall, 1861. Bowsher, 1955b, p. 16.
   Mississippian, Osagean, Burlington Limestone: United States: Iowa

Cusacrinus tuberculosus (Wachsmuth and Springer), 1897. Bowsher, 1955b, p. 17.
   Actinocrinus tuberculosus Wachsmuth and Springer, 1897. Bowsher, 1955b, p. 17.
   Mississippian, Osagean, Burlington Limestone: United States: Iowa

Cusacrinus viaticus (White), 1874. Bowsher, 1955b, p. 17, Pl. 5,figs.3,4.
   Actinocrinus viaticus White, 1874. Bowsher, 1955b, p. 17.
   Lower Mississippian, Monte Cristo Limestone: United States: Nevada

CYATHOCRINITES Miller, 1821. *C. planus* Miller, 1821. Moore and Laudon, 1943b, p. 51, figs. 1, 4; 1944, p. 154, Pl. 52, fig. 12. Moore, 1950, fig. 4r. Ubaghs, 1953, p. 750. Moore, 1962b, p. 8, fig. 1, no. 1. Yakovlev, 1964b, p. 56, figs. 57, 59a. Arendt and Hecker, 1964, p. 83. Lane, 1967a, p. 237.
   Cyathocrinus Cuénot, 1948, p. 46, fig. 59. Termier and Termier, 1947, p. 256, Pl. 19, fig. 50. Yakovlev, 1954b, p. 1088, fig. 1a; 1956a,

p. 90, fig. 36, no. 1;   1958, p. 75, fig. 36, no. 1; 1964a, p. 83, fig. 38, no. 1.
Silurian-Permian

Cyathocrinites acinotubus (Angelin), 1878. Ubaghs, 1953, p. 750, figs. 70, 71.
Cyathocrinus acinotubus Angelin, 1878. Sieverts-Doreck, 1952, p. 422, fig. 3; 1964, p. 7, fig. 1.
Silurian: Sweden

Cyathocrinites arboreus Meek and Worthen, 1865 = *Cyathocrinites multibrachiatus*

Cyathocrinites benedicti (Miller), 1892. Van Sant, *in* Van Sant and Lane, 1964, p. 79.
Mississippian, probably Osagean: United States: exact locality in question

Cyathocrinites brevisacculus (Wachsmuth and Springer), 1890. Ubaghs, 1953, p. 750, fig. 21f.
Mississippian, Osagean, Keokuk Limestone: United States: Indiana

Cyathocrinites bursa (Phillips), 1836. Wright, 1952a, p. 125, Pl. 38, fig. 25.
Carboniferous, Tournaisian: England

Cyathocrinites calcaratus (Phillips), 1836. Wright, 1952a, p. 125, Pl. 38, fig. 27.
Carboniferous, Tournaisian: England

Cyathocrinites conicus (Phillips), 1836. Wright, 1952a, p. 126, Pl. 37, figs. 20, 22, 25, 27; Pl. 38, fig. 28.
Carboniferous, Tournaisian: Scotland; England

?*Cyathocrinites insperatus* (Lyon), 1869. Van Sant, *in* Van Sant and Lane, 1964, p. 79.
Mississippian, Osagean: United States: Indiana

Cyathocrinites longimanus (Angelin), 1878. Ubaghs, 1953, p. 750, fig. 14f.
Silurian: Sweden; England

Cyathocrinites mammillaris (Phillips), 1836. Wright, 1952a, p. 124, Pl. 37, figs. 28-30; Pl. 38, fig. 29.
Carboniferous, Tournaisian-Visean: England

Cyathocrinites multibrachiatus (Lyon and Casseday), 1859. Van Sant, *in* Van Sant and Lane, 1964, p. 80, Pl. 2, figs. 2, 3; Pl. 3, figs. 4, 5, 10, 11; Fig. 21, no. 1.
Cyathocrinites arboreus Meek and Worthen, 1865. Van Sant, *in* Van Sant and Lane, 1964, p. 80.
Mississippian, Osagean, Edwardsville Formation: United States: Indiana

Cyathocrinites opimus (Miller and Gurley), 1890. Van Sant, *in* Van Sant and Lane, 1964, p. 81, Pl. 2, fig. 1.
Mississippian, Osagean, Edwardsville Formation: United States: Indiana

Cyathocrinites patulosus (Wright), 1935. Wright, 1952a, p. 127, Pl. 37, figs. 9, 11-13, 15-19, 21, 24, 26; Pl. 38, fig. 34; Figs. 67-70.
Carboniferous, Tournaisian—lower Visean: Scotland

*Cyathocrinites planus Miller, 1821. Wright, 1952a, p. 121, Pl. 36, fig. 13; Pl. 37, figs. 1-8, 10, 14, 23, 31; Pl. 38, fig. 26. Lane, 1967a, p. 237.
Carboniferous, Tournaisian: England; Ireland

Cyathocrinites poterium (Meek and Worthen), 1870. Van Sant, *in* Van Sant and Lane, 1964, p. 82, Pl. 2, fig. 4.
Mississippian, Osagean, Edwardsville Formation: United States: Indiana

Cyathocrinites ramosus (Angelin), 1878. Ubaghs, 1953, p. 750, fig. 21g.
   Cyathocrinus ramosus Angelin, 1878. Sieverts-Doreck, 1952, p. 422, fig. 4.
   Silurian: Sweden

Cyathocrinites robustus Troost, 1849 = *Gaulocrinus robustus*

Cyathocrinites stubblefieldi Wright, 1952a, p. 128, Pl. 36, fig. 15; Pl. 38, figs. 30-33.
   Carboniferous, Tournaisian: Scotland

Cyathocrinites tuberculatus Miller, 1821. Lane, 1967a, p. 237.
   Age not given

Cyathocrinites tumidus (Hall), 1858. Moore and Laudon, 1944, p. 155, Pl. 57, fig. 27.
   Cyathocrinus tumidus Hall, 1858.
   Mississippian, Osagean, Keokuk Limestone: United States: Illinois

Cyathocrinites wilsoni Springer, 1926. Moore and Laudon, 1944, p. 154, Pl. 54, fig. 21.
   Silurian, Laurel Limestone: United States: Indiana

Cyathocrinites spp.
   Laudon and others, 1952, p. 547, Pl. 65, figs. 3-7.
      Mississippian, Kinderhookian, Banff formation: Canada: Alberta
   Laudon and others, 1952, p. 547, Pl. 65, fig. 8.
      Mississippian, Kinderhookian, Banff formation: Canada: Alberta
   Laudon and Severson, 1953, p. 513, Pl. 51, figs. 13, 14.
      Mississippian, Kinderhookian, Lodgepole Limestone: United States: Montana
   Breimer, 1962a, p. 158, Pl. 15, figs. 9, 10.
      Devonian, Emsian, La Vid Shale: Spain

Cyathocrinus acinotubus Angelin, 1878 = *Cyathocrinites acinotubus*

?*Cyathocrinus dekadactylus* Lyon and Casseday, 1860 = *Histocrinus coreyi*

Cyathocrinus loganensis Müller, 1859 = *Sigambrocrinus laevis*

Cyathocrinus? poterium Meek and Worthen, 1870 = *Cyathocrinites poterium*

Cyathocrinus protuberans Hall, 1858 = *Barycrinus protuberans*

Cyathocrinus ramosus Angelin, 1878 = *Cyathocrinites ramosus*

Cyathocrinus somersi Whitfield, 1882 = *Graffhamicrinus somersi*

Cyathocrinus tumidus Hall, 1858 = *Cyathocrinites tumidus*

Cyathocrinus sp. Huxley and Etheridge, 1878 = *Caleidocrinus turgidulus*

CYDONOCRINUS Bather, 1913. *C. parvulus* Bather, 1913. Moore and Laudon, 1943b, p. 50. Arendt and Hecker, 1964, p. 85. Lane, 1967b, p. 11, fig. 4.
   Lower Carboniferous—Permian

Cydonocrinus parvulus Bather, 1913. Wright, 1952a, p. 134.
   Carboniferous, Visean, Lower Middle Limestone: England

CYDROCRINUS Kirk, 1940. *Poteriocrinus coxanus* Worthen, 1862. Moore and Laudon, 1943b, p. 54; 1944, p. 157.
   Mississippian

Cydrocrinus concinnus (Meek and Worthen), 1870. Van Sant, *in* Van Sant and Lane, 1964, p. 83, Pl. 3, fig. 14; Fig. 31, no. 1.
   Mississippian, Osagean, Edwardsville Formation: United States: Indiana

Cydrocrinus coxanus (Worthen), 1882. Moore and Laudon, 1944, p. 157, Pl. 58, fig. 18. Van Sant, *in* Van Sant and Lane, 1964, p. 84, fig. 31, no. 2.
   Mississippian, Osagean, Keokuk Limestone: United States: Illinois

Cydrocrinus subramulosus (Worthen), 1882. Moore and Laudon, 1944, p. 157, Pl. 58, fig. 2.
    Mississippian, Osagean, Keokuk Limestone: United States: Iowa

CYLICOCRINUS Miller, 1892. *C. canaliculatus* Miller, 1892. Moore and Laudon, 1943b, p. 93; 1944, p. 193. Ubaghs, 1953, p. 740.
    Silurian

Cylicocrinus canaliculatus Miller, 1892. Moore and Laudon, 1943b, p.139, Pl. 10, fig. 15; 1944, p. 193, Pl. 75, fig. 16.
    Silurian, Niagaran, Laurel Limestone: United States: Indiana

CYLIOCRINUS Jaekel, 1918. *Melocrinus? rigidus* Angelin, 1878. Moore and Laudon, 1943b, p. 54.
    Silurian

CYMBIOCRINUS Kirk, 1944. *C. romingeri* Kirk, 1944. Kirk, 1944d, p. 233.
    Mississippian, Meramecian-Pennsylvanian, Morrowan

Cymbiocrinus anatonus Washburn, 1968, p. 120, Pl. 1, figs. 4-6.
    Pennsylvanian, Morrowan, Oquirrh Formation: United States: Utah

Cymbiocrinus anomalos (Wetherby), 1880. Kirk, 1944d, p. 236, Pl. 1, fig.5.
    Poteriocrinus anomalos Wetherby, 1880. Kirk, 1944d, p. 236.
    Mississippian, Chesterian, Glen Dean Group: United States: Kentucky

Cymbiocrinus cuneatus Washburn, 1968, p. 121, Pl. 1, figs. 7-9.
    Pennsylvanian, Morrowan, Oquirrh Formation: United States: Utah

Cymbiocrinus dactylus (Hall), 1860. Kirk, 1944d, p. 237.
    Graphiocrinus dactylus Hall, 1860. Kirk, 1944d, p. 237.
    Mississippian, Meramecian, St. Louis Limestone: United States: Missouri

Cymbiocrinus grandis Kirk, 1944d, p. 238, Pl. 1, figs. 6, 7, 10.
    Mississippian, Chesterian, Glen Dean Formation: United States: Kentucky

Cymbiocrinus gravis Strimple, 1951c, p. 18, Pl. 4, figs. 4-6.
    Mississippian, Chesterian, Fayetteville Formation: United States: Oklahoma

Cymbiocrinus lyoni Kirk, 1944d, p. 240, Pl. 1, figs. 11, 12.
    Mississippian, Chesterian, Glen Dean? Group: United States: Kentucky

Cymbiocrinus pitkini Strimple, 1955, p. 14, fig. 1.
    Mississippian, Chesterian, Pitkin Limestone: United States: Oklahoma

Cymbiocrinus romingeri Kirk, 1944d, p. 241, Pl. 1, figs. 1-4, 8, 9.
    Mississippian, Chesterian, Ste. Genevieve Limestone: United States: Alabama

Cymbiocrinus tumidus Kirk, 1944d, p. 242, Pl. 1, figs. 13, 14.
    Mississippian, Chesterian, Glen Dean Formation, Bangor Limestone: United States: Kentucky, Alabama

CYPHOCRINUS Miller, 1892. *C. gorbyi* Miller, 1892. Moore and Laudon, 1943b, p. 84; 1944, p. 187. Ubaghs, 1950, p. 120, fig. 8; 1953, p. 737.
    Silurian

Cyphocrinus gorbyi Miller, 1892. Moore and Laudon, 1944, p. 187, Pl. 72, fig. 12. Ubaghs, 1953, p. 737, fig. 59b.
    Silurian, Niagaran, Laurel Limestone: United States: Indiana

Cyrtocrinus Kirk, 1943 = *Cytidocrinus*

Cyrtocrinus sculptus (Hall), 1858 = *Cytidocrinus sculptus*

CYTIDOCRINUS Kirk, 1944. *Actinocrinus sculptus* Hall, 1858. Kirk, 1944b, p. 85. Ubaghs, 1953, p. 739. Arendt and Hecker, 1964, p. 96.
    Cyrtocrinus Kirk, 1943a, p. 263.
    Mississippian, Osagean

Cytidocrinus sculptus (Hall), 1858. Kirk, 1944b, p. 85. Laudon, 1948, Pl. 3. Brower, 1965, p. 775, Pl. 91, fig. 16.
  Actinocrinus sculptus Hall, 1858. Kirk, 1943a, p. 263.
  Cyrtocrinus sculptus (Hall), 1858. Kirk, 1943a, p. 263, figs. 3, 5.
  Mississippian, Osagean, Burlington Limestone: United States: Iowa

Cytocrinus Roemer, 1860 = *Melocrinites*

Cytocrinus laevis Roemer, 1860 = *Melocrinites laevis*

CYTTAROCRINUS Goldring, 1923. *Platycrinus eriensis* Hall, 1862. Moore and Laudon, 1943b, p. 101; 1944, p. 203. Ubaghs, 1953, p. 743.
  Devonian

Cyttarocrinus eriensis (Hall), 1862. Moore and Laudon, 1943b, p. 143, Pl. 14, fig. 6; 1944, p. 203, Pl. 77, fig. 22. Goldring, 1945, p. 63. Koenig, 1965, p. 406, figs. 4, 5a-p.
  Devonian, Erian, Coeymans Shale, Ludlowville Shale, Moscow Shale: United States: New York

Cyttarocrinus granopexus Schmidt, 1941, p. 35, Pl. 7, figs. 5-7; Pl. 21, fig. 3.
  Devonian, Eifelian, *Orthocrinus* Beds: Germany

Cyttarocrinus jewetti Goldring, 1923. Goldring, 1954, p. 8, Pl. 1, figs. 6, 7. Breimer, 1962a, p. 142, fig. 32. Koenig, 1965, p. 406, figs. 3D-E.
  Devonian, Erian, Hamilton Group, Ludlowville Shale, Moscow Shale: United States: New York

Cyttarocrinus rauffi Schmidt, 1934 = *Hapalocrinus rauffi*

DACTYLOCRINIS Quenstedt, 1876. *Dimerocrinites oligoptilus* Pacht, 1852. Moore and Laudon, 1943b, p. 74, fig. 10; 1944, p. 179, Pl. 67, fig. 5. Ubaghs, 1953, p. 755. Arendt and Hecker, 1964, p. 93.
  Devonian-Mississippian

Dactylocrinus alpena Springer, 1920. Moore and Laudon, 1944, p. 181, Pl. 69, fig. 14.
  Devonian, Erian, Thunder Bay Limestone: United States: Michigan

Dactylocrinus concavus (Rowley), 1893. Moore and Laudon, 1943b, p. 137, Pl. 8, fig. 4; 1944, p. 181, Pl. 68, fig. 1. Branson, 1944, p. 150, Pl. 21, figs. 10, 11.
  Devonian, Senecan, Snyder Creek Shale: United States: Missouri

Dactylocrinus oligoptilus (Pacht), 1852. Yakovlev, 1941a, p. 325, Pl. 1, figs. 1, 2; 1947a, p. 55, Pl. 10, figs. 1, 2. Arendt and Hecker, 1964, p. 93, Pl. 14, fig. 1.
  Devonian, Frasnian, Chudovo Beds: Russia: Shelon River

Dactylocrinus spiniferus Yakovlev, 1941a, p. 326, Pl. 1, figs. 3, 4.
  Devonian, Frasnian, Chudovo Beds: Russia: Kerest River

DAEDALOCRINUS Ulrich, 1924. *D. kirki* Ulrich, 1924. Moore and Laudon, 1943b, p. 29; 1944, p. 145. Ubaghs, 1953, p. 746. Mocre, 1962b, p. 32, fig. 3, no. 1.
  Ordovician

Daedalocrinus kirki Ulrich, 1924. Moore and Laudon, 1944, p. 145, Pl. 52, fig. 7.
  Ordovician, Champlainian: Canada: Ontario

DASCIOCRINUS Kirk, 1939. *Cyathocrinus florealis* Yandell and Shumard, 1847. Moore and Laudon, 1943b, p. 56; 1944, p. 161. Arendt and Hecker, 1964, p. 88.
  Mississippian

Dasciocrinus aulicus Strimple, 1963b, p. 101, Pl. 1, figs. 8, 9; Figs. 1-4.
  Mississippian, Chesterian, Fayetteville Formation: United States: Oklahoma

*Dasciocrinus florealis (Yandell and Shumard), 1847. Moore and Laudon, 1944, p. 161, Pl. 61, fig. 4.
  Mississippian, Chesterian, Glen Dean Formation: United States: Kentucky

DECADOCRINUS Wachsmuth and Springer, 1897. **Poteriocrinus (Scaphiocrinus) scalaris* Meek and Worthen, 1873. Moore and Laudon, 1943b, p. 59; 1944, p. 169, Pl. 52, fig. 17. Moore, 1962b, p. 15, fig. 9, no. 8. Arendt and Hecker, 1964, p. 89.
  Devonian-Pennsylvanian

Decadocrinus bellus (Miller and Gurley), 1890. Moore and Laudon, 1944, p. 169, Pl. 64, fig. 25.
  Scaphiocrinus bellus Miller and Gurley, 1890.
  Mississippian, Osagean, Keokuk Limestone: United States: Indiana

Decadocrinus brazeauensis Laudon, Parks and Spreng, 1952, p. 556, Pl. 65, figs. 36-38; Pl. 69, fig. 7.
  Mississippian, Kinderhookian, Banff formation: Canada: Alberta

Decadocrinus crassidactylus Laudon, 1936. Kesling, 1964b, p. 140.
  Devonian, Erian, Cedar Valley Formation: United States: Iowa

Decadocrinus depressus (Meek and Worthen), 1870. Van Sant, *in* Van Sant and Lane, 1964, p. 91, Pl. 4, figs. 2, 4, 14.
  Scaphiocrinus granuliferus Miller and Gurley, 1890. Van Sant, *in* Van Sant and Lane, 1964, p. 91.
  Ramulocrinus granuliferus (Miller and Gurley), 1890. Laudon and Severson, 1953, p. 520.
  Mississippian, Osagean, Borden Group: United States: Indiana

Decadocrinus fifensis Wright, 1934. Wright, 1951b, p. 59, Pl. 12, fig. 36; Figs. 21-33.
  Carboniferous, Visean, Lower Limestone Group: Scotland

Decadocrinus gregarius (Williams), 1882. Moore and Laudon, 1944, p. 169, Pl. 60, fig. 18.
  Devonian, Senecan, Chemung Formation: United States: New York

Decadocrinus halli (Hall), 1861. Moore and Laudon, 1944, p. 169, Pl. 59, fig. 8.
  Mississippian, Osagean, upper Burlington Limestone: United States: Iowa

Decadocrinus hughwingi Kesling, 1964b, p. 137, Pl. 1, figs. 1-4; Fig. 1.
  Devonian, Erian, Silica Formation: United States: Ohio

Decadocrinus multinodosus Goldring, 1923. Kesling, 1964b, p. 140.
  Devonian, Erian, Moscow Shale: United States: New York

Decadocrinus nereus (Hall), 1862. Moore and Laudon, 1944, p. 169, Pl. 64, fig. 16. Kesling, 1964b, p. 140.
  Devonian, Erian, Moscow Shale: United States: New York

Decadocrinus oaktrovensis Webby, 1961, p. 539, Pl. 67, fig. 1. Kesling, 1964b, p. 140.
  Devonian, Givetian, Ilfracombe Beds: England

Decadocrinus ornatus Goldring, 1954, p. 30, Pl. 4, figs. 6, 7. Kesling, 1964b, p. 140.
  Devonian, Erian, Moscow Shale: United States: New York

Decadocrinus pachydactylus Laudon, 1936. Kesling, 1964b, p. 140.
  Devonian, Erian, Cedar Valley Formation: United States: Iowa

Decadocrinus repertus (Miller and Gurley), Moore and Laudon, 1944 = *Ramulocrinus repertus*

Decadocrinus spinulifer Laudon, 1936. Moore and Laudon, 1944, p. 169, Pl. 60, fig. 12. Kesling, 1964b, p. 140.
Devonian, Erian, Cedar Valley Formation: United States: Iowa

Decadocrinus stewartae Kier, 1952, p. 73, Pl. 3, figs. 4, 5. Kesling, 1964b, p. 140.
Devonian, Erian, Silica Formation: United States: Ohio

Decadocrinus trymensis Wright, 1951b, p. 60, Pl. 13, fig. 7.
Carboniferous, Tournaisian, K_2: Scotland

Decadocrinus tumidulus (Miller and Gurley), 1894, Moore and Laudon, 1943b, p. 134, Pl. 5, fig. 3; 1944, p. 169, Pl. 59, fig. 7. Van Sant, *in* Van Sant and Lane, 1964, p. 92, Pl. 4, figs. 15, 16.
Mississippian, Osagean, Borden Group: United States: Indiana

Decadocrinus vintonensis Thomas, 1920. Kesling, 1964b, p. 140.
Devonian, Erian, Cedar Valley Formation: United States: Iowa

Decadocrinus wrightae Goldring, 1954, p. 32, Pl. 4, figs. 8, 9. Kesling, 1964b, p. 140.
Devonian, Erian, Arkona shale: Canada: Ontario

Decadocrinus? sp. 1 Wright, 1951b, p. 62, Pl. 13, fig. 9.
Carboniferous: Ireland

Decadocrinus? sp. 2 Wright, 1951b, p. 62, Pl. 13, fig. 8.
Carboniferous: Ireland

DELOCRINUS Miller and Gurley, 1890. *Poteriocrinus hemisphericus* Shumard, 1858. Moore and Laudon, 1943b, p. 60, fig. 8; 1944, p. 171, Pl. 62, fig. 25. Moore, 1962b, p. 8, fig. 1, no. 4.
Mississippian-Permian

Delocrinus abruptus Moore and Plummer, 1940. Moore and Laudon, 1944, p. 173, Pl. 62, fig. 33; Pl. 65, fig. 17.
Permian, Wolfcampian, Beattie Formation, Moran Formation: United States: Kansas, Texas

Delocrinus admirensis Strimple, 1963a, p. 69, Pl. 12, figs. 9-12; Fig. 29.
Permian, Wolfcampian, Falls City Limestone: United States: Kansas

Delocrinus aristatus Strimple, 1949 = *Graffhamicrinus aristatus*

Delocrinus armatura Strimple, 1949 = *Graffhamicrinus armatura*

Delocrinus benthobatus Moore and Plummer, 1940. Moore and Laudon, 1944, p. 171, Pl. 65, fig. 29.
Pennsylvanian, Desmoinesian, Mineral Wells Formation: United States: Texas

Delocrinus beyrichi Wanner, 1916 = *Graphiocrinus? beyrichi*

Delocrinus beyrichi var. *nustoiensis* Wanner, 1916 = *Graphiocrinus verbeeki*

Delocrinus bispinosus Moore and Plummer, 1940 = *Graffhamicrinus bispinosus*

Delocrinus brownvillensis Strimple, 1949d, p. 22, Pl. 4, figs. 1-4.
Pennsylvanian, Virgilian, Brownville Limestone: United States: Oklahoma

Delocrinus bullatus Moore and Plummer, 1940 = *Graffhamicrinus bullatus*

Delocrinus densus Strimple, 1963a, p. 70, Pl. 12, figs. 21-24.
Permian, Wolfcampian, Five Point Limestone: United States: Kansas

Delocrinus depressus Wanner, 1916 = *Graphiocrinus? depressus*

Delocrinus doveri Strimple, 1963a, p. 70, Pl. 12, figs. 13-16.
Pennsylvanian, Virgilian, Dover Limestone: United States: Kansas

Delocrinus cf. *D. dubius* Elias, 1957 = *Paradelocrinus dubius*

Delocrinus excavatissimus Wanner, 1916 = *Graphiocrinus? excavatissimus*

Delocrinus expansus Wanner, 1916 = *Oklahomacrinus expansus*

Delocrinus extraneous Strimple, 1949c, p. 25, Pl. 7, figs. 1-3.
  Pennsylvanian, Virgilian, Stull Shale Member: United States: Kansas

Delocrinus granulosus Moore and Plummer, 1940 = *Graffhamicrinus granulosus*

Delocrinus granulosus var. *moniformis* Moore and Plummer, 1940 = *Graffhamicrinus granulosus* var. *moniformis*

Delocrinus granulosus var. *zonatus* Moore and Plummer, 1940 = *Graffhamicrinus granulosus* var. *zonatus*

Delocrinus graphicus Moore and Plummer, 1940 = *Graffhamicrinus graphicus*

**Delocrinus hemisphericus* (Shumard), 1858. Heuer, 1958, p. 41, Pl. 2, fig. 5a.(Plate printed upside down and incorrectly numbered, originally printed by Plummer and Moore, 1921, as Pl. 23, fig. 19.)
  Pennsylvanian, Virgilian, Harpersville Formation: United States: Texas

Delocrinus magnificus Strimple, 1947 = *Graffhamicrinus magnificus*

Delocrinus? malaianus Wanner, 1916. Wanner, 1949a, p. 31.
  Permian: Timor

Delocrinus matheri Moore and Plummer, 1938 = *Endelocrinus matheri*

Delocrinus cf. *D. matheri* Moore and Plummer, 1938. Washburn, 1968, p. 124, Pl. 3, figs. 1-3.
  Pennsylvanian, Morrowan, Oquirrh Formation: United States: Utah

Delocrinus milleri Croneis, 1930, p. 71, Pl. 18, fig. 15.
  Mississippian, Chesterian, Fayetteville Formation: United States: Arkansas

Delocrinus missouriensis Miller and Gurley, 1890. Moore and Laudon, 1944, p. 173, Pl. 62, fig. 24; Pl. 65, fig. 14.
  Pennsylvanian, Missourian: United States: Missouri

Delocrinus nodosarius Strimple, 1939 = *Graffhamicrinus nodosarius*

Delocrinus papulosus Moore and Plummer, 1940 = *Graffhamicrinus papulosus*

Delocrinus parinodosarius Strimple, 1940 = *Endelocrinus parinodosarius*

Delocrinus paucinodus Moore and Plummer, 1940 = *Graffhamicrinus paucinodus*

Delocrinus pictus Moore and Plummer, 1940 = *Graffhamicrinus pictus*

Delocrinus ponderosus Strimple, 1949d, p. 23, Pl. 4, figs. 5-8.
  Pennsylvanian, Virgilian, Brownville Limestone: United States: Oklahoma

Delocrinus quadratus Moore and Plummer, 1940. Moore and Laudon, 1944, p. 173, Pl. 65, fig. 9.
  Permian, Wolfcampian, Cibolo Formation: United States: Texas

Delocrinus rotundatus Wanner, 1916 = *Graphiocrinus? rotundatus*

Delocrinus rugosus Wanner, 1916, non Shumard, 1858 = *Graphiocrinus? rugosus*

Delocrinus separatus Strimple, 1949c, p. 26, Pl. 7, figs. 4-7. Strimple, 1960b, p. 153, fig. 2G.
  Pennsylvanian, Virgilian, Stull Shale: United States: Kansas

Delocrinus serratomarginatus (Yakovlev), 1930. Arendt and Hecker, 1964, p. 89, Pl. 12, fig. 3.
  Ceriocrinus serratomarginatus Yakovlev, 1930. Yakovlev, *in* Yakovlev

and Ivanov, 1956, p. 64, Pl. 11, fig. 11.
Permian, Artinskian: Russia: Central Urals

Delocrinus somersi (Whitfield), 1882 = *Graffhamicrinus somersi*

Delocrinus stullensis Strimple, 1947 = *Graffhamicrinus stullensis*

Delocrinus subcoronatus Moore and Plummer, 1940 = *Graffhamicrinus subcoronatus*

Delocrinus subhemisphericus Moore and Plummer, 1940. Moore and Laudon, 1944, p. 173, Pl. 62, fig. 30; Pl. 65, fig. 7. Strimple, 1949b, p. 123, Pl. 4, figs. 8-15.
Pennsylvanian, Missourian, Iola Limestone, Lane Shale: United States: Kansas, Missouri

Delocrinus aff. *D. subhemisphericus* Moore and Plummer, 1940. Washburn, 1968, p. 125, Pl. 3, figs. 4-6.
Pennsylvanian, Morrowan, Oquirrh Formation: United States: Utah

Delocrinus tumidus Strimple, 1939 = *Endelocrinus tumidus*

Delocrinus vastus Lane and Webster, 1966, p. 45, Pl. 12, figs. 1, 2, 5. Webster and Lane, 1967, p. 24, Pl. 8, fig. 6.
Permian, Wolfcampian, Bird Spring Formation: United States: Nevada

Delocrinus verbeeki Wanner, 1916 = *Graphiocrinus verbeeki*

Delocrinus verbeeki var. *levis* Wanner, 1916 = *Graphiocrinus verbeeki*

Delocrinus verbeeki var. *pumila* Wanner, 1916 = *Graphiocrinus pumilus*

Delocrinus verus Moore and Plummer, 1940. Moore and Laudon, 1944, p. 173, Pl. 65, fig. 6. Strimple, 1960b, p. 153, fig. 2F.
Pennsylvanian, Missourian, Winterset Limestone, Palo Pinto Limestone: United States: Kansas, Texas

Delocrinus vulgatus Moore and Plummer, 1940. Moore and Laudon, 1944, p. 173, Pl. 65, fig. 19.
Pennsylvanian, Virgilian, Brownville Formation, Harpersville Formation: United States: Oklahoma, Texas

Delocrinus waughi Moore and Strimple, 1941 = *Graffhamicrinus waughi*

Delocrinus wolforum Moore and Plummer, 1940 = *Graffhamicrinus wolforum*

Delocrinus sp. Strimple, 1951a, p. 200, Pl. 36, figs. 1-4.
Pennsylvanian, Missourian, Lake Bridgeport Shale: United States: Texas

DELTACRINUS Ulrich, 1886. *Cheirocrinus clarus* Hall, 1862. Moore and Laudon, 1943b, p. 30; 1944, p. 145. Ubaghs, 1953, p. 746, fig. 140K. Moore, 1962a, p. 28, fig. 5, no. 4. Arendt and Hecker, 1964, p. 81.
Silurian-Devonian

Deltacrinus alleni (Rowley), 1904. Moore, 1962a, p. 28.
Calceocrinus alleni Rowley, 1904. Moore, 1962a, p. 28.
Silurian, Medinian, Edgewood Limestone: United States: Missouri

**Deltacrinus clarus* (Hall), 1862. Moore and Laudon, 1944, p. 145, Pl. 57, fig. 8. Moore, 1962a, p. 28, Pl. 1, fig. 7; Fig. 17.
Devonian, Erian, Moscow Shale: United States: New York

Deltacrinus contractus (Ringueberg), 1889. Moore, 1962a, p. 28.
Silurian, Niagaran, Gasport Limestone: United States: New York

Deltacrinus secundus (Hall), 1872. Moore, 1962a, p. 28.
Calceocrinus (Cheirocrinus) secundus Hall, 1872.
Devonian, Ulsterian, Onondaga Limestone, Tully Limestone: United States: New York

Deltacrinus stigmatus (Hall), 1863. Moore, 1962a, p. 28.
  Cheirocrinus stigmatus Hall, 1863. Moore, 1962a, p. 28.
  Silurian, Niagaran, Waldron Shale: United States: Indiana

DENARIOCRINUS Schmidt, 1941. *D. ferula* Schmidt, 1941. Schmidt, 1941, p. 163.
  Devonian

Denariocrinus ferula Schmidt, 1941, p. 165, Pl. 10, fig. 4; Fig. 45.
  Middle Devonian: Germany

DENDROCRINUS Hall, 1852. *D. longidactylus* Hall, 1852. Moore and Laudon, 1943b, p. 52, figs. 1, 5; 1944, p. 155, Pl. 52, fig. 6. Cuénot, 1948, p. 61. Moore, 1950, p. 34, fig. 4b. Ubaghs, 1953, p. 751. Ramsbottom, 1961, p. 15, fig. 4. Moore, 1962b, p. 15, fig. 9, no. 5. Philip, 1963, p. 268, fig. 2a. Arendt and Hecker, 1964, p. 86. Philip, 1965, fig. 2b.
  Ordovician-Silurian

Dendrocrinus acutidactylus Billings, 1857. Wilson, 1946, p. 37.
  Ordovician, Champlainian, Cobourg or Hull Limestone: Canada: Ontario

Dendrocrinus alternatus (Hall), 1847. Welby, 1962, p. 37, Pl. 14, figs. 25, 26.
  Ordovician, Champlainian, Glen Falls Limestone: United States: Vermont

Dendrocrinus cambriensis Hicks, 1873 = *Ramseyocrinus cambriensis*

Dendrocrinus granditubus Ramsbottom, 1961, p. 15, Pl. 4, figs. 1-7.
  Ordovician, Ashgillian, Upper Drummuck Group: England

Dendrocrinus gregarius Billings, 1857. Wilson, 1946, p. 37, Pl. 6, fig. 4.
  Ordovician, Champlainian, Cobourg Limestone: Canada: Ontario

Dendrocrinus longidactylus Hall, 1852. Moore and Laudon, 1944, p. 155, Pl. 52, fig. 6.
  Silurian, Niagaran, Clinton Group: United States: New York

Dendrocrinus proboscidiatus Billings, 1857. Wilson, 1946, p. 37.
  Ordovician, Champlainian, Cobourg Limestone: Canada: Ontario

Dendrocrinus rugocyathus Ramsbottom, 1961, p. 16, Pl. 3, figs. 1-5.
  Ordovician, Ashgillian, Slade Beds: England

Dendrocrinus rusticus Billings, 1857. Wilson, 1946, p. 37.
  Ordovician, Champlainian, Cobourg Limestone: Canada: Ontario

DEOCRINUS Hudson, 1907. *Rhodocrinus asperatus* Billings, 1859. Moore and Laudon, 1943b, p. 82. Ubaghs, 1953, p. 737. Arendt and Hecker, 1964, p. 95.
  Ordovician

DEPAOCRINUS Wanner, 1937. *D. ottowi* Wanner, 1937. Moore and Laudon, 1943b, p. 55.
  Permian

DERBIOCRINUS Wright, 1951. *D. diversus* Wright, 1951. Wright, 1951b, p. 79.
  Carboniferous

Derbiocrinus diversus Wright, 1951b, p. 79, Pl. 14, figs. 19-24.
  Carboniferous, Visean: England

DESMACRIOCRINUS Strimple, 1966. *Kallimorphocrinus weldenensis* Strimple and Koenig, 1956. Strimple, 1966c, p. 109.
  Mississippian, Kinderhookian-Osagean

Desmacriocrinus moreyi (Peck), 1936. Strimple, 1966c, p. 108.
  Coencystis moreyi Peck, 1936 (sic). Strimple, 1966c, p. 108.
  Mississippian, Osagean, Fern Glen Limestone: United States: Missouri

**Desmacriocrinus weldenensis* (Strimple and Koenig), 1956. Strimple, 1966c, p. 109, fig. 2a.
  Kallimorphocrinus weldenensis Strimple and Koenig, 1956, p. 1234, fig. 2, nos. 29-35. Strimple, 1966c, p. 109.
  Mississippian, Kinderhookian, Welden Limestone: United States: Oklahoma

DESMIDOCRINUS Angelin, 1878. *D. pentadactylus* Angelin, 1878. Moore and Laudon, 1943b, p. 93. Ubaghs, 1953, p. 740.
  Silurian-Devonian

Desmidocrinus laurelianus Springer, 1926 = *Carpocrinus laurelianus*

Desmidocrinus macrodactylus Angelin, 1878. Ubaghs, 1953, p. 740, fig. 93 (misspelled as *D. macrodaotylus*); 1956a, p. 548, figs. 15A, B.
  Silurian: Sweden

*Desmidocrinus pentadactylus Angelin, 1878. Ubaghs, 1956a, p. 548, figs. 15C-E.
  Silurian: Sweden

Desmidocrinus springeri (Thomas), 1924. Strimple, 1963d, p. 93, fig. 29.
  Hexacrinites springeri Thomas, 1924. Strimple, 1963d, p. 93.
  Devonian: United States

DIABOLOCRINUS Wachsmuth and Springer, 1897. *D. perplexus* Wachsmuth and Springer, 1897. Moore and Laudon, 1943b, p. 83, fig. 13; 1944, p. 185. Ubaghs, 1953, p. 737. Ramsbottom, 1961, p. 26, fig. 10. Arendt and Hecker, 1964, p. 95.
  Ordovician

Diabolocrinus asperatus? (Miller and Gurley), 1894. Butts, 1941, p. 91, Pl. 89, figs. 4-8.
  Ordovician, Chazyan, Ottosee Limestone: United States: Virginia

Diabolocrinus craigheadensis Ramsbottom, 1961, p. 27, Pl. 8, figs. 5-9.
  Ordovician, Caradocian, Craighead Limestone: England

Diabolocrinus globularis (Nicholson and Etheridge), 1881. Ramsbottom, 1961, p. 28, Pl. 8, figs. 10-20.
  Glyptocrinus globularis Nicholson and Etheridge, 1881. Ramsbottom, 1961, p. 28.
  Ordovician, Caradocian, Craighead Limestone: England

**Diabolocrinus perplexus* Wachsmuth and Springer, 1897. Butts, 1941, p. 91, Pl. 89, figs. 1-3. Moore and Laudon, 1944, p. 185, Pl. 72, fig. 15.
  Ordovician, Chazyan, Ottosee Limestone: United States: Virginia

Diabolocrinus verperalis (White), 1879. Moore and Laudon, 1943b, p. 138, Pl. 9, fig. 4; 1944, p. 185, Pl. 72, fig. 16.
  Ordovician, Chazyan, Ottosee Limestone: United States: Tennessee

Diabolocrinus spp. Ramsbottom, 1961, p. 29, Pl. 4, fig. 10.
  Ordovician, Caradocian-Ashgillian, Craighead Mudstone, Starfish bed: England

DIALUTOCRINUS Wright, 1955. *D. milleri* Wright, 1955. Wright, 1955a, p. 241.
  Lower Carboniferous

Dialutocrinus aculeatus (Austin and Austin), 1842. Wright, 1955a, p. 247, Pl. 58, fig. 4; Pl. 59, figs. 4, 8, 9; Pl. 60, figs. 1-4, 6; Pl. 61, figs. 4, 9; Pl. 65, figs. 1, 2; Pl. 80, fig. 6; Figs. 118, 122.
  Actinocrinites aculeatus Austin and Austin, 1842. Wright, 1955a, p. 247.
  Carboniferous, Tournaisian: Ireland

Dialutocrinus cataphractus (Austin and Austin), 1842. Wright, 1955a,
p. 248, Pl. 59, figs. 1, 5, 6; Pl. 61, fig. 2; Pl. 64, fig. 3;
Pl. 65, fig. 4; Fig 123.
Actinocrinites cataphractus Austin and Austin, 1842. Wright, 1955a,
p. 248.
Carboniferous, Tournaisian: Ireland

Dialutocrinus elephantinus (Austin and Austin), 1842. Wright, 1955a,
p. 250, Pl. 60, fig. 5; Pl. 61, figs. 1, 7; Pl. 64, figs. 1, 2;
Pl. 65, fig. 3.
Actinocrinites elephantinus Austin and Austin, 1842. Wright, 1943b,
p. 231, Pl. 8, figs. 1, 3; 1955a, p. 250.
Cusacrinus elephantinus (Austin and Austin), 1842. Bowsher, 1955b,
p. 12.
Carboniferous, Tournaisian: Ireland

Dialutocrinus? granulatus Wright, 1955b, p. 256, Pl. 65, fig. 7.
Carboniferous, Tournaisian: Ireland

Dialutocrinus icosidactylus (Portlock), 1843. Wright, 1955b, p. 255,
Pl. 65, fig. 8.
Actinocrinites icosidactylus Portlock, 1843. Wright, 1955b, p. 255.
Carboniferous, Tournaisian: Ireland

Dialutocrinus laevissimus (Austin and Austin), 1842. Wright, 1955a,
p. 252, Pl. 59, fig. 7; Pl. 61, figs. 5, 6; Pl. 64, figs. 7, 8;
Figs. 119, 124.
Actinocrinites laevissimus Austin and Austin, 1842. Wright, 1955a,
p. 252.
Carboniferous, Tournaisian: Ireland

Dialutocrinus longispinosus (Austin and Austin), 1842. Wright, 1955a,
p. 254, Pl. 58, fig. 5; Pl. 59, fig. 2; Pl. 61, fig. 3; Pl. 64,
fig. 10; Fig 125.
Actinocrinites longispinosus Austin and Austin, 1842. Wright, 1955a,
p. 254.
Carboniferous, Tournaisian: Ireland

*Dialutocrinus milleri Wright, 1955a, p. 242, Pl. 58, fig. 1; Pl. 64,
fig. 6; Figs. 117, 120.
Actinocrinites triacontadactylus Miller, 1821, part (Pl. i, fig. 1A
only), Wright, 1955a, p. 242.
Carboniferous, Tournaisian: Ireland

Dialutocrinus polydactylus (Miller), 1821. Wright, 1955a, p. 245, Pl. 58,
figs. 2, 3; Fig. 121.
Actinocrinites polydactylus Miller, 1821. Wright, 1955a, p. 245.
Carboniferous, Tournaisian: England

Dialutocrinus tessellatus (Phillips), 1836. Wright, 1955a, p. 244, Pl. 58,
fig. 9; Pl. 62.
Actinocrinus tessellatus Phillips, 1836. Wright, 1955a, p. 244.
Carboniferous, Tournaisian: England

Dialutocrinus spp.
Wright, 1955b, p. 257, Pl. 60, fig. 7.
Carboniferous, Tournaisian: Ireland
Wright, 1958, Pl. 80, fig. 7.
Carboniferous, Tournaisian: Ireland
Wright, 1955b, p. 257, Pl. 59, fig. 3.
Carboniferous, Tournaisian: Ireland
Wright, 1960, p. 334, Appendix, Pl. A, fig. 11.
Carboniferous, Tournaisian: England

DIAMENOCRINUS Oehlert, 1891. *D. jouani* Oehlert, 1891. Moore and Laudon,
1943b, p. 83. Ubaghs, 1953, p. 735. Arendt and Hecker, 1964, p. 95.
Lower Devonian

Diamenocrinus gonatodes (Müller, *in* Zeiler and Wirtgen), 1855. Schmidt, 1941, p. 204, Pl. 11, fig. 2; Pl. 18, fig. 1; Fig. 56. Lehmann, 1955, p. 137, Pl. 15, fig. 6; Pl. 16, figs. 7-10; Pl. 17, fig. 13; Fig. 1.
  Diamenocrinus grandis Jaekel, 1895. Schmidt, 1941, p. 204. Ubaghs, 1947, p. 3.
  Diamenocrinus pachydactylus Jaekel, 1895. Schmidt, 1941, p. 204. Ubaghs, 1947, p. 3.
  Devonian, upper Coblentzian, Singhofener beds: Germany

Diamenocrinus grandis Jaekel, 1895 = *Diamenocrinus gonatodes*

Diamenocrinus sp. cf. *opitzi* Schmidt, 1934. Sieverts-Doreck, 1957, p. 63, fig. 1.
  Devonian, Coblentzian, Herdorf Beds: Germany

Diamenocrinus pachydactylus Jaekel, 1895 = *Diamenocrinus gonatodes*

Diamenocrinus stellatus Jaekel, 1895. Schmidt, 1941, p. 201, Pl. 11, figs. 3, 4; Fig. 55. Ubaghs, 1947, p. 9, Pl. 1, figs. 5-7; Pl. 2, figs. 10-12; Pl. 3, figs. 14, 15; Fig. 1; 1953, p. 704, fig. 62. Lehmann, 1955, p. 138, Pl. 15, fig. 11; Pl. 16, fig. 12.
  Devonian, Coblentzian, Siegener Beds, Singhofener beds: Germany; Belgium

Diamenocrinus spp.
  Termier and Termier, 1950, p. 83, Pl. 219, fig. 4.
  Lower Devonian: Morocco
  Breimer, 1962a, p. 9, Pl. 1, figs. 7-9.
  Devonian, lower Emsian: Spain

DIATOROCRINUS Wright, 1955. *D. angustus* Wright, 1955. Wright, 1955a, p. 235.
  Carboniferous, Tournaisian

Diatorocrinus adversus Wright, 1955a, p. 237, Pl. 53, figs. 8, 15-20.
  Carboniferous, Tournaisian: Scotland

Diatorocrinus anglicus Wright, 1955a, p. 238, Pl. 54, figs. 9-13; Pl. 56, fig. 1.
  Carboniferous, Tournaisian: Scotland

**Diatorocrinus angustus* Wright, 1955a, p. 235, Pl. 53, figs. 4-7.
  Carboniferous, Tournaisian: Scotland

Diatorocrinus venustus Wright, 1955a, p. 236, Pl. 55, figs. 4, 5.
  Carboniferous, Tournaisian: Scotland

DICHOCRINUS Münster, 1839. *D. radiatus* Münster, 1839. Moore and Laudon, 1943b, p. 96, fig. 16; 1944, p. 197, Pl. 71, fig. 17. Moore, 1950, p. 42, fig. 11. Ubaghs, 1953, p. 740. Yakovlev, 1964b, p. 58, fig. 60. Arendt and Hecker, 1964, p. 96.
  Lower Carboniferous

Dichocrinus bozemanensis (Miller and Gurley), 1897. Laudon and others, 1952, p. 574, Pl. 68, fig. 28. Laudon and Severson, 1953, p. 533, Pl. 54, figs. 15, 16 (*non* fig. 10 as given in text).
  Mississippian, Kinderhookian, Lodgepole Formation, Banff formation: United States: Montana; Canada: Alberta

Dichocrinus campto Laudon, 1933. Laudon and Severson, 1953, p. 534, Pl. 54, fig. 17.
  Mississippian, Kinderhookian, Lodgepole Formation: United States: Montana

Dichocrinus delicatus Wachsmuth and Springer, 1897. Moore and Laudon, 1944, p. 197, Pl. 78, fig. 2.
  Mississippian, Kinderhookian, Hampton Formation: United States: Iowa

Dichocrinus douglassi (Miller and Gurley), 1896. Laudon and Severson, 1953, p. 535, Pl. 54, fig. 10.
Mississippian, Kinderhookian, Lodgepole Formation: United States: Montana

Dichocrinus elegans de Koninck and Le Hon, 1854. Moore, 1948, p. 390, fig. 11.
Carboniferous, Tournaisian: Belgium

Dichocrinus elongatus (Phillips), 1836. Wright, 1956b, p. 304, Pl. 75, figs. 14, 15.
Carboniferous, Tournaisian: Ireland; Scotland

Dichocrinus expansus de Koninck and Le Hon, 1854. Moore, 1948, p. 390, fig. 11.
Carboniferous, Tournaisian: Belgium

Dichocrinus expansus Meek and Worthen, 1868 = *Paradichocrinus polydactylus*

Dichocrinus ficus Casseday and Lyon, 1862. Van Sant, *in* Van Sant and Lane, 1964, p. 121.
Mississippian, Osagean, Borden Group, Keokuk Beds, ?Warsaw Formation: United States: Indiana, Kentucky, Missouri

Dichocrinus ficus Worthen and Meek, 1875 = *Dichocrinus* sp.

Dichocrinus fusiformis Austin and Austin, 1844. Wright, 1956b, p. 305, Pl. 75, figs. 3-5, 10, 11.
Carboniferous, Tournaisian: England; Ireland

Dichocrinus cf. *fusiformis* Austin and Austin, 1844. Wright, 1960, p. 334, Appendix, Pl. A, fig. 4.
Carboniferous, Tournaisian: England

Dichocrinus inornatus Wachsmuth and Springer, 1890. Moore and Laudon, 1943b, p. 141, Pl. 12, fig. 14; 1944, p. 197, Pl. 78, fig. 7.
Mississippian, Kinderhookian, Hampton Formation: United States: Iowa

Dichocrinus liratus Hall, 1872. Moore and Laudon, 1944, p. 199, Pl. 78, fig. 4.
Mississippian, Osagean, Burlington Limestone: United States: Iowa, Missouri

Dichocrinus plicatus Hall, 1872. Moore and Laudon, 1944, p. 197, Pl. 78, fig. 5.
Mississippian, Osagean, Burlington Limestone: United States: Iowa, Missouri

Dichocrinus quadriceptatus Laudon and Severson, 1953, p. 534, Pl. 54, figs. 11-14; Pl. 55, fig. 26.
Mississippian, Kinderhookian, Lodgepole Formation: United States: Montana

Dichocrinus radiatus Münster, 1839. Wright, 1956b, p. 303, Pl. 74, figs. 6, 7.
Carboniferous, Tournaisian: Ireland; Belgium

Dichocrinus cf. *radiatus* Münster, 1839. Termier and Termier, 1950, p. 85, Pl. 217, fig. 15.
Carboniferous, Westphalian: Morocco

Dichocrinus recurvibrachiatus Van Sant, *in* Van Sant and Lane, 1964, p. 121, Pl. 8, fig. 16.
Mississippian, Osagean, Borden Group: United States: Indiana

Dichocrinus rotaii Yakovlev, *in* Yakovlev and Ivanov, 1956, p. 47, Pl. 16, fig. 4; Pl. 17, figs. 1, 2; Pl. 18, fig. 1. Arendt and Hecker, 1964, p. 96, Pl. 15, figs. 2, 3.
Carboniferous, Tournaisian: Russia: Kuznetz Basin

Dichocrinus schmidtii (Stuckenberg), 1875. Yakovlev, *in* Yakovlev and Ivanov, 1956, p. 76.
Platycrinus schmidtii Stuckenberg, 1875 (part, Pl. 1, fig. 8; Pl. 2, fig. 3 only), Yakovlev, *in* Yakovlev and Ivanov, 1956, p. 76.
Permian: Russian: Timan

Dichocrinus sculptus de Koninck and Le Hon, 1854. Wright, 1956b, p. 306, Pl. 75, fig. 16.
Carboniferous, Tournaisian: Ireland

Dichocrinus stelliformis Meek and Worthen, 1868 = *Paradichocrinus polydactylus*

Dichocrinus striatus Owen and Shumard, 1850. Moore and Laudon, 1944, p. 197, Pl. 78, fig. 3.
Mississippian, Osagean, Burlington Limestone: United States: Iowa, Missouri

Dichocrinus tomiensis Yakovlev, *in* Yakovlev and Ivanov, 1956, p. 48, Pl. 16, fig. 1.
Lower Carboniferous: Russia: Roiki

Dichocrinus unicus Wright, 1956b, p. 305, Pl. 75, fig. 12.
Carboniferous, Tournaisian: Ireland

Dichocrinus spp.
Laudon, Parks and Spreng, 1952, p. 574, Pl. 68, figs. 21-24.
Mississippian, Kinderhookian, Banff formation: Canada: Alberta
Laudon, Parks and Spreng, 1952, p. 575, Pl. 68, figs. 25, 26.
Mississippian, Kinderhookian, Banff formation: Canada: Alberta
Laudon, Parks and Spreng, 1952, p. 575, Pl. 68, fig. 27.
Mississippian, Kinderhookian, Banff formation: Canada: Alberta
Laudon, Parks and Spreng, 1952, p. 575, Pl. 68, figs. 29, 30.
Mississippian, Kinderhookian, Banff formation: Canada: Alberta
Van Sant, *in* Van Sant and Lane, 1964, p. 122, Pl. 7, fig. 1.
Dichocrinus ficus Worthen and Meek, 1875. Van Sant, *in* Van Sant and Lane, 1964, p. 122.
Mississippian, Osagean, Borden Group: United States: Indiana

Dichocrinus? sp. Termier and Termier, 1950, p. 212, Pl. 214, fig. 9.
Carboniferous, lower Westphalian: Morocco

Dichocrinus or *Camptocrinus?* sp. Wright, 1960, p. 334, Appendix, Pl. A, fig. 3.
Carboniferous, Tournaisian: England

DICHOSTREBLOCRINUS Weller, 1930. **D. scrobiculus* Weller, 1930. Moore and Laudon, 1943b, p. 50; 1944, p. 154. Arendt and Hecker, 1964, p. 85. Lane, 1967b, p. 13, figs. 2, 3, 7 (fig. 7 given as fig. 8).
Mississippian-Permian

Dichostreblocrinus pyriformis Strimple and Koenig, 1956, p. 1243, fig. 4, nos. 10-15.
Mississippian, Osagean, Nunn Member: United States: New Mexico

**Dichostreblocrinus scrobiculus* Weller, 1930. Moore and Laudon, 1944, p. 154, Pl. 56, fig. 7.
Pennsylvanian, Desmoinesian: United States: Indiana

DICROMYOCRINUS Jaekel, 1918. **Cromyocrinus ornatus* Trautschold, 1879. Arendt and Hecker, 1964, p. 90. Strimple, 1966a, p. 5.
Carboniferous

Dicromyocrinus elongatus Yakovlev and Ivanov, 1956, p. 22, Pl. 4, fig. 4.
Carboniferous: Russia: Moscow Basin

Dicromyocrinus geminatus (Trautschold), 1867. Yakovlev and Ivanov, 1956, p. 19, Pl. 4, figs. 1-3; Fig. 3. Arendt and Hecker, 1964, p. 90,

Pl. 12, figs. 11, 12; Fig. 131.
Cromyocrinus geminatus Trautschold, 1867.
Carboniferous, Moscovian: Russia: Moscow Basin

Dicromyocrinus geminatus Wright, 1952 = *Mantikosocrinus wrighti*

Dicromyocrinus geminatus Strimple, 1961g = *Mooreocrinus geminatus*

Dicromyocrinus granularis Easton, 1962, p. 36, Pl. 4, figs. 4, 5.
Mississippian, or Pennsylvanian, Alaska Bench Limestone: United States: Montana

Dicromyocrinus kumpani Yakovlev, *in* Yakovlev and Ivanov, 1956, p. 44 Pl. 12, figs. 1-5.
Carboniferous, Moscovian: Russia: Donetz Basin

Dicromyocrinus mendesi Lane, 1964 = *Mooreocrinus mendesi*

Dicromyocrinus optimus Strimple, 1951 = *Tarachiocrinus optimus*

*Dicromyocrinus ornatus (Trautschold), 1879. Strimple, 1961g, p. 65.
Yakovlev and Ivanov, 1956, p. 20, Pl. 5, figs. 1-6.
Cromyocrinus ornatus Trautschold, 1879. Yakovlev, 1939a, p. 66, Pl. 11, figs. 7-10; Pl. 12, fig. 10.
Mooreocrinus ornatus (Trautschold), 1879. Wright and Strimple, 1945, p. 222.
Carboniferous, Moscovian: Russia: Moscow Basin

Dicromyocrinus ornatus var. *domgeri* Yakovlev, *in* Yakovlev and Ivanov, 1956, p. 43, Pl. 8, fig. 1.
Carboniferous: Russia: Donetz Basin

Dicromyocrinus periodus Strimple, 1951 = *Tarachiocrinus periodus*

Dicromyocrinus subornatus Yakovlev, *in* Yakovlev and Ivanov, 1956, p. 44, Pl. 8, figs. 2-5; Pl. 21, fig. 1. Arendt and Hecker, 1964, p. 90, Pl. 12, fig. 10.
Carboniferous, Moscovian: Russia: Donetz Basin

Dicromyocrinus tapajosi Strimple, 1960c, p. 75, figs. 1-3. Lane, 1964a, p. 364, Pl. 57, figs. 3-5.
Tarachiocrinus tapajosi (Strimple), 1960. Strimple, 1962c, p. 136.
Pennsylvanian, Tapajos Limestone: Brazil

Dicromyocrinus trautscholdi Yakovlev and Ivanov, 1956, p. 21, Pl. 4, figs. 5-8.
Carboniferous, Moscovian-Uralian: Russia: Moscow Basin

Dicromyocrinus wrighti Yakovlev and Ivanov, 1956 = *Mantikosocrinus wrighti*

DICTENOCRINUS Jaekel, 1918. **Botryocrinus decadactylus* Bather, 1891.
Schmidt, 1941, p. 29. Moore and Laudon, 1943b, p. 54. Ubaghs, 1953, p. 751. Arendt and Hecker, 1964, p. 86. Moore and Jeffords, 1968, p. 11, fig. 1, no. 1.
Silurian-Lower Devonian

Dictenocrinus arthriticus Schmidt, 1941, p. 147, Pl. 13, figs. 3, 4.
Lower Devonian, Stadtfeld Beds, Rittersturz Beds: Germany

Dictenocrinus cyathiformis (Haarmann), 1920. Schmidt, 1941, p. 147, Pl. 5, fig. 6.
Botryocrinus cyathiformis Haarmann, 1920. Schmidt, 1941, p. 147.
Devonian, Coblentzian: Germany

*Dictenocrinus decadactylus (Bather), 1891. Ubaghs, 1953, p. 751, fig. 1.
Botryocrinus decadactylus Bather, 1891. Delpey, 1945, p. 253, fig. 1, no. 7. Stukalina, 1967, p. 203, fig. 4, no. 4 (misspelled as *Blothryocrinus*).
Silurian: Sweden

Dictenocrinus ericius (Schmidt), 1934.
  Botryocrinus (Bathericrinus) ericius Schmidt, 1934. Schmidt, 1941, p. 29.
  Lower Devonian: Germany

Dictenocrinus hystrix (Schmidt), 1934.
  Botryocrinus (Bathericrinus) hystrix Schmidt, 1934. Schmidt, 1941, p. 29.
  Lower Devonian: Germany

Dictenocrinus procerus (Haarmann), 1920. Schmidt, 1941, p. 144, Pl. 5, figs. 7, 8; Fig. 38.
  Botryocrinus procerus Haarmann, 1920. Schmidt, 1941, p. 144.
  Devonian, lower Coblentzian: Germany

Dictenocrinus pumilio Schmidt, 1941, p. 148, fig. 39a.
  Devonian, upper Coblentzian: Germany

Dictenocrinus(?) *pusillus* Schmidt, 1941, p. 150, fig. 40.
  Devonian, upper Coblentzian: Germany

Dictenocrinus pygmaeus Schmidt, 1941, p. 150, fig. 39b.
  Devonian, upper Coblentzian: Germany

Dictenocrinus semipinnulatus (Schmidt), 1934.
  Botryocrinus (Bathericrinus) semipinnulatus Schmidt, 1934. Schmidt, 1941, p. 29.
  Lower Devonian, Hunsrück Slate: Germany

Dictenocrinus speciosus (Schmidt), 1934.
  Botryocrinus (Bathericrinus) speciosus Schmidt, 1934. Schmidt, 1941, p. 29 (misspelled as *B. (D.) spaciosus*)
  Lower Devonian, Hunsrück Slate: Germany

DIEURYOCRINUS Wright, 1954. *Euryocrinus duplex* Wright, 1942. Wright, 1954a, p. 165.
  Carboniferous, Tournaisian

Dieuryocrinus duplex (Wright), 1942. Wright, 1954a, p. 165, Pl. 42, figs. 10, 16.
  Euryocrinus duplex Wright, 1942, p. 280, Pl. 9, figs. 2, 3.
  Carboniferous, Tournaisian: Scotland

DIMEROCRINITES Phillips, 1839. *D. decadactylus* Phillips, 1839. Moore and Laudon, 1943b, p. 83, fig. 13; 1944, p. 187. Ubaghs, 1953, p. 737, fig. 139c.
  Silurian-Devonian

Dimerocrinites arborescens (Talbot), 1905. Moore and Laudon, 1943b, p. 138, Pl. 9, fig. 10; 1944, p. 187, Pl. 72, fig. 5.
  Devonian, Helderbergian, Coeymans Limestone: United States: New York

Dimerocrinites inornatus (Hall), 1863. Moore and Laudon, 1944, p. 187, Pl. 72, fig. 4. Ubaghs, 1953, p. 737, fig. 22a.
  Silurian, Niagaran, Waldron Shale: United States: Indiana

Dimerocrinites nodulosus (Hall), 1862.
  Rhodocrinites nodulosus Le Maître, 1952, p. 96, Pl. 22, figs. 10-12; Fig. 8.
  Devonian, Eifelian: Algeria

Dimerocrinites planus (Springer), 1926. Moore and Laudon, 1943b, p. 138, Pl. 9, fig. 11; 1944, p. 187, Pl. 72, fig. 7.
  Silurian, Niagaran, Beech River Formation: United States: Tennessee

Dinocrinus Wanner, 1916 = *Calceolispongia* Etheridge, 1915

Dinocrinus aculeatus Wanner, 1924 = *Calceolispongia aculeata*

Dinocrinus bifurcatus Wanner, 1937 = *Calceolispongia bifurca*

Dinocrinus cornutus Wanner, 1916 = *Calceolispongia cornuta*
Dinocrinus elegans Oyens, 1940 = *Calceolispongia elegans*
Dinocrinus mammeatus Wanner, 1916 = *Calceolispongia mammeata*
Dinocrinus noetlingi Sieverts-Doreck, 1942 = *Calceolispongia noetlingi*
Dinocrinus? sp. nov. Reed, 1928 = *Calceolispongia?* sp. nov.
DINOTOCRINUS Kirk, 1941. *D. compactus* Kirk, 1941. Moore and Laudon, 1943b, p. 56; 1944, p. 161. Arendt and Hecker, 1964, p. 88. Mississippian

Dinotocrinus albertaensis Laudon, Parks and Spreng, 1952, p. 554, Pl. 65, fig. 32; Pl. 69, fig. 5.
Mississippian, Kinderhookian, Banff formation: Canada: Alberta

Dinotocrinus alvestonensis Wright, 1951b, p. 71, Pl. 13, fig. 5.
Carboniferous, Visean: England

Dinotocrinus compactus Kirk, 1941. Moore and Laudon, 1944, p. 161, Pl. 61, fig. 12.
Mississippian, Meramecian, Ste. Genevieve Limestone: United States: Alabama

Dinotocrinus eirensis Wright, 1951b, p. 70, Pl. 13, fig. 1.
Carboniferous, Tournaisian, Supra-dolomite beds: Ireland

Dinotocrinus logani Laudon and Severson, 1953, p. 516, Pl. 51, figs. 23-27; Pl. 55, figs. 12-14.
Mississippian, Kinderhookian, Lodgepole Formation: United States: Montana

Dinotocrinus? pergracilis Austin, *in* Wright, 1951b, p. 72, Pl. 30, fig. 8.
Carboniferous, ?Tournaisian: England

DIPHUICRINUS Moore and Plummer, 1938. *D. croneisi* Moore and Plummer, 1938. Moore and Laudon, 1943b, p. 60.
Pennsylvanian, Morrowan

Diphuicrinus croneisi Moore and Plummer, 1938. Strimple and Knapp, 1966, p. 312, Pl. 36, figs. 1, 2, 15-17, 24-27; Fig. 1A.
Pennsylvanian, Morrowan, Brentwood Limestone, Bloyd Formation, Wapanucka Limestone: United States: Oklahoma

Diphuicrinus patina Strimple and Knapp, 1966, p. 313, Pl. 36, figs. 18-20, 23; Fig. 1B.
Pennsylvanian, Morrowan, Magoffin Beds: United States: Kentucky

DIZYGOCRINUS Wachsmuth and Springer, 1897. *Actinocrinus indianaensis* Lyon and Casseday, 1860. Moore and Laudon, 1943b, p. 95, fig. 16; 1944, p. 197. Ubaghs, 1953, p. 740. Lane, 1963a, p. 698; 1963b, p. 925.
Mississippian, Osagean-Meramecian

Dizygocrinus biturbinatus (Hall), 1858. Lane, 1963a, p. 699.
  Batocrinus abscissus Rowley and Hare, 1891. Lane, 1963a, p. 699.
  Batocrinus burketi Miller and Gurley, 1895. Lane, 1963a, p. 699.
  Batocrinus gurleyi Rowley and Hare, 1891. Lane, 1963a, p. 699.
  Batocrinus sweeti Rowley and Hare, 1891. Lane, 1963a, p. 699.
  Dizygocrinus montgomeryensis unibrachiatus Wachsmuth and Springer, 1897. Lane, 1963a, p. 699.
  Dizygocrinus whitei var. *didactylus* Wachsmuth and Springer, 1897. Lane, 1963a, p. 699.
Mississippian, Osagean, Keokuk Limestone: United States: Illinois, Iowa, Missouri, Kentucky

Dizygocrinus cantonensis Wachsmuth and Springer, 1897, Lane, 1963a, p. 699.
Middle Mississippian, Harrodsburg Limestone: United States: Indiana

Dizygocrinus caroli (Hall), 1860. Lane, 1963a, p. 699.
  Actinocrinus caroli Hall, 1860. Lane, 1963a, p. 699.

Batocrinus venustulus Miller and Gurley, 1895. Lane, 1963a, p. 699.
Batocrinus veterator Miller and Gurley, 1895. Lane, 1963a, p. 699.
Mississippian, Meramecian, Warsaw Limestone: United States: Illinois, Missouri

Dizygocrinus crawfordsvillensis (Miller), 1891. Lane, 1963a, p. 699. Van Sant, *in* Van Sant and Lane, 1964, p. 111.
Batocrinus decrepitus Miller, 1892. Lane, 1963a, p. 699.
Mississippian, Osagean, Edwardsville Formation: United States: Indiana

*Dizygocrinus indianaensis (Lyon and Casseday), 1860. Moore and Laudon, 1943b, p. 140, Pl. 11, fig. 11 (misspelled as *D. indianensis*); 1944, p. 197, Pl. 75, fig. 3. Laudon, 1948, Pl. 2 (misspelled as *D. indianensis*). Lane, 1963a, p. 699, fig. 3D. Van Sant, *in* Van Sant and Lane, 1964, p. 112, Pl. 8, figs. 5-8; Figs. 19, no. 9; 20, nos. 1, 7; 38-40 (*non* Figs. 13, nos. 1, 7; 38; 40, as given on p. 112).
Dizygocrinus indianensis var. *simplex* Wachsmuth and Springer, 1897. Lane, 1963a, p. 699.
Mississippian, Osagean, Keokuk Limestone: United States: Indiana, Iowa, Missouri

Dizygocrinus indianensis var. *simplex* Wachsmuth and Springer, 1897 (sic) = *Dizygocrinus indianaensis*

Dizygocrinus intermedius (Wachsmuth and Springer), 1881. Lane, 1963a, p. 699.
Eretmocrinus intermedius Wachsmuth and Springer, 1881. Lane, 1963a, p. 699.
Batocrinus nitidulus Miller and Gurley, 1895. Lane, 1963a, p. 699.
Eretmocrinus varsoviensis Worthen, 1882. Lane, 1963a, p. 699.
Middle Mississippian, Harrodsburg Limestone, Warsaw Limestone: United States: Indiana, Illinois, Missouri

Dizygocrinus mediocris (Miller), 1891. Lane, 1963a, p. 700.
Batocrinus mediocris Miller, 1891. Lane, 1963a, p. 700.
Batocrinus boonvillensis Miller, 1891. Lane, 1963a, p. 700.
Batocrinus broadheadi Miller and Gurley, 1895. Lane, 1963a, p. 700.
Batocrinus ignotus Miller and Gurley, 1895. Lane, 1963a, p. 700.
Batocrinus modestus Miller and Gurley, 1895. Lane, 1963a, p. 700.
Batocrinus sampsoni Miller and Gurley, 1895. Lane, 1963a, p. 700.
Mississippian, Meramecian, ?Warsaw Limestone: United States: Missouri

Dizygocrinus montgomeryensis (Worthen), 1884. Lane, 1963a, p. 700. Van Sant, *in* Van Sant and Lane, 1964, p. 114, Pl. 7, figs. 9, 11, 13; Figs. 17, no. 2; 19, no. 7.
Batocrinus subconicus Worthen, 1884. Lane, 1963a, p. 700.
Mississippian, Osagean, Edwardsville Formation, Keokuk Limestone: United States: Indiana, Iowa

Dizygocrinus montgomeryensis unibrachiatus Wachsmuth and Springer, 1897 = *Dizygocrinus biturbinatus*

Dizygocrinus mutabilis Wachsmuth and Springer, 1897. Lane, 1963a, p. 700.
Mississippian, Osagean, Edwardsville Formation: United States: Indiana

Dizygocrinus originarius (Wachsmuth and Springer), 1881. Lane, 1963a, p. 700.
Eretmocrinus adultus Wachsmuth and Springer, 1881. Lane, 1963a, p. 700.
Middle Mississippian, Harrodsburg Limestone, ?Warsaw Limestone: United States: Indiana, Missouri

Dizygocrinus parvus (Shumard), 1855 (*non* Rowley, *in* Greene, 1906). Kirk, 1946e, p. 659.
Actinocrinus parvus Shumard, 1855. Kirk, 1946e, p. 659.
Mississippian, Meramecian, St. Louis Limestone: United States: Missouri

Dizygocrinus persculptus Ulrich, 1917 = *Batocrinus unionensis*

Dizygocrinus rotundus (Yandell and Shumard), 1855. Moore and Laudon, 1944, p. 197, Pl. 75, fig. 2.
  Mississippian, Osagean, Burlington Limestone: United States: Iowa, Missouri

Dizygocrinus superstes Ulrich, 1917 = *Batocrinus unionensis*

Dizygocrinus venustus (Miller), 1891. Lane, 1963a, p. 700.
  Batocrinus venustus Miller, 1891. Lane, 1963a, p. 700.
  Batocrinus heteroclitus Miller and Gurley, 1895. Lane, 1963a, p. 700.
  Batocrinus inconsuetus Miller and Gurley, 1895. Lane, 1963a, p. 700.
  Batocrinus inopinatus Miller and Gurley, 1895. Lane, 1963a, p. 700.
  Batocrinus polydactylus Miller and Gurley, 1895. Lane, 1963a, p. 700.
  Batocrinus serratus Miller and Gurley, 1895. Lane, 1963a, p. 700.
  Batocrinus stelliformis Miller and Gurley, 1896. Lane, 1963a, p. 700.
  Mississippian, Osagean, ?Warsaw Limestone: United States: Missouri

Dizygocrinus whitei (Wachsmuth and Springer), 1881. Lane, 1963a, p. 700.
  Van Sant, *in* Van Sant and Lane, 1964, p. 116.
  Batocrinus facetus Miller and Gurley, 1890. Lane, 1963a, p. 700.
  Batocrinus gorbyi Miller, 1891. Lane, 1963a, p. 700.
  Batocrinus pulchellus Miller, 1891. Lane, 1963a, p. 700.
  Middle Mississippian, Harrodsburg Limestone, ?Warsaw Limestone: United States: Indiana, Missouri

Dizygocrinus whitei var. *didactylus* Wachsmuth and Springer, 1897 = *Dizygocrinus biturbinatus*

DOLATOCRINUS Lyon, 1857. *D. lacus* Lyon, 1857. Moore and Laudon, 1943b, p. 99, fig. 17; 1944, p. 201, Pl. 70, fig. 16. Ubaghs, 1953, p. 742. Lane, 1963b, p. 918. Kesling and Mintz, 1963a, p. 71, figs. 1, 2.
  Stereocrinus Barris, 1878. Moore and Laudon, 1943b, p. 99; 1944, p. 201, Pl. 70, fig. 18. Ubaghs, 1953, p. 742. Kesling and Mintz, 1963b, p. 229.
  Lower—Middle Devonian

Dolatocrinus amplus Miller and Gurley, 1894 = *Dolatocrinus stellifer*

Dolatocrinus aplatus Miller and Gurley, 1896 = *Dolatocrinus stellifer*

Dolatocrinus argutus Miller and Gurley, 1896 = *Dolatocrinus bulbaceous*

Dolatocrinus arrosus Miller and Gurley, 1896 = *Dolatocrinus liratus*

Dolatocrinus arrosus var. *cognatus* Rowley, *in* Greene, 1903 = *Dolatocrinus liratus*

Dolatocrinus asper Miller and Gurley, 1896 = *Dolatocrinus liratus*

Dolatocrinus asperatus Springer, 1921 = *Dolatocrinus liratus*

Dolatocrinus asterias Wood, 1904. Lane, 1963b, p. 918, fig. 1.
  Devonian: United States

Dolatocrinus aureatus Miller and Gurley, 1894 = *Dolatocrinus liratus*

Dolatocrinus barrisi (Wachsmuth and Springer), 1897. Kesling and Mintz, 1963b, p. 232, Pl. 2, figs. 7, 8, 11-15.
  Stereocrinus barrisi Wachsmuth and Springer, 1897. Kesling and Mintz, 1963b, p. 232.
  Devonian, Erian, Thunder Bay Limestone: United States: Michigan

Dolatocrinus basilicus Miller and Gurley, 1896 = *Dolatocrinus liratus*

Dolatocrinus bellarugosus Miller and Gurley, 1896 = *Dolatocrinus liratus*

Dolatocrinus bellulus Miller and Gurley, 1895 = *Dolatocrinus liratus*

Dolatocrinus bethaniensis Goldring, 1950, p. 33, Pl. 2, fig. 9.
  Dolatocrinus springeri Goldring, 1936 (*non D. springeri* Rowley, *in*

Greene, 1903). Goldring, 1950, p. 33. Lane, 1963b, p. 920, fig. 2.
Devonian, Erian, Ludlowville Formation, Moscow Formation: United
States: New York

Dolatocrinus bulbaceous Miller and Gurley, 1894. Moore and Laudon, 1944,
p. 201, Pl. 74, fig. 18. Kesling and Mintz, 1963a, p. 78, Pl. 1,
figs. 1-4; Pl. 7, figs. 1-3.
Dolatocrinus argutus Miller and Gurley, 1896. Moore and Laudon, 1944,
p. 201, Pl. 74, fig. 19. Kesling and Mintz, 1963a, p. 78. Lane,
1963b, p. 918, fig. 2.
Devonian, Erian, Hamilton Formation, Dock Street Clay: United States:
Indiana, Kentucky, Michigan

Dolatocrinus cf. *bulbaceus* Miller and Gurley, 1894. Goldring, 1945, p. 62.
Devonian, Erian, Moscow Shale: United States: New York

Dolatocrinus caelatus Miller and Gurley, 1896 = *Dolatocrinus liratus*

Dolatocrinus canadensis Whiteaves, 188/ = *Dolatocrinus liratus*

Dolatocrinus charlestownensis Miller and Gurley, 1896 = *Dolatocrinus liratus*

Dolatocrinus cistula Miller and Gurley, 1896 = *Dolatocrinus liratus*

Dolatocrinus corbuliformis Rowley, *in* Greene, 1903 = *Dolatocrinus liratus*

Dolatocrinus corporosus Miller and Gurley, 1895 = *Dolatocrinus liratus*

Dolatocrinus corporosus var. *concinnus* Rowley, *in* Greene, 1903 = *Dolatocrinus liratus*

Dolatocrinus corporosus var. *decoratus* Rowley, *in* Greene, 1903 = *Dolatocrinus liratus*

Dolatocrinus costatus Wood, 1904 = *Dolatocrinus stellifer*

Dolatocrinus depressus Miller, 1898 = *Dolatocrinus liratus*

Dolatocrinus dispar Miller and Gurley, 1896 = *Dolatocrinus liratus*

Dolatocrinus dissimilaris Miller and Gurley, 1896 = *Dolatocrinus stellifer*

Dolatocrinus elegantulus Rowley, *in* Greene, 1903 = *Dolatocrinus liratus*

Dolatocrinus exornatus Miller and Gurley, 1895 = *Dolatocrinus liratus*

Dolatocrinus fungiferus Rowley, *in* Greene, 1903. Lane, 1963b, p. 920,
fig. 2.
Devonian: United States

Dolatocrinus glyptus Shumard, 1866 = *Dolatocrinus liratus*

Dolatocrinus glyptus var. *intermedius* Shumard, 1866 = *Dolatocrinus liratus*

Dolatocrinus grabaui Kirk, 1946b, p. 267.
Middle Devonian, Potter Farm Formation: United States: Michigan

Dolatocrinus greenei Miller and Gurley, 1894 = *Dolatocrinus liratus*

Dolatocrinus helderbergensis (Springer), 1921, new comb.
Stereocrinus helderbergensis Springer, 1921. Moore and Laudon, 1944,
p. 201.
Lower Devonian, Linden Group: United States: Tennessee

Dolatocrinus icosodactylus Wachsmuth and Springer, 1897 = *Dolatocrinus liratus*

Dolatocrinus incisus Springer, 1921. Kirk, 1946b, p. 267.
Devonian, Erian, Alpena Limestone, ?Dock Street Clay: United States:
Michigan

Dolatocrinus indianensis Miller and Gurley, 1896 = *Dolatocrinus liratus*

**Dolatocrinus lacus* Lyon, 1857. Moore and Laudon, 1944, p. 201, Pl. 74,
fig. 20.
Lower Devonian: United States

IDENTIFIED CROWNS AND PARTS OF CROWNS

Dolatocrinus lineolatus Miller and Gurley, 1894 = *Dolatocrinus liratus*

Dolatocrinus liratus (Hall), 1862. Whitcomb, 1963, p. 39, fig. 1. Kesling and Mintz, 1963a, p. 81, Pl. 1, figs. 17-22; Pl. 2, figs. 1-13; Pl. 3, figs. 1-16. Kesling and Mintz, 1963b, p. 233, Pl. 1, figs. 13-15; Pl. 2, figs. 16, 17.

Dolatocrinus arrosus Miller and Gurley, 1896. Kesling and Mintz, 1963a, p. 84.

Dolatocrinus arrosus var. *cognatus* Rowley, *in* Greene, 1903. Kesling and Mintz, 1963a, p. 84.

Dolatocrinus asper Miller and Gurley, 1896. Kesling and Mintz, 1963a, p. 84. Lane, 1963b, p. 920, fig. 2.

Dolatocrinus asperatus Springer, 1921. Kesling and Mintz, 1963a, p. 85.

Dolatocrinus aureatus Miller and Gurley, 1894. Kesling and Mintz, 1963a, p. 84.

Dolatocrinus basilicus Miller and Gurley, 1896. Kesling and Mintz, 1963a, p. 84. Lane, 1963b, p. 920, fig. 2.

Dolatocrinus bellarugosus Miller and Gurley, 1896. Kesling and Mintz, 1963a, p. 84. Lane, 1963b, p. 919, fig. 2.

Dolatocrinus bellulus Miller and Gurley, 1895. Moore and Laudon, 1944, p. 201, Pl. 74, fig. 21. Kesling and Mintz, 1963a, p. 84.

Dolatocrinus caelatus Miller and Gurley, 1896. Kesling and Mintz, 1963a, p. 84. Lane, 1963b, p. 920, fig. 2.

Dolatocrinus canadensis Whiteaves, 1887. Kesling and Mintz, 1963a, p. 84.

Dolatocrinus charlestownensis Miller and Gurley, 1896. Kesling and Mintz, 1963a, p. 84. Lane, 1963b, p. 920, fig. 2 (misspelled as *D. charlestonensis*).

Dolatocrinus cistula Miller and Gurley, 1896. Kesling and Mintz, 1963a, p. 84. Lane, 1963b, p. 920, fig. 2.

Dolatocrinus corbuliformis Rowley, *in* Greene, 1903. Kesling and Mintz, 1963a, p. 85. Lane, 1963b, p. 920, fig. 2.

Dolatocrinus corporosus Miller and Gurley, 1895. Kesling and Mintz, 1963a, p. 84. Lane, 1963b, p. 920, fig. 2.

Dolatocrinus corporosus var. *concinnus* Rowley, *in* Greene, 1903. Kesling and Mintz, 1963a, p. 85.

Dolatocrinus corporosus var. *decoratus* Rowley, *in* Greene, 1903. Kesling and Mintz, 1963a, p. 85. Lane, 1963b, p. 920, fig. 2.

Dolatocrinus depressus Miller, 1898. Kesling and Mintz, 1963a, p. 84.

Dolatocrinus dispar Miller and Gurley, 1896. Kesling and Mintz, 1963a, p. 84. Lane, 1963b, p. 920, fig. 2.

Dolatocrinus elegantulus Rowley, *in* Greene, 1903. Kesling and Mintz, 1963a, p. 85. Lane, 1963b, p. 920, fig. 2 (as *D. elegantula*).

Dolatocrinus exornatus Miller and Gurley, 1895. Kesling and Mintz, 1963a, p. 84.

Dolatocrinus glyptus Shumard, 1866. Kesling and Mintz, 1963a, p. 84.

Dolatocrinus glyptus var. *intermedius* Shumard, 1866. Kesling and Mintz, 1963a, p. 84.

Dolatocrinus greenei Miller and Gurley, 1894. Kesling and Mintz, 1963a, p. 84. Lane, 1963b, p. 920, fig. 2.

Dolatocrinus icosodactylus Wachsmuth and Springer, 1897. Kesling and Mintz, 1963a, p. 84.

Dolatocrinus indianensis Miller and Gurley, 1896. Kesling and Mintz, 1963a, p. 84. Lane, 1963b, p. 920, fig. 2.

Dolatocrinus lineolatus Miller and Gurley, 1894. Kesling and Mintz, 1963a, p. 84.

?*Dolatocrinus liratus* var. *parvulus* Goldring, 1923. Kesling and Mintz, 1963a, p. 85.

Dolatocrinus lyoni Miller and Gurley, 1896 (*non* Wachsmuth and Springer, 1897). Moore and Laudon, 1944, p. 201, Pl. 74, fig. 17. Kesling and Mintz, 1963a, p. 84.

Dolatocrinus magnificus Miller and Gurley, 1894. Kesling and Mintz,

1963a, p. 84. Lane, 1963b, p. 918, fig. 1 (as *D. magnificus* group).
Dolatocrinus marshi var. *hamiltonenesis* Wachsmuth and Springer, 1897.
   Kesling and Mintz, 1963a, p. 84.
Dolatocrinus multinodosus Rowley, *in* Greene, 1903. Kesling and Mintz,
   1963a, p. 84. Lane, 1963b, p. 920, fig. 2.
Dolatocrinus nodosus Miller and Gurley, 1895. Kesling and Mintz, 1963a,
   p. 84.
Dolatocrinus noduliferous Rowley, *in* Greene, 1903. Kesling and Mintz,
   1963a, p. 84. Lane, 1963b, p. 920, fig. 2.
Dolatocrinus ornatus Meek, 1871. Kesling and Mintz, 1963a, p. 84.
Dolatocrinus ornatus var. *asperatus* Miller and Gurley, 1894. Kesling
   and Mintz, 1963a, p. 84.
Dolatocrinus preciosus Miller and Gurley, 1896. Kesling and Mintz, 1963a,
   p. 84. Lane, 1963b, p. 920, fig. 2.
Dolatocrinus sacculus Miller and Gurley, 1895. Kesling and Mintz, 1963a,
   p. 84.
Dolatocrinus salebrosus Miller and Gurley, 1895. Kesling and Mintz,
   1963a, p. 84.
Dolatocrinus springeri Rowley, *in* Greene, 1903. Kesling and Mintz, 1963a,
   p. 84.
Dolatocrinus tuberculatus Wachsmuth and Springer, 1897. Kesling and
   Mintz, 1963a, p. 84.
Dolatocrinus venustus Miller and Gurley, 1894. Kesling and Mintz, 1963a,
   p. 84. Lane, 1963b, p. 918, fig. 1 (as *D. venustus* group).
Dolatocrinus welleri Rowley, *in* Greene, 1903. Kesling and Mintz, 1963a,
   p. 84. Lane, 1963b, p. 920, fig. 2.
Cacabocrinus glyptus Hall, 1862. Kesling and Mintz, 1963a, p. 84.
Cacabocrinus glyptus var. *intermedius* Hall, 1862. Kesling and Mintz,
   1963a, p. 84.
Middle Devonian: United States

?*Dolatocrinus liratus* var. *parvulus* Goldring, 1923 = *Dolatocrinus liratus*

Dolatocrinus lyoni Miller and Gurley, 1896 = *Dolatocrinus liratus*

Dolatocrinus lyoni Wachsmuth and Springer, 1897 (*non* Miller and Gurley,
   1896) = *Dolatocrinus stellifer*

Dolatocrinus magnificus Miller and Gurley, 1894 = *Dolatocrinus liratus*

Dolatocrinus marshi Lyon, 1869. Moore and Laudon, 1943b, p. 142, Pl. 13,
   fig. 15; 1944, p. 201, Pl. 74, fig. 22.
   Middle Devonian: United States: Kentucky

Dolatocrinus marshi var. *hamiltonensis* Wachsmuth and Springer, 1897 =
   Dolatocrinus liratus

Dolatocrinus michiganensis Kesling and Mintz, 1963a, p. 89, Pl. 4,
   figs. 4-10.
   Devonian, Erian, Dock Street Clay: United States: Michigan

Dolatocrinus multinodosus Rowley, *in* Greene, 1903 = *Dolatocrinus liratus*

Dolatocrinus nodosus Miller and Gurley, 1895 = *Dolatocrinus liratus*

Dolatocrinus noduliferous Rowley, *in* Greene, 1903 = *Dolatocrinus liratus*

Dolatocrinus ornatus Meek, 1871 = *Dolatocrinus liratus*

Dolatocrinus ornatus var. *asperatus* Miller and Gurley, 1894 = *Dolatocrinus
   liratus*

Dolatocrinus peculiaris Miller and Gurley, 1896 = *Dolatocrinus stellifer*

Dolatocrinus pernodosus Rowley, *in* Greene, 1903 = *Dolatocrinus stellifer*

Dolatocrinus preciosus Miller and Gurley, 1896 = *Dolatocrinus liratus*

Dolatocrinus sacculus Miller and Gurley, 1895 = *Dolatocrinus liratus*

Dolatocrinus salebrosus Miller and Gurley, 1895 = *Dolatocrinus liratus*

Dolatocrinus speciosus (Hall), 1862. Moore and Laudon, 1943b, p. 142,
  Pl. 13, fig. 14; 1944, p. 201, Pl. 74, fig. 16.
Middle Devonian: United States: New York

Dolatocrinus spinosus Miller and Gurley, 1894. Ubaghs, 1953, p. 742,
  fig. 91.
Lower-Middle Devonian: United States

Dolatocrinus springeri Rowley, *in* Greene, 1903 = *Dolatocrinus liratus*

Dolatocrinus springeri Goldring, 1936, *non* Rowley, 1903 = *Dolatocrinus bethaniensis*

Dolatocrinus springeri (Branson and Wilson), 1922, new comb. (This is an invalid combination as the name is twice preoccupied. However, the placement of *Stereocrinus* in the synonymy of *Dolatocrinus* requires the transfer of the species previously recognized under *Stereocrinus* to *Dolatocrinus*. Because *Stereocrinus springeri* may be a junior synonym of a recognized species of *Dolatocrinus*, no new name is proposed.)
Stereocrinus springeri Branson and Wilson, 1922. Branson, 1944, p. 130,
  Pl. 21, figs. 1-3.
Devonian, Erian, Mineola Limestone: United States: Missouri

Dolatocrinus stellifer Miller and Gurley, 1894. Kesling and Mintz, 1963a,
  p. 81, Pl. 1, figs. 5-14; Pl. 2, figs. 14-18; Pl. 4, figs. 11-20;
  Pl. 5, figs. 1-12; Pl. 6, figs. 13-24; Pl. 7, figs. 4-6; 1963b,
  p. 233, Pl. 1, figs. 10-12. Lane, 1963b, p. 918, fig. 1.
Dolatocrinus amplus Miller and Gurley, 1894. Kesling and Mintz, 1963a,
  p. 81. Lane, 1963b, p. 918, figs. 1, 2.
Dolatocrinus aplatus Miller and Gurley, 1896. Lane, 1963b, p. 920,
  fig. 2.
Dolatocrinus costatus Wood, 1904. Kesling and Mintz, 1963a, p. 81.
Dolatocrinus dissimilaris Miller and Gurley, 1896. Lane, 1963b, p. 920,
  fig. 2.
Dolatocrinus lyoni Wachsmuth and Springer, 1897. Cuénot, 1948, p. 52,
  fig. 64. Kesling and Mintz, 1963a, p. 81. Yakovlev, 1964b, p. 65,
  fig. 89.
Dolatocrinus peculiaris Miller and Gurley, 1896. Kesling and Mintz,
  1963a, p. 81. Lane, 1963b, p. 920, fig. 2.
Dolatocrinus pernodosus Rowley, *in* Greene, 1903. Kesling and Mintz,
  1963a, p. 81. Lane, 1963b, p. 920, fig.2.
Dolatocrinus vasculum Miller and Gurley, 1895. Kesling and Mintz, 1963a,
  p. 81. Lane, 1963b, p. 920, fig. 2.
Dolatocrinus wachsmuthi Wood, 1904. Kesling and Mintz, 1963a, p. 81.
Devonian, Erian, Dock Street Clay, numerous other formations: United
  States: Michigan, Indiana, Kentucky

Dolatocrinus triadactylus Barris, 1885 = *Dolatocrinus triangulatus*

Dolatocrinus triangulatus (Barris), 1878. Kesling and Mintz, 1963b, p. 231,
  Pl. 1, figs. 1-9; Pl. 2, figs. 1-6, 9, 10.
Stereocrinus triangulatus Barris, 1878. Moore and Laudon, 1944, p. 201,
  Pl. 74, fig. 23. Kesling and Mintz, 1963b, p. 231.
Stereocrinus triangulatus liratus Barris, 1878. Kesling and Mintz,
  1963b, p. 231.
Dolatocrinus triadactylus Barris, 1885. Kesling and Mintz, 1963a,
  p. 88, Pl. 1, figs. 15, 16; Pl. 4, figs. 1-3. Lane, 1963b, p. 918,
  fig. 1.
Devonian, Erian, Thunder Bay Limestone, Dock Street Clay, Cedar Valley
  Formation: United States: Michigan, Iowa

Dolatocrinus tuberculatus Wachsmuth and Springer, 1897 = *Dolatocrinus liratus*

Dolatocrinus vandiveri (Branson and Wilson), 1922, new comb.
   Stereocrinus vandiveri Branson and Wilson, 1922. Branson, 1944, p. 130, Pl. 21, figs. 7, 8.
   Devonian, Erian, Mineola Limestone: United States: Missouri

Dolatocrinus vasculum Miller and Gurley, 1895 = *Dolatocrinus stellifer*

Dolatocrinus venustus Miller and Gurley, 1894 = *Dolatocrinus liratus*

Dolatocrinus wachsmuthi Wood, 1904 = *Dolatocrinus stellifer*

Dolatocrinus welleri Rowley, in Greene, 1903 = *Dolatocrinus liratus*

Dolatocrinus spp.
   Ubaghs, 1953, p. 742, fig. 92.
      Middle Devonian: United States: Kentucky
   La Rocque and Marple, 1955, p. 94, fig. 233.
      Devonian: United States: Ohio
   Kesling and Mintz, 1963b, p. 234, Pl. 2, figs. 18, 19.
      Middle Devonian, ?Alpena Limestone: United States: Michigan

DORYCRINUS Roemer, 1854. *D. mississippiensis* Roemer, 1854. Moore and Laudon, 1943b, p. 93, fig. 16; 1944, p. 195, Pl. 71, fig. 9. Ubaghs, 1953, p. 740.
   Mississippian

Dorycrinus devonicus Springer, 1911 = *Stamnocrinus devonicus*

Dorycrinus mississippiensis Roemer, 1854. Moore and Laudon, 1944, p. 195, Pl. 76, fig. 6. Laudon, 1948, Pl. 2.
   Mississippian, Osagean, Keokuk Limestone: United States: Indiana, Iowa, Missouri

Dorycrinus missouriensis (Shumard), 1855. Moore and Laudon, 1943b, p. 141, Pl. 12, fig. 6; 1944, p. 195, Pl. 76, fig. 7. Laudon, 1948, Pl. 2.
   Mississippian, Osagean, upper Burlington Limestone: United States: Indiana, Iowa, Missouri

Dorycrinus quinquelobus (Hall), 1860. Laudon, 1948, Pl. 2 (misspelled as *D. quienquelobus*).
   Mississippian, Osagean, Burlington Limestone: United States: Iowa

Dorycrinus unicornis (Owen and Shumard), 1850. Moore and Laudon, 1943b, p. 141, Pl. 12, fig. 5; 1944, p. 195, Pl. 76, fig. 5. Laudon, 1948, Pl. 2.
   Mississippian, Osagean, lower Burlington Limestone: United States: Iowa, Missouri

DRYMOCRINUS Ulrich, 1924. *Heterocrinus geniculatus* Ulrich, 1879. Moore and Laudon, 1943b, p. 29; 1944, p. 145. Ubaghs, 1953, p. 746. Moore, 1962b, p. 12, fig. 3, no. 6.
   Ordovician

Drymocrinus geniculatus (Ulrich), 1879. Moore and Laudon, 1944, p. 145, Pl. 52, fig. 7. Brower, 1966, p. 617, fig. 1I.
   Ordovician, Cincinnatian, Eden Formation: United States: Ohio

DYSTACTOCRINUS Ulrich, 1924. *Heterocrinus (Homocrinus) constrictus* Hall, 1867. Moore and Laudon, 1943b, p. 31, fig. 1; 1944, p. 149. Ubaghs, 1953, p. 747. Moore, 1962b, p. 14, fig. 5, no. 1. Philip, 1963, p. 267, fig. 3d; 1965a, fig. 1e.
   Ordovician

Dystactocrinus constrictus(Hall), 1867. Moore and Laudon, 1944, p. 149, Pl. 52, fig. 11. Moore, 1962b, p. 14, Pl. 1, fig. 1.
   Ordovician, Cincinnatian, Fairview Formation: United States: Ohio

ECTENOCRINUS Miller, 1889. **Heterocrinus simplex* Hall, 1847. Moore and Laudon, 1943b, p. 29, fig. 3; 1944, p. 145, Pl. 52, fig. 7. Moore, 1950, figs. 1e, f. Ubaghs, 1953, p. 746. Moore, 1962a, p. 9, fig. 4. Philip, 1963, fig. 3b.
Ordovician—Early Silurian

Ectenocrinus canadensis (Billings), 1857. Wilson, 1946, p. 32, Pl. 6, fig. 8. Moore, 1962b, p. 31, fig. 3, nos. 2b, c.
Ordovician, Champlainian, Cobourg Limestone, Hull Limestone: Canada: Ontario

Ectenocrinus grandis (Meek), 1873. Moore, 1962b, p. 24, Pl. 1, figs. 2b, c.
Ordovician, Cincinnatian: Ohio

**Ectenocrinus simplex* (Hall), 1847. Butts, 1941, p. 114, Pl. 96, figs. 24, 25. Moore and Laudon, 1943b, p. 130, Pl. 1, fig. 5; 1944, p. 145, Pl. 52, fig. 7; Pl. 53, fig. 8. Moore, 1962b, p. 31, Pl. 1, fig. 2a; Fig. 3, no. 2a. Jillson, 1963, p. 4, fig. 2.
Ordovician, Cincinnatian, Eden Formation, Martinsburg Shale: United States: Ohio, Virginia

ECTOCRINUS Wright, 1955. **Actinocrinus (Amphoracrinus?) olla* McCoy, 1851. Wright, 1955a, p. 204.
Carboniferous, Visean

Ectocrinus expansus Wright, 1955a, p. 206, Pl. 52, figs. 6, 15.
Carboniferous, Visean: England

Ectocrinus macneanensis Wright, 1955a, p. 207, Pl. 52, fig. 9.
Carboniferous, Visean: Ireland

**Ectocrinus olla* (McCoy), 1851. Wright, 1955a, p. 204, Pl. 52, figs. 1-5, 7, 8, 10.
Actinocrinus (Amphoracrinus?) olla McCoy, 1849 [sic]. Wright, 1955a, p. 204.
Carboniferous, Visean: England

EDAPOCRINUS Wright, 1935. **E. rugosus* Wright, 1935. Moore and Laudon, 1943b, p. 50. Lane, 1967b, p. 11, fig. 4.
Edaphocrinus Arendt and Hecker, 1964, p. 85.
Lower Carboniferous

**Edapocrinus rugosus* Wright, 1935. Wright, 1952a, p. 129, Pl. 38, figs. 1-4, 10, 18, 22; Figs. 71-74.
Carboniferous, Tournaisian: Scotland

EDRIOCRINUS Hall, 1858. **E. pocilliformis* Hall, 1859. Moore and Laudon, 1943b, p. 102; 1944, p. 205. Cuénot, 1948, p. 67. Moore and Jeffords, 1968, p. 11, fig. 3, no. 2.
Devonian

Edriocrinus dispansus Kirk, 1911. Moore and Laudon, 1944, p. 205, Pl. 79, fig. 33. Strimple, 1963d, p. 125, Pl. 10, figs. 5, 6.
Devonian, Helderbergian, Linden Group, Haragan Formation: United States: Tennessee, Oklahoma

Edriocrinus holopoides Springer, 1920. Le Maître, 1958b, p. 143, Pl. 3, fig. 8.
Devonian, Ulsterian, Oriskany Sandstone: United States: Maryland

**Edriocrinus pocilliformis* Hall, 1859. Goldring, 1943, p. 164, fig. 30f.
Devonian, Helderbergian, New Scotland Limestone: United States: New York

Edriocrinus cf. *pocilliformis* Hall, 1859. Le Maître, 1958b, p. 143, Pl. 2, figs. 7-9.
Devonian, Couvinian: southern Algeria

Edriocrinus pyriformis Hall, 1862. Moore and Laudon, 1944, p. 205, Pl. 79, fig. 28.
Devonian, Helderbergian, Coeymans Limestone: United States: New York

Edriocrinus sacculus Hall, 1859. Moore and Laudon, 1944, p. 205, Pl. 79, fig. 27. Cuénot, 1948, p. 67, fig. 89.
Devonian, Ulsterian, Oriskany Sandstone: United States: New York, Maryland

EIFELOCRINUS Wanner, 1916. *Ptilocrinus dohmi* Wanner, 1916. Moore and Laudon, 1943b, p. 54.
Devonian

Eifelocrinus bifurcatus Haarmann, 1920. Schmidt, 1941, p. 170, Pl. 5, fig. 1; Fig. 46e.
Devonian, Coblentzian: Germany

Eifelocrinus dohmi (Wanner), 1916. Schmidt, 1941, p. 170, Pl. 5, figs. 2-5; Fig. 46a. Cuénot, 1948, p. 34, fig. 43. Yakovlev, 1964b, p. 67, fig. 97 (misspelled as *E. bohmi*).
Devonian, Coblentzian, Stadtfeld Beds: Germany

Eifelocrinus hefteri Schmidt, 1941, p. 171, Pl. 17, fig. 1, Fig. 46b.
Devonian, Coblentzian: Germany

Eifelocrinus pennula Schmidt, 1941, p. 175, Pl. 13, fig. 6; Fig. 46d.
Devonian, Coblentzian: Germany

Eifelocrinus pulvinatus Schmidt, 1941, p. 172, Pl. 15, figs. 1, 2; Pl. 21, figs. 1, 2; Figs. 46c, 47.
Devonian, Coblentzian: Germany

Eifelocrinus rhenanus (Jaekel), 1921. Schmidt, 1941, p. 176, fig. 48b.
Homocrinus rhenanus Jaekel, 1921. Schmidt, 1941, p. 176.
Devonian, Coblentzian: Germany

Eifelocrinus sp. indet. Schmidt, 1941, p. 178.
Devonian, Coblentzian: Germany

EIREOCRINUS Wright, 1951. *E. ornatus* Wright, 1951. Wright, 1951b, p. 102.
Carboniferous, Tournaisian

Eireocrinus ornatus Wright, 1951b, p. 102, Pl. 31, fig. 2.
Carboniferous, Tournaisian: Ireland

ELIBATOCRINUS Moore, 1940. *E. leptocalyx* Moore, 1940. Moore and Laudon, 1943b, p. 55; 1944, p. 159. Arendt and Hecker, 1964, p. 88.
Pennsylvanian, Desmoinesian-Permian, Wolfcampian

Elibatocrinus catactus Moore, 1940. Moore and Laudon, 1944, p. 159, Pl. 57, fig. 25.
Pennsylvanian, Virgilian, Oread Limestone: United States: Kansas

Elibatocrinus elongatus Webster and Lane, 1967, p. 16, Pl. 5, figs. 1, 2.
Elibatocrinus sp. Lane and Webster, 1966, p. 30, Pl. 10, fig. 7.
Permian, Wolfcampian, Bird Spring Formation: United States: Nevada

Elibatocrinus hoodi Strimple, 1961g, p. 132, Pl. 16, figs. 4, 5; Fig. 23.
Pennsylvanian, Desmoinesian, Holdenville Shale: United States: Oklahoma

Elibatocrinus leptocalyx Moore, 1940. Moore and Laudon, 1944, p. 159, Pl. 57, fig. 28; Pl. 62, fig. 23.
Pennsylvanian, Missourian, Stanton Limestone: United States: Kansas

ELPIDOCRINUS Strimple, 1963. *E. tholiformis* Strimple, 1963. Strimple, 1963d, p. 77.
Silurian, Niagaran

Elpidocrinus exiguus Strimple, 1963d, p. 79, Pl. 4, figs. 1-3; Fig. 21a.
Silurian, Niagaran, Henryhouse Formation: United States, Oklahoma
*Elpidocrinus tholiformis Strimple, 1963d, p. 81, Pl. 5, figs. 4-7; Fig. 22.
Silurian, Niagaran, Henryhouse Formation: United States: Oklahoma
Elpidocrinus tuberosus Strimple, 1963d, p. 78, Pl. 4, figs. 4-7; Figs. 21b, c.
Silurian, Niagaran, Henryhouse Formation: United States: Oklahoma
EMBRYOCRINUS Wanner, 1916. **E. hanieli* Wanner, 1916. Moore and Laudon, 1943b, p. 50. Yakovlev, 1947d, p. 44, fig. 8. Cuénot, 1948, p. 64. Yakovlev, 1956a, p. 80, fig. 28; 1958, p. 65, fig. 28; 1964a, p. 75, fig. 30. Arendt and Hecker, 1964, p. 85. Lane, 1967b, p. 11, figs. 4, 5. Arendt, 1968a, p. 101, fig. 5.
Permian
*Embryocrinus hanieli Wanner, 1916. Cuénot, 1948, p. 64, fig. 83, nos. 2, 3. Ubaghs, 1953, p. 672, figs. 10b, i. Yakovlev, 1964b, p. 56, fig. 53b. Arendt, 1968a, p. 100, fig. 1a.
Permian: Timor

EMPEROCRINUS Miller and Gurley, 1895. **E. indianensis* Miller and Gurley, 1895. Moore and Laudon, 1943b, p. 82; 1944, p. 185. Ubaghs, 1953, p. 737. Arendt and Hecker, 1964, p. 95.
Silurian

*Emperocrinus indianensis Miller and Gurley, 1895. Moore and Laudon, 1944, p. 185, Pl. 72, fig. 14.
Silurian, Niagaran, Laurel Limestone: United States: Indiana

ENALLOCRINUS d'Orbigny, 1849. **Apiocrinites scriptus* Hisinger, 1828. Moore and Laudon, 1943b, p. 50. Bouška, 1950, p. 12, fig. 1. Ubaghs, 1953, p. 750. Arendt and Hecker, 1964, p. 84.
Silurian

Encrinites loricatus Schlotheim, 1820 = *Sampsonocrinus loricatus*

ENDELOCRINUS Moore and Plummer, 1940. **Eupachycrinus fayettensis* Worthen, 1873. Moore and Laudon, 1943b, p. 60; 1944, p. 173. Yakovlev, 1947b, p. 749, fig. 2; 1947d, p. 45, fig. 12; 1956a, p. 82, fig. 32; 1958, p. 67, fig. 32; 1964a, p. 76, fig. 34. Arendt and Hecker, 1964, p. 90.
Corythocrinus Strimple, 1961g (*non* Kirk, 1946), p. 128. Webster and Lane, 1967, p. 24.
Tholiacrinus Strimple, 1962d, p. 136. Webster and Lane, 1967, p. 24.
Pennsylvanian, Morrowan-Permian, Wolfcampian

Endelocrinus allegheniensis (Burke), 1932. Moore and Laudon, 1944, p. 173, Pl. 65, fig. 5.
Pennsylvanian, Conemaughian: United States: Pennsylvania, West Virginia

Endelocrinus bifidus Moore and Plummer, 1940. Moore and Laudon, 1944, p. 173, Pl. 65, fig. 2.
Corythocrinus bifidus (Moore and Plummer), 1940. Strimple, 1961g, p. 129.
Tholiacrinus bifidus (Moore and Plummer), 1940. Strimple, 1962d, p. 136.
Pennsylvanian, Missourian, Graford Formation: United States: Texas

Endelocrinus bransoni Strimple, 1962b, p. 28, figs. 1-3.
Pennsylvanian, Desmoinesian, Perry Farm Shale: United States: Oklahoma

*Endelocrinus fayettensis (Worthen), 1873. Moore and Laudon, 1944, p. 173, Pl. 62, fig. 20; Pl. 65, fig. 4.
Pennsylvanian, Missourian: United States: Illinois

Endelocrinus grafordensis Moore and Plummer, 1940. Moore and Laudon, 1944, p. 173, Pl. 65, fig. 3.
Pennsylvanian, Missourian, Stanton Limestone, Graford Formation: United States: Kansas, Oklahoma, Texas

Endelocrinus kieri Burke, 1966a, p. 459, figs. 1-12.
Pennsylvanian, Conemaughian, Ames Limestone: United States: Ohio, West Virginia

Endelocrinus matheri (Moore and Plummer), 1937. Strimple, 1960b, p.˚151, figs. 1, 2A.
Delocrinus matheri Moore and Plummer, 1937. Moore and Laudon, 1944, p. 171, Pl. 65, fig. 11.
Pennsylvanian, Morrowan, Brentwood Limestone: United States: Oklahoma

Endelocrinus murrysvillensis Burke, 1967a, p. 75, figs. 1-2.
Pennsylvanian, Conemaughian, Brush Creek Limestone: United States: Pennsylvania

Endelocrinus parinodosarius (Strimple), 1940, new comb.
Delocrinus parinodosarius Strimple, 1940. Strimple, 1961g, p. 129.
Corythocrinus parinodosarius (Strimple), 1940. Strimple, 1961g, p. 129.
Tholiacrinus parinodosarius (Strimple), 1940. Strimple, 1962d, p. 136.
Pennsylvanian, Desmoinesian, Wewoka Formation: United States: Oklahoma

Endelocrinus parvus Moore and Plummer, 1940. Moore and Laudon, 1944, p. 173, Pl. 65, fig. 1.
Pennsylvanian, Desmoinesian, Mineral Wells Formation: United States: Texas

Endelocrinus petalosus Strimple, 1949c, p. 24, Pl. 6, figs. 1-4. Strimple, 1960b, p. 154, fig. 2I.
Pennsylvanian, Virgilian, Stull Shale: United States: Kansas

Endelocrinus rectus Moore and Plummer, 1940. Strimple, 1961g, p. 129.
Corythocrinus rectus (Moore and Plummer), 1940. Strimple, 1961g, p. 129.
Tholiacrinus rectus (Moore and Plummer), 1940. Strimple, 1962d, p. 136.
Pennsylvanian, Desmoinesian, Millsap Lake Formation: United States: Texas

Endelocrinus rimulatus (Strimple), 1962, new comb.
Tholiacrinus rimulatus Strimple, 1962e, p. 19, Pl. 7, figs. 11-14.
Pennsylvanian, Desmoinesian, Oologah Formation: United States: Oklahoma

Endelocrinus rotundus Strimple, 1963a, p. 71, Pl. 12, figs. 5-8.
Permian, Wolfcampian, Falls City Limestone: United States: Kansas

Endelocrinus texanus (Weller), 1909. Moore and Laudon, 1944, p. 173, Pl. 65, fig. 10.
Permian, Wolfcampian, Cibolo Formation: United States: Texas

Endelocrinus torus Webster and Lane, 1967, p. 24, Pl. 4, figs. 4-6; Fig. 4, no. 3.
Permian, Wolfcampian, Bird Spring Formation: United States: Nevada

Endelocrinus tumidus (Strimple), 1939. Strimple, 1950a, p. 112.
Pennsylvanian, Missourian, Stanton Limestone: United States: Oklahoma

Endelocrinus tumidus spinosus Strimple, 1950a, p. 112.
Delocrinus tumidus Strimple, 1939 (part, Pl. 2, figs. 1-4 only). Strimple, 1950a, p. 112.
Pennsylvanian, Missourian, Stanton Limestone: United States: Oklahoma

Endelocrinus undulatus (Strimple), 1961, new comb.
Corythocrinus undulatus Strimple, 1961g, p. 129, Pl. 13, figs. 3-6.
Tholiacrinus undulatus (Strimple), 1961. Strimple, 1962d, p. 136.
Pennsylvanian, Desmoinesian, Holdenville Shale: United States: Oklahoma

Endelocrinus sp. Webster and Lane, 1967, p. 25, Pl. 6, figs. 1, 2.
Permian, Wolfcampian, Bird Spring Formation: United States: Nevada

EOMYELODACTYLUS Foerste, 1919. *Myelodactylus (Eomyelodactylus) rotundatus* Foerste, 1919. Moore, 1962b, p. 43.
Silurian, Albian

Eomyelodactylus rotundatus (Foerste), 1919. Moore, 1962b, p. 43, Pl. 4, fig. 1.
Silurian, Albian, Brassfield Limestone: United States: Ohio

EPIHALYSIOCRINUS Arendt, 1965. *Halysiocrinus(?) tuberculatus* Yakovlev, 1927. Arendt, 1965a, p. 90.
Permian, Artinskian

Epihalysiocrinus tuberculatus (Yakovlev), 1927. Arendt, 1965a, p. 91, Pl. 10, figs. 1-9; Fig. 1.
Permian, Artinskian, Boetzkay Suite: Russia: Ural Mountains

EPIPETSCHORACRINUS Yakovlev, *in* Yakovlev and Ivanov, 1956. *E. borealis* Yakovlev, *in* Yakovlev and Ivanov, 1956. Yakovlev, *in* Yakovlev and Ivanov, 1956, p. 81. Arendt and Hecker, 1964, p. 91.
Lower Permian

Epipetschoracrinus borealis Yakovlev, *in* Yakovlev and Ivanov, 1956, p. 81, Pl. 15, fig. 16. Arendt and Hecker, 1964, p. 91, Pl. 13, fig. 6.
Lower Permian: Russia: Arctic Urals

ERATOCRINUS Kirk, 1938. *Zeacrinus elegans* Hall, 1858. Moore and Laudon, 1943b, p. 58; 1944, p. 165. Arendt and Hecker, 1964, p. 89.
Mississippian

Eratocrinus commaticus (Miller), 1892. Moore and Laudon, 1944, p. 165, Pl. 61, fig. 3. Moore, 1962b, p. 24, fig. 14, no. 6.
Mississippian, Meramecian, Warsaw Limestone: United States: Missouri

Eratocrinus elegans (Hall), 1858. Moore and Laudon, 1943b, p. 134, Pl. 5, fig. 9; 1944, p. 165, Pl. 59, fig. 12; Pl. 61, fig. 13. Moore, 1962b, p. 24, fig. 14, no. 5.
Mississippian, Osagean, Burlington Limestone: United States: Iowa

Eratocrinus serratus (Meek and Worthen), 1869. Moore and Laudon, 1944, p. 167, Pl. 60, fig. 11.
Mississippian, Osagean, Burlington Limestone: United States: Iowa

ERETMOCRINUS Lyon and Casseday, 1859. *E. magnificus* Lyon and Casseday, 1859. Moore and Laudon, 1943b, p. 95, fig. 16; 1944, p. 195. Ubaghs, 1953, p. 740. Lane, 1963b, p. 921.
Mississippian

Eretmocrinus adultus Wachsmuth and Springer, 1881 = *Dizygocrinus originarius*

Eretmocrinus commendabilis Miller and Gurley, 1895. Van Sant, *in* Van Sant and Lane, 1964, p. 116, Pl. 7, fig. 12.
Mississippian, Osagean, Borden Group: United States: Indiana

Eretmocrinus granuliferus Wachsmuth and Springer, 1897. Van Sant, *in* Van Sant and Lane, 1964, p. 116.
Mississippian, Osagean, Borden Group: United States: Indiana

Eretmocrinus intermedius Wachsmuth and Springer, 1881 = *Dizygocrinus intermedius*

Eretmocrinus leucosia (Hall), 1861. Laudon, 1948, Pl. 2.
Mississippian, Osagean, Burlington Limestone: United States: Iowa

Eretmocrinus magnificus Lyon and Casseday, 1859. Jillson, 1958, p. 65, unnumbered figure, p. 12.
Mississippian, Osagean, Fort Payne Chert: United States: Kentucky

Eretmocrinus remibrachiatus (Hall), 1861. Moore and Laudon, 1944, p. 195, Pl. 76, fig. 20.
   Mississippian, Osagean, Burlington Limestone: United States: Iowa, Missouri

Eretmocrinus tentor Laudon, 1933. Moore and Laudon, 1944, p. 195, Pl. 76, fig. 12. Laudon and others, 1952, p. 570, Pl. 67, fig. 24.
   Mississippian, Kinderhookian, Gilmore City Limestone, Banff formation: United States: Iowa; Canada: Alberta

Eretmocrinus varsoviensis Worthen, 1882 = *Dizygocrinus intermedius*

Eretmocrinus yandelli (Shumard), 1857. Laudon, 1948, Pl. 2.
   Mississippian, Osagean, Keokuk Limestone: United States: Iowa

ERISOCRINUS Meek and Worthen, 1865. *E. typus* Meek and Worthen, 1865. Moore and Laudon, 1943b, p. 60, fig. 8; 1944, p. 173, Pl. 62, fig. 26. Cuénot, 1948, p. 62. Moore, 1950, fig. 40. Arendt and Hecker, 1964, p. 90.
   Stemmatocrinus Trautschold, 1867. Moore and Laudon, 1943b, p. 60.
   Mississippian-Permian

Erisocrinus antiquus Meek and Worthen, 1869 = *Nactocrinus antiquus*

Erisocrinus araxensis Yakovlev, 1933. Yakovlev, *in* Yakovlev and Ivanov, 1956, p. 82, Pl. 19, figs. 1, 2. Arendt and Hecker, 1964, p. 90, Pl. 12, fig. 4.
   Upper Permian: Russia: Transcaucasian region

Erisocrinus carlopsensis Wright, 1939. Wright, 1951b, p. 101, Pl. 30, figs. 11, 12.
   Carboniferous, Visean, Lower Limestone Group: Scotland

Erisocrinus cernuus (Trautschold), 1867. Yakovlev and Ivanov, 1956, p. 32, Pl. 7, fig. 7; Fig. 7 (misspelled as *E. cernus*). Arendt and Hecker, 1964, p. 90, fig. 130.
   Stemmatocrinus cernus Trautschold, 1867. Yakovlev and Ivanov, 1956, p. 32.
   Middle Carboniferous: Russia: Moscow Basin

Erisocrinus elevatus Moore and Plummer, 1940. Moore and Laudon, 1944, p. 173, Pl. 65, fig. 27.
   Pennsylvanian, Missourian, Palo Pinto Limestone: United States: Texas

Erisocrinus granulatus Wanner, 1924 = *Parastachyocrinus granulatus*

Erisocrinus loczyi Katzer, 1903. Lane, 1964a, p. 365.
   Upper Carboniferous: Brazil

Erisocrinus longwelli Lane and Webster, 1966, p. 43, Pl. 10, figs. 1-3, 5, 6. Webster and Lane, 1967, p. 23, Pl. 1, fig. 13; Pl. 6, fig. 3.
   Permian, Wolfcampian, Bird Spring Formation: United States: Nevada

Erisocrinus lustrum Strimple, 1951b, p. 373, Pl. 56, figs. 8-10.
   Pennsylvanian, Missourian, Raytown Limestone: United States: Kansas

Erisocrinus malaianus Wanner, 1924 = *Parastachyocrinus malaianus*

Erisocrinus mediator Strimple, 1962e, p. 14, Pl. 8, figs. 4-7.
   Pennsylvanian, Desmoinesian, Oologah Limestone: United States: Oklahoma

Erisocrinus obliquus Wanner, 1916 = *Parastachyocrinus obliquus*

Erisocrinus scoticus Wright, 1942 = *?Apographiocrinus scoticus*

Erisocrinus stefaninii Yakovlev, 1934 = *Stachyocrinus stefaninii*

Erisocrinus terminalis Strimple, 1962e, p. 13, Pl. 9, figs. 1-4.
   Pennsylvanian, Desmoinesian, Oologah Limestone: United States: Oklahoma

Erisocrinus typus Meek and Worthen, 1865. Moore and Laudon, 1943b, p. 135, Pl. 6, fig. 3; 1944, p. 173, Pl. 62, fig. 27; Pl. 65, fig. 28. Strimple, 1959a, p. 120, Pl. 1, figs. 14-17; Pl. 2, figs. 2-5; 1960b, p. 155, fig. 2D. Tischler, 1963, p. 1066, figs. 6A, B.
Pennsylvanian, Desmoinesian-Missourian, Oologah Formation, Wann Formation, Madera Formation, and others: United States: Illinois, Colorado, Oklahoma

Erisocrinus sp. Termier and Termier, 1950, p. 90, Pl. 217, figs. 19-21.
Carboniferous, Westphalian: Morocco

ESTHONOCRINUS Jaekel, 1918. *E. laevior* Jaekel, 1918. Moore and Laudon, 1943b, p. 52. Ubaghs, 1953, p. 751. Arendt and Hècker, 1964, p. 86.
Ordovician

Esthonocrinus laevior Jaekel, 1918. Arendt and Hecker, 1964, p. 86, fig. 127.
Middle Ordovician: Estonia

ETHELOCRINUS Kirk, 1937. *Eupachycrinus magister* Miller and Gurley, 1890. Moore and Laudon, 1943b, p. 62; 1944, p. 175.
Pennsylvanian, Atokan-Missourian

Ethelocrinus ardmorensis Strimple, 1949d, p. 11, Pl. 2, figs. 9, 12, 15.
Pennsylvanian, Atokan, Pumpkin Creek Limestone: United States: Oklahoma

Ethelocrinus expansus Strimple, 1938 = *Parethelocrinus expansus*

Ethelocrinus iatani Strimple, 1949 = *Parethelocrinus iatani*

Ethelocrinus magister (Miller and Gurley), 1890. Moore and Laudon, 1943b, p. 135, Pl. 6, fig. 13; 1944, p. 175, Pl. 65, fig. 33.
Pennsylvanian, Missourian, Lane Shale: United States: Missouri

Ethelocrinus magnus Strimple, 1949 = *Parethelocrinus magnus*

Ethelocrinus millsapensis Moore and Plummer, 1940 = *Parethelocrinus millsapensis*

Ethelocrinus oklahomensis Moore and Plummer, 1938 = *Paracromyocrinus oklahomensis*

Ethelocrinus peridous Strimple, 1949 = *Parethelocrinus magnus*

Ethelocrinus plattsburgensis Strimple, 1938 = *Parethelocrinus plattsburgensis*

Ethelocrinus sphaeralis (Miller and Gurley), 1890. Moore and Laudon, 1944, p. 175, Pl. 65, fig. 30.
Pennsylvanian, Missourian, Lane Shale: United States: Missouri, Oklahoma

Ethelocrinus sphaeri Strimple, 1949d, p. 10, Pl. 1, figs. 4-6, 9.
Pennsylvanian, Atokan, Pumpkin Creek Limestone: United States: Oklahoma

Ethelocrinus texasensis Moore and Plummer, 1940 = *Paracromyocrinus oklahomensis*

Ethelocrinus variabilis Strimple, 1949 = *Parethelocrinus variabilis*

Ethelocrinus watkinsi Strimple, 1949 = *Parethelocrinus watkinsi*

EUCALYPTOCRINITES Goldfuss, 1826. *E. rosaceus* Goldfuss, 1831. Moore and Laudon, 1943b, p. 99, fig. 17; 1944, p. 201, Pl. 71, fig. 2. Ubaghs, 1953, p. 742, figs. 116, 117. Arendt and Hecker, 1964, p. 97. Moore and Jeffords, 1968, p. 16, fig. 5, no. 1.
Eucalyptocrinus Cuénot, 1948, p. 69.
Silurian-Devonian

Eucalyptocrinites caelatus (Hall), 1843. Macurda, 1968, p. 115, figs. 9B, 10H (misspelled as *E. coelatus* in figure explanations).
Silurian, Niagaran, Lockport Formation: United States: New York

Eucalyptocrinites crassus (Hall), 1863. Moore and Laudon, 1943b, p. 142, Pl. 13, fig. 16; 1944, p. 201, Pl. 75, fig. 24. La Rocque and Marple, 1955, p. 73, fig. 148. Macurda, 1968, p. 113, figs. 7E; 8C, E-G, I-K; 9C, H, I; 10C (non fig. 8H as misnumbered in figure explanation).
Eucalyptocrinus crassus Wilson, 1949, p. 251, Pl. 25, figs. 1, 2.
Silurian, Niagaran, Waldron Shale, other formations: United States: Indiana, Ohio, Illinois, Wisconsin

Eucalyptocrinites decorus (Phillips), 1839. Macurda, 1968, fig. 9F.
Silurian, Wenlockian: England

Eucalyptocrinites elrodi (Miller), 1891.
Eucalyptocrinus elrodi Wilson, 1949, p. 251, Pl. 25, fig. 10.
Silurian, Niagaran, Waldron Shale: United States: Tennessee

Eucalyptocrinites inchoatus Philip, 1961, p. 150, Pl. 8, figs. 3, 4, 9; Figs. 4-9.
Lower Devonian: Australia: Victoria

Eucalyptocrinites magnus (Worthen), 1875.
Eucalyptocrinus magnus Wilson, 1949, p. 251, Pl. 25, fig. 9.
Silurian, Niagaran, Waldron Shale: United States: Tennessee

Eucalyptocrinites milliganae (Miller and Gurley), 1895. Strimple, 1963d, p. 105, Pl. 7, fig. 3.
Eucalyptocrinus milliganae Cuénot, 1948, p. 38, fig. 47.
Silurian, Niagaran, Henryhouse Formation: United States: Oklahoma

Eucalyptocrinites pernodosus (Springer), 1926. Strimple, 1963d, p. 104, Pl. 7, figs. 1, 2.
Silurian, Niagaran, Henryhouse Formation: United States: Oklahoma

Eucalyptocrinites praerosaceus (Yakovlev), 1940. Arendt and Hecker, 1964, p. 97, Pl. 15, fig. 7. Macurda, 1968, p. 108, fig. 10F.
Eucalyptocrinus praerosaceus Yakovlev, 1940a, p. 103, 2 figs.; 1940b, p. 193, 2 figs.
Lower Devonian: Russia: Ural Mountains

Eucalyptocrinites rosaceus Goldfuss, 1831. Ubaghs, 1953, p. 742, figs. 116, 117. Yakovlev, 1964b, p. 62, fig. 79. Macurda, 1968, p. 108, figs. 8A, B, D.
Middle Devonian: Germany

Eucalyptocrinites springeri (Foerste), 1909. Macurda, 1968, p. 116, fig. 10B.
Silurian, Niagaran, Waldron Shale: United States: Tennessee

Eucalyptocrinites tuberculatus (Miller and Dyer), 1878. Macurda, 1968, p. 115, figs. 8I (non 8H as misnumbered in figure explanation); 9E, G; 10A, D, E, G.
Silurian, Niagaran, Waldron Shale: United States: Indiana

Eucalyptocrinites spp.
Macurda, 1968, p. 115, fig. 9A.
Silurian, Niagaran, Brownsport Formation: United States: Tennessee
Macurda, 1968, fig. 10I.
Silurian, Niagaran, Waldron Shale: United States: Indiana
Eucalyptocrinus sp. Wilson, 1949, p. 251, Pl. 25, figs. 3-8.
Silurian, Niagaran, Waldron Shale: United States: Tennessee

Eucalyptocrinus armosus McChesney, 1861 = *Ochlerocrinus armosus*

Eucalyptocrinus crassus Hall, 1863 = *Eucalyptocrinites crassus*

Eucalyptocrinus elrodi Miller, 1891 = *Eucalyptocrinites elrodi*

Eucalyptocrinus magnus Worthen, 1875 = *Eucalyptocrinites magnus*

Eucalyptocrinus milliganae Miller and Gurley, 1895 = *Eucalyptocrinites milliganae*

Eucalyptocrinus praerosaceus Yakovlev, 1940 = *Eucalyptocrinites praerosaceus*

Eucalyptocrinus sp. Wilson, 1949 = *Eucalyptocrinites* sp.

EUCATILLOCRINUS Springer, 1923. *Catillocrinus bradleyi* Meek and Worthen, 1868. Moore and Laudon, 1943b, p. 30; 1944, p. 149. Ubaghs, 1953, p. 747. Moore, 1962b, p. 16, fig. 12, no. 9. Arendt and Hecker, 1964, p. 81.
Mississippian

Eucatillocrinus bradleyi (Meek and Worthen), 1868. Moore and Laudon, 1944, p. 149, Pl. 55, fig. 3. Van Sant, *in* Van Sant and Lane, 1964, p. 68, Pl. 1, figs. 1, 2; Fig. 24, no. 1.
Catillocrinus bradleyi Cuénot, 1948, p. 62, fig. 80. Yakovlev, 1964b, p. 64, fig. 84.
Mississippian, Osagean, Keokuk Limestone, Edwardsville Formation: United States: Indiana

EUCHEIROCRINUS Meek and Worthen, 1869 = CALCEOCRINUS

Eucheirocrinus anglicus Springer, 1926 = *Calceocrinus anglicus*

Eucheirocrinus chrysalis Moore and Laudon, 1943 = *Calceocrinus chrysalis*

Eucheirocrinus minor Springer, 1926 = *Calceocrinus minor*

Eucheirocrinus ontario Springer, 1919 = *Calceocrinus ontario*

EUCLADOCRINUS Meek, 1871. *Platycrinites (Eucladocrinus) montanaensis* Meek, 1872. Moore and Laudon, 1943b, p. 101; 1944, p. 205. Cuénot, 1948, p. 68. Ubaghs, 1953, p. 743. Arendt and Hecker, 1964, p. 98. Moore and Jeffords, 1968, p. 11, fig. 1, no. 2.
Mississippian

Eucladocrinus millebrachiatus Wachsmuth and Springer, 1878. Moore and Laudon, 1943b, p. 143, Pl. 14, fig. 12; 1944, p. 205, Pl. 78, fig. 29. Ubaghs, 1953, p. 743, fig. 77.
Mississippian, Osagean, Burlington Limestone: United States: Iowa, Missouri

Eucladocrinus pleuroviminus (White), 1862. Moore and Laudon, 1943b, p. 143, Pl. 14, fig. 13 (misspelled as *E. pleurovimineus*); 1944, p. 205, Pl. 78, fig. 28 (misspelled as *E. pleurovimenus*). Ubaghs, 1953, p. 743, fig. 102.
Mississippian, Osagean, Burlington Limestone: United States: Iowa, Missouri

Eucladocrinus tuberosus (Hall), 1858. Moore and Laudon, 1944, p. 205, Pl. 78, fig. 30.
Mississippian, Osagean, Burlington Limestone: United States: Iowa, Missouri

EUDIMEROCRINUS Springer, 1926. *E. multibrachiatus* Springer, 1926. Moore and Laudon, 1943b, p. 83, fig. 14; 1944, p. 187, Pl. 70, fig. 3. Moore, 1950, p. 42, fig. 11. Ubaghs, 1953, p. 737.
Silurian

Eudimerocrinus multibrachiatus Springer, 1926. Moore and Laudon, 1944, p. 187, Pl. 72, fig. 6.
Silurian, Niagaran, Beech River Formation: United States: Tennessee

EUERISOCRINUS Strimple, 1939. **E. waysidensis* Strimple, 1939. Moore and Laudon, 1943b, p. 60.
Pennsylvanian, Missourian

EUMORPHOCRINUS Wright, 1955. **E. erectus* Wright, 1955. Wright, 1955a, p. 232.
Carboniferous, Tournaisian

*Eumorphocrinus erectus Wright, 1955a, p. 232, Pl. 56, figs. 18, 19; Fig. 116.
Carboniferous, Tournaisian: Scotland

Eumorphocrinus excelsus Wright, 1955a, p. 233, Pl. 58, figs. 8, 11.
Carboniferous, Tournaisian: Scotland

Eumorphocrinus hibernicus Wright, 1955a, p. 234, Pl. 56, figs. 4, 5.
Carboniferous, Tournaisian?: Ireland

EUONYCHOCRINUS Strimple, 1940. **E. dubius* Strimple, 1940. Moore and Laudon, 1943b, p. 69.
Pennsylvanian

Euonychocrinus magnus Strimple, 1962e, p. 53, Pl. 5, figs. 1, 2; Pl. 6, figs. 3-5.
Pennsylvanian, Desmoinesian, Oologah Limestone, Wewoka Formation: United States: Oklahoma

Euonychocrinus subservire Strimple, 1951a, p. 206, Pl. 38, figs. 3, 4; Pl. 39, figs. 3, 4, 11.
Pennsylvanian, Missourian, Lake Bridgeport Shale: United States: Texas

EUPACHYCRINUS Meek and Worthen, 1865. **Graphiocrinus quatuordecembrachialis* Lyon, 1867. Moore and Laudon, 1943b, p. 62, fig. 7; 1944, p. 175. Cuénot, 1948, p. 62. Moore, 1950, fig. 4p.
Carboniferous

Eupachycrinus asperatus Worthen, 1883. Moore and Laudon, 1944, p. 175, Pl. 64, fig. 34.
Mississippian, Chesterian: United States: Illinois

Eupachycrinus boydii Meek and Worthen, 1870. Horowitz, 1965, p. 35, Pl. 3, figs. 16-18.
Mississippian, Chesterian, Glen Dean Formation: United States: Indiana

Eupachycrinus? donetzensis Yakovlev, 1961, p. 418, Pl. 2, figs. 8, 9.
Lower Carboniferous: Russia: Donetz Basin

Eupachycrinus germanus Miller, 1879. Horowitz, 1965, p. 34, Pl. 3, figs. 4-6.
Mississippian, Chesterian, Glen Dean Formation: United States: Kentucky

Eupachycrinus macneanensis Wright, 1951b, p. 91, Pl. 31, figs. 12, 14.
Carboniferous, Visean: Ireland

Eupachycrinus modernus Strimple, 1951 = *Intermediacrinus modernus*

Eupachycrinus mooresi Weller, 1898 = *Plaxocrinus mooresi*

*Eupachycrinus quatuordecembrachialis (Lyon), 1867. Moore and Laudon, 1944, p. 175, Pl. 64, fig. 33 (given as *E. 14-brachialis* on p. 175).
Mississippian, Chesterian: United States: Illinois, Kentucky

Eupachycrinus spartarius Miller, 1879. Moore and Laudon, 1943b, p. 135, Pl. 6, fig. 8; 1944, p. 175, Pl. 64, fig. 28. Horowitz, 1965, p. 34, Pl. 3, figs. 10-12.
Mississippian, Chesterian, Glen Dean Formation: United States: Kentucky

Eupachycrinus vapidus Wright, 1951b, p. 92, Pl. 30, figs. 7, 13; Fig. 41.
Carboniferous, Visean: Ireland

EURYOCRINUS Phillips, 1836. *E. concavus Phillips, 1836. Moore and
    Laudon, 1943b, p. 73, fig. 10; 1944, p. 179, Pl. 67, fig. 1.
    Ubaghs, 1953, p. 755. Arendt and Hecker, 1964, p. 94.
Devonian—Lower Carboniferous

Euryocrinus barrisi Springer, 1920. Moore and Laudon, 1943b, p. 137, Pl. 8,
    fig. 8; 1944, p. 179, Pl. 68, fig. 22.
Middle Devonian, Hamilton Group: United States: Iowa, Michigan

*Euryocrinus concavus Phillips, 1836. Wright, 1954a, p. 161, Pl. 44, fig.9.
    Cain, 1968, p. 196, fig. 1b, iv (misspelled as Eurycrinus on p. 196).
Carboniferous, Tournaisian, Limestone Series: Scotland

Euryocrinus duplex Wright, 1942 = Dieuryocrinus duplex

Euryocrinus? granulosus (Phillips), 1836. Wright, 1954a, p. 164, fig. 91.
Carboniferous, Tournaisian: England

Euryocrinus? laddii Stewart, 1940. Kier, 1952, p. 74, Pl. 3, fig. 1.
Middle Devonian, Silica Formation: United States: Ohio

Euryocrinus rofei Bather and Gregory, in Springer, 1920. Wright, 1942,
    p. 280, Pl. 9, fig. 1; 1954a, p. 162, Pl. 42, fig. 13; Pl. 44,
    fig. 3; Fig. 90; 1960, p. 333, Appendix, Pl. B, figs. 5, 6.
Carboniferous, Tournaisian-Visean, Mountain Limestone: Scotland

Euryocrinus tennesseensis Springer, 1920. Moore and Laudon, 1944, p. 179,
    Pl. 68, fig. 23.
Mississippian, Osagean, New Providence Shale: United States: Indiana,
    Kentucky, Tennessee

EUSPIROCRINUS Angelin, 1878. *E. spiralis Angelin, 1878. Moore and
    Laudon, 1943b, p. 51. Ubaghs, 1953, p. 750. Moore, 1962b, p. 15,
    fig. 9, no. 4. Arendt and Hecker, 1964, p. 85.
Ordovician-Silurian

Euspirocrinus cirratus Strimple, 1963d, p. 65, Pl. 3, figs. 1-4.
Silurian, Niagaran, Henryhouse Formation: United States: Oklahoma

Euspirocrinus obconicus Billings, 1885. Wilson, 1946, p. 40.
Ordovician, Champlainian, Cobourg Formation, Sherman Fall Formation or
    Hull Formation: Canada: Ontario, Quebec

*Euspirocrinus spiralis Angelin, 1878. Ubaghs, 1953, p. 750, figs. 21e,
    112. Stukalina, 1967, p. 192, fig. 4, no. 1.
Silurian: Gotland

EUSTENOCRINUS Ulrich, 1924. *E. springeri Ulrich, 1924. Moore and
    Laudon, 1943b, p. 29, fig. 3. Ubaghs, 1953, p. 744, fig. 140A.
    Moore, 1962b, p. 8, figs. 1, no. 9; 2, no. 2.
Ordovician

*Eustenocrinus springeri Ulrich, 1924. Ubaghs, 1953, p. 744, figs. 18a, b.
    Yakovlev, 1964b, p. 66, fig. 92a.
Ordovician: Canada

EUTAXOCRINUS Springer, 1906. *Taxocrinus affinis Müller, 1856. Moore and
    Laudon, 1943b, p. 69, fig. 10; 1944, p. 177, Pl. 67, fig. 14.
    Ubaghs, 1953, p. 755.
Silurian-Mississippian

Eutaxocrinus confer prognatus Schmidt, in Hendriks, 1951, unnumbered figure p. 297.
Devonian, Eifelian: England

Eutaxocrinus fletcheri (Worthen), 1882. Moore and Laudon, 1943b, p. 136,
    Pl. 7, fig. 9; 1944, p. 177, Pl. 68, fig. 10.
Mississippian, Kinderhookian, Hampton Formation: United States: Iowa

Eutaxocrinus fuchsi Schmidt, 1941, p. 188, Pl. 25, fig. 4; Fig. 51.
Lower Devonian: Germany

Eutaxocrinus immersus Dubatolova, 1964, p. 50, Pl. 6, fig. 5.
Lower Devonian: Russia: Kuznetz Basin

Eutaxocrinus ithacensis (Williams), 1882. Moore and Laudon, 1944, p. 177, Pl. 69, fig. 4.
Devonian, Senecan, Portage Formation: United States: New York

Eutaxocrinus maccoyanus (Salter), 1873. Ramsbottom, 1958, p. 108, Pl. 20, figs. 4-6.
Ichthyocrinus McCoyanus Salter, 1873. Ramsbottom, 1958, p. 108.
Silurian, lower Ludlovian, Bannisdale Slates: England

Eutaxocrinus maureri Schmidt, 1941, p. 184, Pl. 25, fig. 5.
Devonian, Coblentzian: Germany

Eutaxocrinus montanensis Springer, 1920. Laudon and Severson, 1953, p. 523, Pl. 54, figs. 18-20.
Mississippian, Kinderhookian, Lodgepole Limestone: United States: Montana

Eutaxocrinus patulus Schmidt, 1941, p. 185, Pl. 25, fig. 1.
Devonian, Coblentzian: Germany

Eutaxocrinus cf. *patulus* Schmidt, 1941, p. 186, Pl. 25, fig. 2.
Devonian, Coblentzian: Germany

Eutaxocrinus procerus Schmidt, 1941, p. 183, Pl. 25, fig. 3.
Devonian, Coblentzian: Germany

Eutaxocrinus pulvinatus Laudon, Parks and Spreng, 1952, p. 560, Pl. 66, figs. 8, 9; Pl. 69, fig. 10.
Mississippian, Kinderhookian, Banff formation: Canada: Alberta

Eutaxocrinus rhenanus (Roemer), 1851. Schmidt, 1941, p. 181, fig. 49.
Devonian, Coblentzian: Germany

Eutaxocrinus sandbergeri Schmidt, 1941, p. 186, fig. 50.
Devonian, Coblentzian: Germany

Eutaxocrinus whiteavesi Springer, 1920. Moore and Laudon, 1943b, p. 136, Pl. 7, fig. 3; 1944, p. 177, Pl. 68, fig. 16.
Devonian, Erian, Hamilton Group: Canada: Ontario

EUTELECRINUS Wanner, 1916. *E. piriformis* Wanner, 1916. Moore and Laudon, 1943b, p. 101. Ubaghs, 1953, p. 743. Arendt and Hecker, 1964, p. 98.
Permian

Eutelecrinus sp. Breimer, 1962a, p. 143, fig. 32.
Permian: Timor

EUTROCHOCRINUS Wachsmuth and Springer, 1897. *Actinocrinus christyi* Shumard, 1855. Moore and Laudon, 1943b, p. 95, fig. 16; 1944, p. 197, Pl. 71, fig. 7. Ubaghs, 1953, p. 740.
Mississippian

Eutrochocrinus christyi (Shumard), 1855. Moore and Laudon, 1943b, p. 141, Pl. 12, fig. 11; 1944, p. 197, Pl. 75, fig. 8; Pl. 76, fig. 21. Laudon, 1948, Pl. 2. Ubaghs, 1953, p. 740, fig. 22c.
Mississippian, Osagean, Burlington Limestone, Fern Glen Limestone, St. Joe Formation, Lake Valley Limestone: United States: Iowa, Illinois, Arkansas, Oklahoma, New Mexico

Eutrochocrinus trochiscus (Meek and Worthen), 1868. Laudon, 1948, Pl. 2.
Mississippian, Osagean, Burlington Limestone: United States: Iowa

EXOCRINUS Strimple, 1949. *E. multirami* Strimple, 1949. Strimple, 1949c, p. 9.
Pennsylvanian, Desmoinesian-Permian, Wolfcampian

Exocrinus desmoinesensis Strimple, 1949c, p. 12, Pl. 2, figs. 3, 4, 8.
Pennsylvanian, Desmoinesian, Altamont Limestone: United States: Oklahoma

Exocrinus moorei (Lane and Webster), 1966. Webster and Lane, 1967, p. 19, Pl. 3, figs. 5, 11.
Marathonocrinus moorei Lane and Webster, 1966, p. 34, Pl. 7, figs. 1, 2, 5; Fig.10. Webster and Lane, 1967, p. 19..
Permian, Wolfcampian, Bird Spring Formation: United States: Nevada

*Exocrinus multirami Strimple, 1949c, p. 10, Pl. 2, figs. 5-7; Pl. 3, figs. 4, 5.
Pennsylvanian, Missourian, Ochelata Group: United States: Oklahoma

Exocrinus pallium Strimple, 1949c, p. 14, Pl. 3, figs. 6-8.
Pennsylvanian, Missourian, Ochelata Group: United States: Oklahoma

Exocrinus virgilensis Strimple, 1949c, p. 13, Pl. 3, figs. 9, 10.
Pennsylvanian, Virgilian, Nelagoney Formation: United States: Oklahoma

Exocrinus wanni Strimple, 1949c, p. 12, Pl. 2, figs. 1, 2; Pl. 3, figs. 1-3.
Pennsylvanian, Missourian, Wann Formation: United States: Oklahoma

FIFEOCRINUS Wright, 1951. **Pachylocrinus tielensis* Wright, 1936. Wright, 1951a, p. 41. Arendt and Hecker, 1964, p. 88.
Carboniferous, Visean

*Fifeocrinus tielensis (Wright), 1936. Wright, 1951a, p. 42, Pl. 11, figs. 2-4, 6, 8; Pl. 32, figs. 7, 11; Figs. 1, 11, 12 (*non* Pl. 18, typographical error).
Pachylocrinus tielensis Wright, 1936. Wright, 1951a, p. 42.
Carboniferous, Visean, Lower Limestone Group: Scotland

FOLLICRINUS Schmidt, 1934. **Taxocrinus grebei* Follmann, 1887. Moore and Laudon, 1943b, p. 52. Ubaghs, 1953, p. 751. Arendt and Hecker, 1964, p. 86.
Devonian

Follicrinus parvestellatus Schmidt, 1941, p. 112, Pl. 19, fig. 1.
Devonian, Coblentzian: Germany

Follicrinus sp. indet. Schmidt, 1941, p. 112, Pl. 21, fig. 5.
Devonian, Coblentzian: Germany

FORBESIOCRINUS de Koninck and Le Hon, 1854. **F. nobilis* de Koninck and Le Hon, 1854. Moore and Laudon, 1943b, p. 73, figs. 10, 11; 1944, p. 179, Pl. 66, fig. 14; Pl. 67, fig. 8. Moore, 1950, p. 37, fig. 40. Ubaghs, 1953, p. 755. Philip, 1963, p. 264, fig. 1c.
Lower Carboniferous

Forbesiocrinus agassizi Hall, 1858. Van Sant, *in* Van Sant and Lane, 1964, p. 43, fig. 10, no. 13.
Mississippian, Osagean, Burlington Limestone: United States: Iowa

Forbesiocrinus burlingtonensis Springer, 1920. Moore and Laudon, 1944, p. 179, Pl. 68, fig. 20.
Mississippian, Osagean, Burlington Limestone: United States: Iowa

Forbesiocrinus communis Hall, 1863. Moore and Laudon, 1944, p. 179, Pl. 68, fig. 19.
Mississippian, Kinderhookian, Cuyahoga Formation: United States: Ohio

Forbesiocrinus multibrachiatus Lyon and Casseday, 1859. Moore and Laudon, 1944, p. 179, Pl. 68, fig. 4. Van Sant, *in* Van Sant and Lane, 1964, p. 100.
Mississippian, Osagean, Keokuk Limestone: United States: Indiana

*Forbesiocrinus nobilis de Koninck and Le Hon, 1854. Ubaghs, 1953, p. 755, figs. 63-69. Van Sant, *in* Van Sant and Lane, 1964, p. 41, figs. 10,

nos. 1-12; 12, nos. 1-4.
Carboniferous, Tournaisian: Belgium

Forbesiocrinus pyriformis Miller and Gurley, 1894. Moore and Laudon, 1943b, p. 137, Pl. 8, fig. 14; 1944, p. 179, Pl. 68, fig. 13. Ubaghs, 1953, p. 755, fig. 13. Yakovlev, 1964b, p. 65, fig. 86.
Mississippian, Osagean, Keokuk Limestone: United States: Kentucky, Tennessee

Forbesiocrinus saffordi Hall, 1860. Moore and Laudon, 1944, p. 179, Pl. 68, fig. 21.
Mississippian, Osagean, Keokuk Limestone: United States: Tennessee

Forbesiocrinus withersi Laudon, Parks and Spreng, 1952, p. 559, Pl. 66, figs. 1-4; Pl. 69, fig. 9.
Mississippian, Kinderhookian, Banff formation: Canada: Alberta

Forbesiocrinus wortheni Hall, 1858. Ubaghs, 1953, p. 755, fig. 15e. Van Sant, *in* Van Sant and Lane, 1964, p. 100.
Mississippian: North America

FORTHOCRINUS Wright, 1942. *F. lepidus* Wright, 1942. Wright, 1942, p. 274. Arendt and Hecker, 1964, p. 88.
Carboniferous, Visean

Forthocrinus lepidus Wright, 1942, p. 275, Pl. 12, figs. 3-6. Wright, 1951b, p. 72, Pl. 35, figs. 1-5, 21.
Carboniferous, Visean, Lower Limestone Group: Scotland

GALATEACRINUS Moore, 1940. *G. stevensi* Moore, 1940. Moore and Laudon, 1943b, p. 58; 1944, p. 165.
Pennsylvanian

Galateacrinus allisoni Moore, 1940. Moore and Laudon, 1944, p. 165, Pl. 64, fig. 13. Strimple, 1959b, p. 195.
Pennsylvanian, Missourian, Iowa Formation, Wann Formation: United States: Kansas, Oklahoma

Galateacrinus ewersi Moore, 1940. Moore and Laudon, 1944, p. 165, Pl. 64, fig. 14.
Pennsylvanian, Missourian, Graford Formation: United States: Texas

Galateacrinus stevensi Moore, 1940. Moore and Laudon, 1943b, p. 134, Pl. 5, fig. 2; 1944, p. 165, Pl. 62, fig. 2; Pl. 64, fig. 15. Strimple, 1962e, p. 39.
Pennsylvanian, Desmoinesian, Oologah Limestone: United States: Oklahoma

Galateacrinus sp. Lane and Webster, 1967, p. 15, fig. 1.
Pennsylvanian: United States

GASTEROCOMA Goldfuss, 1839. *G. antiqua* Goldfuss, 1839. Moore and Laudon, 1943b, p. 50, fig. 4. Ubaghs, 1953, p. 750. Moore, 1962b, p. 8, fig. 1, no. 2.
Devonian

Gasterocoma admota Dubatolova, 1964, p. 18, Pl. 1, figs. 3, 4.
Lower Devonian: Russia: Kuznetz Basin

Gasterocoma(?) *arguta* Dubatolova, 1964, p. 20, Pl. 2, fig. 1.
Lower Devonian: Russia: Kuznetz Basin

Gasterocoma dibapha Dubatolova, 1964, p. 19, Pl. 1, fig. 5.
Lower Devonian: Russia: Kuznetz Basin

Gasterocoma kayseri Schmidt, 1941, p. 106, fig. 28.
Devonian, Eifelian: Germany

Gasterocoma mite Dubatolova, 1964, p. 23, Pl. 2, fig. 2.
Devonian, Eifelian: Russia: Kuznetz Basin

Gasterocoma salairica Dubatolova, 1964, p. 21, Pl. L, figs. 6, 7.
Lower Devonian: Russia: Kuznetz Basin

GASTROCRINUS Jaekel, 1895. **Poteriocrinus patulus* Müller, 1859. Moore and Laudon, 1943b, p. 54. Ubaghs, 1953, p. 751. Arendt and Hecker, 1964, p. 86.
Devonian

Gastrocrinus aldudensis Le Maître and Heddebaut, 1962, p. 2399, figs. 1, 2.
Lower Devonian: France

Gastrocrinus brancai Schmidt, 1914. Schmidt, 1941, p. 135, Pl. 23, fig. 6.
Devonian, Coblentzian: Germany

Gastrocrinus dahmeri Schmidt, 1941, p. 139, Pl. 21, fig. 6.
Devonian, Coblentzian, Hohenrhein Beds: Germany

Gastrocrinus drevermanni Schmidt, 1914. Schmidt, 1941, p. 135, Pl. 23, fig. 4; Fig. 36.
Devonian, Siegenian, Rauhflaser horizon: Germany

Gastrocrinus follmanni Schmidt, 1914. Schmidt, 1941, p. 134, Pl. 23, fig. 1.
Gastrocrinus patulus var. *follmanni* Schmidt, 1914. Schmidt, 1941, p. 134.
Devonian, Coblentzian: Germany

*Gastrocrinus patulus (Müller), 1859. Schmidt, 1941, p. 132, Pl. 23, fig. 2; Pl. 25, fig. 10. Le Maître, 1958b, p. 133, Pl. 2, fig. 11.
Devonian, Coblentzian: Germany

Gastrocrinus patulus var. *follmanni* Schmidt, 1914 = *Gastrocrinus follmanni*

Gastrocrinus rugosus Schmidt, 1914. Schmidt, 1941, p. 139, Pl. 23, fig. 5.
Devonian, Coblentzian: Germany

Gastrocrinus sp. indet. Schmidt, 1941, p. 133, Pl. 23, fig. 3.
Devonian, Coblentzian: Germany

GAULOCRINUS Kirk, 1945. **Stemmatocrinus trautscholdi* Wachsmuth and Springer, 1885. Kirk, 1945b, p. 180.
Mississippian, Osagean

Gaulocrinus bordeni (Springer), 1920. Kirk, 1945b, p. 181.
Mespilocrinus bordeni Springer, 1920. Kirk, 1945b, p. 181.
Mississippian, Osagean, New Providence Shale: United States: Indiana

Gaulocrinus robustus (Troost), 1849. Kirk, 1945b, p. 181.
Cyathocrinites robustus Troost, 1849. Kirk, 1945b, p. 181.
Mississippian, Osagean, Keokuk Formation: United States: Tennessee

*Gaulocrinus trautscholdi (Wachsmuth and Springer), 1885. Kirk, 1945b, p. 182.
Stemmatocrinus trautscholdi Wachsmuth and Springer, 1885. Kirk, 1945b, p. 182.
Mississippian, Osagean, Keokuk Limestone: United States: Tennessee

Gaulocrinus veryi (Rowley, *in* Greene), 1903. Kirk, 1945b, p. 182.
Stemmatocrinus? veryi Rowley, *in* Greene, 1903. Kirk, 1945b, p. 182.
Mississippian, Osagean, Keokuk Limestone: United States: Kentucky

GAUROCRINUS Miller, 1883. **Glyptocrinus nealli* Hall, 1866. Moore and Laudon, 1943b, p. 82. Ubaghs, 1953, p. 735.
Silurian

GAZACRINUS Miller, 1892. **G. inornatus* Miller, 1892. Moore and Laudon, 1943b, p. 84, fig. 14; 1944, p. 189, Pl. 70, fig. 5. Moore, 1950, p. 42, fig. 11. Ubaghs, 1953, p. 737.
Silurian-Devonian

Gazacrinus inornatus Miller, 1892. Moore and Laudon, 1943b, p. 138, Pl. 9, fig. 15; 1944, p. 189, Pl. 72, fig. 11.
Silurian, Niagaran, Laurel Limestone: United States: Indiana

Gazacrinus ramifer (Roemer), 1850. Moore and Laudon, 1944, p. 189, Pl. 72, fig. 10.
Silurian, Niagaran, Beech River Formation: United States: Tennessee

Gazacrinus stellatus Springer, 1926. Moore and Laudon, 1943b, p. 138, Pl. 9, fig. 14; 1944, p. 189, Pl. 72, fig. 9. Strimple, 1963d, p. 88, Pl. 6, figs. 3, 4.
Silurian-Devonian, Niagaran-Helderbergian, Henryhouse Formation, Linden Group: United States: Oklahoma, Tennessee

GENNAEOCRINUS Wachsmuth and Springer, 1881. *Actinocrinus kentuckiensis* Shumard, 1866. Moore and Laudon, 1943b, p. 92, fig. 15; 1944, p. 191, Pl. 70, fig. 14. Ubaghs, 1953, p. 739.
Devonian

Gennaeocrinus arkonensis Whiteaves, 1898. Kesling and Smith, 1962, p. 184. Kesling, 1964c, p. 149.
Devonian: Canada

Gennaeocrinus carinatus Wood, 1901. Moore and Laudon, 1943b, p. 139, Pl.10, fig. 9; 1944, p. 191, Pl. 75, fig. 23. Kesling and Smith, 1962, p. 184. Kesling, 1964c, p. 148.
Devonian, Erian, Hamilton Formation: United States: Indiana, New York

Gennaeocrinus carinatus crassicostatus Goldring, 1923. Kesling and Smith, 1962, p. 184. Kesling, 1964c, p. 148.
Devonian: United States

Gennaeocrinus chilmanae Kesling, 1968a, p. 127, Pl. 1, figs. 1-6; Fig. 1.
Devonian, Erian, Silica Formation: United States: Michigan

Gennaeocrinus comptus Rowley, *in* Greene, 1903. Kesling and Smith, 1962, p. 184. Kesling, 1964c, p. 148.
Devonian: United States

Gennaeocrinus comptus spiniferus Rowley, *in* Greene, 1903. Kesling and Smith, 1962, p. 184. Kesling, 1964c, p. 148.
Devonian: United States

Gennaeocrinus decorus Goldring, 1923. Kesling and Smith, 1962, p. 184. Kesling, 1964c, p. 149.
Devonian: United States

Gennaeocrinus eucharis (Hall), 1862. Moore and Laudon, 1944, p. 191, Pl. 75, fig. 22. Kesling and Smith, 1962, p. 184. Kesling, 1964c, p. 148.
Devonian, Erian, Hamilton Formation: United States: New York

Gennaeocrinus facetus Rowley, *in* Greene, 1903. Kesling and Smith, 1962, p. 184. Kesling, 1964c, p. 148.
Devonian: United States

Gennaeocrinus(?) *germanicus* Schmidt, 1941, p. 98, Pl. 10, figs. 1, 2.
Devonian, Coblentzian, Stadtfeld Beds: Germany

Gennaeocrinus goldringae Ehlers, 1925. Kesling and Smith, 1962, p. 184. Kesling, 1964c, p. 148; 1965b, p. 267, Pl. 1, figs. 1-5; Pl. 2, figs. 1-6; Figs. 1, 2.
Devonian, Erian, Bell Shale: United States: Michigan

?*Gennaeocrinus goldringae* Ehlers, 1925. Kesling, 1965b, p. 270, Pl. 1, figs. 6-22; Pl. 2, figs. 7-19; Pl. 3, figs. 1-78; Pl. 4, figs. 1-56; Pl. 5, figs. 1-72.
Devonian, Erian, Bell Shale: United States: Michigan

Gennaeocrinus kentuckiensis (Shumard), 1866. Goldring, 1948a, p. 1, Pl. 1.
  Kesling and Smith, 1962, p. 184. Kesling, 1964c, p. 148.
  Devonian, Erian, Hamilton Formation: United States: Pennsylvania

Gennaeocrinus mourantae Goldring, 1934. Goldring, 1945, p. 60, Pl. 1,
  fig. 4; 1954, p. 6, Pl. 1, figs. 1-5. Kesling and Smith, 1962,
  p. 185. Kesling, 1964c, p. 149.
  Devonian, Erian, Ludlowville Shale, Arkona shale: Canada: Ontario

Gennaeocrinus nyssa (Hall), 1862. Goldring, 1950, p. 34, Pl. 2, fig. 11.
  Kesling and Smith, 1962, p. 185. Kesling, 1964c, p. 148 (misspelled
  as *G. nyassa*).
  Devonian, Erian, Kashong Member: United States: New York

Gennaeocrinus sp. cf. *G. nyssa* (Hall), 1862. Breimer, 1962a, p. 69.
  Devonian: Spain

Gennaeocrinus peculiaris Goldring, 1923. Kesling and Smith, 1962, p. 185.
  Kesling, 1964c, p. 148.
  Devonian: United States

Gennaeocrinus percarinatus Goldring, 1935. Kesling and Smith, 1962, p. 185.
  Devonian: United States

Gennaeocrinus romingeri Kesling, 1964c, p. 144, Pl. 1, figs. 1-8; Fig. 1.
  Devonian, Erian, Thunder Bay Limestone: United States: Michigan

Gennaeocrinus sculptus Rowley, *in* Greene, 1903. Kesling and Smith, 1962,
  p. 185.
  Devonian: United States

Gennaeocrinus similis Goldring, 1935. Kesling and Smith, 1962, p. 185.
  Devonian: United States

Gennaeocrinus simulans Rowley, *in* Greene, 1904. Kesling and Smith, 1962,
  p. 185.
  Devonian: United States

Gennaeocrinus variabilis Kesling and Smith, 1962, p. 174, Pls. 1-9;
  Figs. 1, 2. Kesling, 1964c, p. 149.
  Devonian, Erian, Bell Shale: United States: Michigan

Gennaeocrinus sp. Fraunfelter, 1965, p. 204, fig. 1.
  Devonian, Bradfordian, Glen Park Formation: United States: Missouri

GERAOCRINUS Ulrich, 1924. *G. sculptus* Ulrich, 1924. Moore and Laudon,
  1943b, p. 31. Ubaghs, 1953, p. 747.
  Ordovician

GEROLDICRINUS Jaekel, 1918. *Lecanocrinus roemeri* Schultze, 1866. Moore
  and Laudon, 1943b, p. 76. Ubaghs, 1953, p. 756. Arendt and Hecker,
  1964, p. 94.
  Devonian

GILBERTSOCRINUS Phillips, 1836. *G. calcaratus* Phillips, 1836. Moore and
  Laudon, 1943b, p. 83; 1944, p. 185, Pl. 70, fig. 4. Cuénot, 1948,
  p. 69. Ubaghs, 1953, p. 737. Van Sant, *in* Van Sant and Lane, 1964,
  p. 104. Arendt and Hecker, 1964, p. 95. Moore and Jeffords, 1968,
  p. 16, fig. 5, no. 3.
  Devonian—lower Carboniferous

Gilbertsocrinus alpenensis Ehlers, 1925. Kier, 1952, p. 70, Pl. 4,
  figs. 14-16.
  Devonian, Erian, Silica Formation, Bell Shale: United States: Ohio,
  Michigan

Gilbertsocrinus bairstowi Wright, 1958, p. 326, Pl. 78, figs. 8-10, 12, 13.
  Carboniferous, Tournaisian: England

Gilbertsocrinus bollandensis Wright, 1958, p. 324, Pl. 76, figs. 12, 16-18; Pl. 77, figs. 9, 10, 15, 21 (*non* Pl. 77, fig. 16, typographical error).
Carboniferous, Tournaisian: Scotland

Gilbertsocrinus bursa Phillips, 1836. Wright, 1958, p. 322, Pl. 77, figs. 1-4.
Carboniferous, Tournaisian: Scotland

Gilbertsocrinus calcaratus Phillips, 1836. Wright, 1958, p. 318, Pl. 77, figs. 17, 23, 24; Pl. 78, figs. 14-16 (*non* Pl. 68, figs. 14-16, typographical error).
Carboniferous, Tournaisian: Scotland

Gilbertsocrinus compactus Wright, 1958, p. 323, Pl. 78, figs. 11, 19-21.
Carboniferous, Tournaisian: Scotland

Gilbertsocrinus coplowensis Wright, 1958, p. 325, Pl. 76, figs. 7, 8.
Carboniferous, Tournaisian: Scotland

Gilbertsocrinus dispansus Wachsmuth and Springer, 1897. Van Sant, *in* Van Sant and Lane, 1964, p. 105.
Mississippian, Osagean, Borden Group: United States: Indiana

Gilbertsocrinus fiscellus (Meek and Worthen), 1866. Moore and Laudon, 1944, p. 187. Pl. 73, fig. 4.
Mississippian, Osagean, Burlington Limestone: United States: Iowa, Missouri

Gilbertsocrinus globosus Wright, 1958, p. 325, Pl. 76, figs. 22-25.
Carboniferous, Tournaisian: Scotland

Gilbertsocrinus konincki Grenfell, 1876. Wright, 1958, p. 321, Pl. 76, fig. 10; Pl. 77, figs. 6, 8, 11-14, 16, 18, 25; Pl. 78, figs. 5, 6; Fig. 132.
Carboniferous, Tournaisian: Scotland

Gilbertsocrinus konincki? Grenfell, 1876. Wright, 1958, p. 321, Pl. 77, figs. 5, 19, 20.
Carboniferous, Tournaisian: Scotland

Gilbertsocrinus mammillaris Phillips, 1836. Wright, 1958, p. 319, Pl. 76, fig. 21; Pl. 77, figs. 7, 22; Pl. 78, figs. 1-4, 7, 17, 18.
Carboniferous, Tournaisian: Scotland

Gilbertsocrinus multicalcaratus Goldring, 1935. Goldring, 1945, p. 60.
Devonian, Erian, Moscow Shale: United States: New York

Gilbertsocrinus ohioensis Stewart, 1940. Kier, 1952, p. 71, Pl. 4, figs. 4-7. La Rocque and Marple, 1955, p. 94, fig. 234.
Devonian, Erian, Silica Formation: United States: Ohio

Gilbertsocrinus robustus (Hall), 1860. Van Sant, *in* Van Sant and Lane, 1964, p. 105.
Trematocrinus robustus Hall, 1860. Van Sant, *in* Van Sant and Lane, 1964, p. 105.
Mississippian, Osagean, Keokuk Limestone: United States: Iowa

Gilbertsocrinus simplex Portlock, 1843. Wright, 1958, p. 326, Pl. 76, fig. 9.
Carboniferous: Ireland

Gilbertsocrinus spinigerus (Hall), 1862. Moore and Laudon, 1944, p. 187, Pl. 73, fig. 5. Goldring, 1950, p. 35, Pl. 2, figs. 12-14.
Devonian, Erian, Hamilton Group, Ludlowville Shale: Canada: Ontario; United States: New York

Gilbertsocrinus tuberosus (Lyon and Casseday), 1859. Moore and Laudon, 1944, p. 187, Pl. 73, figs. 1, 2. Cuénot, 1948, p. 69, fig. 92.

Ubaghs, 1953, p. 737, fig. 57. Lane, 1963c, p. 1004, Pl. 128,
fig. 1. Van Sant, *in* Van Sant and Lane, 1964, p. 105, Pl. 8,
figs. 10, 12, 14, 15; Figs. 20, no. 5; 35.
Mississippian, Osagean, Borden Group, Fort Payne Chert: United States:
Indiana, Kentucky

Gilbertsocrinus typus (Hall), 1860. Moore and Laudon, 1943b, p. 138,
Pl. 9, fig. 9; 1944, p. 187, Pl. 73, fig. 3.
Mississippian, Osagean, Burlington Limestone, Keokuk Limestone: United
States: Iowa, Missouri, Indiana

Gilbertsocrinus sp. Wright, 1958, p. 327, Appendix, Pl. A, fig. 14.
Carboniferous, Lower Limestone: Scotland

GILMOCRINUS Laudon, 1933. *G. iowensis* Laudon, 1933. Moore and Laudon,
1943b, p. 59; 1944, p. 169. Arendt and Hecker, 1964, p. 89.
Mississippian

Gilmocrinus iowensis Laudon, 1933. Moore and Laudon, 1944, p. 169, Pl. 60,
fig. 3.
Mississippian, Kinderhookian, Hampton Formation: United States: Iowa

GISSOCRINUS Angelin, 1878. *G. arthriticus* Angelin, 1878. Moore and
Laudon, 1943b, p. 51; 1944, p. 155. Bouška, 1944, p. 584, fig. 1.
Ramsbottom, 1951a, p. 491, fig. 1. Ubaghs, 1953, p. 750. Yakovlev,
1954b, p. 1088, fig. 1b; 1956a, p. 90, fig. 36, no. 2; 1958, p. 75,
fig. 36, no. 2; 1964a, p. 83, fig. 38, no. 2; 1964b, fig. 59b.
Arendt and Hecker, 1964, p. 83.
Silurian-Devonian

Gissocrinus crassus Bouška, 1944, p. 588, Pl. 2, figs. 1-4.
Upper Silurian: Czechoslovakia: Bohemia

Gissocrinus cyrili Bouška, 1944, p. 587, Pl. 1, figs. 6-9; Fig. 2.
Upper Silurian: Czechoslovakia: Bohemia

Gissocrinus intermedius Bouška, 1944, p. 590, Pl. 2, fig. 9.
Upper Silurian: Czechoslovakia: Bohemia

Gissocrinus cf. *intermedius* Bouška, 1944, p. 591, Pl. 2, fig. 10.
Upper Silurian: Czechoslovakia: Bohemia

Gissocrinus involutus Bouška, 1944, p. 586, Pl. 1, figs. 1-5a.
Upper Silurian: Czechoslovakia: Bohemia

Gissocrinus luculentus Ramsbottom, 1951a, p. 494, Pl. 9, figs. 5-8;
Figs. 6, 7.
Silurian, Wenlockian, Wenlock Limestone: England

Gissocrinus ludensis Sollas (MS) *in* Ramsbottom, 1958, p. 109, Pl. 21,
figs. 8, 9.
Silurian, lower Ludlovian: England

Gissocrinus lyoni Springer, 1926. Moore and Laudon, 1943b, p. 131, Pl. 2,
fig. 1; 1944, p. 155, Pl. 55, fig. 12.
Silurian, Niagaran, Louisville Limestone: United States: Kentucky

Gissocrinus prantli Bouška, 1944, p. 589, Pl. 2, figs. 5-8.
Upper Silurian: Czechoslovakia: Bohemia

Gissocrinus quadratus Springer, 1926. Strimple, 1963d, p. 67, figs. 18a-d.
Silurian, Niagaran, Henryhouse Formation: United States: Oklahoma

Gissocrinus squamiferus (Salter), 1873. Ramsbottom, 1951a, p. 491, Pl. 9,
figs. 1-4; Figs. 2-5.
Silurian, Wenlockian, Wenlock Limestone: England

Gissocrinus typus Bather, 1893. Sieverts-Doreck, 1952, p. 427, fig. 5.
Silurian: Gotland

GLAUCOCRINUS Parks and Alcock, 1912. *G. falconeri Parks and Alcock, 1912. Moore and Laudon, 1943b, p. 31. Ubaghs, 1953, p. 747.
Ordovician

GLAUKOSOCRINUS Strimple, 1951. *Malaiocrinus parvisculus Moore and Plummer, 1940. Strimple, 1951d, p. 191.
Pennsylvanian, Desmoinesian

Glaukosocrinus naturalis Strimple, 1962e, p. 24, Pl. 3, figs. 12-15.
Pennsylvanian, Desmoinesian, Oologah Limestone: United States: Oklahoma

*Glaukosocrinus parvisculus (Moore and Plummer), 1940. Strimple, 1951d, p. 191, figs. 13-16. Strimple, 1962e, p. 23.
Pennsylvanian, Desmoinesian, Oologah Limestone: United States: Oklahoma

GLOBOCRINUS Weller and others, 1920. *Batocrinus unionensis Worthen, 1890. Moore and Laudon, 1943b, p. 95.
Mississippian, Chesterian

*Globocrinus unionensis (Worthen), 1890. Butts, 1941, p. 249, Pl. 132, figs. 9-11.
Mississippian, Chesterian, Gasper limestone: United States: Virginia

GLOBOCRINUS Washburn, 1968 (non Weller and others, 1920). *G. bulbus Washburn, 1968. Washburn, 1968, p. 122.
Pennsylvanian, Morrowan

*Globocrinus bulbus Washburn, 1968, p. 122, Pl. 2, figs. 1-6.
Pennsylvanian, Morrowan, Bridal Veil Falls Member: United States: Utah

GLOSSOCRINUS Goldring, 1923. *G. naplesensis Goldring, 1923. Moore and Laudon, 1943b, p. 54, fig. 6.
Devonian

GLYPTOCRINUS Hall, 1847. *G. decadactylus Hall, 1847. Moore and Laudon, 1943b, p. 96, fig. 13; 1944, p. 199, Pl. 70, fig. 10. Cuénot, 1948, p. 69. Moore, 1950, p. 42, fig. 11. Ubaghs, 1953, p. 741, fig. 139g.
Ordovician-Silurian

Glyptocrinus basalis McCoy, 1850 = Balacrinus basalis

*Glyptocrinus decadactylus Hall, 1847. Bowsher, 1955a, p. 7, Pl. 1, figs. 6-9; Pl. 2, fig. 3.
Ordovician, Cincinnatian, Corryville Shale, Fairmount Shale: United States: Ohio

Glyptocrinus dyeri Meek, 1872. Moore and Laudon, 1943b, p. 142, Pl. 13, fig. 4; 1944, p. 199, Pl. 73, fig. 23. Bowsher, 1955a, p. 7, Pl. 1, figs. 1-5, 10, 11.
Ordovician, Cincinnatian, Waynesville Shale, McMillan Formation, Corryville Shale, Arnheim Formation: United States: Ohio, Indiana

Glyptocrinus globularis Nicholson and Etheridge, 1881 = Diabolocrinus globularis

Glyptocrinus marginatus Billings, 1857 = Archaeocrinus marginatus

Glyptocrinus ottawaensis Wilson, 1946, p. 27.
Glyptocrinus ramulosus Billings, 1859 (non Billings, 1856). Wilson, 1946, p. 27.
Ordovician, Champlainian, Hull Formation: Canada: Quebec

Glyptocrinus quinquepartitus Billings, 1859. Wilson, 1946, p. 28.
Ordovician, Champlainian, Cobourg Formation: Canada: Quebec

Glyptocrinus ramulosus Billings, 1856. Wilson, 1946, p. 28.
Ordovician, Champlainian, Cobourg Formation, Sherman Fall Formation, Hull Formation: Canada: Quebec

Glyptocrinus ramulosus Billings, 1859 (non Billings, 1856) = Glyptocrinus ottawaensis

GNORIMOCRINUS Wachsmuth and Springer, 1879. *Taxocrinus expansus Angelin, 1878. Moore and Laudon, 1943b, p. 69, fig. 10; 1944, p. 177, Pl. 67, fig. 12. Ubaghs, 1953, p. 755.
Silurian

Gnorimocrinus cirrifer Springer, 1920. Moore and Laudon, 1944, p. 177, Pl. 54, fig. 12.
Silurian, Niagaran, Beech River Formation: United States: Tennessee

Gnorimocrinus pontotocensis Strimple, 1952b, p. 78, figs. 10, 11; 1963d, p. 125, Pl. 8, fig. 8.
Silurian, Niagaran, Henryhouse Formation: United States: Oklahoma

GONIOCRINUS Miller and Gurley, 1890. *G. sculptilis Miller and Gurley, 1890. Moore and Laudon, 1943b, p. 55; 1944, p. 158. Ubaghs, 1953, p. 752. Arendt and Hecker, 1964, p. 87.
Mississippian

Goniocrinus angulatus Laudon, Parks and Spreng, 1952, p. 552, Pl. 65, figs. 22, 24; Pl. 69, fig. 4.
Mississippian, Kinderhookian, Banff formation: Canada: Alberta

Goniocrinus harrisi (Miller), 1880. Van Sant, in Van Sant and Lane, 1964, p. 99.
Mississippian, Osagean, Borden Group: United States: Indiana

*Goniocrinus sculptilis Miller and Gurley, 1890. Moore and Laudon, 1944, p. 158, Pl. 57, fig. 7. Laudon and Severson, 1953, p. 514, Pl. 51, fig. 22.
Mississippian, Kinderhookian, Hampton Formation, Lodgepole Limestone: United States: Iowa, Montana

GOTHOCRINUS Bather, 1893. *G. gracilis Bather, 1893. Moore and Laudon, 1943b, p. 54. Ubaghs, 1953, p. 751. Arendt and Hecker, 1964, p. 86.
Silurian

GRAFFHAMICRINUS Strimple, 1961. *G. acutus Strimple, 1961. Strimple, 1961g, p. 123.
Pennsylvanian, Atokan-Virgilian

*Graffhamicrinus acutus Strimple, 1961g, p. 124, Pl. 10, figs. 4-8; Pl. 12, figs. 4-6; Pl. 13, fig. 7; Pl. 15, fig. 6; Pl. 19, fig. 2; Figs. 22a, b.
Pennsylvanian, Desmoinesian, Holdenville Formation: United States: Oklahoma

Graffhamicrinus acutus spicatus Strimple, 1961g, p. 127, Pl. 17, figs. 1-3.
Pennsylvanian, Desmoinesian, Holdenville Shale: United States: Oklahoma

Graffhamicrinus aristatus (Strimple), 1949. Strimple, 1961g, p. 124.
Delocrinus aristatus Strimple, 1949d, p. 18, Pl. 3, figs. 18-21; 1961g, p. 124.
Pennsylvanian, Atokan, Pumpkin Creek Limestone: United States: Oklahoma

Graffhamicrinus armatura (Strimple), 1949. Strimple, 1961g, p. 124.
Delocrinus armatura Strimple, 1949d, p. 18, Pl. 3, figs. 3, 6; 1961g, p. 124.
Pennsylvanian, Atokan, Pumpkin Creek Limestone: United States: Oklahoma

Graffhamicrinus bispinosus (Moore and Plummer), 1940. Strimple, 1961g, p. 123.
Delocrinus bispinosus Moore and Plummer, 1940. Moore and Laudon, 1944, p. 171, Pl. 65, fig. 8. Strimple, 1961g, p. 123.
Pennsylvanian, Missourian, Mineral Wells Formation: United States: Texas

Graffhamicrinus bullatus (Moore and Plummer), 1940. Strimple, 1961g, p. 123.
  Delocrinus bullatus Moore and Plummer, 1940. Strimple, 1961g, p. 123.
  Pennsylvanian, Desmoinesian, Millsap Lake Formation: United States: Texas

Graffhamicrinus granulosus (Moore and Plummer), 1940. Strimple, 1961g, p. 123.
  Delocrinus granulosus Moore and Plummer, 1940. Moore and Laudon, 1944, p. 171, Pl. 65, fig. 13. Strimple, 1961g, p. 123.
  Pennsylvanian, Desmoinesian-Missourian, Wewoka Formation, Mineral Wells Formation: United States: Oklahoma, Texas

Graffhamicrinus granulosus var. *moniformis* (Moore and Plummer), 1940. Strimple, 1961g, p. 124.
  Delocrinus granulosus var. *moniformis* Moore and Plummer, 1940. Strimple, 1961g, p. 124.
  Pennsylvanian: United States

Graffhamicrinus granulosus var. *zonatus* (Moore and Plummer), 1940. Strimple, 1961g, p. 124.
  Delocrinus granulosus var. *zonatus* Moore and Plummer, 1940. Strimple, 1961g, p. 124.
  Pennsylvanian: United States

Graffhamicrinus graphicus (Moore and Plummer), 1940. Strimple, 1961g, p. 124.
  Delocrinus graphicus Moore and Plummer, 1940. Moore and Laudon, 1944, p. 173, Pl. 65, fig. 16. Strimple, 1961g, p. 124.
  Pennsylvanian, Missourian, Graford Formation: United States: Texas

Graffhamicrinus magnificus (Strimple), 1947. Strimple, 1961g, p. 124.
  Delocrinus magnificus Strimple, 1947, p. 3, Pl. 1, figs. 1-4; Pl. 2, fig. 1. Strimple, 1961g, p. 124.
  Pennsylvanian, Virgilian, Haskell Limestone: United States: Kansas

Graffhamicrinus nodosarius (Strimple), 1939. Strimple, 1961g, p. 123.
  Delocrinus nodosarius Strimple, 1939. Strimple, 1961g, p. 123.
  Pennsylvanian: United States

Graffhamicrinus papulosus (Moore and Plummer), 1940. Strimple, 1961g, p. 124.
  Delocrinus papulosus Moore and Plummer, 1940. Moore and Laudon, 1944, p. 171, Pl. 65, fig. 26. Strimple, 1961g, p. 124.
  Pennsylvanian, Missourian, Graford Formation: United States: Texas

Graffhamicrinus paucinodus (Moore and Plummer), 1940. Strimple, 1961g, p. 124.
  Delocrinus paucinodus Moore and Plummer, 1940. Strimple, 1961g, p. 124.
  Pennsylvanian: United States

Graffhamicrinus pictus (Moore and Plummer), 1940. Strimple, 1961g, p. 124.
  Delocrinus pictus Moore and Plummer, 1940. Moore and Laudon, 1944, p. 171, Pl. 65, fig. 12. Strimple, 1961g, p. 124.
  Pennsylvanian, Missourian, Graford Formation: United States: Texas

Graffhamicrinus somersi (Whitfield), 1882. Strimple, 1961g, p. 124.
  Cyathocrinus somersi Whitfield, 1882. Strimple, 1961g, p. 124.
  Delocrinus somersi (Whitfield), 1882. Moore and Laudon, 1944, p. 171, Pl. 65, fig. 15.
  Pennsylvanian, Pottsvillian, Pottsville Group: United States: Ohio

Graffhamicrinus stullensis (Strimple), 1947. Strimple, 1961g, p. 124.
  Delocrinus stullensis Strimple, 1947, p. 5, Pl. 2, figs. 4-6. Strimple, 1961g, p. 124.
  Pennsylvanian, Virgilian, Stull Shale Member: United States: Kansas

Graffhamicrinus subcoronatus (Moore and Plummer), 1940. Strimple, 1961g, p. 124.
  Delocrinus subcoronatus Moore and Plummer, 1940. Strimple, 1961g, p. 124.
  Pennsylvanian: United States
Graffhamicrinus tulsaensis Strimple, 1962e, p. 16, Pl. 4, figs. 1-4.
  Pennsylvanian, Desmoinesian, Oologah Limestone: United States: Oklahoma
Graffhamicrinus variabilis Strimple, 1962e, p. 17, Pl. 4, figs. 5-8.
  Pennsylvanian, Desmoinesian, Oologah Limestone: United States: Oklahoma
Graffhamicrinus waughi (Moore and Strimple), 1941. Strimple, 1961g, p. 124.
  Delocrinus waughi Moore and Strimple, 1941. Strimple, 1961g, p. 124.
  Pennsylvanian: United States
Graffhamicrinus wolforum (Moore and Plummer), 1940. Strimple, 1961g, p. 123.
  Delocrinus wolforum Moore and Plummer, 1940. Hattin, 1958, p. 701, Pl. 98, figs. 1-6. Strimple, 1961g, p. 123.
  Pennsylvanian, Virgilian, Graham Formation: United States: Texas
GRAPHIOCRINUS de Koninck and Le Hon, 1854. *G. encrinoides* de Koninck and Le Hon, 1854. Moore and Laudon, 1943b, p. 60; 1944, p. 171. Cuénot, 1948, p. 61. Arendt and Hecker, 1964, p. 89.
  Mississippian-Permian
Graphiocrinus amplior Wanner, 1924. Wanner, 1949a, p. 17, Pl. 1, fig. 16.
  Permian: Timor
Graphiocrinus? austini Wright, 1952a, p. 104, Pl. 36, figs. 16, 17.
  Carboniferous, Tournaisian: Ireland
Graphiocrinus? beyrichi (Wanner), 1916. Wanner, 1949a, p. 24.
  Delocrinus beyrichi Wanner, 1916. Wanner, 1949a, p. 24.
  Permian: Timor
Graphiocrinus bridgeportensis Strimple, 1951a, p. 201, Pl. 37, figs. 9-12; 1962f, p. 137.
  Pennsylvanian, Missourian, Lake Bridgeport Shale, Brownwood Shale: United States: Texas
Graphiocrinus? crassus (Wanner), 1924. Wanner, 1949a, p. 26.
  Ceriocrinus crassus Wanner, 1924. Wanner, 1949a, p. 26.
  Permian: Timor
Graphiocrinus cristatus Yakovlev, *in* Yakovlev and Ivanov, 1956, p. 50, Pl. 12, fig. 10.
  Middle Carboniferous: Russia: Southern Urals
Graphiocrinus dactylus Hall, 1860 = *Cymbiocrinus dactylus*
Graphiocrinus? declivis Wanner, 1937. Wanner, 1949a, p. 27.
  Permian: Timor
Graphiocrinus deflectus Strimple, 1962e, p. 11, Pl. 7, figs. 8-10.
  Pennsylvanian, Desmoinesian, Oologah Limestone: United States: Oklahoma
Graphiocrinus delicatulus Moore, 1939. Strimple, 1962f, p. 137.
  Pennsylvanian, Virgilian, Brownville Limestone Member: United States: Oklahoma
Graphiocrinus? depressus (Wanner), 1916. Wanner, 1949a, p. 21, Pl. 1, figs. 26, 27; Pl. 2, fig. 1.
  Delocrinus depressus Wanner, 1916. Wanner, 1949a, p. 21.
  Permian: Timor

Graphiocrinus? cf. *depressus* (Wanner), 1916. Wanner, 1949a, p. 22, Pl. 2, figs. 2-4.
Permian: Timor

Graphiocrinus? depressus var. *labiosa* Wanner, 1949a, p. 23, Pl. 2, figs. 5-7.
Permian: Timor

Graphiocrinus encrinoides de Koninck and Le Hon, 1854. Wright, 1952a, p. 103, Pl. 36, fig. 6.
Carboniferous, Tournaisian: Ireland

Graphiocrinus? excavatissimus (Wanner), 1916. Wanner, 1949a, p. 24, Pl. 2, fig. 8.
Delocrinus excavatissimus Wanner, 1916. Wanner, 1949a, p. 24.
Permian: Timor

Graphiocrinus? excavatissimus var. *ornata* Wanner, 1949a, p. 25, Pl. 2, figs. 9-11.
Permian: Timor

Graphiocrinus? fermanaghensis Wright, 1952a, p. 105, Pl. 36, figs. 5, 5a.
Carboniferous, Visean?: Ireland

Graphiocrinus? kansasensis Strimple, 1963a, p. 73, Pl. 12, figs. 1-4.
Permian, Wolfcampian, Bennett Shale: United States: Kansas

Graphiocrinus kingi Moore and Plummer, 1940. Moore and Laudon, 1944, p. 171, Pl. 65, fig. 18. Strimple, 1962f, p. 137.
Pennsylvanian, Missourian, Mineral Wells Formation: United States: Texas

Graphiocrinus lineatus Strimple, 1963c, p. 191, figs. 1-3.
Pennsylvanian, Missourian, Wann Formation: United States: Oklahoma

Graphiocrinus longicirrifer Wachsmuth and Springer, 1889 = *Holcocrinus longicirrifer*

Graphiocrinus mcadamsi (Meek and Worthen), 1873. Van Sant, *in* Van Sant and Lane, 1964, p. 83, Pl. 3, fig. 3.
Mississippian, Osagean, Keokuk Limestone, Borden Group: United States: Illinois, Indiana

Graphiocrinus ovoides Wanner, 1949a, p. 19, Pl. 1, figs. 20-25.
Permian: Timor

Graphiocrinus pumilus (Wanner), 1916. Wanner, 1949a, p. 14, Pl. 1, figs. 9-15; Figs. 8-10.
Delocrinus verbeeki var. *pumila* Wanner, 1916. Wanner, 1949a, p. 14.
Permian: Timor

Graphiocrinus quinquelobus Wanner, 1916. Wanner, 1949a, p. 28, Pl. 2, figs. 18-21; Fig. 12.
Permian: Timor

Graphiocrinus? rotundatus (Wanner), 1916. Wanner, 1949a, p. 27, Pl. 2, fig. 17; Fig. 11.
Delocrinus rotundatus Wanner, 1916. Wanner, 1949a, p. 27.
Permian: Timor

Graphiocrinus? rugosus (Wanner), 1916. Wanner, 1949a, p. 26, Pl. 2, figs. 12-16.
Delocrinus rugosus Wanner, 1916 (*non* Shumard, 1858). Wanner, 1949a, p. 26.
Permian: Timor

Graphiocrinus sampsoni Weller, 1909 = *Acylocrinus sampsoni*

Graphiocrinus scopulus Lane and Webster, 1966, p. 46, Pl. 10, figs. 4, 8. Webster and Lane, 1967, p. 26, Pl. 4, figs. 1-3.
Permian, Wolfcampian, Bird Spring Formation: United States: Nevada

Graphiocrinus smythi Wright, 1934 = *Holcocrinus smythi*

Graphiocrinus stantonensis Strimple, 1939. Strimple, 1962f, p. 137, figs. 1-3.
Pennsylvanian, Missourian, Wann Formation: United States: Oklahoma

Graphiocrinus subamplior Wanner, 1949a, p. 18, Pl. 1, figs. 17-19.
Permian: Timor

Graphiocrinus timanicus Yakovlev, *in* Yakovlev and Ivanov, 1956, p. 74, Pl. 21, fig. 2. Arendt and Hecker, 1964, p. 89, Pl. 12, fig. 2.
Lower Permian: Russia: Timan

Graphiocrinus? timoricus Wanner, 1916. Wanner, 1949a, p. 29, Pl. 2, figs. 22-27.
Permian: Timor

Graphiocrinus? timoricus var. *spinosa* Wanner, 1924. Wanner, 1949a, p. 31.
Permian: Timor

Graphiocrinus treuteri (Yakovlev), 1927. Yakovlev, *in* Yakovlev and Ivanov, 1956, p. 64, Pl. 11, fig. 13.
Permian: Russia: Ural Mountains

Graphiocrinus verbeeki (Wanner), 1916. Wanner, 1949a, p. 12, Pl. 1, figs. 3-8.
Delocrinus verbeeki Wanner, 1916. Wanner, 1949a, p. 12.
Delocrinus beyrichi var. *nustoiensis* Wanner, 1916. Wanner, 1949a, p. 12.
Delocrinus verbeeki var. *levis* Wanner, 1916. Wanner, 1949a, p. 12.
Permian: Timor

GRENPRISIA Moore, 1962. **Ottawacrinus billingsi* Springer, 1911. Moore, 1962b, p. 38.
Ordovician, Champlainian

*Grenprisia billingsi (Springer), 1911. Moore, 1962b, p. 38.
Ottawacrinus billingsi Springer, 1911. Moore and Laudon, 1944, p. 158, Pl. 53, fig. 23. Moore, 1962b, p. 38.
Ordovician, Champlainian, Hull Limestone: Canada: Ontario

Grenprisia springeri Moore, 1962b, p. 38.
Ottawacrinus typus Springer, 1911 (*non* Billings, 1887). Moore, 1962b, p. 38.
Ordovician, Champlainian, Hull Limestone: Canada: Ontario

GRIPHOCRINUS Kirk, 1945. **Rhodocrinus (Acanthocrinus) nodulosus* Hall, 1862. Kirk, 1945a, p. 350. Ubaghs, 1953, p. 737.
Devonian, Erian

Griphocrinus halli (Lyon), 1862. Kirk, 1945a, p. 351.
Rhodocrinus halli Lyon, 1862. Kirk, 1945a, p. 351.
Devonian, Erian, Beechwood Limestone: United States: Kentucky

Griphocrinus insculptus (Goldring), 1935. Kirk, 1945a, p. 352.
Rhodocrinus insculptus Goldring, 1935. Kirk, 1945a, p. 352.
Devonian, Erian, Hamilton Group: United States: New York

*Griphocrinus nodulosus (Hall), 1862. Kirk, 1945a, p. 352. Le Maître, 1958b, p. 128.
Rhodocrinus (Acanthocrinus) nodulosus Hall, 1862. Kirk, 1945a, p. 352.
Devonian, Erian, Tichenor horizon of Hamilton Group: United States: New York; Algeria

Griphocrinus ovetensis Breimer, 1962a, p. 15, Pl. 1, figs. 1, 2; Fig. 1.
Devonian, Couvinian: Spain

Griphocrinus wachsmuthi (Oehlert), 1887. Kirk, 1945a, p. 353.
Raphanocrinus? wachsmuthi Oehlert, 1887. Kirk, 1945a, p. 353.
Middle Devonian: France

GRYPOCRINUS Strimple, 1963. **G. genuinus* Strimple, 1963. Strimple, 1963d, p. 59. Brower, 1966, p. 628.
Silurian, Niagaran

*Grypocrinus genuinus Strimple, 1963d, p. 59, Pl. 2, figs. 7-9; Figs. 16a-c.
Silurian, Niagaran, Henryhouse Formation: United States: Oklahoma

Grypocrinus multibrachiatus Brower, 1966, p. 629, Pl. 75, figs. 14-16, 22; Figs. 2B, F.
Silurian, Niagaran, Beech River Member: United States: Tennessee

Habrocrinus benedicti Foerste, 1917 (*non* Miller), 1892 = *Stiptocrinus* sp.

Habrocrinus benedicti Slocom, 1908, La Rocque and Marple, 1955 = *Saccocrinus benedicti*

Habrocrinus farringtoni Slocom. 1908 = *Stiptocrinus farringtoni*

Habrocrinus sp. Foerste, 1917 = *Stiptocrinus* sp.

HADROCRINUS Lyon, 1869. **H. discus* Lyon, 1869. Moore and Laudon, 1943b, p. 99. Ubaghs, 1953, p. 742. Philip, 1963, fig. 5a.
Devonian

*Hadrocrinus discus Lyon, 1869. Ubaghs, 1956, p. 557, fig. 4.
Devonian: United States

Hadrocrinus hispaniae Schmidt, 1931 = *Trybliocrinus flatheanus*

HAERETOCRINUS Moore and Plummer, 1940. **H. missouriensis* Moore and Plummer, 1940. Moore and Laudon, 1943b, p. 55. Arendt and Hecker, 1964, p. 88.
Pennsylvanian

Haeretocrinus depressus Strimple, 1962e, p. 44, Pl. 1, figs. 9-12; Pl. 7, figs. 5-7.
Pennsylvanian, Desmoinesian, Oologah Limestone: United States: Oklahoma

Haeretocrinus intermedius Strimple, 1961g, p. 97, Pl. 15, figs. 1-3.
Pennsylvanian, Desmoinesian, Holdenville Shale: United States: Oklahoma

Haeretocrinus turbinatus Strimple, 1952d, p. 245, figs. 1-7.
Pennsylvanian, Missourian, Wann Formation: United States: Oklahoma

Haeretocrinus washburni (Beede), 1899. Strimple, 1961g, p. 97.
Scaphiocrinus washburni Beede, 1899. Strimple, 1961g, p. 97.
Pennsylvanian, Virgilian: United States: Kansas

HALLOCRINUS Goldring, 1923. **Cyathocrinus ornatissimus* Hall, 1843. Moore and Laudon, 1943b, p. 54; 1944, p. 157.
Devonian—Lower Carboniferous

Hallocrinus abberrans Schmidt, 1941, p. 167, Pl. 13, fig. 7.
Devonian, Coblentzian: Germany

*Hallocrinus ornatissimus (Hall), 1843. Moore and Laudon, 1943b, p. 132, Pl. 3, fig. 9; 1944, p. 157, Pl. 57, fig. 24.
Devonian, Erian-Senecan, Hamilton Group, Portage Formation: United States: New York; Canada: Ontario

HALYSIOCRINUS Ulrich, 1886. **Cheirocrinus dactylus* Hall, 1860. Moore and Laudon, 1943b, p. 30, fig. 3; 1944, p. 145. Ubaghs, 1953, p. 746, fig. 140L. Moore, 1962a, p. 31, figs. 3, no. 4; 5, no. 8; 20; 1962b, p. 12, fig. 4, no. 3. Strimple, 1963d, p. 48, fig. 12a. Van Sant, *in* Van Sant and Lane, 1964, p. 69, fig. 25, no. 2. Arendt and Hecker, 1964, p. 80.
Devonian-Mississippian, ?Permian

Halysiocrinus bradleyi (Meek and Worthen), 1873. Moore, 1962a, p. 33, fig. 7J. Van Sant, *in* Van Sant and Lane, 1964, p. 69, fig. 25, no. 1.
Mississippian, Osagean, Edwardsville Formation: United States: Indiana

Halysiocrinus carinatus Springer, 1926. Moore, 1962a, p. 33.
  Devonian, Erian, Alpena Limestone, Cedar Valley Formation, Sellersburg
  Limestone: United States: Michigan, Iowa, Indiana

*Halysiocrinus dactylus (Hall), 1860. Moore and Laudon, 1944, p. 145,
  Pl. 55, fig. 5. Moore, 1962a, p. 33, Pl. 2, fig. 1; Fig. 7I. Van
  Sant, *in* Van Sant and Lane, 1964, p. 69, fig. 25, no. 3.
  Mississippian, Osagean, Burlington Limestone: United States: Iowa

Halysiocrinus elephantinus Laudon, 1936. Moore, 1962a, p. 33, fig. 7H.
  Devonian, Erian, Cedar Valley Formation: United States: Iowa

Halysiocrinus granuliferus (Rowley, *in* Greene), 1903. Moore, 1962a, p. 33.
  Calceocrinus? granuliferus Rowley, *in* Greene, 1903. Moore, 1962a, p. 33.
  Mississippian, Osagean, New Providence Shale: United States: Kentucky

Halysiocrinus keyserensis Springer, 1926 = *Synchirocrinus keyserensis*

Halysiocrinus? lamellosus (Hall), 1860. Moore, 1962a, p. 33.
  Cheirocrinus lamellosus Hall, 1860. Moore, 1962a, p. 33.
  Mississippian, Osagean, Burlington Limestone: United States: Iowa

Halysiocrinus marylandensis Springer, 1926 = *Synchirocrinus marylandensis*

Halysiocrinus nodosus (Hall), 1860. Moore and Laudon, 1944, p. 147, Pl. 55,
  fig. 9. Ubaghs, 1953, p. 746, figs. 35a-c. Moore, 1962a, p. 33,
  Pl. 2, fig. 3; Figs. 1, 7K. Yakovlev, 1964b, p. 69, fig. 100. Van
  Sant, *in* Van Sant and Lane, 1964, p. 70, fig. 25, no. 4.
  Mississippian, Osagean, Keokuk Limestone: United States: Indiana, Iowa

Halysiocrinus perplexus (Shumard), 1886. Ubaghs, 1953, p. 746, figs. 35d-f.
  Moore, 1962a, p. 33, Pl. 3, fig. 6.
  Mississippian, Osagean, New Providence Shale: United States: Kentucky,
  Tennessee, Indiana

Halysiocrinus robustus (Worthen), 1890. Moore, 1962a, p. 33.
  Mississippian, Osagean, Keokuk Limestone: United States: Illinois

Halysiocrinus secundus (Hall), 1872 = *Deltacrinus secundus*

Halysiocrinus septarmatus Brower, 1966, p. 630, Pl. 75, figs. 8, 9, 19, 20;
  Fig. 2G.
  Devonian, Erian, Cedar Valley Formation: United States: Iowa

Halysiocrinus springeri Brower, 1966, p. 631, Pl. 75, figs. 1-7, 23.
  Halysiocrinus sp. 2 Springer, 1911. Brower, 1966, p. 631.
  Mississippian, Osagean, New Providence Shale: United States: Indiana,
  Kentucky

Halysiocrinus(?) tuberculatus Yakovlev, 1927. Yakovlev, *in* Yakovlev and
  Ivanov, 1956, p. 54, Pl. 9, fig. 11. Arendt and Hecker, 1964, p. 80,
  Pl. 8, fig. 2.
  Permian, Artinskian: Russia: Ural Mountains

Halysiocrinus tunicatus (Hall), 1860. Moore, 1962a, p. 33. Van Sant, *in*
  Van Sant and Lane, 1964, p. 70.
  Mississippian, Osagean, Keokuk Limestone, Fort Payne Chert: United
  States: Iowa, Illinois, Indiana, Kentucky

Halysiocrinus wachsmuthi (Meek and Worthen), 1869. Moore, 1962a, p. 33.
  Mississippian, Osagean, Burlington Limestone: United States: Iowa

Hapalocrinidae cf. *Culicocrinus nodosus* (Müller), 1852. Breimer, 1962a,
  p. 122, Pl. 11, figs. 1-4.
  Devonian, probably Emsian: Spain

HAPALOCRINUS Jaekel, 1895. **H. elegans* Jaekel, 1895. Moore and Laudon,
  1943b, p. 101; 1944, p. 203. Cuénot, 1948, p. 68. Ubaghs, 1953,
  p. 743.
  Silurian-Devonian

Hapalocrinus cirrifer Springer, 1926. Moore and Laudon, 1944, p. 203, Pl. 77, fig. 21.
Silurian, Niagaran, Beech River Formation: United States: Tennessee

Hapalocrinus frechi imbellis Schmidt, 1934. Schmidt, 1941, p. 22, Pl. 3, fig. 1.
Devonian, Hunsrück Schiefer: Germany

Hapalocrinus frechi var. *rarefurcata* Schmidt, 1934. Schmidt, 1941, p. 21, Pl. 1, fig. 2; Pl. 2, fig. 4.
Devonian, Hunsrück Schiefer: Germany

Hapalocrinus gracilis Springer, 1926. Moore and Laudon, 1944, p. 203, Pl. 77, fig. 15.
Silurian, Niagaran, Beech River Formation: United States: Tennessee

Hapalocrinus penniger Schmidt, 1941, p. 18, Pl. 2, figs. 1-3; Fig. 1.
Devonian, Hunsrück Schiefer: Germany

Hapalocrinus quinquepennis Salter, *in* Ramsbottom, 1958, p. 113, Pl. 20, figs. 1-3.
Silurian, Ludlovian: England

Hapalocrinus rauffi (Schmidt), 1934. Schmidt, 1941, p. 23.
Cyttarocrinus rauffi Schmidt, 1934. Schmidt, 1941, p. 23.
Devonian, Hunsrück Schiefer: Germany

Hapalocrinus sp. Seilacher, 1961, p. 15, fig. 1, title cover.
Devonian, Hunsrück Schiefer: Germany

HAPLOCRINITES Steininger, 1834. *Eugeniacrinites mespiliformis* Goldfuss, 1831. Moore and Laudon, 1943b, p. 30, fig. 3; 1944, p. 147, Pl. 52, fig. 2. Bouška, 1946, p. 1, Pl. 1, fig. 5. Ubaghs, 1953, p. 746. Moore, 1962b, p. 10, fig. 3, no. 3. Strimple, 1963d, p. 29, fig. 8c.
Asterocrinites Roemer, 1851. Moore and Laudon, 1943b, p. 30.
Asterocrinus Münster, 1838. Moore and Laudon, 1943b, p. 30.
Haplocrinus Steininger, 1834. Cuénot, 1948, p. 60.
Devonian-Carboniferous

Haplocrinites clio (Hall), 1862. Moore and Laudon, 1943b, p. 130, Pl. 1, fig. 6; 1944, p. 147, Pl. 56, fig. 14. Termier and Termier, 1950, p. 88, Pl. 228, figs. 3-14. Strimple, 1963d, figs. 10b-d.
Devonian, Erian, Givetian, Cherry Valley Limestone, Marcellus Shale: United States: New York; Morocco

Haplocrinites aff. *clio* (Hall), 1862. Le Maître, 1958b, p. 132, Pl. 3, fig. 7.
Devonian, Givetian: Algeria

Haplocrinites granatus (de Koninck), 1869. Wright, 1954a, p. 149.
Carboniferous: England

Haplocrinites mespiliformis (Goldfuss), 1831. Ubaghs, 1953, p. 746, figs. 21c, d. Yakovlev, 1964b, p. 60, fig. 70.
Devonian: Germany

HELIOSOCRINUS Strimple, 1951. *H. aftonensis* Strimple, 1951. Strimple, 1951f, p. 675.
Mississippian, Chesterian

Heliosocrinus aftonensis Strimple, 1951f, p. 676, Pl. 98, figs. 3, 4, 7, 8.
Mississippian, Chesterian, Fayetteville Formation: United States: Oklahoma

HEMIINDOCRINUS Yakovlev, 1926. *H. fredericksi* Yakovlev, 1926. Yakovlev, 1939e, p. 832; 1939f, p. 832. Moore and Laudon, 1943b, p. 64. Arendt and Hecker, 1964, p. 92.
Lower Permian

*Hemiindocrinus fredericksi Yakovlev, 1926. Yakovlev, in Yakovlev and
Ivanov, 1956, p. 60, Pl. 11, figs. 1-5.
Permian: Russia: Ural Mountains

Hemiindocrinus sp. Yakovlev, 1949b, p. 897, fig. 1, II; 1952b, p. 150,
fig. 4.
Lower Permian: Russia

HEMIMOLLOCRINUS Yakovlev, 1930. *H. uralensis Yakovlev, 1930. Moore and
Laudon, 1943b, p. 55. Arendt and Hecker, 1964, p. 92.
Permian, Artinskian

*Hemimollocrinus uralensis Yakovlev, 1930. Yakovlev, in Yakovlev and
Ivanov, 1956, p. 59, Pl. 10, figs. 27, 33. Arendt and Hecker, 1964,
p. 92, Pl. 13, fig. 11.
Permian, Artinskian: Russia: Ural Mountains

HEMISTREPTACRON Yakovlev, 1926. *H. abrachiatum Yakovlev, 1926. Moore
and Laudon, 1943b, p. 50. Yakovlev, 1947d, p. 45, fig. 10; 1956a,
p. 81, figs. 30; 38, nos. 2, 3; 1958, p. 66, figs. 30; 38, nos. 2, 3;
1964a, p. 76, figs. 32; 40, nos. 2, 3. Arendt and Hecker, 1964, p. 84.
Lane, 1967b, p. 13, figs. 3, 7 (fig. 7 listed as fig. 8).
Lower Permian

*Hemistreptacron abrachiatum Yakovlev, 1926. Yakovlev, 1939b, p. 60,
Pl. 10, figs. 3, 4; Fig. 14. Yakovlev, in Yakovlev and Ivanov, 1956,
p. 59, Pl. 10, figs. 4-6; Fig. 16. Arendt and Hecker, 1964, p. 84,
Pl. 9, figs. 13-15; Fig. 126. Arendt, 1964, p. 111, figs. 1, 2;
1968a, p. 100, figs. 1b, 4, 8.
Lower Permian: Russia: Ural Mountains

HERCOCRINUS Hudson, 1907. *H. elegans Hudson, 1907. Moore and Laudon,
1943b, p. 82. Ubaghs, 1953, p. 737. Arendt and Hecker, 1964, p. 95.
Ordovician

HERPETOCRINUS Salter, 1873. *H. fletcheri Salter, 1873. Moore, 1962b,
p. 42. Moore and Jeffords, 1968, p. 12, fig. 4, no. 4.
Silurian

Herpetocrinus ammonis alternicirrus Bather, 1893. Moore, 1962b, p. 42.
Silurian, Gotlandian, Wenlock Limestone, Niagaran, Beech River Formation,
Waldron Shale: Sweden; England; United States: Tennessee

Herpetocrinus ammonis ammonis (Bather), 1893. Moore, 1962b, p. 42.
Herpetocrinus ammonis bijugicirrus Bather, 1893. Moore, 1962b, p. 42.
Silurian, Gotlandian, Wenlock Limestone, Beech River Formation, Waldron
Shale: Sweden; England; United States: Tennessee

Herpetocrinus ammonis bijugicirrus Bather, 1893 = Herpetocrinus ammonis
ammonis

?Herpetocrinus (Myelodactylus) dicirrocrinus Ehrenberg, 1930 = Dicirrocrinus? dicirrocrinus

*Herpetocrinus fletcheri Salter, 1873. Moore, 1962b, p. 42, Pl. 4, fig. 3.
Silurian, Wenlockian, Wenlock Limestone: England

HETEROCRINUS Hall, 1847. *H. heterodactylus Hall, 1847. Moore and Laudon,
1943b, p. 31, figs. 1, 3; 1944, p. 149, Pl. 52, fig. 11. Ubaghs,
1953, p. 747, fig. 140M. Ramsbottom, 1961, p. 10, fig. 1. Moore,
1962b, p. 8, figs. 1, no. 6; 5, no. 3.
Ordovician

Heterocrinus? Huxley and Etheridge, 1878 = Cupulocrinus sepulchrum

*Heterocrinus heterodactylus Hall, 1847. Moore and Laudon, 1944, p. 149,
Pl. 52, fig. 11. Ubaghs, 1953, p. 747, fig. 18c. Yakovlev, 1964b,
p. 66, fig. 92b.

Ordovician, Cincinnatian, Lorraine Formation, Eden Formation: United
States: New York, Ohio

Heterocrinus? milleri Wetherby, 1880 = *Peniculocrinus milleri*

Heterocrinus pachydactylus Sandberger and Sandberger, 1855 = *Botryocrinus pachydactylus*

Heterocrinus propinquus Meek, 1873. Moore, 1962b, p. 14, Pl. 1, fig. 4.
Ordovician, Cincinnatian, Eden Formation: United States: Ohio

Heterocrinus tenuis Billings, 1857. Wilson, 1946, p. 32.
Ordovician, Champlainian, Cobourg Limestone, Sherman Fall Formation, Hull Formation: Canada: Quebec

Undetermined Heterocrinid 1, Ramsbottom, 1961, p. 10, Pl. 6, fig. 5.
Ordovician, Ashgillian, Upper Drummuck Group: England

Undetermined Heterocrinid 2, Ramsbottom, 1961, p. 10, Pl. 3, fig. 8.
Heterocrinus? Strahan and others, 1909. Ramsbottom, 1961, p. 10.
Ordovician, Ashgillian: England

HEXACRINITES Austin and Austin, 1843. *Platycrinites interscapularis*
Phillips, 1841. Moore and Laudon, 1943b, p. 95, fig. 16; 1944, p. 197. Ubaghs, 1953, p. 740, fig. 139f. Strimple, 1963d, p. 89, fig. 27 (part). Xu, 1963, p. 108, fig. 1a. Arendt and Hecker, 1964, p. 96.
Hexacrinus [sic] Austin and Austin, 1843. Wanner, 1942b, p. 209, fig. 5c. Cuénot, 1948, p. 69.
Silurian-Carboniferous

Hexacrinites adaensis Strimple, 1952b, p. 75, figs. 4, 5.
Silurian, Niagaran, Henryhouse Formation: United States: Oklahoma

Hexacrinites anaglypticus var. *stellaris* (Schultze), 1866. Xu, 1963, p. 112, Pl. 1, figs. 12, 13; Fig. 4 (incorrectly ascribes authorship to Goldfuss).
Middle Devonian: China: Guansi Province

Hexacrinites carboniferus (Yakovlev and Ivanov), 1956. Arendt and Hecker, 1964, p. 96, Pl. 15, fig. 1.
Hexacrinus carboniferus Yakovlev and Ivanov, 1956, p. 34, Pl. 4, fig. 10.
Middle Carboniferous (C_2pd): Russia: Moscow Basin

Hexacrinites carinatus Strimple, 1963d, p. 94, Pl. 6, figs. 1, 2; Fig. 30.
Silurian, Niagaran, Henryhouse Formation: United States: Oklahoma

Hexacrinites confragosus Dubatolova, 1964, p. 35, Pl. 4, figs. 1, 2.
Lower Devonian: Russia: Kuznetz Basin

Hexacrinites crispus Dubatolova, 1964 (*non* Quenstedt, 1861), p. 34, Pl. 4, figs. 3, 4.
Lower Devonian: Russia: Kuznetz Basin

Hexacrinites elongatus (Goldfuss), 1839.
Hexacrinus elongatus Sieverts-Doreck, 1950, p. 80, fig. 1.
Devonian, Eifelian: Germany

Hexacrinites invitabilis Dubatolova, 1964, p. 36, Pl. 4, fig. 5.
Lower Devonian: Russia: Kuznetz Basin

Hexacrinites mui Xu, 1963, p. 111, Pl. 1, figs. 7-11; Fig. 3.
Middle Devonian: China: Kwangsi Province

Hexacrinites occidentalis (Wachsmuth and Springer), 1897. Moore and Laudon, 1943b, p. 141, Pl. 12, fig. 12; 1944, p. 197, Pl. 78, fig. 8.
Devonian, Erian, Cedar Valley Limestone: United States: Iowa

Hexacrinites pateraeformis (Schultze), 1866. Yakovlev, 1964b, p. 60, fig. 72.
Devonian, Eifelian: Germany

Hexacrinites yui Xu, 1963, p. 109, Pl. 1, figs. 1-6; Fig. 2.
Middle Devonian: China: Kwangsi Province

Hexacrinites? sp. Philip, 1961, p. 157, Pl. 8, figs. 5-7; Fig. 10.
Lower Devonian: Australia: Victoria

Hexacrinus springeri Thomas, 1924 = *Desmidocrinus springeri*

Hexacrinus? sp. Termier and Termier, 1950, p. 84, Pl. 211, figs. 23-26.
Devonian, Eifelian: Morocco

HIMEROCRINUS Springer, 1921. **Hadrocrinus plenissimus* Lyon, 1869. Moore and Laudon, 1943b, p. 99. Ubaghs, 1953, p. 742.
Coronocrinus Hall, 1859. Moore and Laudon, 1943b, p. 99.
Devonian

HISTOCRINUS Kirk, 1940. **Poteriocrinus coreyi* Worthen, 1875. Moore and Laudon, 1943b, p. 59; 1944, p. 169. Arendt and Hecker, 1964, p. 89.
Mississippian

*Histocrinus coreyi (Worthen), 1875. Moore and Laudon, 1943b, p. 132, Pl. 3, fig. 1; 1944, p. 169, Pl. 60, fig. 22. Van Sant, *in* Van Sant and Lane, 1964, p. 93, Pl. 4, figs. 5, 8, 9; Pl. 5, figs. 8, 9, 12.
?*Cyathocrinus dekadactylus* Lyon and Casseday, 1860. Van Sant, *in* Van Sant and Lane, 1964, p. 93.
Mississippian, Osagean, Borden Group, Keokuk Limestone: United States: Indiana

Histocrinus graphicus (Miller and Gurley), 1890. Van Sant, *in* Van Sant and Lane, 1964, p. 94, Pl. 5, figs. 2, 5.
Mississippian, Osagean, Borden Group: United States: Indiana

HOLCOCRINUS Kirk, 1945. **Graphiocrinus longicirrifer* Wachsmuth and Springer, *in* Miller, 1889. Kirk, 1945c, p. 517.
Mississippian, Kinderhookian-Osagean

*Holcocrinus longicirrifer (Wachsmuth and Springer, *in* Miller), 1889. Kirk, 1945c, p. 519. Laudon and Severson, 1953, p. 522, Pl. 51, figs. 31-35; Pl. 55, fig. 2.
Graphiocrinus longicirrifer Wachsmuth and Springer, *in* Miller, 1889. Kirk, 1945c, p. 519.
Mississippian, Kinderhookian, Hampton Formation, Lodgepole Limestone: United States: Iowa, Montana

Holcocrinus? nodobrachiatus (Hall), 1861. Kirk, 1945c, p. 519.
Scaphiocrinus nodobrachiatus Hall, 1861. Kirk, 1945c, p. 519.
Mississippian, Osagean, Keokuk Limestone: United States: Indiana

Holcocrinus smythi (Wright), 1934. Kirk, 1945c, p. 520. Wright, 1952a, p. 105, Pl. 36, figs. 1, 4, 14, 20; Fig. 42.
Graphiocrinus smythi Wright, 1934. Kirk, 1945c, p. 520.
Carboniferous, Tournaisian, Supra-dolomite beds: Ireland

Holcocrinus spinobrachiatus (Hall), 1861. Kirk, 1945c, p. 520.
Scaphiocrinus spinobrachiatus Hall, 1861. Kirk, 1945c, p. 520.
Mississippian, Kinderhookian, lower part Burlington Limestone: United States: Iowa

Holocrinus wachsmuthi (Meek and Worthen), 1861. Kirk, 1945c, p. 520.
Poteriocrinus (Scaphiocrinus) wachsmuthi Meek and Worthen, 1861. Kirk, 1945c, p. 520.
Mississippian, Kinderhookian, lower part Burlington Limestone: United States: Iowa

HOLYNOCRINUS Bouška, 1948. **H. moorei* Bouška, 1948. Bouška, 1948, p. 521, fig. 8. Ubaghs, 1953, p. 747.
Middle Devonian

Holynocrinus moorei Bouška, 1948, p. 522, figs. 1-5.
   Middle Devonian, Hlubočepy Limestone: central Bohemia

Holynocrinus spinifer Bouška, 1948, p. 523, figs. 6, 7.
   Middle Devonian, Hlubočepy Limestone: central Bohemia

HOMALOCRINUS Angelin, 1878. *H. parabasalis* Angelin, 1878. Moore and
   Laudon, 1943b, p. 74. Ubaghs, 1953, p. 756. Arendt and Hecker,
   1964, p. 94.
   Silurian

HOMOCRINUS Hall, 1852. *H. parvus* Hall, 1852. Moore and Laudon, 1943b,
   p. 29; 1944, p. 145, Pl. 52, fig. 7. Ubaghs, 1953, p. 746, fig. 140C.
   Moore, 1962b, p. 8, figs. 1, no. 8; 3, no. 4.
   Silurian-Devonian

Homocrinus cylindricus Hall, 1852 = *Botryocrinus? cylindricus*

Homocrinus parvus Hall, 1852. Moore and Laudon, 1943b, p. 130, Pl. 1,
   fig. 4; 1944, p. 145, Pl. 52, fig. 7; Pl. 53, fig. 4.
   Silurian, Niagaran, Rochester Shale, Clinton Formation: United States:
   New York

Homocrinus rhenanus Jaekel, 1921 = *Eifelocrinus rhenanus*

HOPLOCRINUS Grewingk, 1867. *Apiocrinus dipentas* Leuchtenberg, 1843.
   Moore and Laudon, 1943b, p. 32. Ubaghs, 1953, p. 749. Moore, 1962b,
   p. 14, fig. 8, no. 1. Arendt and Hecker, 1964, p. 82.
   Ordovician

Hoplocrinus dalecarlicus Regnell, 1948, p. 3, Pl. 1, figs. 1, 2; Figs. 1, 2.
   Ubaghs, 1953, p. 749, fig. 80a.
   Ordovician, Llandeilian, Lower Chasmops Limestone: Sweden

Hoplocrinus dipentas (Leuchtenberg), 1843. Arendt and Hecker, 1964, p. 82,
   fig. 116.
   Middle Ordovician: Russia: Estonia

Hoplocrinus (cf.) *dipentas* Öpik, 1935 = *Hoplocrinus symmetricus* Miannil,
   1959.

Hoplocrinus estonus Öpik, 1935. Miannil, 1959, p. 82, figs. 1i-m, 2.
   Arendt and Hecker, 1964, p. 82, Pl. 8, fig. 3.
   Middle Ordovician: Russia: Estonia

Hoplocrinus grewingki Öpik, 1925. Kaljo, 1958, p. 43, fig. 21. Miannil,
   1959, p. 83, Pl. 2, fig. 3; Fig. 1e.
   Middle Ordovician: Russia: Estonia

Hoplocrinus heckeri Miannil, 1959, p. 87, Pl. 1, fig. 1.
   Middle Ordovician, Ukhakuskian horizon: Russia: Estonia

Hoplocrinus laevis Miannil, 1959, p. 89, Pl. 1, figs. 3, 4; Figs. 1v, g.
   Middle Ordovician, Keylaskian horizon: Russia: Estonia

Hoplocrinus oanduensis Miannil, 1959, p. 91, Pl. 1, figs. 5, 6.
   Middle Ordovician, Oanduskian horizon: Russia: Estonia

Hoplocrinus pseudodicyclicus Öpik, 1935. Miannil, 1959, p. 83, fig. 1b.
   Middle Ordovician: Russia: Estonia

Hoplocrinus symmetricus Miannil, 1959, p. 88, Pl. 1, fig. 2; Fig. 1a.
   Hoplocrinus (cf.) *dipentas* Öpik, 1935. Miannil, 1959, p. 88.
   Hybocrinus sp. Öpik, 1925. Miannil, 1959, p. 88.
   Middle Ordovician, Ukhakuskian horizon: Russia: Estonia

Hoplocrinus tallinnensis Öpik, 1935. Miannil, 1959, p. 83, fig. 1zh.
   Middle Ordovician: Russia: Estonia

Hoplocrinus tuberculatus Miannil, 1959, p. 90, Pl. 2, figs. 4, 5; Figs. 1d, 3.
   Middle Ordovician, Keylaskian horizon: Russia: Estonia

Hoplocrinus vasalemmaensis Miannil, 1959, p. 92, Pl. 2, figs. 2, 3; Fig. 1z.
  Middle Ordovician, Oanduskian horizon: Russia: Estonia

HORMOCRINUS Springer, 1920. *Centrocrinus tennesseensis* Worthen, 1890.
  Moore and Laudon, 1943b, p. 74, fig. 11; 1944, p. 181, Pl. 66,
  fig. 2. Ubaghs, 1953, p. 756. Arendt and Hecker, 1964, p. 94.
  Silurian

Hormocrinus tennesseensis (Worthen), 1890. Moore and Laudon, 1943b,
  p. 137, Pl. 8, fig. 2; 1944, p. 181, Pl. 54, fig. 6.
  Silurian, Niagaran, Beech River Formation, Laurel Limestone: United
  States: Tennessee, Indiana

HOSIEOCRINUS Wright, 1952. *Tribrachiocrinus caledonicus* Wright, 1936.
  Wright, 1952a, p. 137.
  Carboniferous, Visean

Hosieocrinus caledonicus (Wright), 1936. Wright, 1952a, p. 137, Pl. 38,
  figs. 5-9, 11-17, 19-21, 23, 24.
  Carboniferous, Visean, Lower Limestone Group: Scotland

HYBOCHILOCRINUS Weller, 1930. *Allagecrinus americanus* Rowley, 1895.
  Moore and Laudon, 1943b, p. 30; 1944, p. 147. Ubaghs, 1953, p. 747.
  Strimple and Koenig, 1956, p. 1227. Arendt and Hecker, 1964, p. 81.
  Moore and Jeffords, 1968, p. 11, fig. 3, no. 3.
  Upper Devonian-Mississippian

Hybochilocrinus americanus (Rowley), 1895. Moore and Laudon, 1944, p. 147,
  Pl. 56, fig. 2.
  Upper Devonian, Louisiana Formation: United States: Missouri

Hybochilocrinus rowleyi (Peck), 1936. Moore and Laudon, 1944, p. 147,
  Pl. 56, fig. 1. Ubaghs, 1953, p. 747, fig. 19.
  Mississippian, Kinderhookian, Chouteau Group: United States: Missouri

HYBOCRINUS Billings, 1857. *H. conicus* Billings, 1857. Moore and Laudon,
  1943b, p. 32, fig. 3; 1944, p. 151, Pl. 52, fig. 3. Termier and
  Termier, 1947, Pl. 19, fig. 39. Cuénot, 1948, p. 56, fig. 70.
  Ubaghs, 1953, p. 749. Moore, 1962b, p. 14, fig. 7, no. 1. Philip,
  1963, fig. 3e. Yakovlev, 1964b, p. 64, fig. 82. Arendt and Hecker,
  1964, p. 82.
  Ordovician-Silurian

Hybocrinus conicus Billings, 1857. Moore and Laudon, 1944, p. 151, Pl. 53,
  fig. 22. Wilson, 1946, p. 30. Ubaghs, 1953, p. 749, figs. 141, 142.
  Ordovician, Champlainian, Cobourg Limestone, Trenton Group: Canada:
  Ontario; United States: Kentucky

Hybocrinus crinerensis Strimple and Watkins, 1949, p. 132, Pl. 1, figs. 4-8.
  Ordovician, Champlainian, Bromide Formation: United States: Oklahoma

Hybocrinus nitidus Sinclair, 1945, p. 713, Pl. 2, figs. 1-4.
  Ordovician, Champlainian, Bromide Formation: United States: Oklahoma

Hybocrinus pyxidatus Sinclair, 1945, p. 713, Pl. 2, figs. 5-7.
  Ordovician, Champlainian, Bromide Formation: United States: Oklahoma

Hybocrinus tumidus Billings, 1859. Moore and Laudon, 1944, p. 151, Pl. 53,
  fig. 3. Wilson, 1946, p. 31.
  Ordovician, Champlainian, Cobourg Limestone, Sherman Fall Formation,
  Hull Formation, Trenton Group: Canada: Ontario; United States:
  Kentucky

Hybocrinus sp. Öpik, 1925 = *Hoplocrinus symmetricus* Miannil, 1959

HYBOCYSTITES Wetherby, 1880. *H. problematicus* Wetherby, 1880. Moore and
  Laudon, 1943b, p. 32; 1944, p. 151, Pl. 52, fig. 3. Ubaghs, 1953,
  p. 749. Moore, 1962b, p. 14, fig. 8, no. 3.

Hybocystis Cuénot, 1948, p. 60.
Ordovician

Hybocystites eldonensis (Parks), 1908. Moore and Laudon, 1943b, p. 130,
Pl. 1, fig. 8; 1944, p. 151, Pl. 53, fig. 1. Ubaghs, 1953, p. 749,
figs. 143, 144.
Hybocystis eldonensis Cuénot, 1948, p. 55, fig. 69.
Ordovician, Champlainian, Hull Limestone, Trenton Group: Canada:
Ontario; United States: Kentucky

Hybocystites problematicus Wetherby, 1880. Regnell, 1948, p. 9, fig. 3B.
Ubaghs, 1953, p. 749, fig. 80c.
Ordovician, Champlainian: Canada; United States

HYDREIONOCRINUS de Koninck, 1858. *Woodocrinus goniodactylus* de Koninck
and Wood, 1858. Moore and Laudon, 1943b, p. 56, fig. 6.
Carboniferous, Tournaisian-Namurian

Hydreionocrinus amplus Wright, 1951b, p. 84, Pl. 12, figs. 1-3; Pl. 16,
figs. 1, 3, 4, 6, 8, 9; Pl. 17, figs. 3, 5, 10-14, 16, 17, 19, 20;
Fig. 40. Moore, 1962b, p. 24, fig. 14, no. 7.
Hydreionocrinus woodianus Wright, 1937, 1939. Wright, 1951b, p. 84.
Hydreionocrinus cf. *woodianus* Wright, 1935. Wright, 1951b, p. 84.
Hydreionocrinus No. 1 sp., ?*H. woodianus* Wright, 1925. Wright, 1951b,
p. 84.
Hydreionocrinus sp. Bather, 1912; Wright, 1914, 1920, 1927 (part).
Wright, 1951b, p. 84.
Woodocrinus goniodactylus Wright, 1936. Wright, 1951b, p. 84.
Carboniferous, Visean, Lower Limestone Group: Scotland

Hydreionocrinus amplus? Wright, 1951b, p. 87, Pl. 16, figs. 7, 11.
Carboniferous, Visean, Lower Limestone Group: Scotland

Hydreionocrinus cf. *amplus* Wright, 1951b, p. 87, Pl. 14, figs. 17, 18.
Poteriocrinus granulosus Phillips, McCoy, *in* Sedgwick and McCoy, 1851.
Wright, 1951b, p. 87.
Carboniferous, Visean: England

Hydreionocrinus artus Wright, 1945, p. 118, Pl. 2, figs. 16-24; 1951b,
p. 88, Pl. 12, figs. 15-21, 27-29, 34.
Carboniferous, Visean, Lower Limestone Group: Scotland

Hydreionocrinus balladoolensis Wright, 1942, p. 277, Pl. 11, figs. 18, 22.
Moore, 1948, p. 390, fig. 11 (misspelled as *H. balladooensis*).
Wright, 1951b, p. 90, Pl. 15, figs. 9, 11.
Carboniferous, Visean: England

Hydreionocrinus depressus Cuénot, 1948 = *Tholocrinus spinosus*

Hydreionocrinus deweyensis Strimple, 1939 = *Bathronocrinus deweyensis*

Hydreionocrinus formosus Wright, 1939. Wright, 1951b, p. 89, Pl. 16,
figs. 2, 5, 10.
Hydreionocrinus No. 2 sp. Wright, 1925. Wright, 1951b, p. 89.
Hydreionocrinus sp. Wright, 1927 (part); Tait and Wright, 1923. Wright,
1951b, p. 89.
Carboniferous, Visean, Lower Limestone Group: Scotland

Hydreionocrinus goniodactylus (de Koninck and Wood), 1858. Wright, 1951b,
p. 82, Pl. 15, figs. 1-3, 13, 14.
Hydreionocrinus woodianus de Koninck, 1958. Wright, 1945, p. 119, Pl. 2,
figs. 1-4. Moore, 1948, p. 390, fig. 11. Ubaghs, 1953, p. 700,
fig. 56a. Wright, 1951b, p. 82.
Carboniferous, Visean-Namurian, Lower Limestone Group, red beds above
Main Limestone: Scotland; England

Hydreionocrinus kansasensis Weller, 1898 = *Plaxocrinus kansasensis*

Hydreionocrinus nitidus Wright, 1942, p. 278, Pl. 11, figs. 19, 23; 1951b, p. 90, Pl. 15, figs. 7, 10.
Lower Carboniferous, Visean: England

Hydreionocrinus parkinsoni Wright, 1942, p. 277, Pl. 9, fig. 6; Pl. 11, figs. 20, 21; 1951b, p. 89, Pl. 15, figs. 4-6; 1952c, p. 406, Pl. 11, figs. 1-11; Figs. 1-3; 1960, p. 331, Appendix, Pl. A, figs. 5, 6.
Carboniferous, Tournaisian: Scotland

Hydreionocrinus scoticus de Koninck, 1858 = *Phanocrinus scoticus*

Hydreionocrinus uddeni Weller, 1909 = *Neozeacrinus uddeni*

Hydreionocrinus verrucosus Bather, 1917. Wright, 1951b, p. 91, Pl. 15, figs. 8, 12.
Carboniferous, Visean: England

Hydreionocrinus woodianus de Koninck, 1858 = *Hydreionocrinus goniodactylus*

Hydreionocrinus woodianus Wright, 1937 and 1939 = *Hydreionocrinus amplus*

Hydreionocrinus woodianus Wright and Strimple, 1945 = *Hydreionocrinus amplus*

Hydreionocrinus No. 1 sp., ?*H. woodianus* Wright, 1925 = *Hydreionocrinus amplus*

Hydreionocrinus cf. *woodianus* Wright, 1935 = *Hydreionocrinus amplus*

Hydreionocrinus sp. Bather, 1912; Wright, 1914, 1920, and 1927 (part) = *Hydreionocrinus amplus*

Hydreionocrinus sp. Tait and Wright, 1923; Wright, 1927 (part) = *Hydreionocrinus formosus*

Hydreionocrinus No. 2 sp. Wright, 1925 = *Hydreionocrinus formosus*

HYDRIOCRINUS Trautschold, 1867. **H. pusillus* Trautschold, 1867. Moore and Laudon, 1943b, p. 55. Arendt and Hecker, 1964, p. 88.
Carboniferous

Hydriocrinus pusillus Trautschold, 1867. Yakovlev and Ivanov, 1956, p. 30, Pl. 8, fig. 8; Fig. 5. Arendt and Hecker, 1964, p. 88, Pl. 11, fig. 2; Fig. 129.
Middle-Upper Carboniferous: Russia: Moscow Basin

Hydriocrinus rosei Moore and Plummer, 1938. Strimple, 1961f, p. 306.
Pennsylvanian, Morrowan, Brentwood Limestone: United States: Oklahoma

HYLODECRINUS Kirk, 1941. **H. sculptus* Kirk, 1941. Moore and Laudon, 1943b, p. 56; 1944, p. 161. Arendt and Hecker, 1964, p. 88.
Mississippian

Hylodecrinus asper Kirk, 1941. Moore and Laudon, 1944, p. 161, Pl. 60, fig. 4.
Mississippian, Osagean, Keokuk Limestone: United States: Indiana

*Hylodecrinus sculptus Kirk, 1941. Moore and Laudon, 1944, p. 161, Pl. 60, fig. 2.
Mississippian, Osagean, Keokuk Limestone: United States: Indiana

Hylodecrinus sp. Laudon and others, 1952, p. 555, Pl. 65, fig. 35.
Mississippian, Kinderhookian, Banff formation: Canada: Alberta

HYPERMORPHOCRINUS Arendt, 1968. **H. magnospinosus* Arendt, 1968. Arendt, 1968b, p. 99.
Lower Permian

*Hypermorphocrinus magnospinosus Arendt, 1968b, p. 101, fig. 1.
Lower Permian, Artinskian: Russia: Ural Mountains

HYPOCRINUS Beyrich, 1862. **H. schneideri* Beyrich, 1862. Moore and Laudon, 1943b, p. 50, fig. 4. Yakovlev, 1951, p. 577, fig. 1I. Arendt and Hecker, 1964, p. 85. Lane, 1967b, p. 11, figs. 4, 5. Arendt, 1968a, p. 98, fig. 1.
Permian

**Hypocrinus schneideri* Beyrich, 1862. Haaf, 1950, p. 891, figs. 1-4. Ubaghs, 1953, p. 672, figs. 10c,d.
Permian: Timor

Hypocrinus sp. Yakovlev, 1947d, p. 41, fig. 1 (misspelled as *Hyocrinus*).
Permian

HYPSELOCRINUS Kirk, 1940. **Poteriocrinus hoveyi* Worthen, 1875. Moore and Laudon, 1943b, p. 59; 1944, p. 169. Arendt and Hecker, 1964, p. 89.
Mississippian, Kinderhookian-Pennsylvanian, Missourian

Hypselocrinus arcanus (Miller and Gurley), 1890. Moore and Laudon, 1944, p. 169, Pl. 58, fig. 17.
Mississippian, Osagean, Keokuk Limestone: United States: Indiana

Hypselocrinus ardrossensis Wright, 1951a, p. 33, Pl. 11, fig. 7.
Carboniferous, Visean, Calciferous Sandstone series: Scotland

Hypselocrinus campanulus Horowitz, 1965, p. 27, Pl. 2, figs. 11-14; Fig. 3.
Mississippian, Chesterian, Glen Dean Formation: United States: Kentucky

Hypselocrinus(?) *cavus* Washburn, 1968, p. 129, Pl. 2, figs. 13-15.
Pennsylvanian, Morrowan, Oquirrh Formation: United States: Utah

Hypselocrinus defendus Washburn, 1968, p. 127, Pl. 3, figs. 7-15.
Pennsylvanian, Morrowan, Oquirrh Formation: United States: Utah

Hypselocrinus douglassi (Miller and Gurley), 1896. Moore and Laudon, 1944, p. 169, Pl. 60, fig. 8.
Mississippian, Kinderhookian, Gilmore City Limestone: United States: Iowa

**Hypselocrinus hoveyi* (Worthen), 1875. Moore and Laudon, 1944, p. 169, Pl. 60, fig. 21. Van Sant, *in* Van Sant and Lane, 1964, p. 95, Pl. 5, fig. 16 (*non* Pl. 5, fig. 7 as given in text).
Mississippian, Osagean, Borden Group: United States: Indiana

Hypselocrinus indianensis (Meek and Worthen), 1865. Van Sant, *in* Van Sant and Lane, 1964, p. 95, Pl. 5, figs. 3, 7, 14.
  Poteriocrinus indianensis Meek and Worthen, 1865. Van Sant, *in* Van Sant and Lane, 1964, p. 95.
  Poteriocrinus crawfordsvillensis Miller and Gurley, 1890. Van Sant, *in* Van Sant and Lane, 1964, p. 95.
  Paracosmetocrinus crawfordsvillensis (Miller and Gurley), 1890. Strimple, 1967, p. 81.
Mississippian, Osagean, Borden Group: United States: Indiana

Hypselocrinus? lasallensis (Worthen), 1875. Moore and Laudon, 1944, p. 169, Pl. 57, fig. 17.
  Poteriocrinus lasallensis Worthen, 1875.
Pennsylvanian, Missourian: United States: Illinois

Hypselocrinus maccabei (Miller and Gurley), 1894. Moore and Laudon, 1944, p. 169, Pl. 60, fig. 9. Laudon and Severson, 1953, p. 519, Pl. 51, figs. 2-5.
Mississippian, Kinderhookian, Hampton Formation, Lodgepole Limestone: United States: Iowa, Montana

Hypselocrinus? macoupinensis (Worthen), 1873. Moore and Laudon, 1944, p. 169, Pl. 57, fig. 19.

Poteriocrinites macoupinensis Worthen, 1873.
Pennsylvanian, Desmoinesian: United States: Illinois

Hypselocrinus monensis (Wright), 1938. Wright, 1951a, p. 32, Pl. 8, figs. 4, 9; Fig. 10.
Pachylocrinus monensis Wright, 1938. Wright, 1951a, p. 32.
Carboniferous, Visean: England

Hypselocrinus sansabensis (Moore and Plummer), 1940. Moore and Laudon, 1944, p. 169, Pl. 60, fig. 10.
Scytalocrinus sansabensis Moore and Plummer, 1940.
Pennsylvanian, Morrowan, Marble Falls Formation: United States: Texas

Hypselocrinus(?) *superus* Washburn, 1968, p. 128, Pl. 2, figs. 10-12.
Pennsylvanian, Morrowan, Oquirrh Formation: United States: Utah

HYPSOCRINUS Springer and Slocum, 1906. *H. fieldi* Springer and Slocum, 1906. Moore and Laudon, 1943b, p. 30. Strimple, 1963d, p. 22, fig. 1a. Ubaghs, 1953, p. 747.
Devonian

IBEROCRINUS Sieverts-Doreck, 1951. *I. multibrachiatus* Sieverts-Doreck, 1951. Sieverts-Doreck, 1951b, p. 105, fig. 1. Ubaghs, 1953, p. 739.
Carboniferous, Westphalian

Iberocrinus multibrachiatus Sieverts-Doreck, 1951b, p. 109, Pl. 8, figs. 1, 2; Figs. 2d, 3. Breimer, 1962a, p. 75, Pl. 8, figs. 1-4; Fig. 13.
Carboniferous, Westphalian, Cotarazzo Limestone: Spain

ICHTHYOCRINUS Conrad, 1842. *I. laevis* Conrad, 1842. Moore and Laudon, 1943b, p. 76, fig. 11; 1944, p. 183, Pl. 66, figs. 11, 15. Termier and Termier, 1947, p. 256, Pl. 19, fig. 52. Cuénot, 1948, p. 67. Moore, 1950, p. 37, fig. 40. Ubaghs, 1953, p. 756.
Silurian-Devonian

Ichthyocrinus corbis Winchell and Marcy, 1865. Moore and Laudon, 1944, p. 183, Pl. 68, fig. 11. Strimple, 1963d, p. 112, Pl. 9, fig. 6.
Silurian, Niagaran, Racine Dolomite, Henryhouse Formation: United States: Illinois, Oklahoma

Ichthyocrinus devonicus Springer, 1920. Moore and Laudon, 1944, p. 183, Pl. 68, fig. 17.
Devonian, Ulsterian, Linden Group: United States: Tennessee

Ichthyocrinus gotlandicus Wachsmuth and Springer, 1880. Ubaghs, 1953, p. 756, fig. 12.
Silurian: Sweden

Ichthyocrinus laevis Conrad, 1842. Moore and Laudon, 1943b, p. 136, Pl. 7, fig. 1; 1944, p. 183, Pl. 68, fig. 3.
Silurian, Niagaran, Rochester Shale: United States: New York; Canada: Ontario

Ichthyocrinus magniradialis Weller, 1903 = *Lecanocrinus* (?*Miracrinus*) *magniradialis*

Ichthyocrinus mccoyanus Salter, 1873 = *Eutaxocrinus maccoyanus*

Ichthyocrinus pyriformis (Phillips, *in* Murchison), 1839. Cuénot, 1948, p. 67, fig. 88 (misspelled as *Ichtyocrinus pyriformis*).
Silurian: Europe; North America

Ichthyocrinus subangularis Hall, 1863. Moore and Laudon, 1943b, p. 136, Pl. 7, fig. 7; 1944, p. 183, Pl. 54, fig. 5.
Silurian, Niagaran, Rochester Shale, Waldron Shale, Racine Dolomite: United States: New York, Indiana, Illinois

IDOSOCRINUS Wright, 1954. **I. bispinosus* Wright, 1954. Wright, 1954b, p. 167.
Lower Carboniferous

**Idosocrinus bispinosus* Wright, 1954b, p. 167, Pl. 3, figs. 1, 2; Fig. 1; 1960, p. 332, Appendix, Pl. B, fig. 1.
Carboniferous, Visean, Calciferous Sandstone series: Scotland

Idosocrinus tumidus Wright, 1954b, p. 169, Pl. 3, fig. 3; 1960, p. 332, Appendix, Pl. B, fig. 7.
Carboniferous, Visean, Calciferous Sandstone series: Scotland

IMITATOCRINUS Schmidt, 1934. **Cyathocrinus gracilior* Roemer, 1863. Moore and Laudon, 1943b, p. 54. Ubaghs, 1953, p. 752. Arendt and Hecker, 1964, p. 86.
Devonian

INDOCRINUS Wanner, 1916. **I. elegans* Wanner, 1916. Yakovlev, 1939e, p. 832; 1939f, p. 832. Moore and Laudon, 1943b, p. 64. Yakovlev, 1956a, p. 24, figs. 3, nos. 2, 3; 37, nos. 2-4; 1958, p. 13, figs. 3, nos. 2, 3; 37, nos. 2-4; 1964a, p. 24, figs. 3, nos. 2, 3; 39, nos. 2-4. Arendt and Hecker, 1964, p. 92.
Proindocrinus Yakovlev, 1927. Yakovlev, 1939e, p. 832; 1939f, p. 832.
Permian

Indocrinus crassus Wanner, 1916. Yakovlev, 1950, p. 94, fig. 1IV; 1957, p. 12, fig. 2g. Arendt and Hecker, 1964, p. 92, Pl. 13, fig. 16.
Permian: Timor

**Indocrinus elegans* Wanner, 1916. Yakovlev, 1957, p. 12, fig. 2d. Arendt and Hecker, 1964, p. 92, fig. 134.
Permian: Timor

Indocrinus piszowi Yakovlev, 1926. Yakovlev, 1951, p. 578, fig. kIII; 1957, p. 12, fig. 2v.
Proindocrinus piszowi (Yakovlev), 1926. Yakovlev, 1939b, p. 62, Pl. 10, figs. 7-12; Fig. 16; 1949b, p. 897, figs. 1III,IV.
Lower Permian: Russia: Krasnoufimsk

Indocrinus(?) *piszowi* Yakovlev, *in* Yakovlev and Ivanov, 1956, p. 61, Pl. 11, figs. 6-10; Fig. 17. Yakovlev, 1964b, p. 59, fig. 69 (misspelled as *I.*(?) *piszovi*). Arendt and Hecker, 1964, p. 92, Pl. 13, fig. 17.
Permian, Artinskian: Russia: Krasnoufimsk

Indocrinus rimosus Wanner, 1916. Yakovlev, 1950, p. 94, fig. 1V.
Permian: Timor

Indocrinus sp. Yakovlev, 1952b, p. 150, figs. 3, 4 (part).
Permian: Russia

INTERMEDIACRINUS Sutton and Winkler, 1940. **Eupachycrinus asperatus* Worthen, 1882. Strimple, 1961g, p. 81.
Mississippian, Chesterian

**Intermediacrinus asperatus* (Worthen), 1882. Strimple, 1961g, p. 80.
Mississippian, Chesterian: United States: Illinois

Intermediacrinus modernus (Strimple), 1951. Strimple, 1961g, p. 81.
Eupachycrinus modernus Strimple, 1951g, p. 291, figs. 6-8; 1961g, p. 81.
Mississippian, Chesterian, unnamed limestone below Fayetteville Formation: United States: Oklahoma

IOCRINUS Hall, 1866. **Heterocrinus crassus* Meek and Worthen, 1865. Moore and Laudon, 1943b, p. 29, figs. 1, 3; 1944, p. 143, Pl. 52, fig. 15. Moore, 1950, p. 34, figs. 1c, d. Ubaghs, 1953, p. 744, fig. 140B. Ramsbottom, 1961, p. 3, fig. 3. Moore, 1962b, p. 8, figs. 1, no. 7; 11, no. 1. Philip, 1963, p. 268, fig. 3a. Arendt and Hecker, 1964,

p. 80. Philip, 1965a, p. 148, fig. 1b.
Ordovician-Early Silurian

Iocrinus brithdirensis Bates, 1965, p. 355, Pl. 45.
Ordovician, Llanvirnian: England

Iocrinus? cambriensis Ramsbottom, 1961 = *Ramseyocrinus cambriensis*

Iocrinus crassus (Meek and Worthen), 1865. Moore and Laudon, 1944, p. 143, Pl. 52, fig. 15; Pl. 53, fig. 2. Moore, 1962b, p. 40, Pl. 2, figs. 3, 4; Pl. 3, fig. 4; Fig. 16, no. 2.
Ordovician, Cincinnatian, Maquoketa Formation: United States: Illinois

Iocrinus shelvensis Ramsbottom, 1961, p. 3, Pl. 1, figs. 3-8.
cf. *Caleidocrinus* Whittard, 1931. Ramsbottom, 1961, p. 3.
Ordovician, Llanvirnian, Weston Stage: England

Iocrinus similis (Billings), 1857. Wilson, 1946, p. 33, Pl. 5, fig. 3; Fig. 2.
Ordovician, Champlainian, Cobourg Limestone: Canada: Ontario

Iocrinus subcrassus (Meek and Worthen), 1865. Moore and Laudon, 1943b, p. 130, Pl. 1, fig. 11; 1944, p. 143, Pl. 53, figs. 10, 11. Wilson, 1946, p. 34. Ubaghs, 1953, p. 744, fig. 18d. Moore, 1962b, p. 40, Pl. 2, figs. 1, 2; Pl. 3, fig. 3; Fig. 16, no. 1.
Ordovician, Champlainian-Cincinnatian, Richmond Group, Trenton Group, Cobourg Limestone, Hull Limestone: United States: Ohio, New York; Canada: Ontario

Iocrinus cf. *subcrassus* (Meek and Worthen), 1868. Wilson, 1946, p. 34, Pl. 6, fig. 7.
Ordovician, Champlainian, Cobourg Limestone: Canada: Ontario

Iocrinus whitteryi Ramsbottom, 1961, p. 5, Pl. 1, fig. 9.
Ordovician, Caradocian, Whittery Beds: England

ISOALLAGECRINUS Strimple, 1966. *Allagecrinus bassleri* Strimple, 1938.
Strimple, 1966c, p. 105.
Pennsylvanian, Desmoinesian-lower Permian, Wolfcampian

Isoallagecrinus bassleri (Strimple), 1938. Strimple, 1966c, p. 104.
Allagecrinus bassleri Strimple, 1938. Moore and Laudon, 1943b, p. 130, Pl. 1, fig. 19; 1944, p. 149, Pl. 56, fig. 9. Ubaghs, 1953, p. 747, fig. 50. Strimple, 1959a, p. 116, Pl. 1, figs. 18-23; 1966c, p. 104.
Allagecrinus bassleri nodosus Strimple, 1951a, p. 204, Pl. 37, figs. 1-4.
Strimple, 1966c, p. 101.
Pennsylvanian, Missourian, Wann Formation, Lake Bridgeport Shale: United States: Oklahoma, Texas

Isoallagecrinus sp. cf. *I. bassleri* (Moore), 1940. Strimple, 1966c, p. 104.
Allagecrinus sp. cf. *A. bassleri* Moore, 1940. Strimple, 1966c, p. 104.
Pennsylvanian, Desmoinesian, Wewoka Formation: United States: Oklahoma

Isoallagecrinus constellatus (Moore), 1940. Strimple, 1966c, p. 104.
Allagecrinus constellatus Moore, 1940. Ubaghs, 1953, p. 747, fig. 49.
Strimple, 1962e, p. 10; 1966c, p. 104.
Pennsylvanian, Desmoinesian, Oologah Limestone: United States: Oklahoma

Isoallagecrinus copani (Strimple), 1949. Strimple, 1966c, p. 104.
Allagecrinus copani Strimple, 1949c, p. 21, Pl. 5, figs. 1-7.
Pennsylvanian, Missourian, unnamed shale 30 feet above Torpedo Sandstone: United States: Oklahoma

Isoallagecrinus dignatus (Moore), 1940. Strimple, 1966c, p. 104.
Allagecrinus dignatus Moore, 1940. Moore and Laudon, 1944, p. 149, Pl. 56, fig. 10. Strimple, 1962e, p. 10; 1966c, p. 104.
Pennsylvanian, Desmoinesian, Oologah Limestone: United States: Oklahoma

Isoallagecrinus? donetzensis (Yakovlev), 1930. Strimple, 1966c, p. 104.
Allagecrinus donetzensis Yakovlev, 1930. Strimple, 1966c, p. 104.
Kallimorphocrinus donetzensis (Yakovlev), 1930. Yakovlev, in Yakovlev and Ivanov, 1956, p. 43, Pl. 12, fig. 7.
Permian: Russia

Isoallagecrinus eaglei Strimple, 1966c, p. 105, Pl. 1, figs. 1-3. Webster and Lane, 1967, p. 11, Pl. 3, figs. 9, 10.
Permian, Wolfcampian, Red Eagle Limestone, Bird Spring Formation: United States: Oklahoma, Nevada

Isoallagecrinus graffhami (Strimple), 1948. Strimple, 1966c, p. 104.
Allagecrinus graffhami Strimple, 1948c, p. 3, Pl. 1, figs. 1-11; 1966c, p. 104.
Pennsylvanian, Virgilian, Stull Shale: United States: Kansas

Isoallagecrinus illinoisensis (Weller), 1930. Strimple, 1966c, p. 109.
Kallimorphocrinus illinoisensis Weller, 1930. Strimple, 1966c, p. 109.
Pennsylvanian: United States: Illinois, Missouri

Isoallagecrinus kylensis (Strimple), 1948. Strimple, 1966c, p. 104.
Allagecrinus kylensis Strimple, 1948c, p. 4, Pl. 1, figs. 12-20; 1966c, p. 104.
Pennsylvanian, Missourian, Brownwood Shale: United States: Texas

Isoallagecrinus pocillus (Weller), 1930. Strimple, 1966c, p. 109.
Kallimorphocrinus pocillus Weller, 1930. Strimple, 1966c, p. 109.
Pennsylvanian: United States: Indiana

Isoallagecrinus status (Strimple), 1951. Strimple, 1966c, p. 104.
Allagecrinus bassleri status Strimple, 1951a, p. 204, Pl. 37, figs. 5-8; 1966c, p. 104.
Pennsylvanian, Missourian, Lake Bridgeport Shale: United States: Texas

Isoallagecrinus strimplei (Kirk), 1936. Strimple, 1966c, p. 104.
Allagecrinus strimplei Kirk, 1936. Moore and Laudon, 1944, p. 149, Pl. 56, fig. 8. Ubaghs, 1953, p. 747, fig. 51. Strimple, 1966c, p. 104.
Pennsylvanian, Missourian, Dewey Limestone: United States: Oklahoma

ISOCATILLOCRINUS Wanner, 1937. *I. indicus* Wanner, 1937. Moore and Laudon, 1943b, p. 31, fig. 3. Ubaghs, 1953, p. 747. Moore, 1962b, p. 16, fig. 12, no. 12. Arendt and Hecker, 1964, p. 81.
Permian

ISOTOMOCRINUS Ulrich, in Foerste, 1924. *I. typus* Ulrich, in Foerste, 1924. Moore and Laudon, 1943b, p. 31; 1944, p. 149. Ubaghs, 1953, p. 747. Moore 1962b, p. 14, fig. 5, no. 2. Philip, 1965a, p. 148, fig. 1d.
Ordovician

Isotomocrinus typus Ulrich, in Foerste, 1924. Moore and Laudon, 1944, p. 149, Pl. 52, fig. 11.
Ordovician, Champlainian, Trenton Group: Canada: Ontario

ITEACRINUS Goldring, 1923. *I. flagellum* Goldring, 1923. Moore and Laudon, 1943b, p. 55. Ubaghs, 1953, p. 752. Arendt and Hecker, 1964, p. 87.
Devonian

Iteacrinus dactylus (Schmidt), 1934. Schmidt, 1941, p. 30.
Rhadinocrinus dactylus Schmidt, 1934. Schmidt, 1941, p. 30.
Devonian: Germany

Iteacrinus nanus (Roemer), 1863. Schmidt, 1941, p. 30.
Poteriocrinus nanus Roemer, 1863. Schmidt, 1941, p. 30.
Devonian: Germany

Ivanovicrinus Yakovlev, in Arendt and Hecker, 1964 = *Trautscholdicrinus*

Ivanovicrinus miloradovitschi (Yakovlev), 1939 = *Trautscholdicrinus miloradovitschi*

JAEKELICRINUS Yakovlev, 1949. *J. baskiricus* Yakovlev, 1949. Yakovlev, 1947c, p. 609, fig. 1, nomen nudum; 1949d, p. 435; 1964b, p. 65, fig. 85. Arendt and Hecker, 1964, p. 81.
Upper Devonian

Jaekelicrinus baskiricus Yakovlev, 1949d, p. 435, fig. 1.
Upper Devonian: Russia: Ural Mountains

JAHNOCRINUS Jaekel, 1918. *J. minutus* Jaekel, 1918. Moore and Laudon, 1943b, p. 54.
Middle Devonian

JIMBACRINUS Teichert, 1954. *J. bostocki* Teichert, 1954. Teichert, 1954, p. 71. Yakovlev, 1956a, p. 124, fig. 46; 1958, p. 108, fig. 46; 1964a, p. 109, fig. 48; 1964b, p. 69, figs. 102, 103.
Permian, Artinskian

Jimbacrinus bostocki Teichert, 1954, p. 71, Pls. 13, 14. Yakovlev, 1957, p. 11, figs. 1a, b.
Permian, Artinskian: Australia

KALLIMORPHOCRINUS Weller, 1930. *K. astrus* Weller, 1930. Moore and Laudon, 1943b, p. 30; 1944, p. 147, Pl. 52, fig. 4. Ubaghs, 1953, p. 747. Moore, 1962b, p. 16, fig. 12, no. 1. Arendt and Hecker, 1964, p. 81. Strimple, 1966c, p. 107.
Mississippian-Permian

Kallimorphocrinus angulatus Strimple and Koenig, 1956, p. 1233, figs. 1c; 2, nos. 13-28. Regnell, 1960, p. 780, fig. 6.
Mississippian, Osagean, Nunn Member, St. Joe Formation: United States: New Mexico, Oklahoma

Kallimorphocrinus astrus Weller, 1930. Moore and Laudon, 1944, p. 149, Pl. 56, fig. 3. Strimple, 1966c, p. 108, fig. 2c.
Pennsylvanian, Desmoinesian: United States: Indiana

Kallimorphocrinus donetzensis (Yakovlev), 1930 = *Isoallagecrinus? donetzensis*

Kallimorphocrinus elongatus (Wright), 1932. Wright, 1952a, p. 146, Pl. 40, figs. 7, 8, 11, 18, 28.
Carboniferous, Visean, Lower Limestone Group: Scotland

Kallimorphocrinus expansus Weller, 1930. Moore and Laudon, 1943b, p. 130, Pl. 1, fig. 7; 1944, p. 149, Pl. 56, fig. 4.
Pennsylvanian, Desmoinesian, Piasa Limestone: United States: Illinois

Kallimorphocrinus extensus Wright, 1952a, p. 146, Pl. 40, fig. 10.
Carboniferous, Visean, Lower Limestone Group: Scotland

Kallimorphocrinus illinoisensis Weller, 1930 = *Isoallagecrinus illinoisensis*

Kallimorphocrinus indianensis Weller, 1930. Moore and Laudon, 1944, p. 149, Pl. 56, fig. 5.
Pennsylvanian, Desmoinesian: United States: Indiana

Kallimorphocrinus multibrachiatus (Yakovlev), 1927. Yakovlev, *in* Yakovlev and Ivanov, 1956, p. 54, Pl. 9, fig. 4. Arendt and Hecker, 1964, p. 81, Pl. 8, fig. 5.
Permian, Artinskian: Russia: Central Urals

Kallimorphocrinus pocillus Weller, 1930 = *Isoallagecrinus pocillus*

Kallimorphocrinus pristinus (Peck), 1936. Strimple and Koenig, 1956, p. 1231, fig.1a.
Mississippian, Osagean, Fern Glen Limestone: United States: Missouri

Kallimorphocrinus puteatus (Peck), 1936. Moore and Laudon, 1944, p. 149, Pl. 56, fig. 6. Strimple and Koenig, 1956, p. 1231, fig. 1b.
  Mississippian, Osagean, Fern Glen Limestone: United States: Missouri

Kallimorphocrinus scoticus (Wright), 1932. Wright, 1952a, p. 143, Pl. 40, figs. 1, 4-6; Fig. 75.
  Carboniferous, Visean, Lower Limestone Group: Scotland

Kallimorphocrinus tintinabulum Strimple and Koenig, 1956, p. 1235, figs. 1d; 2, nos. 36-39.
  Mississippian, Kinderhookian, shale in Welden Limestone: United States: Oklahoma

Kallimorphocrinus uralensis (Yakovlev), 1939. Yakovlev, 1939b, p. 59, Pl. 10, fig. 2; Fig. 12. Yakovlev, *in* Yakovlev and Ivanov, 1956, p. 53, Pl. 9, figs. 1-3; Fig. 13. Arendt and Hecker, 1964, p. 81, Pl. 8, fig. 4; Fig. 115.
  Middle Carboniferous-Permian: Russia: Don Basin, Northern Urals

Kallimorphocrinus weldenensis Strimple and Koenig, 1956 = *Desmacriocrinus weldenensis*

KALPIDOCRINUS Goldring, 1954. *K. eriensis* Goldring, 1954. Goldring, 1954, p. 17.
  Devonian, Erian

Kalpidocrinus eriensis Goldring, 1954, p. 18, Pl. 4, figs. 10-12.
  Devonian, Erian, Ludlowville Group: United States: New York

KOPFICRINUS Goldring, 1954. *K. pustuliferus* Goldring, 1954. Goldring, 1954, p. 11.
  Devonian, Ulsterian

Kopficrinus pustuliferus Goldring, 1954, p. 12, Pl. 2, figs. 3, 4.
  Devonian, Ulsterian, Onondaga Limestone: United States: New York

KOPHINOCRINUS Goldring, 1954. *K. spiniferus* Goldring, 1954. Goldring, 1954, p. 36.
  Devonian, Erian

Kophinocrinus spiniferus Goldring, 1954, p. 37, Pl. 5, figs. 2-6.
  Devonian, Erian, Kashong Member: United States: New York

Lagarocrinus anglicus Jaekel, 1900 = *Cicerocrinus anglicus*

Lagarocrinus tenuis Jaekel, 1900 = *Cicerocrinus elegans*

LAGENIOCRINUS de Koninck and Le Hon, 1854. *L. seminulum* de Koninck and Le Hon, 1854. Moore and Laudon, 1943b, p. 50. Yakovlev, 1947d, p. 44, fig. 9. Cuénot, 1948, p. 64. Yakovlev, 1956a, p. 80, figs. 29; 38, no. 1; 1958, p. 65, figs. 29; 38, no. 1; 1964a, p. 75, fig. 40, no. 1. Arendt and Hecker, 1964, p. 85. Lane, 1967b, p. 13, fig. 2.
  Lower Carboniferous-Permian

Lageniocrinus jacksoni (Austin and Austin), 1842. Wright, 1952a, p. 132, Pl. 36, figs. 11, 12.
  Sycocrinites jacksoni Austin and Austin, 1842. Wright, 1952a, p. 132.
  Carboniferous, Visean?: Scotland

Lageniocrinus seminulum de Koninck and Le Hon, 1854. Ubaghs, 1953, p. 672, fig. 10j. Yakovlev, 1964b, p. 56, fig. 56.
  Carboniferous, Visean: Belgium

Lahuseniocrinus Tschernyschew, 1892 = *Spyridiocrinus*

Lahuseniocrinus tirlensis Tschernyschew, 1892 = *Spyridiocrinus tirlensis*

LAMPADOSOCRINUS Strimple and Koenig, 1956. **Dichostreblocrinus minutus* Peck, 1936. Strimple and Koenig, 1956, p. 1244. Lane, 1967b, p. 13, figs. 2, 7 (fig. 7 listed as fig. 8).
Mississippian, Osagean

**Lampadosocrinus minutus* (Peck), 1936. Strimple and Koenig, 1956, p. 1245, fig. 4, nos. 19-21.
Dichostreblocrinus minutus Peck, 1936. Strimple and Koenig, 1956, p. 1245.
Mississippian, Osagean, St. Joe Formation: United States: Oklahoma

Lampadosocrinus obtusus Strimple and Koenig, 1956, p. 1245, fig. 4, nos. 16-18.
Mississippian, Osagean, Nunn Member: United States: New Mexico

LAMPTEROCRINUS Roemer, 1860. **L. tennesseensis* Roemer, 1860. Moore and Laudon, 1943b, p. 84, fig. 14; 1944, p. 187, Pl. 70, fig. 2. Ubaghs, 1953, p. 737. Strimple, 1963d, p. 82, fig. 26 (part).
Silurian

Lampterocrinus fatigatus Strimple, 1963d, p. 83, Pl. 5, figs. 1-3; Figs. 23-25.
Silurian, Niagaran, Henryhouse Formation: United States: Oklahoma

Lampterocrinus roemeri Springer, 1926. Moore and Laudon, 1944, p. 187, Pl. 72, fig. 17.
Silurian, Niagaran, Beech River Formation: United States: Tennessee

**Lampterocrinus tennesseensis* Roemer, 1860. Moore and Laudon, 1943b, p. 138, Pl. 9, fig. 13; 1944, p. 187, Pl. 72, fig. 18.
Silurian, Niagaran, Beech River Formation: United States: Tennessee

LASANOCRINUS Moore and Plummer, 1940. **Hydreionocrinus daileyi* Strimple, 1940. Moore and Laudon, 1943b, p. 58; 1944, p. 163.
Pennsylvanian

Lasanocrinus altamontensis Strimple, 1950b, p. 573, Pl. 77, figs. 5-8; 1962e, p. 35.
Pennsylvanian, Desmoinesian, Altamont Limestone: United States: Oklahoma

Lasanocrinus cornutus Moore and Plummer, 1940. Moore and Laudon, 1944, p. 163, Pl. 62, fig. 6.
Pennsylvanian, Morrowan, Marble Falls Formation: United States: Texas

**Lasanocrinus daileyi* (Strimple), 1940. Moore and Laudon, 1944, p. 163, Pl. 63, fig. 14.
Pennsylvanian, Morrowan, Wapanucka Limestone, Brentwood Limestone: United States: Oklahoma

LASIOCRINUS Kirk, 1914. **Homocrinus scoparius* Hall, 1859. Moore and Laudon, 1943b, p. 55, fig. 7; 1944, p. 158, Pl. 52, fig. 5. Ubaghs, 1953, p. 752. Moore, 1962b, p. 15, fig. 10, no. 5. Arendt and Hecker, 1964, p. 86. Philip, 1965a, fig. 2c.
Devonian-Mississippian

Lasiocrinus expressus Laudon, 1933. Moore and Laudon, 1944, p. 158, Pl. 53, fig. 14.
Mississippian, Kinderhookian, Hampton Formation: United States: Iowa

Lasiocrinus konkolodesae Yakovlev, 1956, *nom. correct*.
Lasiocrinus kon-kolodesae Yakovlev, 1956b, p. 91, Pl. 1, fig. 3.
Arendt and Hecker, 1964, p. 86, Pl. 10, fig. 1.
Devonian, Frasnian, Evlanovsk horizon: Russia: Voronez Region

Lasiocrinus(?) scaber Schmidt, 1941, p. 101, Pl. 13, fig. 5.
Lower Devonian, *Cultrijugatus* Zone: Germany

**Lasiocrinus scoparius* (Hall), 1859. Moore and Laudon, 1943b, p. 132, Pl. 3, fig. 4; 1944, p. 158, Pl. 52, fig. 5; Pl. 53, fig. 15.
Devonian, Ulsterian, Coeymans Limestone: United States: New York

Lasiocrinus(?) *vastus* Schmidt, 1941, p. 102, Pl. 12, fig. 4.
Devonian, Coblentzian: Germany

Lasiocrinus? sp. Breimer, 1962a, p. 169, Pl. 15, figs. 11, 12.
Devonian, Emsian, La Vid Shale: Spain

LAUDONOCRINUS Moore and Plummer, 1940. *Hydreionocrinus subsinuatus* Miller and Gurley, 1894. Moore and Laudon, 1943b, p. 58; 1944, p. 163.
Pennsylvanian

Laudonocrinus catillus Moore and Plummer, 1940. Moore and Laudon, 1944, p. 165, Pl. 64, fig. 9. Strimple, 1962e, p. 25, Pl. 1, figs. 13,14; Pl. 4, figs. 9-11.
Pennsylvanian, Desmoinesian, Oologah Limestone, Millsap Lake Formation: United States: Oklahoma, Texas

**Laudonocrinus subsinuatus* (Miller and Gurley), 1894. Moore and Laudon, 1944, p. 165, Pl. 62, fig. 10; Pl. 64, fig. 10.
Pennsylvanian, Missourian: United States: Missouri

Laudonocrinus sp. Strimple, 1957, p. 369, figs. 1a, b.
Pennsylvanian, Missourian, Avant Limestone: United States: Oklahoma

LAURELOCRINUS Springer, 1926. *L. paulensis* Springer, 1926. Moore and Laudon, 1943b, p. 98; 1944, p. 200. Ubaghs, 1953, p. 742.
Silurian, Niagaran

**Laurelocrinus paulensis* Springer, 1926. Moore and Laudon, 1944, p. 200, Pl. 73, fig. 20.
Silurian, Niagaran, Laurel Limestone: United States: Indiana

Laurelocrinus wilsoni Springer, 1926. Moore and Laudon, 1943b, p. 142, Pl. 13, fig. 13; 1944, p. 200, Pl. 73, fig. 19.
Silurian, Niagaran, Laurel Limestone: United States: Indiana

LEBETOCRINUS Kirk, 1940. *L. grandis* Kirk, 1940. Moore and Laudon, 1943b, p. 55. Arendt and Hecker, 1964, p. 88.
Middle Mississippian

**Lebetocrinus grandis* Kirk, 1940. Moore, 1962b, p. 24, fig. 15. Van Sant, *in* Van Sant and Lane, 1964, p. 83.
Mississippian, Osagean, Keokuk Limestone: United States: Indiana

LECANOCRINUS Hall, 1852. *L. macropetalus* Hall, 1852. Moore and Laudon, 1943b, p. 76, fig. 11; 1944, p. 181, Pl. 66, figs. 6, 13. Cuénot, 1948, p. 67. Moore, 1950, p. 37, fig. 40. Ubaghs, 1953, p. 756. Philip, 1963, fig. 1b. Arendt and Hecker, 1964, p. 94.
Silurian-Devonian

Lecanocrinus angulatus Springer, 1920. Strimple, 1963d, p. 120.
Silurian, Wenlockian: Sweden: Isle of Gotland

Lecanocrinus brevis Strimple, 1952f p. 319, figs. 13-17; 1963d, p. 122, Pl. 9, figs. 1, 2.
Silurian, Niagaran, Henryhouse Formation: United States: Oklahoma

Lecanocrinus erectus Strimple, 1952f, p. 319, figs. 9, 10; 1963d, p. 123.
Silurian, Niagaran, Henryhouse Formation: United States: Oklahoma

Lecanocrinus huntonensis Strimple, 1952f, p. 323, figs. 11, 12; 1963d, p. 124.
Devonian, Ulsterian, Haragan Formation: United States: Oklahoma

Lecanocrinus invaginatus Strimple, 1952f, p. 322, figs. 1-4; 1963d, p. 121.
Silurian, Niagaran, Henryhouse Formation: United States: Oklahoma

Lecanocrinus lindenensis Strimple, 1952f, p. 320, figs. 5-8.
Silurian, Niagaran, Lobelville Formation: United States: Tennessee

Lecanocrinus macropetalus Hall, 1852. Moore and Laudon, 1943b, p. 137, Pl. 8, fig. 7; 1944, p. 181, Pl. 68, fig. 15. Bowsher, 1953, p. 2, fig. 1f, g.
Silurian, Niagaran, Rochester Formation: United States: New York

Lecanocrinus (Miracrinus) magniradialis (Weller), 1903. Bowsher, 1953 = *Miracrinus magniradialis*

Lecanocrinus papilloseous Strimple, 1954b, p. 281, figs. 1-4, 9, 10; 1963d, p. 121, Pl. 10, fig. 1.
Silurian, Niagaran, Henryhouse Formation: United States: Oklahoma

Lecanocrinus (Miracrinus) perdewi Bowsher, 1953 = *Miracrinus perdewi*

Lecanocrinus pisiformis (Roemer), 1860. Moore and Laudon, 1944, p. 181, Pl. 54, fig. 2. Amsden, 1949, p. 74, Pl. 12, figs. 18-20. Ubaghs, 1953, p. 756, fig. 15d.
Silurian, Niagaran, Brownsport Formation: United States: Tennessee

Lecanocrinus pusillus (Hall), 1863. Moore and Laudon, 1944, p. 181, Pl. 54, fig. 1. Wilson, 1949, p. 251, Pl. 25, figs. 18-20.
Silurian, Niagaran, Waldron Shale: United States: Indiana, Tennessee

Lecanocrinus waukoma (Hall), 1865. La Rocque and Marple, 1955, p. 73, fig. 149.
Silurian, Niagaran, Cedarville Limestone, Guelph formation: United States: Ohio

LECYTHIOCRINUS White, 1879. *L. olliculaeformis* White, 1879. Moore and Laudon, 1943b, p. 50; 1944, p. 154. Ubaghs, 1953, p. 751. Moore, 1962b, p. 8, fig. 1, no. 3. Arendt and Hecker, 1964, p. 85. Lane, 1967b, p. 11, fig. 4.
Pennsylvanian

Lecythiocrinus adamsi Worthen, 1882. Moore and Laudon, 1943b, p. 131, Pl. 2, fig. 2; 1944, p. 154, Pl. 53, fig. 20.
Pennsylvanian, Desmoinesian: United States: Illinois

Lecythiocrinus fusiformis Strimple, 1949d, p. 20, Pl. 3, figs. 11, 13, 14.
Pennsylvanian, Atokan, Pumpkin Creek Limestone: United States: Oklahoma

Lecythiocrinus olliculaeformis White, 1879. Moore and Laudon, 1944, p. 154, Pl. 53, fig. 21.
Pennsylvanian, Virgilian, Oread Limestone: United States: Kansas

Lecythiocrinus optimus Strimple, 1951d, p. 194, figs. 6-8; 1962e, p. 48.
Pennsylvanian, Desmoinesian, Oologah Limestone: United States: Oklahoma

Lecythiocrinus? problematicus Springer, 1926 = *Thyridocrinus problematicus*

Lecythiocrinus tubiformis Strimple, 1951b, p. 376, Pl. 56, figs. 15, 16.
Pennsylvanian, Missourian, Raytown Limestone: United States: Kansas

LECYTHOCRINUS Müller, 1858. *L. eifelianus* Müller, 1858. Moore and Laudon, 1943b, p. 51. Ubaghs, 1953, p. 750.
Devonian

Lecythocrinus eifelianus Müller, 1858. Wanner, 1942a, p. 26, fig. 1.
Devonian, Eifelian: Germany

LENNEOCRINUS Jaekel, 1921. *L. cirratus* Jaekel, 1921. Breimer, 1962a, p. 30.
Devonian

Lenneocrinus cirratus Jaekel, 1921. Breimer, 1962a, p. 30.
Devonian, Givetian: Germany

Lenneocrinus ventanillensis Breimer, 1962a, p. 30, Pl. 2, figs. 1, 2; Pl. 3, figs. 11, 13.
Devonian, Frasnian: Spain

Lindstroemiocrinus Jaekel, 1918 = *Melocrinites*

LINOBRACHIOCRINUS Goldring, 1939. *L. kindlei* Goldring, 1939. Moore and Laudon, 1943b, p. 59. Arendt and Hecker, 1964, p. 89.
Devonian

LINOCRINUS Kirk, 1938. *L. wachsmuthi* Kirk, 1938. Moore and Laudon, 1943b, p. 58; 1944, p. 167. Arendt and Hecker, 1964, p. 89.
Mississippian

Linocrinus arboreus (Worthen), 1873. Moore and Laudon, 1944, p. 167, Pl. 59, fig. 5.
Pachylocrinus arboreus Worthen, 1873. Cuénot, 1948, p. 61, fig. 78.
Mississippian, Chesterian: United States: Alabama

Linocrinus asper (Meek and Worthen), 1869. Moore and Laudon, 1944, p. 167, Pl. 64, fig. 19.
Mississippian, Osagean, Burlington Limestone: United States: Iowa

Linocrinus compactus (Laudon), 1933. Moore and Laudon, 1944, p. 167, Pl. 64, fig. 23.
Mississippian, Kinderhookian, Gilmore City Limestone: United States: Iowa

Linocrinus faculensis Laudon, Parks and Spreng, 1952, p. 555, Pl. 65, fig. 16; Pl. 69, fig. 6.
Mississippian, Kinderhookian, Banff formation: Canada: Alberta

Linocrinus penicillus (Meek and Worthen), 1869. Moore and Laudon, 1944, p. 167, Pl. 64, fig. 20.
Mississippian, Osagean, Burlington Limestone: United States: Iowa

Linocrinus praemorsus (Miller and Gurley), 1890. Moore and Laudon, 1944, p. 167, Pl. 64, fig. 22.
Mississippian, Osagean, Keokuk Limestone: United States: Indiana

Linocrinus scobina (Meek and Worthen), 1869. Moore and Laudon, 1944, p. 167, Pl. 64, fig. 24.
Mississippian, Osagean, Burlington Limestone: United States: Iowa

*Linocrinus wachsmuthi Kirk, 1938. Moore and Laudon, 1944, p. 167, Pl. 59, fig. 3.
Mississippian, Chesterian, Ste. Genevieve Limestone: United States: Alabama

Linocrinus wilsallensis Laudon and Severson, 1953, p. 517, Pl. 51, fig. 28; Pl. 55, fig. 15.
Mississippian, Kinderhookian, Lodgepole Limestone: United States: Montana

LIOMOLGOCRINUS Strimple, 1963. *L. dissutus* Strimple, 1963. Strimple, 1963d, p. 103.
Devonian, Ulsterian

*Liomolgocrinus dissutus Strimple, 1963d, p. 103.
Devonian, Ulsterian, Haragan Formation: United States: Oklahoma

LIPAROCRINUS Goldring, 1923. *L. batheri* Goldring, 1923. Moore and Laudon, 1943b, p. 54.
Devonian

LITHOCRINUS Wachsmuth and Springer, 1880. *Forbesiocrinus divaricatus* Angelin, 1878. Moore and Laudon, 1943b, p. 73. Ubaghs, 1953, p. 755.
Silurian

Lobocrinus spiniferus Wachsmuth and Springer, 1897 = *Uperocrinus marinus*

Lobolithus Waagen and Jahn, 1899 = *Scyphocrinites*

LODANELLA Kayser, 1885. *L. mira* Kayser, 1885.
Lower Devonian

Lodanella mira Kayser, 1885. Schmidt, 1941, p. 190, figs. 52, 53. Gross, 1948, p. 40, fig. 1.
Devonian, Coblentzian: Germany

LOGOCRINUS Goldring, 1923. *L. geniculatus* Goldring, 1923. Moore and Laudon, 1943b, p. 59; 1944, p. 169. Arendt and Hecker, 1964, p. 89.
Devonian

Logocrinus conicus Kesling, 1968d, p. 164, Pl. 1, figs. 1-8; Fig. 1.
Devonian, Erian, Bell Shale: United States: Michigan

Logocrinus geniculatus Goldring, 1923. Moore and Laudon, 1944, p. 169, Pl. 58, fig. 1.
Devonian, Erian, Hamilton Group: United States: New York

Logocrinus infundibuliformis Goldring, 1923. Moore and Laudon, 1944, p. 169, Pl. 58, fig. 13.
Devonian, upper Chemung Formation: United States: New York

Logocrinus kopfi Goldring, 1946, p. 37, Pl. 1, figs. 1-6 (misspelled as *Loganocrinus*).
Devonian, Senecan, Alfred Shale: United States: New York

LOPADIOCRINUS Wanner, 1916. *L. granulatus* Wanner, 1916. Moore and Laudon, 1943b, p. 60. Arendt and Hecker, 1964, p. 90.
Permian

Lopadiocrinus broweri Wanner, 1916. Wanner, 1949a, p. 51, Pl. 3, figs. 26-28; Figs. 14, 15.
Permian: Timor

Lopadiocrinus granulatus Wanner, 1916. Wanner, 1949a, p. 50, Pl. 3, figs. 23-25.
Permian: Timor

LOPHOCRINUS von Meyer, 1858. *Poteriocrinus minutus* Roemer, 1850. Moore and Laudon, 1943b, p. 54.
Devonian-lower Carboniferous

Lophocrinus aff. *minutus* (Roemer), 1850.
Poteriocrinus aff. *minutus* Roemer, 1850. Larrauri y Mercadillo, 1944, p. 50, Pl. 6, fig. 1; Pl. 8, fig. 1.
Devonian: Spanish Sahara

LOVENIOCRINUS Jaekel, 1918. *L. gotlandicus* Jaekel, 1918. Moore and Laudon, 1943b, p. 98. Ubaghs, 1953, p. 742.
Silurian

LOXOCRINUS Wanner, 1916. *L. globulus* Wanner, 1916. Moore and Laudon, 1943b, p. 76. Arendt and Hecker, 1964, p. 94.
Permian

LYONICRINUS Springer, 1926. *Coccocrinus bacca* Roemer, 1860. Moore and Laudon, 1943b, p. 101; 1944, p. 201. Ubaghs, 1953, p. 743.
Silurian

Lyonicrinus bacca (Roemer), 1860. Moore and Laudon, 1943b, p. 143, Pl. 14, fig. 1; 1944, p. 201, Pl. 77, fig. 19. Breimer, 1962a, p. 141, fig. 32. Koenig, 1965, p. 407, figs. 3A-C.
Silurian, Niagaran, Beech River Formation: United States: Tennessee

LYRIOCRINUS Hall, 1852. *Marsupiocrinites? dactylus* Hall, 1843. Moore and Laudon, 1943b, p. 82; 1944, p. 185. Ubaghs, 1953, p. 737.

Arendt and Hecker, 1964, p. 95.
Silurian

Lyriocrinus britannicus Ramsbottom, 1950, p. 653, Pl. 9, figs. 1-4; Fig. 2.
Silurian, Wenlockian, Upper Wenlock Limestone: England

Lyriocrinus melissa (Hall), 1863. Moore and Laudon, 1944, p. 185, Pl. 72, fig. 28. Wilson, 1949, p. 251, Pl. 25, fig. 15. Ramsbottom, 1950, p. 651, fig. 1.
Silurian, Niagaran, Waldron Shale: United States: Indiana

MACAROCRINUS Jaekel, 1895. *M. springeri* Jaekel, 1895. Moore and Laudon, 1943b, p. 93. Ubaghs, 1953, p. 737.
Devonian

Macarocrinus terfurcatus Schmidt, 1934. Schmidt, 1941, p. 32, Pl. 4, fig. 3.
Devonian, Coblentzian, Hunsrück Schiefer: Germany

Macarocrinus? sp. Breimer, 1962a, p. 13, Pl. 1, figs. 10-12.
Devonian: Spain

MACROCRINUS Wachsmuth and Springer, 1897. *Actinocrinus konincki* Shumard, 1855. Moore and Laudon, 1943b, p. 95, fig. 16; 1944, p. 197. Ubaghs, 1953, p. 740. Lane, 1963b, p. 922.
Mississippian, Osagean

Macrocrinus konincki (Shumard), 1855. Moore and Laudon, 1943b, p. 140, Pl. 11, fig. 12 (misspelled as *M. kononcki*); 1944, p. 197, Pl. 76, fig. 17.
Mississippian, Osagean, Burlington Limestone: United States: Iowa, Missouri, Illinois

Macrocrinus mundulus (Hall), 1860. Van Sant, *in* Van Sant and Lane, 1964, p. 117, Pl. 8, figs. 6, 13.
Actinocrinus mundulus Hall, 1860. Van Sant, *in* Van Sant and Lane, 1964, p. 117.
Actinocrinus lagunculus Hall, 1860. Van Sant, *in* Van Sant and Lane, 1964, p. 117.
Batocrinus jucundus Miller and Gurley, 1890. Van Sant, *in* Van Sant and Lane, 1964, p. 117.
Macrocrinus jucundus (Miller and Gurley), 1890. Moore and Laudon, 1944, p. 197, Pl. 76, fig. 18. Laudon, 1948, Pl. 2.
Mississippian, Osagean, Keokuk Limestone: United States: Illinois, Indiana, Iowa

Macrocrinus verneuilianus (Shumard), 1855. Moore and Laudon, 1944, p. 197, Pl. 76, fig. 19. Laudon, 1948, Pl. 2 (misspelled as *M. verneulianus*). Lane, 1963d, p. 922, fig. 3.
Mississippian, Osagean, Burlington Limestone: United States: Illinois, Iowa, Missouri

MACROSTYLOCRINUS Hall, 1852. *M. ornatus* Hall, 1852. Moore and Laudon, 1943b, p. 98; 1944, p. 200, Pl. 70, fig. 15. Moore, 1950, p. 42, fig. 11. Ubaghs, 1953, p. 742. Ramsbottom, 1961, p. 20, fig. 8.
Silurian-Devonian

Macrostylocrinus bornholmensis Laursen, 1940, p. 32, Pl. 5, fig. 1.
Middle Silurian: Denmark

Macrostylocrinus cirrifer Ramsbottom, 1961, p. 20, Pl. 6, figs. 6-13.
Ordovician, Ashgillian, Upper Drummuck Group: England

Macrostylocrinus fasciatus (Hall), 1877. Moore and Laudon, 1944, p. 200, Pl. 74, fig. 7.
Silurian, Niagaran, Waldron Shale: United States: Indiana

Macrostylocrinus granulosus Hall, 1879. Moore and Laudon, 1944, p. 200, Pl. 74, fig. 6.
Silurian, Niagaran, Waldron Shale: United States: Indiana

Macrostylocrinus laevis Springer, 1926. Moore and Laudon, 1944, p. 200, Pl. 74, fig. 4. Strimple, 1963d, p. 108, Pl. 8, fig. 1.
Silurian, Niagaran, Beech River Formation, Henryhouse Formation: United States: Tennessee, Oklahoma

**Macrostylocrinus ornatus* Hall, 1852. Moore and Laudon, 1944, p. 200, Pl. 74, fig. 13. Bowsher, 1955a, p. 8, Pl. 1, fig. 13.
Silurian, Niagaran, Rochester Shale: United States: New York

Macrostylocrinus striatus Hall, 1864. Moore and Laudon, 1943b, p. 142, Pl. 13, fig. 11; 1944, p. 200, Pl. 74, fig. 5. Strimple, 1963d, p. 109, Pl. 8, fig. 2.
Silurian, Niagaran, Waldron Shale, Henryhouse Formation: United States: Indiana, Oklahoma

MALAIOCRINUS Wanner, 1923. **Zeacrinus? sundaicus* Wanner, 1916. Moore and Laudon, 1943b, p. 55.
Pennsylvanian-Permian

Malaiocrinus azygous Strimple, 1949 = *Schistocrinus azygous*

Malaiocrinus crassitesta Wanner, 1924. Moore, 1962b, p. 24, fig. 13, no. 4.
Permian: Timor

**Malaiocrinus sundaicus* (Wanner), 1916. Moore, 1962b, p. 24, fig. 13, no. 3.
Permian: Timor

MALIGNEOCRINUS Laudon, Parks and Spreng, 1952. **M. medicinensis* Laudon, Parks and Spreng, 1952. Laudon and others, 1952, p. 564.
Mississippian, Kinderhookian

**Maligneocrinus medicinensis* Laudon, Parks and Spreng, 1952, p. 566, Pl. 67, figs. 12-15; Pl. 69, fig. 26.
Mississippian, Kinderhookian, Banff formation: Canada: Alberta

MANTIKOSOCRINUS Strimple, 1951. **M. castus* Strimple, 1951. Strimple, 1951f, p. 673.
Carboniferous, Visean, ?basal Namurian

**Mantikosocrinus castus* Strimple, 1951f, p. 674, Pl. 98, figs. 11-13. Strimple, 1961a, p. 23, fig. 1.
Mississippian, Chesterian, Fayetteville Formation: United States: Oklahoma

Mantikosocrinus wrighti (Yakovlev and Ivanov), 1956. Strimple, 1966a, p. 5.
Dicromyocrinus wrighti Yakovlev and Ivanov, 1956, p. 20. Strimple, 1966a, p. 5.
Dicromyocrinus geminatus Wright, 1952a, p. 111, Pl. 12, figs. 23-25.
Carboniferous, Visean, Calciferous Sandstone series: Scotland

MAQUOKETOCRINUS Slocum, 1924. **M. ornatus* Slocum, 1924. Moore and Laudon, 1943b, p. 82. Ubaghs, 1953, p. 737. Arendt and Hecker, 1964, p. 95.
Early Silurian

MARAGNICRINUS Whitfield, 1905. **M. portlandicus* Whitfield, 1905. Moore and Laudon, 1943b, p. 54; 1944, p. 157.
Devonian

**Maragnicrinus portlandicus* Whitfield, 1905. Moore and Laudon, 1944, p. 157, Pl. 57, fig. 12.
Devonian, Senecan, Portage Shale: United States: New York

MARATHONOCRINUS Moore and Plummer, 1940. **M. bakeri* Moore and Plummer, 1940. Moore and Laudon, 1943b, p. 56.
Pennsylvanian

Marathonocrinus moorei Lane and Webster, 1966 = *Exocrinus moorei*

Mariacrinus Hall, 1859 = *Melocrinites* Goldfuss

Mariacrinus carleyi (Hall), 1877 = *Alisocrinus carleyi*

Mariacrinus nobilissimus Hall, 1859 = *Melocrinites nobilissimus*

Mariacrinus pachydactylus Wells, 1963 = *Melocrinites pachydactylus*

Mariacrinus paucidactylus Hall, 1859 = *Melocrinites paucidactylus*

Mariacrinus plumosus Hall, 1859 = *Melocrinites plumosus*

Mariacrinus? rotundus Springer, 1926 = *Abathocrinus rotundus*

MARSUPIOCRINUS Morris, 1843. *Marsupiocrinites coelatus* Phillips, 1839. Moore and Laudon, 1943b, p. 99, fig. 18; 1944, p. 201, Pl. 71, fig. 16. Cuénot, 1948, p. 68. Ubaghs, 1953, p. 743.
Silurian-Lower Devonian

Marsupiocrinus coelatus (Phillips), 1839. Bowsher, 1955a, p. 3, Pl. 1, fig. 12.
Silurian, Wenlockian, Wenlock Limestone: England

Marsupiocrinus inflatus (Troost), 1849. Ubaghs, 1953, p. 743, fig. 59a.
Silurian: United States

Marsupiocrinus praematurus (Hall and Whitfield), 1875. La Rocque and Marple, 1955, p. 73, fig. 146.
Silurian, Niagaran, Cedarville Limestone, Guelph formation: United States: Ohio

Marsupiocrinus radiatus Angelin, 1878. Ubaghs, 1953, p. 743, fig. 22b.
Silurian: Sweden: Isle of Gotland

Marsupiocrinus rosaeformis (Troost), 1849. Moore and Laudon, 1943b, p. 143, Pl. 14, fig. 7; 1944, p. 201, Pl. 77, fig. 25.
Silurian, Niagaran, Beech River Formation: United States: Tennessee

Marsupiocrinus stellatus communis Strimple, 1963d, p. 105, Pl. 6, figs. 5-9.
Silurian, Niagaran, Henryhouse Formation: United States: Oklahoma

Marsupiocrinus striatissimus (Springer), 1926. Moore and Laudon, 1943b, p. 143, Pl. 14, fig. 8; 1944, p. 201, Pl. 77, fig. 24.
Silurian, Niagaran, Beech River Formation: United States: Tennessee

MASTIGOCRINUS Bather, 1892. *M. loreus* Bather, 1892. Moore and Laudon, 1943b, p. 54. Ubaghs, 1953, p. 751. Arendt and Hecker, 1964, p. 86.
Silurian

Mastigocrinus bravoniensis Ramsbottom, 1958, p. 110, Pl. 21, figs. 6, 7.
Silurian, Ludlovian, Lower Ludlow Beds: England

Mastigocrinus loreus Bather, 1892. Ubaghs, 1953, p. 751, fig. 24.
Silurian: England

MEGALIOCRINUS Moore and Laudon, 1942. *M. aplatus* Moore and Laudon, 1942. Moore and Laudon, 1942, p. 68; 1943b, p. 92, fig. 15. Sieverts-Doreck, 1951b, p. 105, fig. 2c. Ubaghs, 1953, p. 739.
Pennsylvanian, Morrowan

Megaliocrinus aplatus Moore and Laudon, 1942, p. 68, figs. 1-3.
Pennsylvanian, Morrowan: United States: Oklahoma

Megaliocrinus exotericus Strimple, 1951c, p. 14, Pl. 3, figs. 5-8.
Pennsylvanian, Morrowan, Brentwood Limestone: United States: Oklahoma

MEGISTOCRINUS Owen and Shumard, 1852. *Actinocrinus evansii* Owen and Shumard, 1850. Moore and Laudon, 1943b, p. 92, fig. 15; 1944, p. 191, Pl. 71, fig. 3. Moore, 1950, fig. 1g. Ubaghs, 1953, p. 739.
Devonian-Mississippian

Megistocrinus? bifrons Schmidt, 1931 = *Pyxidocrinus? bifrons*

Megistocrinus broadheadi Branson and Wilson, 1922. Branson, 1944, p. 130, Pl. 21, fig. 4.
Devonian, Erian, Mineola Limestone: United States: Missouri

Megistocrinus concavus Wachsmuth, 1886. Moore and Laudon, 1944, p. 191, Pl. 75, fig. 12.
Devonian, Erian, Traverse Formation: United States: Michigan

Megistocrinus depressus (Hall), 1862. Moore and Laudon, 1942, p. 74, fig. 4F. Goldring, 1945, p. 61.
Devonian, Erian, Kashong Member: United States: New York

**Megistocrinus evansii* (Owen and Shumard), 1850. Moore and Laudon, 1942, p. 74, fig. 4E; 1944, p. 191, Pl. 75, fig. 11.
Mississippian, Osagean, Burlington Limestone: United States: Iowa, Missouri

Megistocrinus globosus (Phillips), 1836. Wright, 1955a, p. 191, Pl. 48, figs. 1-4, 8.
Carboniferous, Tournaisian?-Visean: Scotland

Megistocrinus mineolaensis Branson and Wilson, 1922. Branson, 1944, p. 130, Pl. 21, fig. 9.
Devonian, Erian, Mineola Limestone: United States: Missouri

Megistocrinus missouriensis Branson and Wilson, 1922. Branson, 1944, p. 130, Pl. 21, figs. 15-17.
Devonian, Erian, Mineola Limestone: United States: Missouri

Megistocrinus nobilis Wachsmuth and Springer, *in* Miller, 1889. Moore and Laudon, 1943b, p. 139, Pl. 10, fig. 8; 1944, p. 191, Pl. 75, fig. 9.
Mississippian, Osagean, Hampton Formation: United States: Iowa

Megistocrinus nodosus Barris, 1878. Moore and Laudon, 1944, p. 191, Pl. 75, fig. 10.
Devonian, Erian, Traverse Formation, Cedar Valley Formation: United States: Michigan, Iowa

Megistocrinus parvus Wachsmuth and Springer, *in* Miller, 1889 = *Aryballocrinus parvus*

Megistocrinus waliszewskii Oehlert, 1896 = *Pithocrinus waliszewskii*

MELBACRINUS Strimple, 1939. **M. americanus* Strimple, 1939. Moore and Laudon, 1943b, p. 55. Arendt and Hecker, 1964, p. 88.
Upper Pennsylvanian

MELOCRINITES Goldfuss, 1826. **M. hieroglyphicus* Goldfuss, 1831. Moore and Laudon, 1943b, p. 96, fig. 17; 1944, p. 200. Ubaghs, 1953, p. 741, fig. 82f; 1958b, p. 295, fig. 16. Kesling, 1964a, p. 91. Schewtschenko, 1966, p. 133, fig. 4.
Astrocrinites Conrad, 1841. Moore and Laudon, 1943b, p. 96. Kesling, 1964a, p. 91.
Astrocrinus Bather, 1900. Moore and Laudon, 1943b, p. 96. Kesling, 1964a, p. 92.
Clonocrinus Oehlert, 1879 (*non* Quenstedt, 1876). Ubaghs, 1958b, p. 279.
Ctenocrinites Steininger, 1849. Kesling, 1964a, p. 91.
Ctenocrinus Bronn, 1840. Moore and Laudon, 1943b, p. 96; 1944, p. 200, Pl. 70, fig. 12. Ubaghs, 1953, p. 741, figs. 82d,e. Le Maître, 1958b, p. 131, Pl. 2, fig. 5. Ubaghs, 1958b, p. 294, fig. 16, Kesling, 1964a, p. 91.
Cytocrinus Roemer, 1960. Moore and Laudon, 1943b, p. 96; 1944, p. 200. Kesling, 1964a, p. 91.
Lindstroemiocrinus Jaekel, 1918. Moore and Laudon, 1943b, p. 92 (as a junior synonym of *Ctenocrinus*). Ubaghs, 1958b, p. 279.

Mariacrinus Hall, 1959. Cuénot, 1948, p. 69. Kesling, 1964a, p. 91.
Trichotocrinus Olsson, 1912. Ubaghs, 1953, p. 741, fig. 82g; 1958b, p. 295, fig. 16. Kesling, 1964a, p. 92.
Turbinicrinites Troost, *in* Wood, 1909. Kesling, 1964a, p. 92.
Turbinocrinus Wachsmuth and Springer, 1881. Moore and Laudon, 1943b, p. 96. Kesling, 1964a, p. 92.
Xenocrinus Jahn, 1892 (*non* Miller, 1881). Kesling, 1964a, p. 92.
Zenkericrinus Waagen and Jahn, 1899. Kesling, 1964a, p. 92.
Silurian-Mississippian

Melocrinites acicularis (Follmann), 1887, new comb.
   Ctenocrinus acicularis Follmann, 1887. Schmidt, 1941, p. 60, Pl. 22, fig. 3; Fig. 12. Ubaghs, 1945a, p. 7, Pl. 1, fig. 9.
   Ctenocrinus sculptus Jaekel, 1895. Schmidt, 1941, p. 60.
   Devonian, lower Coblentzian: Germany

Melocrinites acidepressus (Schmidt), 1941, new comb.
   Ctenocrinus acidepressus Schmidt, 1941, p. 62, Pl. 20, fig. 3; Fig. 13. Ubaghs, 1945a, p. 14.
   Devonian, lower Coblentzian: Germany; Belgium

Melocrinites aculeatus (Ubaghs), 1945, new comb.
   Ctenocrinus aculeatus Ubaghs, 1945a, p. 5, Pl. 1, figs. 1-8; 1945b, p. 3, Pl. 1, figs. 5-7. Van Sant, *in* Van Sant and Lane, 1964, p. 59, fig. 20, no. 3.
   Devonian, lower Emsian: Belgium

Melocrinites acutior (Schmidt), 1941, new comb.
   Ctenocrinus acutior Schmidt, 1941, p. 63, Pl. 8, fig. 3; Fig. 14.
   Devonian, Coblentzian, Nellenköptchen Beds: Germany

Melocrinites aequus Schmidt, 1941 (*non* Miller, 1892), p. 88, Pl. 20, fig. 1.
   Devonian, lower Eifelian: Germany

Melocrinites bainbridgensis (Hall and Whitfield), 1875. Moore and Laudon, 1943b, p. 142, Pl. 13, fig. 9; 1944, p. 200, Pl. 74, fig. 9.
   Devonian, Senecan, Huron Shale: United States: Ohio

Melocrinites clathratus (Schmidt), 1941, new comb.
   Ctenocrinus clathratus Schmidt, 1941, p. 78, Pl. 8, figs. 5-8; Fig. 20.
   Devonian, lower Coblentzian, Rittersturz Beds: Germany

Melocrinites? constrictus (Wanner), 1942, new comb.
   Melocrinus constrictus Wanner, 1942a, p. 35, Pl. 1, figs. 5a-c; Fig. 4.
   Devonian, Eifelian: Germany

Melocrinites decadactylus (Tannenberg, *in* Goldfuss), 1839, new comb.
   Ctenocrinus decadactylus (Goldfuss), 1839. Schmidt, 1941, p. 65, fig. 15.
   Devonian, upper Coblentzian: Germany

Melocrinites cf. *decadactylus* (Tannenberg, *in* Goldfuss), 1839, new comb.
   Ctenocrinus cf. *decadactylus* (Goldfuss), 1839. Ubaghs, 1945a, p. 15, fig. 2.
   Devonian, upper Emsian: Belgium

Melocrinites decadactylus densalternans (Schmidt), 1941, new comb.
   Ctenocrinus decadactylus densalternans Schmidt, 1941, p. 69, fig. 16. Ubaghs, 1945a, p. 16.
   Devonian, upper Coblentzian: Germany; Belgium

Melocrinites gottlandicus (Pander, *in* Helmersen), 1858, new comb.
   Ctenocrinus gottlandicus (Angelin), 1878. Ubaghs, 1953, p. 741, fig. 82d (misspelled as *C. gothlandicus* and authorship should be attributed to Pander).
   Ctenocrinus gottlandicus (Pander), 1858. Ubaghs, 1958b, p. 279, Pls. 3, 4; Figs. 8-13, 16, 17B.
   Silurian, calcareous oolite and sandstone of Burgsvik: Sweden: Isle of Gotland

Melocrinites aff. *gracilis* (Jaekel), 1895 (*non* Wachsmuth and Springer, 1897), new comb.
   Ctenocrinus aff. *gracilis* Jaekel, 1895. Ubaghs, 1945a, p. 9, Pl. 2, figs. 1-5; 1945b, p. 4, Pl. 1, fig. 8.
   Devonian, Emsian: Belgium

Melocrinites gregeri (Rowley), 1893. Branson, 1944, p. 150, Pl. 21, fig. 14 (incorrectly given as *Melocrinus gregeri*).
   Devonian, Senecan, Snyder Creek Shale: United States: Missouri

Melocrinites harrisi (Olsson), 1912. Kesling, 1964a, p. 93.
   Trichotocrinus harrisi Olsson, 1912. Ubaghs, 1953, p. 741, fig. 82g. Kesling, 1964a, p. 93.
   Devonian, Senecan, Ithaca Shale, Portage Formation: United States: New York

Melocrinites laevis Roemer, 1860 (*non M. laevis* Goldfuss, 1826). Kesling, 1964a, p. 94.
   Cytocrinus laevis Roemer, 1860. Moore and Laudon, 1943b, p. 142, Pl. 13, fig. 6; 1944, p. 200, Pl. 73, fig. 15. Kesling, 1964a, p. 94.
   Silurian, Niagaran, Beech River Formation: United States: Tennessee

Melocrinites loricatus (Schmidt), 1941, new comb.
   Ctenocrinus loricatus Schmidt, 1941, p. 80, Pl. 8, figs. 1, 2; Fig. 21.
   Devonian, Coblentzian, Koblenz-Quarzit: Germany

Melocrinites lylii (Rowley), 1894. Branson, 1944, p. 150, Pl. 21, figs. 5, 6 (incorrectly designated as *Melocrinus lylii*).
   Devonian, Senecan, Snyder Creek Shale: United States: Missouri

Melocrinites melocrinoides (Waagen and Jahn), 1899. Kesling, 1964a, p. 93.
   Zenkericrinus melocrinoides Waagen and Jahn, 1899. Kesling, 1964a, p. 93.
   Silurian (Div. E1, E2): Bohemia

Melocrinites michiganensis Kesling, 1964a, p. 94, Pls. 1, 2; Figs. 1, 2.
   Devonian, Erian, Bell Shale: United States: Michigan

Melocrinites nobilissimus (Hall), 1859. Kesling, 1964a, p. 92.
   Mariacrinus nobilissimus Hall, 1859. Kesling, 1964a, p. 92.
   Ctenocrinus nobilissimus (Hall), 1859. Moore and Laudon, 1944, p. 200, Pl. 74, fig. 15. Ubaghs, 1953, p. 741, fig. 82e.
   Devonian, Helderbergian, Coeymans Limestone: United States: New York

Melocrinites nodiferus (Follmann), 1882, new comb.
   Ctenocrinus nodiferus Follmann, 1882. Schmidt, 1941, p. 86, fig. 24.
   Devonian, upper Coblentzian: Germany

Melocrinites nodosus (Hall), 1861. Moore and Laudon, 1944, p. 200, Pl. 74, fig. 10.
   Devonian, Senecan, Traverse Group, Cedar Valley Limestone: United States: Michigan, Iowa, Wisconsin

Melocrinites pachydactylus (Conrad), 1841, new comb.
   Astrocrinites pachydactylus Conrad, 1841.
   Actinocrinites polydactylus Bonny (*non* Miller), 1835. Wells, 1963, p. 59, Pl. 11, fig. 1.
   Ctenocrinus pachydactylus (Conrad), 1841. Moore and Laudon, 1943b, p. 142, Pl. 13, fig. 8; 1944, p. 200, Pl. 74, figs. 11, 14.
   Mariacrinus pachydactylus (Conrad) Hall, 1861. Wells, 1963, p. 59, Pl. 11, fig. 2. Stukalina, 1967, fig. 4, no. 2.
   Melocrinus pachydactylus (Conrad) Goldring, 1923. Wells, 1963, p. 59, Pl. 11, fig. 3.
   Devonian, Helderbergian, Coeymans Limestone: United States: New York

Melocrinites paucidactylus (Hall), 1859, new comb.
   Mariacrinus paucidactylus Hall, 1859.

Ctenocrinus paucidactylus (Hall), 1859. Moore and Laudon, 1943b, p. 142, Pl. 13, fig. 7; 1944, p. 200, Pl. 74, fig. 12.
Devonian, Helderbergian, Coeymans Limestone: United States: New York

Melocrinites plumosus (Hall), 1859. Kesling, 1964a, p. 92.
Mariacrinus plumosus Hall, 1859. Kesling, 1964a, p. 92.
Devonian, Helderbergian, Coeymans Limestone: United States: New York

Melocrinites powelli (Goldring), 1923. Goldring, 1945, p. 62, Pl. 1, fig. 5.
Devonian, Erian, Moscow Shale: United States: New York

Melocrinites pulcher (Spriestersbach), 1919. Schmidt, 1941, p. 88, Pl. 20, fig. 2; Figs. 25a-f. Ubaghs, 1958b, p. 297, fig. 17E.
Middle Devonian, *Cultrijugatus* Zone: Germany

Melocrinites pulcher annulifera (Spriestersbach), 1919. Schmidt, 1941, p. 89, fig. 25g.
Middle Devonian: Germany

Melocrinites pulvinatus (Luttig), 1951, new comb.
Ctenocrinus pulvinatus Luttig, 1951, p. 122, Pl. 9, figs. 1-3.
Devonian, Siegenian: Germany

Melocrinites pyramidalis (Ubaghs), 1945 (*non* Goldfuss, 1839), new comb.
Ctenocrinus pyramidalis Ubaghs, 1945a, p. 17, Pl. 2, figs. 6, 7.
Devonian, upper Emsian: Belgium

Melocrinites rhenanoides (Schmidt), 1941, new comb.
Ctenocrinus rhenanoides Schmidt, 1941, p. 84, Pl. 22, fig. 1.
Devonian, upper Coblentzian: Germany

Melocrinites rhenanoides demissa (Schmidt), 1941, new comb.
Ctenocrinus rhenanoides demissa Schmidt, 1941, p. 85, Pl. 22, fig. 2; Fig. 23.
Devonian, upper Coblentzian: Germany

Melocrinites rhenanus (Follmann), 1887, new comb.
Ctenocrinus rhenanus Follmann, 1887. Schmidt, 1941, p. 81, fig. 22. Ubaghs, 1945b, p. 6, Pl. 1, figs. 1, 2.
Devonian, Coblentzian: Germany; Belgium

Melocrinites rhenanus multiplex (Schmidt), 1941, new comb.
Ctenocrinus rhenanus multiplex Schmidt, 1941, p. 83, Pl. 18, fig. 3; Pl. 20, fig. 5.
Devonian, upper Coblentzian: Germany

Melocrinites rhenanus simplex (Schmidt), 1941, new comb.
Ctenocrinus rhenanus simplex Schmidt, 1941, p. 83, Pl. 20, fig. 4.
Lower Devonian, Hohenrhein Beds: Germany

Melocrinites rhenanus ulcifer (Schmidt), 1941, new comb.
Ctenocrinus rhenanus ulcifer Schmidt, 1941, p. 83, Pl. 24, fig. 3.
Devonian, upper Coblentzian: Germany

Melocrinites rugosus (Schmidt), 1941, new comb.
Ctenocrinus rugosus Schmidt, 1941, p. 74, fig. 19.
Devonian, upper Coblentzian: Germany

Melocrinites signatus (Schmidt), 1941, new comb.
Ctenocrinus signatus Schmidt, 1941, p. 72, Pl. 17, fig. 2; Fig. 18.
Devonian, upper Coblentzian: Germany

Melocrinites spectabilis (Angelin), 1878.
Melocrinus spectabilis Angelin, 1878. Ubaghs, 1958b, p. 288.
Ctenocrinus spectabilis (Angelin), 1878. Ubaghs, 1958b, p. 288, Pl. 5; Figs. 14, 15.
Melocrinus volborthi Angelin, 1878. Ubaghs, 1958b, p. 288.
Upper Silurian: Sweden

Melocrinites splendens (Goldring), 1923. Ubaghs, 1953, p. 741, fig. 82f.
  Devonian: United States: New York

Melocrinites stellifer (Follmann), 1887, new comb.
  Ctenocrinus stellifer Follman, 1887. Schmidt, 1941, p. 70, Pl. 8,
    fig. 4; Fig. 17.
  Devonian, lower Coblentzian: Germany

Melocrinites tersus (Rowley), 1894. Branson, 1944, p. 150, Pl. 21,
    figs. 12, 13 (incorrectly designated as *Melocrinus tersus*).
  Devonian, Senecan, Snyder Creek Shale: United States: Missouri

Melocrinites(?) *triformis* Dubatolova, 1964, p. 47, Pl. 5, figs. 8, 9;
    Pl. 6, figs. 2-4; Pl. 7, fig. 1; Fig. 6.
  Devonian, Frasnian: Russia: Kuznetz Basin

Melocrinites tumidus Dubatolova, 1964, p. 44, Pl. 5, fig. 7.
  Lower Devonian: Russia: Kuznetz Basin

Melocrinites typus (Bronn), 1840. Kesling, 1964a, p. 93.
  Ctenocrinus typus Bronn, 1840. Schmidt, 1941, p. 56, Pl. 20, fig. 6;
    Fig. 11. Kesling, 1964a, p. 93.
  Devonian, Siegenian: Germany

Melocrinites umbraculum (Luttig), 1951, new comb.
  Ctenocrinus umbraculum Luttig, 1951, p. 123, Pl. 9, figs. 6, 7.
  Devonian, Eifelian, Unnenbergen Beds: Germany

Melocrinites unnenbergensis (Luttig), 1951, new comb.
  Ctenocrinus unnenbergensis Luttig, 1951, p. 124, Pl. 9, figs. 4, 5.
  Devonian, Eifelian, Unnenbergen Sandstone: Germany

Melocrinites verneuili (Troost, *in* Wood), 1909. Kesling, 1964a, p. 93.
  Turbinicrinites verneuili Troost, *in* Wood, 1909. Kesling, 1964a, p. 93.
  Silurian, Niagaran, Beech River Formation: United States: Tennessee

Melocrinites sp.
  Ctenocrinus sp. Ubaghs, 1958b, p. 299, figs. 17C, D.
  Devonian, Siegenian-Emsian transition: Belgium

Melocrinites sp. indet.
  Ctenocrinus sp. indet. Ubaghs, 1945a, p. 3, fig. 1; 1945b, p. 8, Pl. 1,
    figs. 3, 4, 9, 10.
  Devonian, Siegenian-Emsian transition: Belgium

Melocrinites sp. nov.
  Ctenocrinus n. sp. Sieverts-Doreck, 1957, p. 65.
    Devonian, Herdorf Beds: Germany
  Ctenocrinus sp. nov. Ubaghs, 1945a, p. 19, fig. 3.
    Devonian, upper Emsian: Belgium

Melocrinus constrictus Wanner, 1942 = *Melocrinites*? *constrictus*

Melocrinus pachydactylus Wells, 1963 = *Melocrinites pachydactylus*

Melocrinus volborthi Angelin, 1878 (in part) = *Melocrinites spectabilis*

Melocrinus volborthi Angelin, 1878 (in part) = *Paramelocrinus angelini*

MERISTOCRINUS Springer, 1906. *Cyathocrinus interbrachiatus* Angelin, 1878.
  Moore and Laudon, 1943b, p. 73. Ubaghs, 1953, p. 755.
  Silurian

Meristocrinus orbignyi (McCoy), 1850. Ramsbottom, 1958, p. 107, Pl. 20,
    fig. 7.
  Silurian, Ludlovian, Bannisdale Slates: England

MEROCRINUS Walcott, 1883. *M. typus* Walcott, 1883. Moore and Laudon,
  1943b, p. 52, figs. 1, 6; 1944, p. 155. Moore, 1950, p. 34,
  fig. 4e. Ramsbottom, 1961, p. 11, fig. 5. Moore, 1962b, p. 8,

figs. 1, no. 5; 11, no. 2. Philip, 1963, p. 264, fig. 4c. 1965a, fig. 2a.
Ordovician

Merocrinus corroboratus Walcott, 1883. Moore and Laudon, 1944, p. 155, Pl. 53, fig. 17.
Ordovician, Champlainian, Trenton Formation: United States: New York

Merocrinus salopiae Bather, 1896. Ramsbottom, 1961, p. 11, Pl. 3, fig. 6.
Ordovician, Llandeilian, Meadowtown Beds: England

Merocrinus typus Walcott, 1883. Moore and Laudon, 1943b, p. 132, Pl. 3, fig. 3; 1944, p. 155, Pl. 53, fig. 16.
Ordovician, Champlainian, Trenton Limestone: United States: New York

MESPILOCRINUS de Koninck and Le Hon, 1853. *M. forbesianus* de Koninck and Le Hon, 1853. Moore and Laudon, 1943b, p. 76, fig. 11; 1944, p. 181, Pl. 66, fig. 7. Ubaghs, 1953, p. 756, figs. 32, 97. Yakovlev, 1956a, p. 86, fig. 35, no. 5 (misspelled as *Moepillocrinus*); 1958, p. 72, fig. 35, no. 5; 1964a, p. 79, fig. 37, no. 5. Arendt and Hecker, 1964, p. 94. Moore and Jeffords, 1968, p. 11, fig. 1, no. 3.
Lower Carboniferous

Mespilocrinus blairi (Miller and Gurley), 1895. Moore and Laudon, 1943b, p. 137, Pl. 8, fig. 13; 1944, p. 181, Pl. 68, fig. 12. Branson, 1944, p. 211, Pl. 38, figs. 6-9.
Mississippian, Kinderhookian, Chouteau Group: United States: Missouri

Mespilocrinus bordeni Springer, 1920 = *Gaulocrinus bordeni*

Mespilocrinus depressus Wright, 1936. Wright, 1954a, p. 180, Pl. 47, figs. 8, 10, 13, 17, 18, 25; Figs. 100, 101, 104.
Carboniferous, Visean, Lower Limestone Group: Scotland

Mespilocrinus forbesianus de Koninck and Le Hon, 1853. Ubaghs, 1943, p. 2, figs. 1-11; 1953, p. 756, fig. 32. Wright, 1954a, p. 177, Pl. 47, fig. 11. Yakovlev, 1964b, p. 58, fig. 65.
Carboniferous, Tournaisian: Belgium; Scotland

Mespilocrinus konincki Hall, 1860. Moore and Laudon, 1943b, p. 137, Pl. 8, fig. 10; 1944, p. 181, Pl. 68, fig. 18. Ubaghs, 1953, p. 756, fig. 97. Moore and others, 1968a, p. 20, Pl. 3, fig. 1.
Mississippian, Osagean, Burlington Limestone: United States: Iowa

Mespilocrinus cf. *konincki* Hall, Wright, 1936 = *Mespilocrinus pringlei*

Mespilocrinus pringlei Wright, 1954a, p. 178, Pl. 47, figs. 7, 9, 14-16, 19-24; Figs. 97-99, 102, 103, 105.
Mespilocrinus cf. *konincki* Hall, Wright, 1936 and 1939. Wright, 1954a, p. 178.
Carboniferous, Visean, Lower Limestone Group: Scotland

MESPILOCYSTITES Barrande, 1887. *M. bohemicus* Barrande, 1887. Fay, 1962b, p. 156.
Ordovician

Mespilocystites bohemicus Barrande, 1887. Fay, 1962b, p. 156, Pls. 1, 2; Figs. 1-3.
Ordovician, Llandeilian, Letná Beds: Czechoslovakia

METACATILLOCRINUS Moore and Strimple, 1942. *M. bulbosus* Moore and Strimple, 1942. Moore and Strimple, 1942, p. 79. Moore and Laudon, 1943b, p. 31. Ubaghs, 1953, p. 747. Moore, 1962b, p. 16, fig. 12, no. 6. Arendt and Hecker, 1964, p. 81.
Pennsylvanian, Desmoinesian

Metacatillocrinus bulbosus Moore and Strimple, 1942, p. 80, figs. 1-6. Strimple, 1962e, p. 11.
Pennsylvanian, Desmoinesian, Marmaton Group: United States: Oklahoma

METACROMYOCRINUS Strimple, 1961. **M. holdenvillensis* Strimple, 1961.
Strimple, 1961g, p. 68; 1966a, p. 6.
Pennsylvanian, Morrowan-Desmoinesian

Metacromyocrinus gillumi Strimple, 1966a, p. 6, Pl. 1, figs. 1-6; Pl. 2, figs. 5, 6.
Pennsylvanian, Morrowan, Bloyd Formation: United States: Arkansas

*Metacromyocrinus holdenvillensis Strimple, 1961g, p. 69, Pl. 6, figs. 1-3; Figs. 16-18. Branson, 1962, p. 163, fig. 4.
Pennsylvanian, Desmoinesian, Holdenville Formation: United States: Oklahoma

Metacromyocrinus minutus Strimple, 1962e, p. 39, Pl. 4, figs. 12-15.
Pennsylvanian, Desmoinesian, Oologah Formation: United States: Oklahoma

Metacromyocrinus oklahomensis Strimple and Knapp, 1966 = *Paracromyocrinus oklahomensis*

METAINDOCRINUS Strimple, 1966. **M. cooperi* Strimple, 1966. Strimple, 1966b, p. 81.
Permian, Guadalupian

*Metaindocrinus cooperi Strimple, 1966b, p. 83, figs. 1-4.
Permian, Guadalupian, Word Formation: United States: Texas

METALLAGECRINUS Strimple, 1966. **Allagecrinus quinquebrachiatus* Wanner, 1929. Strimple, 1966c, p. 109.
Permian

Metallagecrinus acutus (Wanner), 1929. Strimple, 1966c, p. 105.
Allagecrinus acutus Wanner, 1929. Strimple, 1966c, p. 105.
Permian: Timor

Metallagecrinus dux (Wanner), 1930. Strimple, 1966c, p. 105.
Allagecrinus dux Wanner, 1930. Strimple, 1966c, p. 105.
Permian: Timor

Metallagecrinus excavatus (Wanner), 1929. Strimple, 1966c, p. 105.
Allagecrinus excavatus Wanner, 1929. Strimple, 1966c, p. 105.
Permian: Timor

Metallagecrinus indoaustralicus (Wanner), 1916. Strimple, 1966c, p. 105.
Allagecrinus indoaustralicus Wanner, 1916. Strimple, 1966c, p. 105.
Permian: Timor

Metallagecrinus inflatus (Wanner), 1929. Strimple, 1966c, p. 105.
Allagecrinus inflatus Wanner, 1929. Strimple, 1966c, p. 105.
Permian: Timor

Metallagecrinus multibrachiatus (Yakovlev), 1927. Strimple, 1966c, p. 104.
Allagecrinus multibrachiatus Yakovlev, 1927. Strimple, 1966c, p. 104.
Permian: Russia

Metallagecrinus ornatus (Wanner), 1929. Strimple, 1966c, p. 105.
Allagecrinus ornatus Wanner, 1929. Strimple, 1966c, p. 105.
Permian: Timor

Metallagecrinus procerus (Wanner), 1929. Strimple, 1966c, p. 105.
Allagecrinus procerus Wanner, 1929. Strimple, 1966c, p. 105.
Permian: Timor

*Metallagecrinus quinquebrachiatus (Wanner), 1929. Strimple, 1966c, p. 109, fig. 2e.
Allagecrinus quinquebrachiatus Wanner, 1929. Strimple, 1966c, p. 109.
Permian: Timor

Metallagecrinus quinquelobus (Wanner), 1929. Strimple, 1966c, p. 105.
   Allagecrinus quinquelobus Wanner, 1929. Strimple, 1966c, p. 105.
   Permian: Timor

Metallagecrinus uralensis (Yakovlev), 1927. Strimple, 1966c, p. 105.
   Allagecrinus uralensis Yakovlev, 1927. Strimple, 1966c, p. 105.
   Permian: Russia

Metallagecrinus uralensis var. *nodocarinatus* (Yakovlev), 1927. Strimple, 1966c, p. 105.
   Allagecrinus uralensis var. *nodocarinatus* Yakovlev, 1927. Strimple, 1966c, p. 105.
   Permian: Russia

METAPERIMESTOCRINUS Strimple, 1961. *M. spiniferus* Strimple, 1961. Strimple, 1961g, p. 36.
   Pennsylvanian, Desmoinesian

Metaperimestocrinus spiniferus Strimple, 1961g, p. 37, Pl. 4, figs. 3-7; Pl. 18, fig. 5; Fig. 13a, b. Branson, 1962, p. 162, fig. 3.
   Pennsylvanian, Desmoinesian, Holdenville Formation: United States: Oklahoma

Metaperimestocrinus trapezoidalis Strimple, 1962e, p. 35, Pl. 7, figs. 15-18.
   Pennsylvanian, Desmoinesian, Oologah Limestone: United States: Oklahoma

METASYCOCRINUS Wanner, 1920. *Hypocrinus? piriformis* Rothpletz, 1892. Moore and Laudon, 1943b, p. 50. Arendt and Hecker, 1964, p. 85. Lane, 1967b, p. 12, fig. 6.
   Permian

METHABOCRINUS Jaekel, 1918. *M. erraticus* Jaekel, 1918. Moore and Laudon, 1943b, p. 54. Ubaghs, 1953, p. 740; 1958a, p. 53, fig. 4.
   Silurian

Methabocrinus erraticus Jaekel, 1918. Ubaghs, 1958a, p. 53, figs. 1-3.
   Upper Silurian, Gotlandian: Germany

METICHTHYOCRINUS Springer, 1906. *Cyathocrinus tiaraeformis* Troost, in Hall, 1858. Moore and Laudon, 1943b, p. 76, fig. 11; 1944, p. 183, Pl. 66, fig. 12. Ubaghs, 1953, p. 756.
   Mississippian

Metichthyocrinus burlingtonensis (Hall), 1858. Moore and Laudon, 1943b, p. 136, Pl. 7, fig. 2; 1944, p. 183, Pl. 68, fig. 6.
   Mississippian, Osagean, Burlington Limestone: United States: Iowa, Missouri

Metichthyocrinus clarkensis (Miller and Gurley), 1894. Moore and Laudon, 1944, p. 183, Pl. 68, fig. 9. Branson, 1944, p. 165, Pl. 30, figs. 12, 15.
   Mississippian, Osagean, Northview Formation, New Providence Shale: United States: Missouri, Indiana, Kentucky, Tennessee

MICTOCRINUS Goldring, 1923. *M. robustus* Goldring, 1923. Moore and Laudon, 1943b, p. 50. Ubaghs, 1953, p. 750.
   Devonian

Mictocrinus robustus Goldring, 1923. Goldring, 1954, p. 10, Pl. 2, figs. 1, 2.
   Devonian, Ulsterian, Onondaga Limestone: United States: New York

MIRACRINUS Bowsher, 1953. *Lecanocrinus (Miracrinus) perdewi* Bowsher, 1953. Strimple, 1963d, p. 117.
   Lecanocrinus (Miracrinus) Bowsher, 1953, p. 3.
   Devonian

Miracrinus magniradialis (Weller), 1903.
   Ichthyocrinus magniradialis Weller, 1903. Bowsher, 1953, p. 7.

Lecanocrinus (Miracrinus) magniradialis (Weller), 1903. Bowsher, 1953, p. 7.
Devonian, Helderbergian, New Scotland Limestone: United States: New Jersey

*Miracrinus perdewi (Bowsher), 1953. Strimple, 1963d, p. 117.
Lecanocrinus (Miracrinus) perdewi Bowsher, 1953, p. 5, Pl. 1, figs. 1-11; Figs. 1a-e.
Devonian, Ulsterian, New Scotland Limestone: United States: Maryland

MISSOURICRINUS Miller, 1891. *M. admonitus* Miller, 1891. Moore and Laudon, 1943b, p. 60. Arendt and Hecker, 1964, p. 93.
Mississippian, Kinderhookian

MOAPACRINUS Lane and Webster, 1966. *M. rotundatus* Lane and Webster, 1966. Lane and Webster, 1966, p. 30.
Permian, Wolfcampian

Moapacrinus inornatus Webster and Lane, 1967, p. 18, Pl. 2, figs. 1, 2, 5-7, 9, 10.
Permian, Wolfcampian, Bird Spring Formation: United States: Nevada

*Moapacrinus rotundatus Lane and Webster, 1966, p. 31, Pl. 6, figs. 1-10; Fig. 9. Webster and Lane, 1967, p. 17, Pl. 3, fig. 12.
Permian, Wolfcampian, Bird Spring Formation: United States: Nevada

MOLLOCRINUS Wanner, 1916. *M. poculum* Wanner, 1916. Moore and Laudon, 1943b, p. 55. Arendt and Hecker, 1964, p. 92.
Permian

*Mollocrinus poculum Wanner, 1916. Wanner, 1954, p. 232, figs. 2, 3.
Permian: Timor

MONOBRACHIOCRINUS Wanner, 1916. *M. ficiformis* Wanner, 1916. Moore and Laudon, 1943b, p. 50. Cuénot, 1948, p. 64. Yakovlev, 1951, p. 578, fig. 1, no. II. Arendt and Hecker, 1964, p. 84. Lane, 1967b, p. 12, fig. 6.
Permian

*Monobrachiocrinus ficiformis Wanner, 1916. Arendt and Hecker, 1964, p. 84, Pl. 9, fig. 6; Fig. 124.
Permian: Timor

Monobrachiocrinus ficiformis granulatus Wanner, 1920. Ubaghs, 1953, p. 672, figs. 10a, g, h. Yakovlev, 1964b, p. 56, figs. 53a, 54.
Monobrachiocrinus granulatus Cuénot, 1948, p. 64, fig. 83, no. 1.
Permian: Timor

Monobrachiocrinus oviformis Yakovlev, 1926. Yakovlev, 1939b, p. 60, Pl. 10, fig. 13; Fig. 13. Yakovlev, *in* Yakovlev and Ivanov, 1956, p. 56, Pl. 10, figs. 1-3; Fig. 14. Arendt and Hecker, 1964, p. 84, Pl. 9, figs. 3-5; Fig. 123. Arendt, 1968a, p. 104, fig. 7.
Lower Permian: Russia: Krasnoufimsk

MONSTROCRINUS Schmidt, 1941. *M. securifer* Schmidt, 1941. Schmidt, 1941, p. 213. Ubaghs, 1953, p. 737.
Lower Devonian

Monstrocrinus granosus Schmidt, 1941, p. 217, fig. 61.
Devonian, Coblentzian, Mandeln Beds: Germany

*Monstrocrinus securifer Schmidt, 1941, p. 215, Pl. 16, figs. 1-7; Fig. 60.
Lower Devonian, *Orthocrinus* Beds: Germany

Mooreocrinus Wright and Strimple, 1945. **Cromyocrinus geminatus* Trautschold, 1867. Wright and Strimple, 1945, p. 221. Strimple, 1966a, p. 5.
Carboniferous

Mooreocrinus bowsheri Strimple, 1949 = *Cryphiocrinus bowsheri*

*Mooreocrinus geminatus (Trautschold), 1867. Wright and Strimple, 1945,
p. 222. Strimple, 1966a, p. 5.
Cromyocrinus geminatus Trautschold, 1867. Yakovlev, 1939a, p. 66,
Pl. 12, fig. 7. Wright and Strimple, 1945, p. 222.
Dicromyocrinus geminatus (Trautschold), 1867. Strimple, 1961g, p. 65.
Carboniferous: Russia; Scotland

Mooreocrinus hemisphericus (Worthen), 1882. Wright and Strimple, 1945,
p. 222.
Agassizocrinus hemisphericus Worthen, 1882. Wright and Strimple, 1945,
p. 222.
Mississippian, Chesterian: United States: Illinois

Mooreocrinus meadowensis Strimple, 1949 = *Tarachiocrinus meadowensis*

Mooreocrinus mendesi (Lane), 1964. Strimple, 1966a, p. 6.
Dicromyocrinus mendesi Lane, 1964a, p. 362, Pl. 57, figs. 1, 2, 6-10.
Middle Pennsylvanian, Itaituba Series: Brazil

Mooreocrinus ornatus (Trautschold), Wright and Strimple, 1945 = *Dicromyocrinus ornatus*

Mooreocrinus papillatus (Worthen), 1882. Wright and Strimple, 1945, p. 222.
Agassizocrinus papillatus Worthen, 1882. Wright and Strimple, 1945,
p. 222.
Mississippian, Chesterian: United States: Illinois

MORROWCRINUS Moore and Plummer, 1938. **M. fosteri* Moore and Plummer, 1938.
Moore and Laudon, 1943b, p. 59. Arendt and Hecker, 1964, p. 89.
Pennsylvanian, Morrowan

MOSCOVICRINUS Jaekel, 1918. **Poteriocrinus multiplex* Trautschold, 1867.
Moore and Laudon, 1943b, p. 55. Yakovlev, 1947d, p. 45, fig. 11;
1956a, p. 81, fig. 31; 1958, p. 66, fig. 31; 1964a, p. 76, fig. 33;
1964b, p. 58, fig. 66. Arendt and Hecker, 1964, p. 87.
Carboniferous-Permian

Moscovicrinus bijugus (Trautschold), 1867. Yakovlev and Ivanov, 1956,
p. 11, Pl. 3, fig. 1; Fig. 1. Ivanova, 1958, p. 132, frontise.
Poteriocrinus bijugus Trautschold, 1867. Yakovlev and Ivanov, 1956,
p. 11.
Carboniferous, Moscovian: Russia: Moscow Basin

Moscovicrinus bipinnatus Lane and Webster, 1966, p. 29, Pl. 5, figs. 1, 3,
6-8.
Permian, Wolfcampian, Bird Spring Formation: United States: Nevada

*Moscovicrinus multiplex (Trautschold), 1867. Yakovlev, 1939a, p. 65,
Pl. 11, figs. 5, 6; 1947b, p. 748, fig. 1. Yakovlev and Ivanov,
1956, p. 10, Pl. 1, figs. 1, 2. Arendt and Hecker, 1964, p. 87,
Pl. 11, fig. 1.
Carboniferous, Visean: Russia: Moscow Basin, Central Kazakhstan

MYCOCRINUS Schultze, 1866. **M. boletus* Schultze, 1866. Moore and Laudon,
1943b, p. 30. Ubaghs, 1953, p. 747, fig. 11c. Moore, 1962b, p. 16,
fig. 12, no. 2. Arendt and Hecker, 1964, p. 81.
Devonian, Eifelian

*Mycocrinus boletus Schultze, 1866. Ubaghs, 1953, p. 747, figs. 40-42.
Devonian, Eifelian: Germany

MYELODACTYLUS Hall, 1852. **M. convolutus* Hall, 1852. Moore and Laudon,
1943b, p. 29; 1944, p. 143. Cuénot, 1948, p. 60. Ubaghs, 1953,
p. 744, figs. 123, 124, 128. Yakovlev, 1964b, p. 69, fig. 105. Arendt and
Hecker, 1964, p. 80. Moore and Jeffords, 1968, p. 11, fig. 1, no. 6.
Silurian-Lower Devonian

Myelodactylus ammonis (Bather), 1893. Ubaghs, 1953, p. 744, figs. 124, 128.
  Silurian: Europe; North America
Myelodactylus brachiatus Hall, 1852 = *Crinobrachiatus brachiatus*
Myelodactylus canaliculatus (Goldfuss), 1826. Sieverts-Doreck, 1954, p. 46, figs. 1-4.
  Middle Devonian: Germany
**Myelodactylus convolutus* Hall, 1852. Moore and Laudon, 1944, p. 143, Pl. 54, fig. 26. Moore, 1962b, p. 41, Pl. 4, fig. 2.
  Silurian, Albion-Niagaran, Brassfield Limestone, Laurel Limestone, Racine Formation, Clinton Formation, Rochester Shale, Lockport Formation: United States: Ohio, Indiana, Illinois, New York
Myelodactylus extensus Springer, 1926. Amsden, 1949, p. 78, Pl. 12, fig. 17. Strimple, 1963d, p. 27, Pl. 1, figs. 9, 10.
  Silurian, Niagaran, Brownsport Formation, Henryhouse Formation: United States: Tennessee, Oklahoma
Myelodactylus extensus alternicirrus Springer, 1926 = *Myelodactylus extensus extensus*
Myelodactylus extensus bijugicirrus Springer, 1926. Moore, 1962b, p. 42.
  Silurian, Niagaran, Beech River Formation: United States: Tennessee
Myelodactylus extensus extensus Springer, 1926. Moore, 1962b, p. 42.
Myelodactylus extensus alternicirrus Springer, 1926. Moore, 1962b, p. 42.
  Silurian, Niagaran, Beech River Formation: United States: Tennessee
Myelodactylus fletcheri (Salter), 1873. Ubaghs, 1953, p. 744, figs. 125-127. Yakovlev, 1964b, p. 69, fig. 106.
  Silurian: Europe
Myelodactylus keyserensis Springer, 1926. Moore and Laudon, 1943b, p. 134, Pl. 5, fig. 7; 1944, p. 145, Pl. 54, fig. 27. Ubaghs, 1953, p. 744, fig. 140N. Moore, 1962b, p. 41, Pl. 3, fig. 1.
  Silurian, Niagaran-Cayugan, Rochester Shale, Keyser Limestone: United States: New York, West Virginia
Myelodactylus nodosarius (Hall), 1859. Moore and Laudon, 1944, p. 145, Pl. 54, fig. 25. Strimple, 1963d, p. 26, Pl. 1, figs. 1-4; Fig. 3a.
  Devonian, Helderbergian, New Scotland Limestone, Haragan Formation, Bois d'Arc Limestone: United States: New York, Oklahoma
Myelodactylus (*Eomyelodactylus*) *rotundatus* Foerste, 1919 = *Eomyelodactylus rotundatus*

Myelodactylus spp.
  Termier and Termier, 1949, p. 42, Pl. 2, fig. 7.
    Silurian: Sweden
  Breimer, 1962a, p. 145.
    Devonian, Santa Lucia Limestone, Portilla Limestone: Spain
  Strimple, 1963d, p. 28, Pl. 1, fig. 5.
    Silurian, Niagaran, Henryhouse Formation: United States: Oklahoma
MYRTILLOCRINUS Sandberger and Sandberger, 1856. **M. elongatus* Sandberger and Sandberger, 1856. Moore and Laudon, 1943b, p. 50; 1944, p. 153. Ubaghs, 1953, p. 750.
  Devonian
Myrtillocrinus americanus Hall, 1862. Moore and Laudon, 1943b, p. 131, Pl. 2, fig. 3; 1944, p. 154, Pl. 57, fig. 16.
  Devonian, Ulsterian, Onondaga Limestone: United States: New York
Myrtillocrinus(?) *curtus* Schmidt, 1913. Schmidt, 1941, p. 107, fig. 29.
  Devonian, lower Eifelian: Germany

Myrtillocrinus orbiculatus Dubatolova, 1964, p. 24, Pl. 2, figs. 3-5; Fig. 2.
  Devonian, Eifelian: Russia: Kuznetz Basin

MYSTICOCRINUS Springer, 1918. *M. wilsoni* Springer, 1918. Moore and Laudon, 1943b, p. 50; 1944, p. 153. Ubaghs, 1953, p. 750. Xu, 1962, p. 45, 50, fig. 1.
  Silurian, Niagaran, ?Carboniferous, Namurian

Mysticocrinus wilsoni Springer, 1918. Moore and Laudon, 1944, p. 153, Pl. 54, fig. 23.
  Silurian, Niagaran, Laurel Limestone: United States: Indiana

Mysticocrinus? sp. Termier and Termier, 1950, p. 88, Pl. 216, figs. 9-11.
  Carboniferous, Namurian: Morocco

NACTOCRINUS Kirk, 1947. *N. nitidus* Kirk, 1947. Kirk, 1947, p. 288.
  Mississippian, Kinderhookian-Osagean

Nactocrinus antiquus (Meek and Worthen), 1869. Kirk, 1947, p. 289.
  Erisocrinus antiquus Meek and Worthen, 1869. Kirk, 1947, p. 289.
  Mississippian, Osagean, lower Burlington Limestone: United States: Iowa

Nactocrinus nitidus Kirk, 1947, p. 290, Pl. 1, figs. 1-6.
  Mississippian, Osagean, lower Burlington Limestone: United States: Iowa

Nactocrinus notatus (Miller and Gurley), 1896. Kirk, 1947, p. 292.
  Scaphiocrinus notatus Miller and Gurley, 1896. Kirk, 1947, p. 292.
  Mississippian, Kinderhookian, Hampton Formation: United States: Iowa

NANOCRINUS Müller, 1856. *N. paradoxus* Müller, 1856. Moore and Laudon, 1943b, p. 50. Ubaghs, 1953, p. 750.
  Devonian, Eifelian

NASSOVIOCRINUS Jaekel, 1918. *Heterocrinus pachydactylus* Sandberger and Sandberger, 1855. Moore and Laudon, 1943b, p. 54.
  Devonian, Coblentzian-Frasnian

Nassoviocrinus rarifissus Jaekel, 1921 = *Antihomocrinus rarifissus*

NEBRASKACRINUS Moore, 1939. *N. tourteloti* Moore, 1939. Moore and Laudon, 1943b, p. 55; 1944, p. 159. Arendt and Hecker, 1964, p. 88.
  Permian, Wolfcampian

Nebraskacrinus tourteloti Moore, 1939. Moore and Laudon, 1944, p. 159, Pl. 60, fig. 13.
  Permian, Wolfcampian, Winfield Limestone: United States: Nebraska, Kansas

NEOARCHAEOCRINUS Strimple and Watkins, 1955. *Thysanocrinus (Rhodocrinus) pyriformis* Billings, 1857. Strimple and Watkins, 1955, p. 350.
  Ordovician-Silurian, Trenton-Niagaran

Neoarchaeocrinus necopinus Strimple, 1963d, p. 70, Pl. 3, fig. 8; Fig. 20b.
  Silurian, Niagaran, Henryhouse Formation: United States: Oklahoma

Neoarchaeocrinus pyriformis (Billings), 1857. Strimple, 1963d, p. 70, fig. 19b.
  Thysanocrinus pyriformis Billings, 1857. Strimple, 1963d, p. 70.
  Ordovician, Champlainian, Hull Limestone: Canada: Quebec

NEOCATILLOCRINUS Wanner, 1937. *N. incissus* Wanner, 1937. Moore and Laudon, 1943b, p. 31. Ubaghs, 1953, p. 747, figs. 11g, 140H. Moore, 1962b, p. 16, fig. 12, no. 7. Arendt and Hecker, 1964, p. 81.
  Permian

NEODICHOCRINUS Wanner, 1937. *N. nanus* Wanner, 1937. Moore and Laudon, 1943b, p. 96. Ubaghs, 1953, p. 740. Arendt and Hecker, 1964, p. 97.
  Permian

NEOPLATYCRINUS Wanner, 1916. *N. dilatatus Wanner, 1916. Moore and Laudon, 1943b, p. 101. Ubaghs, 1953, p. 743. Arendt and Hecker, 1964, p. 98.
Permian

*Neoplatycrinus dilatatus Wanner, 1916. Breimer, 1962a, p. 136, figs. 32, 33.
Permian: Timor

Neoplatycrinus somoholensis Wanner, 1916. Meyer, 1965, p. 1207, fig. 1G.
Permian: Timor

Neoplatycrinus sp. Meyer, 1965, fig. 1H (typographical error as Neoplatycrinites sp.).
Permian: Timor

NEOZEACRINUS Wanner, 1937. *N. peramplus Wanner, 1937. Moore and Laudon, 1943b, p. 58; 1944, p. 167, Pl. 52, fig. 13. Arendt and Hecker, 1964, p. 89.
Pennsylvanian, Desmoinesian-Permian, Wolfcampian

Neozeacrinus coronulus Webster and Lane, 1967, p. 28, Pl. 7, fig. 4.
Permian, Wolfcampian, Bird Spring Formation: United States: Nevada

Neozeacrinus praecursor Moore and Plummer, 1940 = Anobasicrinus praecursor

Neozeacrinus uddeni (Weller), 1909. Moore and Laudon, 1944, p. 167, Pl. 62, fig. 15; Pl. 63, fig. 22. Webster and Lane, 1967, p. 27.
Plaxocrinus uddeni (Weller), 1909. Strimple, 1961g, p. 47.
Permian, Wolfcampian, Cibolo Formation: United States: Texas

Neozeacrinus wanneri Webster and Lane, 1967, p. 27, Pl. 7, fig. 3; Pl. 8, figs. 1-4; Fig. 4, nos. 4, 5.
Permian, Wolfcampian, Bird Spring Formation: United States: Nevada

NEREOCRINUS Wanner, 1924. *N. antiquus Wanner, 1924. Moore and Laudon, 1943b, p. 50. Arendt and Hecker, 1964, p. 84. Lane, 1967b, p. 11, fig. 5.
Permian

*Nereocrinus antiquus Wanner, 1924. Arendt and Hecker, 1964, p. 84, Pl. 9, fig. 11; Fig. 125.
Permian: Timor

Nereocrinus jemeljantzewi Yakovlev, 1937. Yakovlev, in Yakovlev and Ivanov, 1956, p. 55, Pl. 11, fig. 19. Arendt and Hecker, 1964, p. 84, Pl. 9, fig. 12.
Permian, Artinskian: Russia: Krasnoufimsk

NEVADACRINUS Lane and Webster, 1966. *N. geniculatus Lane and Webster, 1966. Lane and Webster, 1966, p. 15.
Permian, Wolfcampian

*Nevadacrinus geniculatus Lane and Webster, 1966, p. 16, Pl. 1, figs. 2, 4, 7; Figs. 3-5. Webster and Lane, 1967, p. 9, fig. 2. Moore and others, 1968a, p. 20, Pl. 4, fig. 6.
Permian, Wolfcampian, Bird Spring Formation: United States: Nevada

Nevadacrinus sp. Lane and Webster, 1967, p. 14, fig. 5.
Permian, Wolfcampian, Bird Spring Formation: United States: Nevada

NIPTEROCRINUS Wachsmuth, in Meek and Worthen, 1868. *N. wachsmuthi Meek and Worthen, 1868. Moore and Laudon, 1943b, p. 74, fig. 11; 1944, p. 181, Pl. 66, fig. 5. Ubaghs, 1953, p. 756. Arendt and Hecker, 1964, p. 94.
Mississippian, Kinderhookian-Osagean

*Nipterocrinus wachsmuthi Meek and Worthen, 1868. Moore and Laudon, 1943b, p. 137, Pl. 8, fig. 1; 1944, p. 181, Pl. 69, fig. 2. Laudon and others, 1952, p. 559, Pl. 66, figs. 5-7.

Mississippian, Kinderhookian-Osagean, Banff formation, lower Burlington Limestone: Canada: Alberta; United States: Iowa

NOTIOCRINUS Wanner, 1924. *N. timoricus Wanner, 1924. Moore and Laudon, 1943b, p. 59.
Permian

NUNNACRINUS Bowsher, 1955. *N. mammillatus Bowsher, 1955. Bowsher, 1955b, p. 17.
Mississippian

Nunnacrinus armatus (de Koninck and Le Hon), 1854. Bowsher, 1955b, p. 21.
Actinocrinus armatus de Koninck and Le Hon, 1854. Bowsher, 1955b, p. 21.
Carboniferous: Belgium

Nunnacrinus dalayanus (Miller), 1881. Bowsher, 1955b, p. 18.
Actinocrinus dalayanus Miller, 1881. Bowsher, 1955b, p. 18.
Mississippian, Osagean, Lake Valley Limestone: United States: New Mexico

Nunnacrinus deornatus (de Koninck and Le Hon), 1854. Bowsher, 1955b, p. 21.
Actinocrinus deornatus de Koninck and Le Hon, 1854. Bowsher, 1955b, p. 21.
Carboniferous: Belgium

Nunnacrinus dorsatus (de Koninck and Le Hon), 1854. Bowsher, 1955b, p. 21.
Actinocrinus dorsatus de Koninck and Le Hon, 1854. Bowsher, 1955b, p. 21.
Carboniferous: Belgium

Nunnacrinus foveatus (Miller and Gurley), 1895. Bowsher, 1955b, p. 18.
Actinocrinus foveatus Miller and Gurley, 1895. Bowsher, 1955b, p. 18.
Mississippian, Osagean, Burlington Limestone: United States: Missouri

Nunnacrinus jessieae (Miller and Gurley), 1896. Bowsher, 1955b, p. 19.
Actinocrinus jessieae Miller and Gurley, 1896. Bowsher, 1955b, p. 19.
Mississippian, Osagean, Burlington Limestone: United States: Missouri

Nunnacrinus locellus (Hall), 1861. Bowsher, 1955b, p. 19, Pl. 3, fig. 4.
Actinocrinus locellus Hall, 1861. Bowsher, 1955b, p. 19.
Mississippian, Osagean, lower Burlington Limestone: United States: Iowa

*Nunnacrinus mammillatus Bowsher, 1955b, p. 19, Pl. 3, figs. 6, 7; Pl. 6, figs. 2, 3; Figs. 3A-D.
Mississippian, Osagean, Nunn Member: United States: New Mexico

Nunnacrinus pallubrum (Miller and Gurley), 1896. Bowsher, 1955b, p. 20.
Actinocrinus pallubrum Miller and Gurley, 1896. Bowsher, 1955b, p. 20.
Mississippian, Osagean, Burlington Limestone: United States: Missouri

Nunnacrinus pettisensis (Miller and Gurley), 1896. Bowsher, 1955b, p. 21.
Actinocrinus pettisensis Miller and Gurley, 1896. Bowsher, 1955b, p. 21.
Mississippian, Osagean, Burlington Limestone: United States: Missouri

Nunnacrinus puteatus (Rowley and Hare), 1896. Bowsher, 1955b, p. 22.
Actinocrinus puteatus Rowley and Hare, 1891. Bowsher, 1955b, p. 22.
Mississippian, Osagean, lower Burlington Limestone: United States: Missouri

Nunnacrinus rubra (Weller), 1909. Bowsher, 1955b, p. 21.
Actinocrinites rubra (Weller), 1909. Moore and Laudon, 1944, p. 193, Pl. 76, fig. 2 (misspelled as A. ruber). Laudon, 1948, Pl. 3.
Actinocrinus rubra Weller, 1909. Bowsher, 1955b, p. 21.
Mississippian, Osagean, Fern Glen Limestone, Lake Valley Limestone: United States: Missouri, New Mexico

Nunnacrinus sampsoni (Miller and Gurley), 1896. Bowsher, 1955b, p. 21.
 Actinocrinus sampsoni Miller and Gurley, 1896. Bowsher, 1955b, p. 21.
 Mississippian, Osagean, Burlington Limestone: United States: Missouri

Nunnacrinus stellaris (de Koninck and Le Hon), 1854. Bowsher, 1955b, p. 22.
 Actinocrinus stellaris de Koninck and Le Hon, 1854. Bowsher, 1955b, p. 22.
 Nunnacrinus? stellaris (de Koninck and Le Hon), 1854. Breimer, 1962a, p. 78, Pl. 8, fig. 7.
 Carboniferous: Belgium

Nunnacrinus sp. A. Bowsher, 1955b, p. 21, Pl. 3, fig. 5.
 Mississippian, Osagean, Burlington Limestone: United States: Iowa

NYCTOCRINUS Springer, 1926. *N. magnitubus* Springer, 1926. Moore and Laudon, 1943b, p. 84, fig. 14; 1944, p. 189. Moore, 1950, p. 42, fig. 11. Ubaghs, 1953, p. 738.
 Silurian, Niagaran

Nyctocrinus magnitubus Springer, 1926. Moore and Laudon, 1943b, p. 138, Pl. 9, fig. 16; 1944, p. 189, Pl. 77, fig. 23.
 Silurian, Niagaran, Beech River Formation: United States: Tennessee

OCHLEROCRINUS Strimple, 1963. *Eucalyptocrinus armosus* McChesney, 1861. Strimple, 1963d, p. 87.
 Silurian, Niagaran

Ochlerocrinus armosus (McChesney), 1861. Strimple, 1963d, p. 87.
 Siphonocrinus armosus (McChesney), 1861. Strimple, 1963d, p. 87.
 Silurian, Niagaran, Racine Dolomite: United States: Wisconsin

Ochlerocrinus pentagonus (Wachsmuth and Springer), 1897. Strimple, 1963d, p. 87.
 Siphonocrinus pentagonus Wachsmuth and Springer, 1897. Bolton, 1957, p. 54, Pl. 7, figs. 1-4. Strimple, 1963d, p. 87.
 Silurian, Niagaran, Racine Dolomite, Amabel Formation: United States: Wisconsin; Canada: Ontario

Octocrinus Peck, 1936 = *Tytthocrinus*

Octocrinus inconsuetus Peck, 1936 = *Tytthocrinus inconsuetus*

OENOCHOACRINUS Breimer, 1962. *O. princeps* Breimer, 1962. Breimer, 1962a, p. 124.
 Devonian

Oenochoacrinus pileatus Breimer, 1962a, p. 127, Pl. 12, figs. 7-9; Figs. 30, 32.
 Devonian, lower Couvinian, Santa Lucia Limestone: Spain

Oenochoacrinus princeps Breimer, 1962a, p. 124, Pl. 12, figs. 1-6; Fig. 29.
 Devonian, horizon unknown: Spain

Oenochoacrinus scaber Breimer, 1962a, p. 129, Pl. 13, figs. 3-5, 10, 11; Fig. 31.
 Devonian, lower Emsian, La Vid Shale: Spain

Oenochoacrinus? sp. Breimer, 1962a, p. 131, Pl. 12, figs. 10, 11; Pl. 13, figs. 1, 2.
 Devonian, lower Couvinian, Santa Lucia Limestone: Spain

OHIOCRINUS Wachsmuth and Springer, 1886. *Homocrinus laxus* Hall, 1867. Moore and Laudon, 1943b, p. 31. Ubaghs, 1953, p. 747. Moore, 1962b, p. 14, fig. 5, no. 4.
 Ordovician, Champlainian-Cincinnatian

Ohiocrinus laxus (Hall), 1867. Moore, 1962b, p. 14, Pl. 1, fig. 5.
   Ordovician, Cincinnatian, Maysville Limestone: United States: Ohio

OKLAHOMACRINUS Moore, 1939. *O. supinus* Moore, 1939. Moore and Laudon, 1943b, p. 59; 1944, p. 167. Arendt and Hecker, 1964, p. 89.
   Pennsylvanian-Desmoinesian, Permian-Wolfcampian

Oklahomacrinus abruptus Strimple, 1961g, p. 107, Pl. 13, figs. 1, 2.
   Pennsylvanian, Desmoinesian, Holdenville Shale: United States: Oklahoma

Oklahomacrinus bowsheri Moore, 1939. Burke, 1966b, p. 466, fig. 3.
   Pennsylvanian, Virgilian, Nelagoney Formation: United States: Oklahoma

Oklahomacrinus cirriferous Strimple, 1963a, p. 71, Pl. 12, figs. 17-20.
   Permian, Wolfcampian, Five Point Limestone: United States: Kansas

Oklahomacrinus discus Strimple, 1947, p. 6, Pl. 2, figs. 2, 3. Burke, 1966b, p. 466, fig. 3.
   Pennsylvanian, Virgilian, Stull Shale: United States: Kansas

Oklahomacrinus expansus (Wanner), 1916. Burke, 1966b, p. 467, figs. 4-6.
   Delocrinus expansus Wanner, 1916. Burke, 1966b, p. 467.
   Permian: Timor

Oklahomacrinus loeblichi Moore, 1939. Moore and Laudon, 1944, p. 167, Pl. 56, fig. 25.
   Pennsylvanian, Missourian, Francis Formation: United States: Oklahoma

Oklahomacrinus ohioensis Burke, 1966b, p. 465, figs. 1-3.
   Pennsylvanian, Conemaughian, upper Ames Limestone: United States: Ohio

Oklahomacrinus regularis Strimple, 1951c, p. 26, Pl. 4, figs. 8, 9.
   Pennsylvanian, Missourian, Wann Formation: United States: Oklahoma

*Oklahomacrinus supinus Moore, 1939. Moore and Laudon, 1943b, p. 135, Pl. 6, fig. 5; 1944, p. 167, Pl. 56, fig. 24; Pl. 62, fig. 19. Burke, 1966b, p. 466, fig. 3.
   Pennsylvanian, Virgilian, Brownville Limestone: United States: Oklahoma

Ollulocrinus Bouška, 1956 = *Parapisocrinus* Mu, 1954

Ollulocrinus ollula elegans Bouška, 1956 = *Pisocrinus (Parapisocrinus) ollula elegans*

Ollulocrinus ollula grandis Bouška, 1956 = *Pisocrinus (Parapisocrinus) ollula grandis*

Ollulocrinus ollula hlubocepensis Bouška, 1956 = *Pisocrinus (Parapisocrinus) ollula hlubocepensis*

Ollulocrinus ollula ollula (Angelin), 1878 = *Pisocrinus (Parapisocrinus) ollula ollula*

Ollulocrinus pribyli Bouška, 1956 = *Pisocrinus (Parapisocrinus) pribyli*

Ollulocrinus quinquelobus (Bather), 1893 = *Pisocrinus (Parapisocrinus) quinquelobus*

Ollulocrinus rimosus Bouška, 1956 = *Pisocrinus (Parapisocrinus) rimosus*

Ollulocrinus sphericus (Rowley), 1904 = *Pisocrinus (Parapisocrinus) sphericus*

Ollulocrinus tennesseensis (Roemer), 1860 = *Pisocrinus (Parapisocrinus) tennesseensis*

Ollulocrinus yassensis (Etheridge), 1904 = *Pisocrinus (Parapisocrinus) yassensis*

Ollulocrinus cf. *yassensis* (Etheridge), 1904 = *Pisocrinus (Parapisocrinus)* cf. *yassensis*

Ollulocrinus sp. Bouška, 1956 = *Pisocrinus (Parapisocrinus)* sp.

ONYCHOCRINUS Lyon and Casseday, 1860. *O. exsculptus* Lyon and Casseday, 1860. Moore and Laudon, 1943b, p. 69, fig. 10; 1944, p. 179, Pl. 67, fig. 13. Cuénot, 1948, p. 66. Ubaghs, 1953, p. 755, fig. 26. Van Sant, *in* Van Sant and Lane, 1964, p. 48, fig. 15. Moore and Jeffords, 1968, p. 11, fig. 1, no. 4.
 Mississippian

Onychocrinus asteriaeformis (Hall), 1861. Moore and Laudon, 1944, p. 179, Pl. 69, fig. 10.
 Mississippian, Osagean, Burlington Limestone: United States: Iowa

Onychocrinus diversus Meek and Worthen, 1866. Moore and others, 1968a, p. 20, Pl. 4, fig. 2.
 Mississippian, Osagean, Burlington Limestone: United States: Iowa

**Onychocrinus exsculptus* Lyon and Casseday, 1860. Moore and Laudon, 1943b, p. 136, Pl. 7, fig. 14; 1944, p. 179, Pl. 69, fig. 19. Van Sant, *in* Van Sant and Lane, 1964, p. 100, Pl. 7, fig. 10; Figs. 14; 34, no. 2.
 Mississippian, Osagean, Edwardsville Formation, Keokuk Limestone: United States: Indiana, Illinois, Iowa

Onychocrinus hibernicus Wright, 1934. Wright, 1954a, p. 174, Pl. 44, fig. 4; Pl. 45, figs. 2, 3, 5-11, 13-17; Pl. 46, figs. 3, 5; Figs. 93-96.
 Carboniferous, Tournaisian, Supra-dolomite beds: Ireland; England

Onychocrinus liddelensis Wright, 1954a, p. 171, Pl. 41, fig. 8; Pl. 44, fig. 2.
 Carboniferous, Visean, Lower Limestone Group: Scotland

Onychocrinus polydactylus (McCoy), 1844. Wright, 1954a, p. 172, Pl. 45, fig. 12; Pl. 46, fig. 4; Fig. 92.
 Carboniferous, ?Tournaisian: Ireland

Onychocrinus pulaskiensis Miller and Gurley, 1895. Moore and Laudon, 1944, p. 179, Pl. 69, fig. 3. Horowitz, 1965, p. 39, Pl. 4, figs. 11, 12. Moore and others, 1968a, p. 20, Pl. 4, fig. 3.
 Mississippian, Chesterian, Glen Dean Formation: United States: Illinois, Indiana, Kentucky, Alabama

Onychocrinus ramulosus (Lyon and Casseday), 1859. Van Sant, *in* Van Sant and Lane, 1964, p. 101, Pl. 6, figs. 2, 7, 9; Fig. 34, no. 3.
 Mississippian, Osagean, Borden Group: United States: Indiana, Kentucky

Onychocrinus ulrichi Miller and Gurley, 1890. Moore and Laudon, 1943b, p. 136, Pl. 7, fig. 13; 1944, p. 179, Pl. 69, fig. 15. Cuénot, 1948, p. 66, fig. 87. Ubaghs, 1953, p. 755, fig. 26. Van Sant, *in* Van Sant and Lane, 1964, p. 102, Pl. 6, figs. 6, 8; Fig. 34, no. 1.
 Mississippian, Osagean, Borden Group: United States: Indiana

Onychocrinus wrighti Springer, 1920. Wright, 1954a, p. 170, Pl. 41, fig. 3; Pl. 44, figs. 6, 10.
 Carboniferous, Visean, Lower Limestone Group: Scotland

Onychocrinus sp. Laudon and others, 1952, p. 561, Pl. 66, fig. 15.
 Mississippian, Kinderhookian, Banff formation: Canada: Alberta

OPHIUROCRINUS Jaekel, 1918. **Poteriocrinus originarius* Trautschold, 1867. Moore and Laudon, 1943b, p. 59. Arendt and Hecker, 1964, p. 89.
 Carboniferous

Ophiurocrinus arbiglandensis (Wright), 1934. Wright, 1950, p. 22, Pl. 6, fig. 8.

Scytalocrinus arbiglandensis Wright, 1934. Wright, 1950, p. 22.
Carboniferous, Calciferous Sandstone series: Scotland; England

Ophiurocrinus beggi (Wright), 1938. Wright, 1950, p. 22, Pl. 6, fig. 6.
Scytalocrinus beggi Wright, 1938. Wright, 1950, p. 22.
Carboniferous, Visean, C_2: England

Ophiurocrinus dactyloides (Austin and Austin), 1842. Wright, 1950, p. 20, Pl. 6, figs. 1-3, 7.
Poteriocrinus dactyloides Austin and Austin, 1842. Wright, 1950, p. 20.
Carboniferous, Tournaisian: Ireland

Ophiurocrinus cf. *dactyloides* (Austin and Austin), 1842. Wright, 1960, p. 329, Appendix, Pl. A, fig. 13.
Carboniferous, Tournaisian: Ireland

Ophiurocrinus gowerensis Wright, 1960, p. 329, Appendix, Pl. A, figs. 7, 8.
Carboniferous, Tournaisian, Z: England

Ophiurocrinus hebdenensis Wright, 1950, p. 23, Pl. 6, figs. 4, 5.
Carboniferous, Namurian, Millstone Grit: England

OPSIOCRINUS Kier, 1952. *O. mariana* Kier, 1952. Kier, 1952, p. 65.
Middle Devonian

**Opsiocrinus mariana* Kier, 1952, p. 66, Pl. 3, figs. 2, 3. Kier, 1958, p. 201, Pl. 1, figs. 1-6; Figs. 1, 2.
Devonian, Erian, Silica Formation: United States: Ohio

OROCRINUS Sieverts-Doreck, 1951. *O. hercynicus* Sieverts-Doreck, 1951. Sieverts-Doreck, 1951a, p. 119.
Lower Carboniferous

Orocrinus granulosus Sieverts-Doreck, 1951a, p. 129, fig. 5.
Lower Carboniferous, Erdbecker Limestone: Germany

Orocrinus grundensis Sieverts-Doreck, 1951a, p. 127, Pl. 9, figs. 1, 2, 4.
Woodocrinus? sp. indet. Schmidt, 1934, fig. 11b, b_1. Sieverts-Doreck, 1951a, p. 127.
Lower Carboniferous, Erdbacker Limestone: Germany

**Orocrinus hercynicus* Sieverts-Doreck, 1951a, p. 124, Pl. 8, figs. 1-7; Fig. 1.
Woodocrinus? sp. indet. Schmidt, 1934, fig. 11a. Sieverts-Doreck, 1951a, p. 124.
Lower Carboniferous, Erdbacker Limestone: Germany

Orocrinus winterbergensis Sieverts-Doreck, 1951a, p. 126, Pl. 8, figs. 8-10; Pl. 9, fig. 3?.
Woodocrinus? sp. indet. Schmidt, 1934, fig. 11c. Sieverts-Doreck, 1951, p. 126.
Lower Carboniferous, Erdbacker Limestone: Germany

Orocrinus? sp. A. Sieverts-Doreck, 1951a, p. 130, Pl. 9, figs. 6, 7; Fig. 2.
Lower Carboniferous, Erdbacker Limestone: Germany

Orocrinus? sp. B. Sieverts-Doreck, 1951a, p. 134.
Lower Carboniferous, Erdbacker Limestone: Germany

ORTHOCRINUS Jaekel, 1895. *O. simplex* Jaekel, 1895. Ubaghs, 1953, p. 737.
Devonian

Orthocrinus elongatus Breimer, 1962a, p. 21, Pl. 3, fig. 10; Fig. 2B.
Devonian, upper Emsian, Santa Lucia Formation: Spain

Orthocrinus sp. cf. *O. elongatus* Breimer, 1962a, p. 22, Pl. 1, figs. 4-6.
Devonian: Spain

Orthocrinus plannus Mellado, 1949 = *Orthocrinus robustus*

Orthocrinus planus Schmidt, 1932. Schmidt, 1952, p. 162, Pl. 9, figs. 1a, b; Fig. 11. Breimer, 1962a, p. 19; 1962b, p. 325. China, 1965, p. 45 (nom. conserv.).
Devonian: Spain

Orthocrinus robustus Breimer, 1962a, p. 20, fig. 2C.
Orthocrinus plannus Mellado, 1949, p. 657, Pls. 29, 30; Fig. 1 (misspelling of *O. planus*).
Devonian, lower Couvinian, Santa Lucia Formation: Spain

**Orthocrinus simplex* Jaekel, 1895. Schmidt, 1941, p. 195, Pl. 21, fig. 7.
Poteriocrinus rhenanus Müller, *in* Zeiler and Wirtgen, 1855 (Pl. 7, fig. 1 only). Schmidt, 1941, p. 195.
Middle Devonian, Kieselgallen Beds: Germany

Orthocrinus tuberculatus Schmidt, 1913. Schmidt, 1941, p. 196, Pl. 11, fig. 5; Pl. 21, fig. 4; Fig. 54. Sieverts-Doreck, 1952, p. 428, fig. 1.
Middle Devonian, *Cultrijugatus* Zone: Germany

Orthocrinus tuberculatus modesta Schmidt, 1941, p. 200, Pl. 11, fig. 6.
Middle Devonian, *Cultrijugatus* Zone: Germany

Orthocrinus sp. (nov.?) Breimer, 1962a, p. 23, Pl. 3, fig. 1; Fig. 2A.
Devonian, lower Emsian: Spain

Orthocrinus sp. Breimer, 1962a, p. 19, Pl. 3, fig. 12.
Devonian, lower Couvinian: Spain

OTTAWACRINUS Billings, 1887. **O. typus* Billings, 1887. Moore and Laudon, 1943b, p. 55, fig. 7; 1944, p. 158, Pl. 52, fig. 5. Moore, 1950, p. 34, fig. 4f. Ubaghs, 1953, p. 752, fig. 9. Arendt and Hecker, 1964, p. 87.
Ordovician, Champlainian

Ottawacrinus billingsi Springer, 1911 = *Grenprisia billingsi*

**Ottawacrinus typus* Billings, 1887. Moore and Laudon, 1943b, p. 132, Pl. 3, fig. 7; 1944, p. 158, Pl. 53, fig. 24. Wilson, 1946, p. 38. Ubaghs, 1953, p. 752, fig. 9. Sieverts-Doreck, 1957b, p. 152, figs. 3, 4. Moore, 1962b, p. 36, Pl. 1, fig. 3.
Ordovician, Champlainian, Cobourg Formation, Hull Formation: Canada: Ontario

Ottawacrinus typus Springer, 1911 (*non* Billings), 1887 = *Grenprisia springeri*

PACHYLOCRINUS Wachsmuth and Springer, 1880. **Scaphiocrinus aequalis* Hall, 1861. Moore and Laudon, 1943b, p. 56; 1944, p. 161. Cuénot, 1948, p. 61. Arendt and Hecker, 1964, p. 88.
Mississippian

**Pachylocrinus aequalis* (Hall), 1861. Moore and Laudon, 1944, p. 161, Pl. 61, fig. 10. Van Sant, *in* Van Sant and Lane, 1964, p. 87, Pl. 4, figs. 1, 3, 6, 10; Fig. 33.
Scaphiocrinus gurleyi White, 1878. Van Sant, *in* Van Sant and Lane, 1964, p. 87.
Mississippian, Osagean, Edwardsville Formation: United States: Indiana

Pachylocrinus arboreus Springer, 1926, Cuénot, 1948 = *Linocrinus arboreus*

Pachylocrinus baschmakowae Yakovlev and Ivanov, 1956, p. 25, Pl. 6, figs. 5, 6. Arendt and Hecker, 1964, p. 88, Pl. 11, fig. 3.
Upper Carboniferous (C_3ks): Russia: Moscow Basin

Pachylocrinus bellirugosus Moore. 1939 = *Plummericrinus bellirugosus*

Pachylocrinus boonvillensis (Miller), 1891. Van Sant, *in* Van Sant and Lane, 1964, p. 88.
   Mississippian: United States: Indiana?

Pachylocrinus cirrifer Laudon, 1933 = *Paracosmetocrinus cirrifer*

Pachylocrinus clavatus Wright, 1937 = *Pedinocrinus clavatus*

Pachylocrinus colubrosus Moore, 1939 = *Plummericrinus colubrosus*

Pachylocrinus conicus (Phillips), 1836. Wright, 1942, p. 274, Pl. 10, figs. 9, 10, 12. Wright, 1947, p. 102, Pl. 3, figs. 8-10.
   Poteriocrinus conicus Phillips, 1836. Wright, 1947, p. 103.
   Carboniferous, Tournaisian (C_1): Scotland

Pachylocrinus? crassimanus (McCoy), 1849. Wright, 1951b, p. 67, Pl. 31, fig. 11.
   Poteriocrinus crassimanus McCoy, 1849. Wright, 1951b, p. 67.
   Carboniferous, Tournaisian: Ireland

Pachylocrinus cravenensis Wright, 1942 = *Bollandocrinus cravenensis*

Pachylocrinus dunlopi Wright, 1936 = *Aphelecrinus dunlopi*

Pachylocrinus gibsoni (White), 1878. Van Sant, *in* Van Sant and Lane, 1964, p. 89, Pl. 4, figs. 7, 11.
   Mississippian, Osagean, Edwardsville Formation: United States: Indiana

Pachylocrinus globosus (Wachsmuth and Springer), 1889. Moore and Laudon, 1944, p. 161, Pl. 60, fig. 17.
   Mississippian, Osagean, Keokuk Limestone: United States: Indiana

Pachylocrinus lacunosus (Miller and Gurley), 1890. Moore and Laudon, 1944, p. 161, Pl. 64, fig. 21.
   Mississippian, Osagean, Keokuk Limestone: United States: Iowa

Pachylocrinus latifrons Wright, 1934 (part) = *Pachylocrinus? wexfordensis*

Pachylocrinus? latifrons (Austin and Austin), 1847. Wright, 1951b, p. 65, Pl. 31, fig. 5.
   Carboniferous, Tournaisian: England

Pachylocrinus lutanus Boos, 1939 = *Plummericrinus lutanus*

Pachylocrinus manus (Miller and Gurley), 1890. Van Sant, *in* Van Sant and Lane, 1964, p. 89, Pl. 4, fig. 12.
   Mississippian, Osagean, Indian Creek crinoid beds: United States: Indiana

Pachylocrinus mcguirei Moore, 1939 = *Plummericrinus mcguirei*

Pachylocrinus monensis Wright, 1938 = *Hypselocrinus monensis*

Pachylocrinus ogarai Moore and Plummer, 1940 = *Plummericrinus ogarai*

Pachylocrinus pachypinnularis Yakovlev and Ivanov, 1956, p. 25, Pl. 7, fig. 1.
   Upper Carboniferous (C_3ks): Russia: Moscow Basin

Pachylocrinus? skinneri Wright, 1951b, p. 68, Pl. 31, fig. 4.
   Carboniferous, Tournaisian: Ireland

Pachylocrinus tenuiramosus Yakovlev, 1939a, p. 67, Pl. 12, fig. 5. Yakovlev and Ivanov, 1956, p. 24, Pl. 6, fig. 4. Arendt and Hecker, 1964, p. 88, Pl. 11, fig. 4.
   Upper Carboniferous (C_3ks): Russia: Moscow Basin

Pachylocrinus tielensis Wright, 1936 = *Fifeocrinus tielensis*

Pachylocrinus twenhofeli Moore, 1939 = *Plummericrinus twenhofeli*

Pachylocrinus tyriensis Wright, 1937 = **Scotiacrinus tyriensis*

Pachylocrinus uddeni Moore and Plummer, 1940 = *Texacrinus uddeni*

Pachylocrinus? wexfordensis Wright, 1951b, p. 66, Pl. 31, figs. 1, 10, 13.
  Pachylocrinus latifrons Wright, 1934 (part, Pl. 15, figs. 9, 10, 16 only).
    Wright, 1951b, p. 66.
  Carboniferous, Tournaisian: Ireland

Pachylocrinus spp.
  Laudon and others, 1952, p. 555, Pl. 65, fig. 34.
    Mississippian, Kinderhookian, Banff formation: Canada: Alberta
  Yakovlev and Ivanov, 1956, p. 24, Pl. 6, fig. 7.
    Upper Carboniferous (C_3ks): Russia: Moscow Basin

PAGECRINUS Kirk, 1929. *P. gracilis* Kirk, 1929. Moore and Laudon, 1943b,
  p. 55; 1944, p. 158. Arendt and Hecker, 1964, p. 86.
  Devonian

Pagecrinus gracilis Kirk, 1929. Moore and Laudon, 1944, p. 158, Pl. 57,
  fig. 4.
  Devonian, Ulsterian, Jeffersonville Limestone: United States: Indiana

Pagecrinus heckeri Yakovlev, 1941a, p. 327, Pl. 2, fig. 1; 1947a, p. 56,
  Pl. 10, fig. 3. Arendt and Hecker, 1964, p. 86, Pl. 10, fig. 2.
  Devonian, Frasnian, Pskov Beds: Russia: Syas River, Main Devonian Field

PAIANOCRINUS Strimple, 1951. *P. durus* Strimple, 1951. Strimple, 1951f,
  p. 669.
  Mississippian, Chesterian

Paianocrinus aptus Strimple, 1951f, p. 670, Pl. 99, figs. 3-5; 1961g, p. 22,
  Pl. 3, figs. 4, 5; Fig. 6.
  Mississippian, Chesterian, Pitkin Limestone: United States: Oklahoma

Paianocrinus durus Strimple, 1951f, p. 669, Pl. 99, figs. 6-8.
  Mississippian, Chesterian, Pitkin Limestone: United States: Oklahoma

PALAEOCRINUS Billings, 1859. *P. striatus* Billings, 1859. Moore and Laudon,
  1943b, p. 50, fig. 4; 1944, p. 153, Pl. 52, fig. 22. Ubaghs, 1953,
  p. 750, fig. 33c. Xu, 1962, p. 45, 50, fig. 1. Moore, 1962b, p. 15,
  fig. 9, no. 7.
  Ordovician

Palaeocrinus angulatus (Billings), 1857. Wilson, 1946, p. 40, Pl. 6,
  fig. 1.
  Ordovician, Champlainian, Cobourg Formation: Canada: Ontario

Palaeocrinus hudsoni Sinclair, 1945, p. 715, Pl. 2, figs. 11-13.
  Ordovician, Champlainian, Bromide Formation: United States: Oklahoma

Palaeocrinus pulchellus Billings, 1859. Wilson, 1946, p. 40, Pl. 6, fig. 3.
  Ordovician, Champlainian, Cobourg Formation: Canada: Ontario

Palaeocrinus rhombiferus Billings, 1859. Wilson, 1946, p. 40, Pl. 6, fig. 2.
  Ordovician, Champlainian, Cobourg Formation: Canada: Ontario

Palaeocrinus striatus Billings, 1859. Moore and Laudon, 1943b, p. 131,
  Pl. 2, fig. 10; 1944, p. 153, Pl. 53, fig. 19.
  Ordovician, Champlainian: Canada: Ontario, Quebec

Palaeocrinus aff. *P. striatus* Billings, 1859. Butts, 1941, p. 91, Pl. 89,
  figs. 9-11.
  Ordovician, Champlainian, Ottosee Limestone: United States: Virginia

PALAEOHOLOPUS Wanner, 1916. *P. pretiosus* Wanner, 1916. Moore and Laudon,
  1943b, p. 76. Ubaghs, 1953, p. 756. Arendt and Hecker, 1964, p. 94.
  Permian

Palaeoholopus pretiosus Wanner, 1916. Ubaghs, 1953, p. 756, fig. 135.
  Permian: Timor

PANDORACRINUS Jaekel, 1918.   *P. pinnulatus Jaekel, 1918.   Moore and
    Laudon, 1943b, p. 54.
    Ordovician

Pandoracrinus mincopensis Ramsbottom, 1961, p. 18, Pl. 3, fig. 7.
    Ordovician, Llandeilian, Meadowtown Beds: England

PARABOTRYOCRINUS Yakovlev, 1941.   *P. tschudovensis Yakovlev, 1941.
    Yakovlev, 1941a, p. 328. Arendt and Hecker, 1964, p. 86.
    Upper Devonian

*Parabotryocrinus tschudovensis Yakovlev, 1941a, p. 328, Pl. 2, fig. 2.
    Yakovlev, 1947a, p. 57, Pl. 10, fig. 4 (misspelled as P. tschudo-
    wensis). Arendt and Hecker, 1964, p. 86, Pl. 9, fig. 17 (misspelled
    as P. tschudowensis).
    Upper Devonian, Chudovo Beds: Russia: Leningrad district

PARABURSACRINUS Wanner, 1924.   *Bursacrinus procerus Wanner, 1916. Moore
    and Laudon, 1943b, p. 59.
    Permian

Parabursacrinus? gracilis Wanner, 1949a, p. 38, Pl. 3, figs. 8, 9; Fig. 13.
    Permian: Timor

*Parabursacrinus magnificus (Wanner), 1916. Wanner, 1949a, p. 34, Pl. 2,
    figs. 28, 29.
    Permian: Timor

Parabursacrinus magnificus var. granulata Wanner, 1949a, p. 35, Pl. 3,
    figs. 1-3.
    Permian: Timor

Parabursacrinus pyramidatus (Wanner), 1916. Wanner, 1949a, p. 36, Pl. 3,
    figs. 4-7.
    Permian: Timor

PARACATILLOCRINUS Wanner, 1916.   *P. granulatus Wanner, 1916. Moore and
    Laudon, 1943b, p. 30. Ubaghs, 1953, p. 747. Moore, 1962b, p. 16,
    fig. 12, no. 4. Arendt and Hecker, 1964, p. 81.
    Permian

PARACOSMETOCRINUS Strimple, 1967.   *P. strakai Strimple, 1967. Strimple,
    1967, p. 82.
    Mississippian, Kinderhookian

Paracosmetocrinus cirrifer (Laudon), 1933. Strimple, 1967, p. 81.
    Pachylocrinus cirrifer Laudon, 1933. Strimple, 1967, p. 81.
    Mississippian, Kinderhookian, Gilmore City Formation: United States:
    Iowa

Paracosmetocrinus crawfordsvillensis (Miller and Gurley), 1890. Strimple,
    1967 = Hypselocrinus indianensis

Paracosmetocrinus madisonensis (Laudon and Severson), 1953. Strimple,
    1967, p. 81.
    Aphelecrinus madisonensis Laudon and Severson, 1953, p. 521, Pl. 51,
    figs. 20, 21; Pl. 55, figs. 10, 11 (misspelled as Amphelecrinus
    madisonensis). Strimple, 1967, p. 81.
    Mississippian, Kinderhookian, Lodgepole Formation: United States:
    Montana

*Paracosmetocrinus strakai Strimple, 1967, p. 84, Pl. 1, figs. 1-5.
    Mississippian, Kinderhookian, Wassonville Limestone: United States:
    Iowa

PARACROMYOCRINUS Strimple, 1966.   *Parulocrinus vetulus Lane, 1964.
    Strimple, 1966a, p. 4.
    Mississippian, Chesterian-Pennsylvanian, Missourian

Paracromyocrinus compactus (Moore and Plummer), 1940. Strimple, 1966a, p. 5.
   Parulocrinus compactus Moore and Plummer, 1940. Moore and Laudon, 1943b, p. 135, Pl. 6, figs. 11, 12; 1944, p. 175, Pl. 64, fig. 36. Strimple, 1966a, p. 5.
   Aglaocrinus compactus (Moore and Plummer), 1940. Strimple, 1961g, p. 86.
   Pennsylvanian, Missourian, Brad Formation, Winterset Limestone, Iola Limestone, Ochelata Formation: United States: Texas, Kansas, Oklahoma

Paracromyocrinus marquisi (Moore and Plummer), 1940. Strimple, 1966a, p. 5.
   Parulocrinus marquisi Moore and Plummer, 1940. Strimple, 1966a, p. 5.
   Pennsylvanian, probably Missourian: United States: Texas

Paracromyocrinus oklahomensis (Moore and Plummer), 1938. Strimple, 1966a, p. 5.
   Ethelocrinus oklahomensis Moore and Plummer, 1938. Moore and Laudon, 1944, p. 175. Strimple, 1966a, p. 5.
   Metacromyocrinus oklahomensis (Moore and Plummer), 1938. Strimple and Knapp, 1966, p. 311, Pl. 36, figs. 21, 22.
   Ethelocrinus texasensis Moore and Plummer, 1940. Moore and Laudon, 1944, p. 175, Pl. 65, fig. 31. Plummer, 1950, p. 53, Pl. 11, fig. 8. Strimple, 1962i, p. 270; 1966a, p. 5.
   Pennsylvanian, Morrowan, Marble Falls Formation, Kendrick Shale, Brentwood Limestone Member: United States: Texas, Kentucky, Oklahoma, Arkansas

Paracromyocrinus pustulosus (Moore and Plummer), 1940. Strimple, 1966a, p. 5.
   Parulocrinus pustulosus Moore and Plummer, 1940. Moore and Laudon, 1944, p. 175, Pl. 64, fig. 31. Strimple, 1966a, p. 5.
   Aglaocrinus pustulosus (Moore and Plummer), 1940. Strimple, 1961g, p. 86.
   Pennsylvanian, Missourian, Mineral Wells Formation: United States: Texas

**Paracromyocrinus vetulus* (Lane), 1964. Strimple, 1966a, p. 5.
   Parulocrinus vetulus Lane, 1964b, p. 681, Pl. 112, figs. 11-13. Strimple, 1966a, p. 5.
   Mississippian, Chesterian, Indian Springs Formation: United States: Nevada

PARACTOCRINUS Jaekel, 1918.   **P. tuberculatus* Jaekel, 1918.
   Ordovician

**Paractocrinus tuberculatus* Jaekel, 1918. Ubaghs, 1953, p. 669, fig. 8.
   Ordovician: Russia

PARADELOCRINUS Moore and Plummer, 1938.   **P. aequabilis* Moore and Plummer, 1938. Moore and Laudon, 1943b, p. 60; 1944, p. 173.
   Carboniferous

**Paradelocrinus aequabilis* Moore and Plummer, 1938. Moore and Laudon, 1944, p. 173, Pl. 62, fig. 3; Pl. 65, fig. 20. Strimple, 1960b, p. 154, fig. 2B.
   Pennsylvanian, Morrowan, Brentwood Limestone: United States: Arkansas

Paradelocrinus atoka Strimple, 1961d, p. 228, Pl. 1, figs. 1-3, 13-15.
   Pennsylvanian, Atokan, Griley Limestone: United States: Oklahoma

Paradelocrinus brachiatus Moore and Plummer, 1940. Moore and Laudon, 1944, p. 173, Pl. 62, fig. 34; Pl. 65, fig. 25.
   Pennsylvanian, Desmoinesian, Millsap Lake Formation: United States: Texas

Paradelocrinus cranei Strimple, 1949d, p. 6, Pl. 1, figs. 1-3, 7.
  Pennsylvanian, Atokan, Pumpkin Creek Limestone: United States: Oklahoma

Paradelocrinus disculus Strimple, 1949d, p. 7, Pl. 3, figs. 15-16.
  Pennsylvanian, Atokan, Otterville Limestone: United States: Oklahoma

Paradelocrinus dubius (Mather), 1915. Moore and Laudon, 1944, p. 173, Pl. 62, fig. 31; Pl. 65, fig. 21. Strimple, 1960b, p. 154, fig. 2H.
  Delocrinus dubius Mather, 1915. Croneis, 1930, p. 85, Pl. 21, figs.21, 22.
  Pennsylvanian, Morrowan, Brentwood Limestone: United States: Arkansas

Paradelocrinus cf. *P. dubius* (Mather), 1915.
  Delocrinus cf. *D. dubius* (Mather), 1915. Elias, 1957, p. 385, Pl. 41, fig. 3.
  Mississippian, Chesterian, Redoak Hollow Formation: United States: Oklahoma

Paradelocrinus iolaensis Strimple, 1949b, p. 124, Pl. 4, figs. 5-7; Fig. 1a.
  Pennsylvanian, Missourian, Raytown Limestone: United States: Kansas

Paradelocrinus johnstonensis Strimple, 1961d, p. 226, Pl. 1, figs. 4-6, 10-12.
  Pennsylvanian, Atokan, Pumpkin Creek Limestone: United States: Oklahoma

Paradelocrinus obovatus Moore and Plummer, 1940. Moore and Laudon, 1943b, p. 135, Pl. 6, fig. 2; 1944, p. 173, Pl. 65, fig. 22.
  Pennsylvanian, Missourian, Graford Formation: United States: Texas

Paradelocrinus planus (White), 1879. Moore and Laudon, 1944, p. 173, Pl. 65, fig. 24.
  Pennsylvanian, Virgilian: United States: Kansas

Paradelocrinus protensus Moore and Plummer, 1940. Moore and Laudon, 1944, p. 173, Pl. 65, fig. 23. Strimple, 1960b, p. 154, fig. 2C.
  Pennsylvanian, Missourian, Graford Formation: United States: Texas

Paradelocrinus regulatus Strimple, 1949d, p. 7, Pl. 1, figs. 13, 15, 17; Pl. 2, figs. 2, 5.
  Pennsylvanian, Atokan, Pumpkin Creek Limestone: United States: Oklahoma

Paradelocrinus subplanus Moore and Plummer, 1940. Strimple, 1962e, p. 21, Pl. 3, figs. 16, 17; Pl. 6, figs. 1, 2; Pl. 8, fig. 1.
  Pennsylvanian, Desmoinesian, Oologah Limestone: United States: Oklahoma

Paradelocrinus wapanucka Strimple, 1961d, p. 225, Pl. 1, figs. 7-9.
  Pennsylvanian, Morrowan, Wapanucka Limestone: United States: Oklahoma

Paradelocrinus sp. 1. Breimer, 1962a, p. 172.
  Carboniferous, upper Visean, Griotte Limestone: Spain

Paradelocrinus sp. 2. Breimer, 1962a, p. 172.
  Carboniferous, Moscovian: Spain

Paradelocrinus sp. Frederickson and Waddell, 1960, p. 172, Pl. 1.
  Pennsylvanian, Atokan, Pumpkin Creek Limestone: United States: Oklahoma

PARADICHOCRINUS Springer, 1926. *Dichocrinus polydactylus* Casseday and Lyon, 1862. Moore and Laudon, 1943b, p. 96; 1944, p. 199. Ubaghs, 1953, p. 740. Arendt and Hecker, 1964, p. 97.
  Mississippian, Osagean

*Paradichocrinus polydactylus (Casseday and Lyon), 1862. Moore and Laudon, 1943b, p. 142, Pl. 13, fig. 1; 1944, p. 199, Pl. 78, figs. 1, 6. Van Sant, in Van Sant and Lane, 1964, p. 124, Pl. 8, figs. 7, 11; Figs. 17, nos. 6, 7; 19, nos. 5, 6, 8.
Dichocrinus expansus Meek and Worthen, 1868. Van Sant, in Van Sant and Lane, 1964, p. 124.
Dichocrinus stelliformis Meek and Worthen, 1868. Van Sant, in Van Sant and Lane, 1964, p. 124.
Mississippian, Osagean, Keokuk Limestone: United States: Indiana

PARADOXOCRINUS Wanner, 1937. *P. patella Wanner, 1937. Wanner, 1942b, p. 201, figs. 4, 5B. Moore and Laudon, 1943b, p. 102.
Permian

*Paradoxocrinus patella Wanner, 1937. Wanner, 1942b, p. 201, figs. 1-3, 6.
Permian: Timor

PARAGARICOCRINUS Yakovlev, 1934. *P. mediterraneus Yakovlev, 1934. Moore and Laudon, 1943b, p. 92, fig. 15. Sieverts-Doreck, 1951b, p. 105, figs. 1, 2a. Ubaghs, 1953, p. 739.
Permian

*Paragaricocrinus mediterraneus Yakovlev, 1934. Moore and Laudon, 1942, p. 75, figs. 5A-C.
Permian: Sicily

PARAGASSIZOCRINUS Moore and Plummer, 1940. *Agassizocrinus tarri Strimple, 1939. Moore and Laudon, 1943b, p. 62; 1944, p. 177. Arendt and Hecker, 1964, p. 91.
Pennsylvanian, Morrowan-Desmoinesian

Paragassizocrinus asymmetricus Strimple, 1960a, p. 15, Pl. 1, figs. 1-5.
Pennsylvanian, Morrowan, Wapanucka Limestone: United States: Oklahoma

Paragassizocrinus atoka Strimple and Blythe, 1960, p. 25, Pl. 3, figs. 4-9; Fig. 2.
Pennsylvanian, Atokan, Atoka Formation: United States: Oklahoma

Paragassizocrinus bulbosus Strimple, 1961e, p. 296, Pl. 1, figs. 7-10.
Pennsylvanian, Atokan, Bostwick Limestone: United States: Oklahoma

Paragassizocrinus caliculus (Moore and Plummer), 1938. Strimple, 1961b, p. 158.
Agassizocrinus caliculus Moore and Plummer, 1938.
Pennsylvanian, Morrowan, Brentwood Limestone: United States: Arkansas, Oklahoma

Paragassizocrinus deltoideus Strimple, 1960a, p. 20, Pl. 1, figs. 11-17.
Pennsylvanian, Morrowan, Wapanucka Limestone: United States: Oklahoma

Paragassizocrinus disculus Strimple, 1960a, p. 23, Pl. 3, figs. 1-3. Strimple, 1961e, p. 296.
Pennsylvanian, Morrowan, Wapanucka Limestone: United States: Oklahoma

Paragassizocrinus cf. P. disculus Strimple, 1960. Strimple and Knapp, 1966, p. 310, Pl. 36, figs. 6-8.
Pennsylvanian, Morrowan, Kendrick Shale: United States: Kentucky

Paragassizocrinus elevatus Strimple, 1961e, p. 297, Pl. 1, figs. 1-6.
Pennsylvanian, Atokan, Bostwick Limestone: United States: Oklahoma

Paragassizocrinus ellipticus Strimple, 1961e, p. 297, Pl. 1, figs. 11-13.
Pennsylvanian, Missourian, Chanute Formation: United States: Oklahoma

Paragassizocrinus elongatus Strimple, 1960a, p. 18, Pl. 2, figs. 1, 2.
Pennsylvanian, Missourian, Hogshooter Limestone: United States: Oklahoma

Paragassizocrinus hoodi Strimple, 1960a, p. 22, Pl. 1, figs. 9, 10.
  Pennsylvanian, Desmoinesian, Sand Creek Formation: United States: Oklahoma

Paragassizocrinus kendrickensis Strimple and Knapp, 1966, p. 310, Pl. 36, figs. 3-5, 9-11.
  Pennsylvanian, Morrowan, Kendrick Shale: United States: Kentucky

Paragassizocrinus magnus (Moore and Plummer), 1938. Strimple, 1961b, p. 158.
  Agassizocrinus magnus Moore and Plummer, 1938.
  Pennsylvanian, Morrowan, Kessler Limestone: United States: Arkansas

Paragassizocrinus mcguirei (Strimple), 1939. Strimple, 1960a, p. 13, Pl. 2, figs. 3, 8-11. Strimple, 1961e, p. 294.
  Pennsylvanian, Missourian, Hogshooter Limestone, Checkerboard Limestone, Hoxbar Formation, Seminole Formation: United States: Oklahoma

Paragassizocrinus springeri Strimple, 1960a, p. 24, Pl. 2, figs. 6, 7.
  Pennsylvanian, Desmoinesian, Pumpkin Creek Limestone: United States: Oklahoma

**Paragassizocrinus tarri* (Strimple), 1939. Moore and Laudon, 1943b, p. 135, Pl. 6, fig. 9; 1944, p. 177, Pl. 56, fig. 31. Strimple, 1959a, p. 123, Pl. 2, figs. 7, 10, 13; 1960a, p. 9, Pl. 2, figs. 4, 5, 12, 13. Yakovlev, 1964b, p. 56, fig. 58.
  Pennsylvanian, Missourian, Wann Formation, Nellie Bly Formation, Inola Limestone, Eudora Shale, Graford Formation: United States: Oklahoma, Kansas, Texas

Paragassizocrinus turris Strimple, 1960a, p. 19, Pl. 1, figs. 6-8.
  Pennsylvanian, Morrowan, Wapanucka Limestone: United States: Oklahoma

Paragassizocrinus cf. *P. turris* Strimple, 1960. Strimple and Knapp, 1966, p. 311, Pl. 36, figs. 12-14.
  Pennsylvanian, Morrowan, Kendrick Shale: United States: Kentucky

PARAGAZACRINUS Springer, 1926. **P. rotundus* Springer, 1926. Moore and Laudon, 1943b, p. 82; 1944, p. 185. Ubaghs, 1953, p. 737. Arendt and Hecker, 1964, p. 95.
  Silurian, Niagaran

**Paragazacrinus rotundus* Springer, 1926. Moore and Laudon, 1943b, p. 138, Pl. 9, fig. 5; 1944, p. 185, Pl. 72, fig. 25.
  Silurian, Niagaran, Laurel Formation: United States: Indiana

PARAGRAPHIOCRINUS Wanner, 1937. **Graphiocrinus exornatus* Wanner, 1916. Moore and Laudon, 1943b, p. 60. Arendt and Hecker, 1964, p. 90.
  Permian

PARAHEXACRINUS Schewtschenko, 1967. **P. fungiformis* Schewtschenko, 1967. Schewtschenko, 1967, p. 77; 1968b, p. 72.
  Lower Devonian

**Parahexacrinus fungiformis* Schewtschenko, 1967, p. 77, Pl. 9, figs. 1-3; Figs. 1, 2; 1968b, p. 73, Pl. 9, figs. 1-3; Figs. 1, 2.
  Lower Devonian, Kshtut horizon: Russia: Zeravshan Range

PARAMELOCRINUS Ubaghs, 1958. **P. angelini* Ubaghs, 1958. Ubaghs, 1958b, p. 261.
  Silurian

**Paramelocrinus angelini* Ubaghs, 1958b, p. 262, Pl. 1, fig. 1; Figs. 1, 2.
  Melocrinus volborthi Angelin, 1878 (part, p. 20, Pl. XVIII, fig. 16). Ubaghs, 1958b, p. 262.
  Melocrinus volborthi Wachsmuth and Springer, 1881. Ubaghs, 1958b, p. 262.
  Silurian, level and strata indeterminate: Sweden

PARAPERNEROCRINUS Yakovlev, 1949. *P. sibiricus Yakovlev, 1949. Yakovlev, 1949c, p. 17. Arendt and Hecker, 1964, p. 84.
Devonian, Coblentzian

*Parapernerocrinus sibiricus Yakovlev, 1949c, p. 17, Pl. 1, figs. 4-8; Pl. 2, figs. 1, 2. Arendt and Hecker, 1964, p. 84, Pl. 8, figs. 13-15.
Devonian, Coblentzian: Russia: Central Urals

PARAPLASOCRINUS Moore and Plummer, 1938. *Cibolocrinus transitorius Wanner, 1916. Moore and Laudon, 1943b, p. 60. Wanner, 1949a, p. 49. Arendt and Hecker, 1964, p. 90.
Permian

PARARCHAEOCRINUS Strimple and Watkins, 1955. *P. decoratus Strimple and Watkins, 1955. Strimple and Watkins, 1955, p. 351.
Ordovician, Champlainian

*Pararchaeocrinus decoratus Strimple and Watkins, 1955, p. 351, figs. 2b-f, 9, 10.
Ordovician, Champlainian, Bromide Formation: United States: Oklahoma

PARASTACHYOCRINUS Wanner, 1949. *Erisocrinus malaianus Wanner, 1924. Wanner, 1949a, p. 42.
Permian

Parastachyocrinus granulatus (Wanner), 1924. Wanner, 1949a, p. 42.
Erisocrinus granulatus Wanner, 1924. Wanner, 1949a, p. 42.
Permian: Timor

Parastachyocrinus inflatus Wanner, 1949a, p. 46, Pl. 3, figs. 20, 21.
Permian: Timor

*Parastachyocrinus malaianus (Wanner), 1924. Wanner, 1949a, p. 44, Pl. 3, figs. 10-17.
Erisocrinus malaianus Wanner, 1924. Wanner, 1949a, p. 44.
Permian: Timor

Parastachyocrinus malaianus var. ornata Wanner, 1949a, p. 45, Pl. 3, figs. 18, 19.
Permian: Timor

Parastachyocrinus obliquus (Wanner), 1916. Wanner, 1949a, p. 42.
Erisocrinus obliquus Wanner, 1916. Wanner, 1949a, p. 42. Van Sant, in Van Sant and Lane, 1964, p. 65, fig. 23, no. 1.
Permian: Timor

PARASYCOCRINUS Marez Oyens, 1940. *P. fastigateplicatus Marez Oyens, 1940. Lane, 1967b, p. 12, fig. 6.
Permian

PARAZOPHOCRINUS Strimple, 1963. *P. callosus Strimple, 1963. Strimple, 1963d, p. 61.
Silurian, Niagaran

*Parazophocrinus callosus Strimple, 1963d, p. 61, Pl. 2, figs. 10, 11.
Silurian, Niagaran, Henryhouse Formation: United States: Oklahoma

PARETHELOCRINUS Strimple, 1961. *P. ellipticus Strimple, 1961. Strimple, 1961g, p. 81.
Aglaocrinus Strimple, 1961g, p. 86. Webster and Lane, 1967, p. 21.
Pennsylvanian, Morrowan-Permian, Wolfcampian

Parethelocrinus beedei (Moore and Plummer), 1940 = Parulocrinus beedei

*Parethelocrinus ellipticus Strimple, 1961g, p. 83, Pl. 8, figs. 4, 5; Pl. 18, figs. 1, 2; Pl. 19, fig. 5. Branson, 1962, p. 163, fig. 2.
Pennsylvanian, Desmoinesian, Holdenville Shale: United States: Oklahoma

Parethelocrinus expansus (Strimple), 1938, new comb.
  Aglaocrinus expansus (Strimple), 1938. Strimple, 1961g, p. 86.
  Ethelocrinus expansus Strimple, 1938. Strimple, 1959a, p. 127, Pl. 2,
    figs. 8, 11 (*non* Pl. 2, fig. 10, as given on p. 127).
  Pennsylvanian, Missourian, Wann Formation: United States: Oklahoma

Parethelocrinus iatani (Strimple), 1949, new comb.
  Aglaocrinus iatani (Strimple), 1949. Strimple, 1961g, p. 86.
  Ethelocrinus iatani Strimple, 1949d, p. 25, Pl. 4, rigs. 9-11.
  Pennsylvanian, Missourian, Iatan Limestone: United States: Nebraska

Parethelocrinus magnus (Strimple), 1949, new comb.
  Aglaocrinus magnus (Strimple), 1949. Strimple, 1961g, p. 87, Pl. 6,
    fig. 4; Pl. 8, figs. 1-3; Fig. 19.
  Ethelocrinus magnus Strimple, 1949d, p. 12, Pl. 2, figs. 1, 3, 4, 6.
  Ethelocrinus peridous Strimple, 1949d, p. 13, Pl. 2, figs. 11, 13, 14,
    16. Strimple, 1961g, p. 87.
  Pennsylvanian, Morrowan-Desmoinesian, Holdenville Shale, Pumpkin Creek
    Limestone, Otterville Limestone: United States: Oklahoma

Parethelocrinus millsapensis (Moore and Plummer), 1940. Strimple, 1961g,
    p. 82.
  Ethelocrinus millsapensis Moore and Plummer, 1940. Strimple, 1961g,
    p. 82.
  Pennsylvanian, Desmoinesian, Millsap Lake Formation: United States:
    Texas

Parethelocrinus plattsburgensis (Strimple), 1938. Strimple, 1961g, p. 82.
  Ethelocrinus plattsburgensis Strimple, 1938. Strimple, 1959a, p. 126,
    Pl. 2, figs. 6, 9, 12; 1961g, p. 82.
  Pennsylvanian, Missourian, Plattsburg Limestone, Wann Formation: United
    States: Oklahoma

Parethelocrinus rectilatus (Lane and Webster), 1966. Webster and Lane,
    1967, p. 22, Pl. 5, fig. 6.
  Aglaocrinus? rectilatus Lane and Webster, 1966, p. 39, Pl. 7, figs. 3,
    4, 7.
  Permian, Wolfcampian, Bird Spring Formation: United States: Nevada

Parethelocrinus variabilis (Strimple), 1949. Strimple, 1961g, p. 82.
  Ethelocrinus variabilis Strimple, 1949d, p. 14, Pl. 1, figs. 8, 10, 11.
    Strimple, 1961g, p. 82.
  Pennsylvanian, Morrowan, Otterville Limestone: United States:
    Oklahoma

Parethelocrinus watkinsi (Strimple), 1949. Strimple, 1961g, p. 82.
  Ethelocrinus watkinsi Strimple, 1949d, p. 8, Pl. 3, figs. 10, 17;
    Fig. 1; 1961g, p. 82.
  Pennsylvanian, Desmoinesian, Arnold Limestone: United States: Oklahoma

Parethelocrinus sp. Strimple, 1962e, p. 47.
  Pennsylvanian, Desmoinesian, Oologah Limestone: United States: Oklahoma

PARICHTHYOCRINUS Springer, 1902. *Ichthyocrinus nobilis* Wachsmuth and
    Springer, 1878. Moore and Laudon, 1943b, p. 69, fig. 10; 1944,
    p. 177, Pl. 67, fig. 9. Ubaghs, 1953, p. 755.
  Mississippian, Kinderhookian-Osagean

Parichthyocrinus? crawfordsvillensis (Miller and Gurley), 1894. Van Sant,
    in Van Sant and Lane, 1964, p. 102.
  Mississippian, Osagean, Borden Group: United States: Indiana

Parichthyocrinus meeki (Hall), 1858. Moore and Laudon, 1944, p. 179,
    Pl. 69, fig. 12.
  Mississippian, Osagean, Keokuk Limestone: United States: Illinois,
    Iowa

Parichthyocrinus nobilis (Wachsmuth and Springer), 1878. Moore and
Laudon, 1943b, p. 136, Pl. 7, fig. 8; 1944, p. 177, Pl. 69, fig. 8.
Mississippian, Osagean, Burlington Limestone: United States: Iowa

Parichthyocrinus seversoni Laudon, Parks and Spreng, 1952, p. 560, Pl. 66,
figs. 10-14; Pl. 69, fig. 11 (misspelled as *Paraichthyocrinus seversoni*).
Mississippian, Kinderhookian, Banff formation: Canada: Alberta

PARINDOCRINUS Wanner, 1937. *P. oyensi* Wanner, 1937. Moore and Laudon,
1943b, p. 64. Arendt and Hecker, 1964, p. 92.
Permian

PARISANGULOCRINUS Schmidt, 1934. *Poteriocrinus zeaeformis* Schultze, 1866.
Moore and Laudon, 1943b, p. 52. Ubaghs, 1953, p. 751. Arendt and
Hecker, 1964, p. 86.
Lower Devonian

Parisangulocrinus cucumis Schmidt, 1941, p. 26, Pl. 4, fig. 2.
Parisangulocrinus minax? Schmidt, 1934. Schmidt, 1941, p. 26.
Parisangulocrinus? sp. Schmidt, 1934. Schmidt, 1941, p. 26.
?*Parisocrinus zeaformis* Opitz, 1932. Schmidt, 1941, p. 26.
?*Bactrocrinus nanus* Opitz, 1932. Schmidt, 1941, p. 26.
Lower Devonian, Hunsrück Schiefer: Germany

Parisangulocrinus minax Schmidt, 1934. Schmidt, 1941, p. 27.
Lower Devonian, Hunsrück Schiefer: Germany

Parisangulocrinus minax? Schmidt, 1934 = *Parisangulocrinus cucumis*

Parisangulocrinus schmidti Lehmann, 1939, p. 3, Pl. 1, fig. 1. Schmidt,
1941, p. 25.
Lower Devonian, Hunsrück Schiefer: Germany

Parisangulocrinus zeaeformis (Schultze), 1866. Lehmann, 1939, p. 4,
Pl. 1, fig. 2. Schmidt, 1941, p. 28.
Parisocrinus zeaeformis Schultze, 1866. Stürmer, 1965, p. 222, Pl. 6.
Lower Devonian, Hunsrück Schiefer: Germany

Parisangulocrinus? sp. Schmidt, 1934 = *Parisangulocrinus cucumis*

PARISOCRINUS Wachsmuth and Springer, 1880. *Poteriocrinites perplexus*
Meek and Worthen, 1869. Moore and Laudon, 1943b, p. 51; 1944,
p. 154. Ubaghs, 1953, p. 750. Philip, 1963, fig. 2c. Yakovlev,
1964b, p. 65, fig. 90v. Arendt and Hecker, 1964, p. 85.
Silurian-lower Carboniferous

Parisocrinus asiaticus Yakovlev, *in* Yakovlev and Ivanov, 1956, p. 51,
Pl. 17, figs. 3, 4. Arendt and Hecker, 1964, p. 85, Pl. 9, fig. 16.
Carboniferous, Visean: Russia: Kuznetz Basin

Parisocrinus canaliculatus Jaekel, 1895. Schmidt, 1941, p. 220.
No age or locality given.

Parisocrinus crawfordsvillensis (Miller), 1882. Van Sant, *in* Van Sant
and Lane, 1964, p. 72, Pl. 1, figs. 7, 8, 11; Fig. 27.
Parisocrinus subramosus (Miller and Gurley), 1890. Moore and Laudon,
1944, p. 154, Pl. 57, fig. 20. Ubaghs, 1953, p. 750, figs. 14e, 33i.
Poteriocrinites circumtextus Miller and Gurley, 1894. Van Sant, *in*
Van Sant and Lane, 1964, p. 72.
Poteriocrinus subramosus Miller and Gurley, 1890. Van Sant, *in* Van
Sant and Lane, 1964, p. 72.
Mississippian, Osagean, Keokuk Limestone: United States: Indiana

Parisocrinus radiatus (Austin and Austin), 1842. Wright, 1950, p. 17,
Pl. 5, figs. 1, 5, 9.
Carboniferous, Tournaisian: Ireland; England

Parisocrinus siluricus Springer, 1926. Moore and Laudon, 1944, p. 154, Pl. 55, fig. 13.
  Silurian, Niagaran, Laurel Limestone: United States: Indiana

Parisocrinus subramosus (Miller and Gurley), 1890 = *Parisocrinus crawfordsvillensis*

?*Parisocrinus zeaformis* Opitz, 1932 = *Parisangulocrinus cucumis*

Parisocrinus zeaeformis Stürmer, 1965 = *Parisangulocrinus zeaeformis*

Parisocrinus sp. Seilacher, 1961, p. 15, fig. 1, frontise (probably a *Parisangulocrinus*, GDW).
  Lower Devonian, Hunsrück Schiefer: Germany

PARULOCRINUS Moore and Plummer, 1940. *Ulocrinus blairi* Miller and Gurley, 1894. Moore and Laudon, 1943b, p. 62; 1944, p. 175. Strimple, 1966a, p. 4.
  Pennsylvanian, Missourian-Permian, Wolfcampian

Parulocrinus americanus (Weller), 1909. Strimple, 1966a, p. 4.
  Ulocrinus americanus (Weller), 1909. Strimple, 1966a, p. 4.
  Permian, Wolfcampian, Cibolo Formation: United States: Texas

Parulocrinus beedei Moore and Plummer, 1940. Strimple, 1966a, p. 4.
  Parethelocrinus beedei (Moore and Plummer), 1940. Strimple, 1961g, p. 82.
  Pennsylvanian, Missourian, Palo Pinto Limestone: United States: Texas

**Parulocrinus blairi* (Miller and Gurley), 1894. Moore and Laudon, 1944, p. 175, Pl. 64, fig. 37. Wright and Strimple, 1945, p. 227, fig. 2. Strimple, 1961g, p. 75; 1966a, p. 4.
  Ulocrinus blairi Miller and Gurley, 1894. Strimple, 1961g, p. 75.
  Pennsylvanian, Missourian, Argentine Limestone: United States: Missouri

Parulocrinus caverna (Strimple), 1949. Strimple, 1966a, p. 4.
  Ulocrinus caverna Strimple, 1949c, p. 23, Pl. 6, figs. 5-8. Strimple, 1966a, p. 4.
  Pennsylvanian, Missourian, Iola Limestone: United States: Kansas

Parulocrinus compactus Moore and Plummer, 1940 = *Paracromyocrinus compactus*

Parulocrinus marquisi Moore and Plummer, 1940 = *Paracromyocrinus marquisi*

Parulocrinus pustulosus Moore and Plummer, 1940 = *Paracromyocrinus pustulosus*

Parulocrinus vetulus Lane, 1964 = *Paracromyocrinus vetulus*

Parulocrinus wallacei Termier and Termier, 1950, p. 91, Pl. 222, figs. 6-8, 14-21 (*non* figs. 4, 5 as given on p. 91); Pl. 217, figs. 5-14, 17, 18 (*non* fig. 16 as given on p. 91).
  Carboniferous, Westphalian: Morocco

Parulocrinus sp. Termier and Termier, 1950, p. 228, Pl. 222, figs. 4, 5.
  Carboniferous, lower Westphalian: Morocco

Parulocrinus? sp. Newell and others, 1953, p. 164, Pl. 35, figs. 5, 6.
  Middle Pennsylvanian, Tarma Group: Peru

PASSALOCRINUS Peck, 1936. **P. triangularis* Peck, 1936.
  Mississippian, Osagean

**Passalocrinus triangularis* Peck, 1936. Strimple and Koenig, 1956, p. 1246, fig. 4, nos. 22-27.
  Mississippian, Osagean, Nunn Member, St. Joe Formation, Welden Limestone: United States: New Mexico, Oklahoma

PATELLIOCRINUS Angelin, 1878. **P. pachydactylus* Angelin, 1878. Moore and Laudon, 1943b, p. 98, fig. 18; 1944, p. 200.
Silurian

Patelliocrinus ornatus Springer, 1926. Moore and Laudon, 1943b, p. 142, Pl. 13, fig. 10; 1944, p. 200, Pl. 73, fig. 16.
Silurian, Niagaran, Laurel Limestone: United States: Indiana

Patelliocrinus rugosus Springer, 1926. Moore and Laudon, 1944, p. 200, Pl. 73, fig. 22.
Silurian, Niagaran, Laurel Limestone: United States: Indiana

PAULOCRINUS Springer, 1926. **P. biturbinatus* Springer, 1926. Moore and Laudon, 1943b, p. 82; 1944, p. 183. Ubaghs, 1953, p. 737. Arendt and Hecker, 1964, p. 95.
Silurian, Niagaran

*Paulocrinus biturbinatus Springer, 1926. Moore and Laudon, 1944, p. 185, Pl. 72, fig. 20. Ubaghs, 1950, p. 120, fig. 8 (misspelled as *Pauliocrinus biturbinatus*).
Silurian, Niagaran, Laurel Limestone: United States: Indiana

PEDINOCRINUS Wright, 1951. **Pachylocrinus clavatus* Wright, 1937. Wright, 1951b, p. 77. Arendt and Hecker, 1964, p. 88.
Carboniferous, Visean

*Pedinocrinus clavatus (Wright), 1937. Wright, 1951b, p. 77, Pl. 30, figs. 4, 6.
Pachylocrinus clavatus Wright, 1937. Wright, 1951b, p. 77.
Carboniferous, Visean, Lower Limestone Group: Scotland

PEGOCRINUS Kirk, 1940. **Poteriocrinus bijugis* Trautschold, 1867. Moore and Laudon, 1943b, p. 59. Arendt and Hecker, 1964, p. 89.
Carboniferous, Moscovian

PELECOCRINUS Kirk, 1941. **P. insignis* Kirk, 1941. Moore and Laudon, 1943b, p. 55, fig. 6; 1944, p. 157.
Mississippian, Kinderhookian, Osagean

Pelecocrinus aqualis (Hall), 1860. Moore and Laudon, 1944, p. 157, Pl. 55, fig. 19.
Mississippian, Osagean, Burlington Limestone: United States: Iowa

Pelecocrinus banffensis Laudon, Parks and Spreng, 1952, p. 547, Pl. 65 figs. 13-15; Pl. 69, fig. 1.
Mississippian, Kinderhookian, Banff formation: Canada: Alberta

*Pelecocrinus insignis Kirk, 1941. Moore and Laudon, 1943b, p. 133, Pl. 4, fig. 3; 1944, p. 157, Pl. 58, fig. 10.
Mississippian, Osagean, Keokuk Limestone, Burlington Limestone: United States: Iowa

Pelecocrinus primordialis Laudon, Parks and Spreng, 1952, p. 548, Pl. 65, fig. 17; Pl. 69, fig. 2.
Mississippian, Kinderhookian, Banff formation: Canada: Alberta

Pelecocrinus sp. Laudon and others, 1952, p. 550, Pl. 65, figs. 11, 12.
Mississippian, Kinderhookian, Banff formation: Canada: Alberta

Pelecocrinus sp. Laudon and Severson, 1953, p. 515, Pl. 51, fig. 15; Pl. 55, fig. 6.
Mississippian, Kinderhookian, Lodgepole Limestone: United States: Montana

PELLECRINUS Kirk, 1929. **Cyathocrinus hexadactylus* Lyon and Casseday, 1860. Moore and Laudon, 1943b, p. 51. Ubaghs, 1953, p. 750.
Mississippian, Kinderhookian-Osagean

Pellecrinus hexadactylus (Lyon and Casseday), 1860. Van Sant, *in* Van
Sant and Lane, 1964, p. 78, Pl. 1, fig. 15; Pl. 2, figs. 6, 10, 11;
Fig. 21, no. 3.
Mississippian, Osagean, Edwardsville Formation, Fort Payne Chert:
United States: Indiana, Kentucky

Pellecrinus sp. Laudon and others, 1952, p. 546, Pl. 65, figs. 1, 2.
Mississippian, Kinderhookian, Banff formation: Canada: Alberta

PENICULOCRINUS Moore, 1962. *Heterocrinus? milleri* Wetherby, 1880.
Moore, 1962b, p. 33. Philip, 1965a, p. 149, fig. 1a.
Ordovician, Trentonian

Peniculocrinus milleri (Wetherby), 1880. Moore, 1962b, p. 34, Pl. 1,
fig. 7; Fig. 2, no. 3.
Heterocrinus? milleri Wetherby, 1880. Moore, 1962b, p. 34.
Ordovician, Trentonian, Tyrone Limestone: United States: Kentucky

PENTARAMICRINUS Sutton and Winkler, 1940. *Cromyocrinus gracilis* Wetherby,
1880. Moore and Laudon, 1943b, p. 60; 1944, p. 171. Arendt and
Hecker, 1964, p. 90.
Mississippian, Chesterian

Pentaramicrinus gracilis (Wetherby), 1880. Moore and Laudon, 1944,
p. 171, Pl. 64, fig. 30.
Mississippian, Chesterian, Golconda Formation: United States: Kentucky,
Illinois

PENTECECRINUS Koenig and Niewoehner, 1959. *P. parvus* Koenig and Niewoehner, 1959. Koenig and Niewoehner, 1959, p. 464. Lane, 1967b, p. 13,
fig. 2.
Devonian-Mississippian transition

Pentececrinus parvus Koenig and Niewoehner, 1959, p. 467, figs. 1, 3.
Devonian-Mississippian transition, Louisiana Formation: United States:
Missouri

Periechocrinites Austin and Austin, 1943 = *Sagenocrinites* Austin and Austin,
1842

Periechocrinites articulosus Austin and Austin, 1943 = *Sagenocrinites
expansus*

Periechocrinites costatus Austin and Austin, 1843 = *Periechocrinus costatus*

Periechocrinites marcouanus (Winchell and Marcy), 1865 = *Periechocrinus
marcouanus*

Periechocrinites tennesseensis (Hall), 1875 = *Periechocrinus tennesseensis*

Periechocrinites whitei (Hall), 1861 = *Periechocrinus whitei*

Periechocrinites sp. 1. Laudon and Severson, 1953 = *Aryballocrinus* sp.

Periechocrinites sp. Laudon and Severson, 1953 = *Periechocrinus* sp.

PERIECHOCRINUS Morris, 1843. *Periechocrinites costatus* Austin and Austin,
1843. Ramsbottom, 1954, p. 687.
Periechocrinites Moore and Laudon, 1943b, p. 92, fig. 15; 1944, p. 189,
Pl. 70, fig. 17. Moore, 1950, p. 42, fig. 11. Ubaghs, 1953, p. 739.
Silurian, Mississippian

Periechocrinus? awthornsensis Wright, 1955 = *Aryballocrinus awthornsensis*

Periechocrinus chicagoensis Weller, 1900 = *Stiptocrinus chicagoensis*

Periechocrinus costatus (Austin and Austin), 1843. Ramsbottom, 1954,
p. 687.
Periechocrinites costatus Austin and Austin, 1943. Ramsbottom, 1951b,
p. 1042.
Silurian, Wenlockian, Wenlock Limestone: England

Periechocrinus marcouanus (Winchell and Marcy), 1865.
  Periechocrinites marcouanus (Winchell and Marcy), 1865. Moore and Laudon, 1942, p. 74, fig. 4D.
  Silurian, Niagaran: United States: Illinois

Periechocrinus radiatus Angelin, 1878 = *Promelocrinus radiatus*

Periechocrinus sanmigueli Astre, 1925 = *Pyxidocrinus sanmigueli*

Periechocrinus tennesseensis (Hall), 1875.
  Periechocrinites tennesseensis (Hall), 1875. Moore and Laudon, 1943b, p. 139, Pl. 10, fig. 4; 1944, p. 189, Pl. 75, fig. 15. La Rocque and Marple, 1955, p. 73, fig. 147.
  Silurian, Niagaran, Beech River Formation: United States: Tennessee

Periechocrinus whitei (Hall), 1861.
  Periechocrinites whitei (Hall), 1861. Moore and Laudon, 1944, p. 189, Pl. 75, fig. 14.
  Mississippian, Osagean, Hampton Formation, Burlington Limestone: United States: Iowa, Missouri

Periechocrinus sp.
  Periechocrinites sp. Laudon and Severson, 1953, p. 527, Pl. 52, figs. 1, 2.
  Mississippian, Kinderhookian, Lodgepole Limestone: United States: Montana

PERIGLYPTOCRINUS Wachsmuth and Springer, 1897. *P. billingsi* Wachsmuth and Springer, 1897. Moore and Laudon, 1943b, p. 96; 1944, p. 199. Ubaghs, 1953, p. 741.
  Ordovician, Trentonian

Periglyptocrinus billingsi Wachsmuth and Springer, 1897. Moore and Laudon, 1944, p. 199, Pl. 73, fig. 23. Wilson, 1946, p. 29.
  Ordovician, Trentonian, Trenton Limestone, Hull Limestone: Canada: Ontario

Periglyptocrinus priscus (Billings), 1857. Wilson, 1946, p. 30.
  Ordovician, Trentonian, Rockland Formation or Leray beds: Canada: Ontario

PERIMESTOCRINUS Moore and Plummer, 1938. *Hydreionocrinus noduliferus* Miller and Gurley, 1894. Moore and Laudon, 1943b, p. 58; 1944, p. 161.
  Pennsylvanian, Morrowan-Permian, Wolfcampian

Perimestocrinus? bulbosus Strimple, 1962e, p. 26, Pl. 2, figs. 1-4.
  Pennsylvanian, Desmoinesian, Oologah Limestone: United States: Oklahoma

Perimestocrinus calyculus Moore and Plummer, 1940. Moore and Laudon, 1944, p. 161, Pl. 64, fig. 2.
  Pennsylvanian, Desmoinesian, Mineral Wells Formation: United States: Texas

Perimestocrinus fabulosus (Strimple), 1950. Strimple, 1961g, p. 24.
  Utharocrinus fabulosus Strimple, 1950b, p. 572, Pl. 77, figs. 1-4; 1961g, p. 24.
  Pennsylvanian, Missourian, Stoner Limestone: United States: Kansas

Perimestocrinus facilis (Strimple), 1950. Strimple, 1961g, p. 24.
  Utharocrinus facilis Strimple, 1950b, p. 571, Pl. 77, figs. 11, 12, 18-20.
  Pennsylvanian, Virgilian, Stull Shale: United States: Kansas

Perimestocrinus formosus Moore and Plummer, 1940. Moore and Laudon, 1944, p. 161, Pl. 64, fig. 3.
  Pennsylvanian, Missourian, Palo Pinto Group: United States: Texas

Perimestocrinus granulosus (Strimple), 1939. Strimple, 1961g, p. 24; 1963b, p. 106, Pl. 1, figs. 1-7.
  Utharocrinus granulosus Strimple, 1939. Strimple, 1961g, p. 24.
  Pennsylvanian, Missourian, Wann Formation: United States: Oklahoma

Perimestocrinus habitus (Strimple), 1950. Strimple, 1961g, p. 24.
  Utharocrinus habitus Strimple, 1950b, p. 572, Pl. 77, figs. 13-16; 1961g, p. 24.
  Pennsylvanian, Missourian, Wann Formation: United States: Oklahoma

Perimestocrinus hexagonus Strimple, 1952 = *Stenopecrinus hexagonus*

Perimestocrinus impressus Moore and Plummer, 1940. Moore and Laudon, 1943b, p. 134, Pl. 5, fig. 5; 1944, p. 161, Pl. 62, fig. 8; Pl. 64, fig. 12. Strimple, 1948a, p. 115, Pl. 10, fig. 1 (misspelled as *Peremistocrinus impressus*). Strimple, 1962e, p. 29, Pl. 2, figs. 9-12.
  Pennsylvanian, Desmoinesian-Missourian, Mineral Wells Formation, Dewey Limestone, Oologah Limestone: United States: Texas, Oklahoma, Missouri, Kansas

Perimestocrinus (*Triceracrinus*) *moorei* (Bramlette), 1943 = *Triceracrinus moorei*

Perimestocrinus moseleyi Strimple, 1951 = *Stenopecrinus moseleyi*

Perimestocrinus nevadensis Lane and Webster, 1966, p. 51, Pl. 12, figs. 7, 8. Webster and Lane, 1967, p. 30, Pl. 5, fig. 3.
  Permian, Wolfcampian, Bird Spring Formation: United States: Nevada

Perimestocrinus noduliferus (Miller and Gurley), 1894. Moore and Laudon, 1944, p. 163, Pl. 62, fig. 9; Pl. 64, fig. 5. Strimple, 1961g, p. 20, figs. 1, 2.
  Pennsylvanian, Missourian, Argentine Limestone: United States: Missouri

Perimestocrinus oasis Webster and Lane, 1967, p. 30, Pl. 5, figs. 9, 10.
  Permian, Wolfcampian, Bird Spring Formation: United States: Nevada

Perimestocrinus oreadensis (Moore), 1939. Strimple, 1961g, p. 24.
  Utharocrinus oreadensis Moore, 1939. Strimple, 1961g, p. 24.
  Pennsylvanian, Virgilian, Oread Limestone: United States: Kansas

Perimestocrinus papillatus Strimple, 1962e, p. 28, Pl. 2, figs. 5-8.
  Pennsylvanian, Desmoinesian, Oologah Formation: United States: Oklahoma

Perimestocrinus planus Strimple, 1952 = *Stenopecrinus planus*

Perimestocrinus pumilis Moore and Plummer, 1938. Moore and Laudon, 1944, p. 161, Pl. 64, fig. 4.
  Pennsylvanian, Morrowan, Brentwood Limestone: United States: Oklahoma

Perimestocrinus quinquacutus (Moore), 1939. Strimple, 1961g, p. 24.
  Utharocrinus quinquacutus Moore, 1939. Moore and Laudon, 1944, p. 163, Pl. 62, fig. 14; Pl. 64, fig. 17. Strimple, 1961g, p. 24.
  Pennsylvanian, Virgilian, Brownville Limestone: United States: Oklahoma

Perimestocrinus spinosus (Strimple), 1950. Strimple, 1961g, p. 24.
  Utharocrinus spinosus Strimple, 1950b, p. 573, Pl. 77, figs. 9, 10, 17. Strimple, 1961g, p. 24.
  Pennsylvanian, Missourian, Ochelata Group: United States: Oklahoma

Perimestocrinus teneris Moore and Plummer, 1938. Moore and Laudon, 1944, p. 161, Pl. 64, fig. 11.
  Pennsylvanian, Morrowan, Brentwood Limestone: United States: Arkansas

Perimestocrinus topekensis (Moore), 1939. Strimple, 1961g, p. 24.
  Utharocrinus topekensis Moore, 1939. Strimple, 1961g, p. 24.
  Pennsylvanian, Virgilian, Topeka Limestone: United States: Kansas

Perimestocrinus sp. Strimple, 1948a, p. 115, Pl. 10, figs. 2-14 (misspelled as *Peremistocrinus*).
Pennsylvanian, Missourian, Dewey Limestone: United States: Oklahoma

PERISSOCRINUS Goldring, 1936. *P. papillatus* Goldring, 1936. Moore and Laudon, 1943b, p. 30. Ubaghs, 1953, p. 747.
Devonian, Erian

**Perissocrinus papillatus* Goldring, 1936. Strimple, 1963d, p. 22, fig. 1b.
Devonian, Erian, Centerfield Limestone: United States: New York

PERITTOCRINUS Jaekel, 1902. **Porocrinus radiatus* Beyrich, 1879. Moore and Laudon, 1943b, p. 50. Ubaghs, 1953, p. 750. Xu, 1962, p. 45, 50, fig. 1.
Ordovician

Perittocrinus sp. Ubaghs, 1953, p. 691, fig. 33a.
Ordovician: Russia

PERMIOCRINUS Wanner, 1949. **P. immaturus* Wanner, 1949. Wanner, 1949a, p. 53.
Permian

**Permiocrinus immaturus* Wanner, 1949a, p. 53, Pl. 3, figs. 29, 30; Fig. 16.
Permian: Timor

PERNEROCRINUS Bouška, 1950. **P. paradoxus* Bouška, 1950. Bouška, 1950, p. 18. Ubaghs, 1953, p. 750. Arendt and Hecker, 1964, p. 84.
Lower Devonian

**Pernerocrinus paradoxus* Bouška, 1950, p. 19, Pl. 3, figs. 1-4b; Pl. 4, figs. 1-3; Fig. 6. Ubaghs, 1953, p. 750, fig. 84. Yakovlev, 1964b, p. 62, fig. 78.
Lower Devonian, Koněprusy Limestone: central Bohemia

Pernerocrinus? sp. indet. Bouška, 1950, p. 21, Pl. 4, figs. 4, 4a.
Lower Devonian, Koněprusy Limestone: central Bohemia

PETALOCRINUS Weller and Davidson, 1896. **P. mirabilis* Weller and Davidson, 1896. Moore and Laudon, 1943b, p. 50, fig. 4; 1944, p. 153. Cuénot, 1948, p. 64. Moore, 1950, p. 37, fig. 40. Ubaghs, 1953, p. 750, fig. 85. Moore and Jeffords, 1968, p. 16, fig. 5, no. 2.
Silurian

Petalocrinus chiai Mu, 1950, p. 95, figs. 3, nos. 7-9; 11.
Silurian, Shihniulan Limestone: China: Kueichou

Petalocrinus inferior Bather, 1898. Mu, 1950, p. 94, figs. 3, nos. 1-3; 12.
Silurian, Shihniulan Limestone: China: Kueichou

Petalocrinus inflatus Mu, 1950, p. 94, figs. 2; 3, nos. 4-6; 10; 13.
Silurian, Shihniulan Limestone: China: Kueichou

**Petalocrinus mirabilis* Weller and Davidson, 1896. Moore and Laudon, 1943b, p. 131, Pl. 2, fig. 8; 1944, p. 153, Pl. 54, fig. 31. Cuénot, 1948, p. 64, fig. 82. Mu, 1950, p. 93, fig. 1. Ubaghs, 1953, p. 750, figs. 85a, b, d. Norford, 1962, p. 25, Pl. 4, figs. 25-27. Yakovlev, 1964b, p. 62, fig. 77.
Silurian, Niagaran, Hopkinton Dolomite, Caralline Member: United States: Iowa; Canada: British Columbia

Petalocrinus visbycensis Bather, 1898. Ubaghs, 1953, p. 750, fig. 85c.
Silurian: Sweden

Petalocrinus? sp. Pinna, 1963, p. 108, fig. II, no. 4.
Silurian, Wenlockian, Woolhope Limestone: England

PETROCRINUS Wanner, 1924. **P. beyrichi* Wanner, 1924. Moore and Laudon, 1943b, p. 76. Ubaghs, 1953, p. 756. Arendt and Hecker, 1964, p. 94.
Permian

PETSCHORACRINUS Yakovlev, 1928. *P. variabilis Yakovlev, 1928. Yakovlev, 1939e, p. 832; 1939f, p. 832. Moore and Laudon, 1943b, p. 62. Yakovlev, 1947d, p. 43, figs. 5, 6; 1956a, p. 78, figs. 25, 26; 1958, p. 63, figs. 25, 26; 1964a, p. 74, figs. 27, 28. Arendt and Hecker, 1964, p. 91.
Lower Permian

*Petschoracrinus variabilis Yakovlev, 1928. Yakovlev, 1939b, p. 62, Pl. 10, figs. 15-18; Fig. 15. Yakovlev, in Yakovlev and Ivanov, 1956, p. 79, Pl. 15, figs. 2-15, 17. Arendt and Hecker, 1964, p. 91, Pl. 13, figs. 2-5.
Lower Permian: Russia: Arctic Urals

Petschoracrinus? sp. Webster and Lane, 1967, p. 21, Pl. 3, figs. 7, 8.
Permian, Wolfcampian, Bird Spring Formation: United States: Nevada

PHACELOCRINUS Kirk, 1940. *Poteriocrinus wetherbyi Miller, 1879. Moore and Laudon, 1943b, p. 59; 1944, p. 171. Arendt and Hecker, 1964, p. 89.
Mississippian

Phacelocrinus dactyliformis (Hall), 1858. Moore and Laudon, 1944, p. 171, Pl. 60, fig. 7.
Mississippian, Meramecian, St. Louis Limestone: United States: Missouri

Phacelocrinus decabrachiatus (Hall), 1858. Moore and Laudon, 1944, p. 171, Pl. 60, fig. 6.
Mississippian, Chesterian: United States: Illinois

Phacelocrinus intermodius (Hall), 1858. Moore and Laudon, 1944, p. 171, Pl. 60, fig. 16.
Mississippian, Chesterian: United States: Illinois

Phacelocrinus longidactylus (McChesney), 1859. Horowitz, 1965, p. 28, Pl. 2, figs. 19-21.
Mississippian, Chesterian, Glen Dean Formation: United States: Indiana

Phacelocrinus? pentagonus (Austin and Austin), 1843. Wright, 1951a, p. 32, Pl. 9.
Cladocrinites pentagonus Austin and Austin, 1843. Wright, 1951a, p. 32.
Carboniferous, Tournaisian: England

Phacelocrinus rostratus (Austin and Austin), 1842. Wright, 1951a, p. 29, Pl. 8, figs. 1-3, 7, 8, 10; Pl. 10, fig. 4; Fig. 9; 1960, p. 330, Appendix, Pl. A, fig. 1.
Poteriocrinus rostratus Austin and Austin, 1842. Wright, 1951a, p. 29.
Carboniferous, Tournaisian: Ireland

Phacelocrinus vanhornei (Worthen), 1875. Moore and Laudon, 1944, p. 171, Pl. 60, fig. 5.
Mississippian, Meramecian, St. Louis Formation: United States: Illinois

Phacelocrinus? sp. Wright, 1960, p. 330, Appendix, Pl. A, fig. 9.
Carboniferous, Visean, S_2: England

PHANOCRINUS Kirk, 1937. *Poteriocrinites (Zeacrinus) formosus Worthen, in Meek and Worthen, 1873. Moore and Laudon, 1943b, p. 60, fig. 8; 1944, p. 171. Ubaghs, 1953, p. 677, figs. 16g-i. Moore, 1962b, p. 15, fig. 10, no. 1. Arendt and Hecker, 1964, p. 90.
Mississippian, Kinderhookian-Permian, Wolfcampian

Phanocrinus alexanderi Strimple, 1948b, p. 493, Pl. 77, figs. 1-6.
Mississippian, Chesterian, Fayetteville Formation: United States: Oklahoma

Phanocrinus altus Wright, 1942, p. 276, Pl. 12, figs. 11, 13-16; 1951b, p. 101, Pl. 17, figs. 2, 4.
Carboniferous, Visean, Lower Limestone Group: Scotland

Phanocrinus ardrossensis (Wright), 1934. Wright, 1951b, p. 97, Pl. 33, figs. 3-5, 7-12; 1952b, p. 324, Pl. 13, fig. 8; 1954b, p. 169, Pl. 3, fig. 4.
Carboniferous, Visean, Calciferous Sandstone series: Scotland

Phanocrinus calyx (McCoy), 1849. Wright, 1951b, p. 94, Pl. 17, figs. 15, 21; Pl. 18, figs. 1-19, 22, 23, 29; Pl. 33, figs. 6, 14.
Carboniferous, Visean, Lower Limestone Group: Scotland

Phanocrinus compactus Sutton and Winkler, 1940. Horowitz, 1965, p. 32, Pl. 3, figs. 7-9.
Mississippian, Chesterian, Glen Dean Formation: United States: Indiana

Phanocrinus cylindricus (Miller and Gurley), 1894. Strimple, 1948b, p. 491, Pl. 77, fig. 7; 1951g, p. 291, fig. 11.
Mississippian, Chesterian, Fayetteville Formation: United States: Oklahoma

**Phanocrinus formosus* (Worthen, *in* Meek and Worthen), 1873. Moore and Laudon, 1943b, p. 135, Pl. 6, fig. 1; 1944, p. 171, Pl. 65, fig. 32. Termier and Termier, 1950, p. 90, Pl. 221, figs. 1-8.
Mississippian, Chesterian, Glen Dean Formation, upper Visean: United States: Kentucky; Morocco

Phanocrinus cf. *P. formosus* (Worthen, *in* Meek and Worthen), 1873. Horowitz, 1965, p. 32, Pl. 3, figs. 13-15.
Mississippian, Chesterian, Glen Dean Formation: United States: Indiana

Phanocrinus gordoni Wright, 1939. Wright, 1951b, p. 99, Pl. 17, figs. 6-9, 22, 23; Pl. 19, figs. 11, 12.
Carboniferous, Visean, Lower Limestone Group: Scotland

Phanocrinus? insolitus Webster and Lane, 1967, p. 26, Pl. 5, figs. 4, 5.
Permian, Wolfcampian, Bird Spring Formation: United States: Nevada

Phanocrinus irregularis Strimple, 1951e, p. 262, figs. 5-9.
Mississippian, Chesterian, Pitkin Limestone: United States: Oklahoma

Phanocrinus modulus Strimple, 1951e, p. 263, figs. 10-12.
Mississippian, Chesterian, Pitkin Limestone: United States: Oklahoma

Phanocrinus nitidus (Miller and Gurley), 1894. Moore and Laudon, 1944, p. 171, Pl. 64, fig. 35.
Mississippian, Chesterian: United States: Illinois

Phanocrinus ornatus Wright, 1951b, p. 98, Pl. 18, fig. 28; Pl. 19, figs. 14, 17.
Carboniferous, Visean, Lower Limestone Group: Scotland

Phanocrinus parvaramus Sutton and Winkler, 1940. Horowitz, 1965, p. 30, Pl. 3, figs. 1-3.
Mississippian, Chesterian, Glen Dean Limestone: United States: Indiana

Phanocrinus scoticus (de Koninck), 1858. Wright, 1951b, p. 96, Pl. 18, fig. 21.
Hydreionocrinus scoticus de Koninck, 1858. Wright, 1951b, p. 96.
Carboniferous, Visean, Lower Limestone Group: Scotland

Phanocrinus stellaris (Wright), 1934. Wright, 1951b, p. 98, Pl. 18, figs. 24-27.
Carboniferous, Visean, Lower Limestone Group: Scotland

Phanocrinus vadosus Washburn, 1968, p. 126, Pl. 1, figs. 10-12.
Pennsylvanian, Morrowan, Oquirrh Formation: United States: Utah

PHIALOCRINUS Eichwald, 1856. **P. impressus*, 1856.
Ordovician-Permian

Phialocrinus americanus Weller, 1909 = *Parulocrinus americanus*

Phialocrinus elenae Yakovlev, 1930. Yakovlev, *in* Yakovlev and Ivanov, 1956, Pl. 11, fig. 12.
Lower Permian: Russia: Krasnoufimsk

PHILOCRINUS de Koninck, 1863. *P. cometa* de Koninck, 1863. Moore and Laudon, 1943b, p. 55. Arendt and Hecker, 1964, p. 88.
Carboniferous

PHIMOCRINUS Schultze, 1866. *P. laevis* Schultze, 1866. Moore and Laudon, 1943b, p. 31. Ubaghs, 1953, p. 746.
Devonian

Phimocrinus jouberti Oehlert, 1882 = *Theloreus jouberti*

PHYSETOCRINUS Meek and Worthen, 1869. *Actinocrinus ventricosus* Hall, 1858. Moore and Laudon, 1943b, p. 93, fig. 15; 1944, p. 193, Pl. 71, fig. 15. Ubaghs, 1953, p. 739. Arendt and Hecker, 1964, p. 96.
Lower Carboniferous

Physetocrinus brightoni Wright, 1955a, p. 239, Pl. 52, figs. 16-19.
Carboniferous, ?Tournaisian: ?Ireland

Physetocrinus costus (McCoy), 1844. Wright, 1955a, p. 240.
Carboniferous, Tournaisian: Scotland

Physetocrinus lobatus Wachsmuth and Springer, 1897. Moore and Laudon, 1944, p. 193, Pl. 77, fig. 8. Laudon, 1948, Pl. 3. Bowsher, 1954, p. 113, figs. 1-3.
Mississippian, Osagean, Lake Valley Limestone, Redwall Limestone: United States: New Mexico, Arizona

Physetocrinus ventricosus (Hall), 1858. Moore and Laudon, 1943b, p. 140, Pl. 11, fig. 4; 1944, p. 193, Pl. 77, fig. 7. Laudon, 1948, Pl. 3.
Mississippian, Osagean, Burlington Limestone: United States: Iowa, Missouri

PILIDIOCRINUS Wanner, 1937. *P. permicus* Wanner, 1937. Moore and Laudon, 1943b, p. 50. Arendt and Hecker, 1964, p. 85. Lane, 1967b, p. 13, fig. 3.
Permian

Pilidiocrinus permicus Wanner, 1937. Arendt, 1968a, p. 99, fig. 2.
Permian: Timor

PIMLICOCRINUS Wright, 1943. *Amphoracrinus clitheroensis* Wright, 1942. Wright, 1943a, p. 89. Ubaghs, 1953, p. 739.
Carboniferous

Pimlicocrinus brevicalix (Rofe), 1865. Wright, 1955a, p. 211, Pl. 51, fig. 27; Fig. 111.
Actinocrinus (Amphoracrinus) brevicalix Rofe, 1865. Wright, 1943b, p. 234; 1955a, p. 211.
Carboniferous, ?Tournaisian: Scotland

Pimlicocrinus clitheroensis (Wright), 1942. Wright, 1943a, p. 90, Pl. 3, figs. 12, 13, 18-20; 1955a, p. 208, Pl. 51, figs. 1-4, 11, 12.
Amphoracrinus clitheroensis Wright, 1942, p. 272, Pl. 10, figs. 1-4. Wright, 1943a, p. 90.
Carboniferous, Tournaisian: Scotland

Pimlicocrinus latus Wright, 1943a, p. 91, Pl. 3, figs. 1-11, 14-17, 21, 22; 1955a, p. 210, Pl. 51, figs. 6-10, 13, 14, 21-25, 33-37. Breimer, 1962a, p. 80, Pl. 8, figs. 8-10; Fig. 14.
Carboniferous, Tournaisian-Namurian, Rabanal Limestone: Scotland; Spain

Pimlicocrinus cf. *latus* Wright, 1943. Wright, 1955a, p. 210, Pl. 61, fig. 8.
Carboniferous, Tournaisian: England

Pimlicocrinus sp. 1. Breimer, 1962a, p. 82, figs. 15A, B.
   Carboniferous, Westphalian: Spain
Pimlicocrinus sp. 2. Breimer, 1962a, p. 83, Pl. 8, fig. 11; Fig. 15C.
   Age and locality unknown in Spain
Pimlicocrinus sp. Wright, 1955a, p. 212, Pl. 53, fig. 21.
   Carboniferous, Tournaisian: Scotland

PIRASOCRINUS Moore and Plummer, 1940. *P. scotti* Moore and Plummer, 1940.
   Moore and Laudon, 1943b, p. 58, fig. 7; 1944, p. 165. Moore, 1950,
   p. 34, fig. 4j.
   Pennsylvanian, Desmoinesian-Missourian

Pirasocrinus depressus Strimple, 1961g, p. 57, Pl. 14, figs. 1, 2.
   Pennsylvanian, Desmoinesian, Holdenville Formation: United States:
   Oklahoma

Pirasocrinus invaginatus Strimple, 1951a, p. 202, Pl. 36, figs. 8-11;
   Pl. 38, figs. 5, 6.
   Pennsylvanian, Missourian, Lake Bridgeport Shale: United States: Texas

*Pirasocrinus scotti Moore and Plummer, 1940. Moore and Laudon, 1943b,
   p. 134, Pl. 5, fig. 11; 1944, p. 165, Pl. 62, fig. 18; Pl. 63,
   fig. 23.
   Pennsylvanian, Desmoinesian, Millsap Lake Formation: United States:
   Texas

PISOCRINUS de Koninck, 1858. *P. pilula* de Koninck, 1858. Moore and
   Laudon, 1943b, p. 30, fig. 3; 1944, p. 147, Pl. 54, fig. 19. Cuénot,
   1948, p. 60. Ubaghs, 1953, p. 746, figs. 11a, 140E. Mu, 1954,
   p. 325, fig. 1. Bouška, 1956a, p. 12, fig. 1. Moore, 1962b, p. 14,
   figs. 6, no. 1; 7, no. 2. Strimple, 1963d, p. 37, fig. 9c. Arendt
   and Hecker, 1964, p. 81. Moore and Jeffords, 1968, p. 11, fig. 3,
   no. 4.
   Silurian-Lower Devonian

Pisocrinus baccula Miller and Gurley, 1895 = *Pisocrinus (Pisocrinus) baccula*

Pisocrinus benedicti Miller, 1891 = *Pisocrinus (Pisocrinus) benedicti*

Pisocrinus bogdani Yeltyschewa, 1968, p. 32, Pl. 4, figs. 9-12.
   Silurian, Borschorskian horizon: Russia: Podolsky

Pisocrinus bohemicus Bouška, 1956a, p. 20, Pl. 2, figs. 10-12; Fig. 4.
   Silurian, middle Ludlovian, Přidoli Beds: Bohemia

Pisocrinus campana Miller, 1891 = *Pisocrinus (Pisocrinus) campana*

Pisocrinus crassiortabelaris Bouška, 1956a, p. 25, Pl. 3, figs. 1-3.
   Lower Devonian, Prokop Limestone: Bohemia

Pisocrinus gemmiformis Miller, 1879 = *Pisocrinus (Pisocrinus) gemmiformis*

Pisocrinus gemmiformis var. *globosus* (Ringueberg), 1884 = *Pisocrinus
   (Pisocrinus) gemmiformis* var. *globosus*

Pisocrinus gorbyi Miller, 1891 = *Pisocrinus (Pisocrinus) gorbyi*

Pisocrinus cf. *gorbyi* Miller, 1891. Bouška, 1956a, p. 19, Pl. 2, figs. 1-4.
   Silurian, lower Ludlovian, Kopanina Limestone: Bohemia

Pisocrinus granulosus Rowley, 1904 = *Pisocrinus (Pisocrinus) granulosus*

Pisocrinus kolihai Bouška, 1956a, p. 23, Pl. 2, fig. 14.
   Lower Devonian, Zlichov Limestone: Bohemia

Pisocrinus kosovensis Bouška, 1956a, p. 18, Pl. 1, figs. 16-18.
   Silurian, lower Ludlovian, Kopanina Limestone: Bohemia

Pisocrinus lobata Etheridge, 1904 = *Pisocrinus (Parapisocrinus) lobata*

Pisocrinus minutus Bouška, 1956a, p. 19, Pl. 2, figs. 5-9.
   Silurian, lower Ludlovian, Kopanina Limestone: Bohemia
Pisocrinus morinensis Bouška, 1956a, p. 26, Pl. 3, figs. 4-6.
   Silurian, Ludlovian, Kopanina Limestone: Bohemia
Pisocrinus ollula Angelin, 1878 = *Pisocrinus (Parapisocrinus) ollula*
Pisocrinus (Parapisocrinus) Mu, 1954, p. 326, fig. 2. Weyer, 1965, p. 969.
   Ollulocrinus Bouška, 1956a, p. 28. Arendt and Hecker, 1964, p. 81.
      Weyer, 1965, p. 969.
   Silurian-Lower Devonian
Pisocrinus (Parapisocrinus) lobata Etheridge, 1904. Mu, 1954, p. 326.
   Upper Silurian: Australia: Yass District
Pisocrinus (Parapisocrinus) malobatschatensis (Dubatolova), 1964, new comb.
   Ollulocrinus malobatschatensis Dubatolova, 1964, p. 14, Pl. 1, fig. 1.
   Lower Devonian: Russia: Kuznetz Basin
Pisocrinus (Parapisocrinus) ollula Angelin, 1878. Mu, 1954, p. 326, Pl. 1,
   figs. 8, 9.
   Silurian: Sweden
Pisocrinus (Parapisocrinus) ollula elegans (Bouška), 1956, new comb.
   Ollulocrinus ollula elegans Bouška, 1956a, p. 31, Pl. 4, fig. 1.
   Lower Devonian, Prokop Limestone: Bohemia
Pisocrinus (Parapisocrinus) ollula grandis (Bouška), 1956, new comb.
   Ollulocrinus ollula grandis Bouška, 1956a, p. 30, Pl. 4, fig. 5; Fig. 6.
   Silurian, Ludlovian, Přidoli Beds: Bohemia
Pisocrinus (Parapisocrinus) ollula hlubocepensis (Bouška), 1956, new comb.
   Ollulocrinus ollula hlubocepensis Bouška, 1956a, p. 31, Pl. 4, fig. 6;
      Fig. 7.
   Lower Devonian, Zlichov Limestone: Bohemia
Pisocrinus (Parapisocrinus) ollula ollula Angelin, 1878.
   Ollulocrinus ollula ollula (Angelin), 1878. Bouška, 1956a, p. 29,
      Pl. 3, fig. 15. Strimple, 1963d, fig. 10k.
   Silurian, lower Ludlovian-middle Wenlockian, Kopanina Limestone: Bohemia
Pisocrinus (Parapisocrinus) pribyli (Bouška), 1956, new comb.
   Ollulocrinus pribyli Bouška, 1956a, p. 36, Pl. 4, figs. 8-10.
   Silurian, Ludlovian, Kopanina Limestone: Bohemia
Pisocrinus (Parapisocrinus) quinquelobus Bather, 1893. Mu, 1954, p. 326.
   Pisocrinus quinquelobus Bather, 1893. Moore and Laudon, 1943b, p. 130,
      Pl. 1, fig. 13; 1944, p. 147, Pl. 54, fig. 15. Cuénot, 1948, p. 60,
      fig. 75. Amsden, 1949, p. 77, Pl. 12, figs. 14-16. Ubaghs, 1953,
      p. 746, fig. 36.
   Ollulocrinus quinquelobus (Bather), 1893. Bouška, 1956a, p. 33, Pl. 5,
      figs. 1-14. Strimple, 1963d, p. 47, figs. 10a, 11a, b.
   Silurian, Niagaran-Ludlovian, Beech River Formation, Racine Formation,
      Bainbridge Formation, and Henryhouse Formation, Přidoli Beds: United
      States: Tennessee, Oklahoma; Bohemia
Pisocrinus (Parapisocrinus) rimosus (Bouška), 1956, new comb.
   Ollulocrinus rimosus Bouška, 1956a, p. 35, Pl. 6, fig. 1; Figs. 8, 9.
      Strimple, 1963d, p. 45 (misspelled as *O. rinosus*).
   Lower Devonian, Koněprusy Limestone: Bohemia
Pisocrinus (Parapisocrinus) sphericus Rowley, 1904. Mu, 1954, p. 326.
   Pisocrinus sphericus Rowley, 1904. Moore and Laudon, 1944, p. 147,
      Pl. 54, fig. 16.
   Ollulocrinus sphericus (Rowley), 1904. Bouška, 1956a, p. 32.
   Silurian, Niagaran, Brownsport Formation, Bainbridge Formation: United
      States: Tennessee, Missouri

Pisocrinus cf. *sphericus* Rowley, 1904. Ramsbottom, 1958, p. 113, Pl. 21, fig. 1.
  Silurian, lower Ludlovian: England

Pisocrinus (*Parapisocrinus*) *tennesseensis* (Roemer), 1860. Mu, 1954, p. 326.
  Pisocrinus tennesseensis (Roemer), 1860. Moore and Laudon, 1944, p. 147, Pl. 54, fig. 18. Amsden, 1949, p. 78, Pl. 12, figs. 10, 11. Ubaghs, 1953, p. 746, figs. 37-39.
  Ollulocrinus tennesseensis (Roemer), 1860. Strimple, 1963d, p. 46, fig. 11d.
  Silurian, Niagaran, Brownsport Formation, Henryhouse Formation: United States: Tennessee, Oklahoma

Pisocrinus (*Parapisocrinus*) *yassensis* Etheridge, 1904. Mu, 1954, p. 326.
  Ollulocrinus yassensis (Etheridge), 1904. Bouška, 1956a, p. 32.
  Silurian: Australia: Yass District

Pisocrinus (*Parapisocrinus*)? cf. *yassensis* Etheridge, 1904, new comb.
  Ollulocrinus? cf. *yassensis* (Etheridge), 1904. Bouška, 1956a, p. 33, Pl. 4, figs. 2-4.
  Silurian, lower Ludlovian, Kopanina Limestone: Bohemia

Pisocrinus (*Parapisocrinus*) sp. new comb.
  Ollulocrinus sp. Bouška, 1956a, p. 37, Pl. 4, fig. 7.
  Silurian, Ludlovian, Přidoli Beds: Bohemia

Pisocrinus pilula var. *ornatus* Angelin, 1878 = *Pisocrinus* (*Pisocrinus*) *pilula* var. *ornatus*

Pisocrinus (*Pisocrinus*) *baccula* Miller and Gurley, 1895. Mu, 1954, p. 325.
  Pisocrinus baccula Miller and Gurley, 1895. Moore and Laudon, 1944, p. 147, Pl. 54, fig. 17 (misspelled as *P. bacula*).
  Silurian, Niagaran, Laurel Limestone: United States: Indiana

Pisocrinus (*Pisocrinus*) *benedicti* Miller, 1891. Mu, 1954, p. 325.
  Silurian, Niagaran: United States

Pisocrinus (*Pisocrinus*) *campana* Miller, 1891. Mu, 1954, p. 325.
  Pisocrinus campana Miller, 1891. Amsden, 1949, p. 76, Pl. 12, figs. 12, 13. Mu, 1954, p. 325.
  Silurian, Niagaran, Brownsport Formation: United States: Tennessee

Pisocrinus (*Pisocrinus*) *gemmiformis* Miller, 1879. Mu, 1954, p. 325.
  Pisocrinus gemmiformis Miller, 1879. Moore and Laudon, 1944, p. 147, Pl. 54, fig. 14.
  Silurian, Niagaran, Laurel Limestone, Osgood Formation, Racine Formation, Bainbridge Formation: United States: Indiana, Tennessee, Illinois, Missouri

Pisocrinus (*Pisocrinus*) *gemmiformis* var. *globosus* (Ringueberg), 1884. Mu, 1954, p. 325.
  Silurian: United States

Pisocrinus (*Pisocrinus*) *gorbyi* Miller, 1891. Mu, 1954, p. 325.
  Silurian: United States

Pisocrinus (*Pisocrinus*) *granulosus* Rowley, 1904. Mu, 1954, p. 325.
  Silurian: United States

*Pisocrinus (*Pisocrinus*) *pilula* de Koninck, 1858. Mu, 1954, p. 325, Pl. 1, figs. 6, 7.
  Pisocrinus pilula de Koninck, 1858. Bouška, 1956a, p. 13, Pl. 1, figs. 1-15; Fig. 3. Strimple, 1963d, fig. 10*l*.
  Silurian, Ludlovian: Bohemia

Pisocrinus (*Pisocrinus*) *pilula* var. *flagellifer* Angelin, 1878. Mu, 1954, p. 325, Pl. 1, fig. 10.
  Silurian: Sweden

Pisocrinus (Pisocrinus) pilula var. *ornatus* de Koninck, 1858. Mu, 1954, p. 325.
  Silurian: Sweden

Pisocrinus (Pisocrinus) pilula var. *yini* Mu, 1954, p. 326, Pl. 1, figs. 1-5; Fig. 3.
  Middle Silurian, Lojopingian Series: China: North Szechuan

Pisocrinus (Pisocrinus) pocillum Angelin, 1878. Mu, 1954, p. 325.
  Silurian: Sweden

Pisocrinus pocillum Angelin, 1878 = *Pisocrinus (Pisocrinus) pocillum*

Pisocrinus quinquelobus Bather, 1893 = *Pisocrinus (Parapisocrinus) quinquelobus*

Pisocrinus spatulatus Strimple, 1954b, p. 281, figs. 5-8; 1963d, p. 41.
  Silurian, Niagaran, Henryhouse Formation: United States: Oklahoma

Pisocrinus sphericus Rowley, 1904 = *Pisocrinus (Parapisocrinus) sphericus*

Pisocrinus tenuis Bouška, 1956a, p. 24, Pl. 2, fig. 15.
  Silurian, Ludlovian, Kopanina Limestone: Bohemia

Pisocrinus ubaghsi Bouška, 1956a, p. 27, Pl. 3, figs. 7-14. Strimple, 1963d, fig. 10i. Yeltyschewa, 1968, p. 31, Pl. 4, figs. 1-8.
  Silurian, Ludlovian, Pridoli Beds, Borschovskian horizon: Bohemia; Russia: Podolsky

Pisocrinus varus Strimple, 1963d, p. 42, Pl. 1, fig. 11; Figs. 11c, e-l.
  Silurian, Niagaran, Henryhouse Formation: United States: Oklahoma

Pisocrinus yakovlevi Bouška, 1956a, p. 22, Pl. 2, fig. 13; Fig. 5.
  Lower Devonian, Prokop Limestone: Bohemia

Pisocrinus? yassensis Etheridge, 1904 = *Pisocrinus (Parapisocrinus) yassensis*

PITHOCRINUS Kirk, 1945. *P. cooperi* Kirk, 1945. Kirk, 1945a, p. 341. Ubaghs, 1953, p. 739.
  Lower-Middle Devonian

Pithocrinus bifrons (Schmidt), 1931, Kirk, 1945 = *Pyxidocrinus? bifrons*

Pithocrinus cooperi Kirk, 1945a, p. 343, Pl. 1, figs. 4-7. Breimer, 1962a, p. 46.
  Devonian, Erian, Alpena Limestone: United States: Michigan

Pithocrinus intrastigmatus (Schmidt), 1931, Kirk, 1945 = *Stamnocrinus intrastigmatus*

Pithocrinus ovatus Breimer, 1962a, p. 47, Pl. 4, figs. 1, 2; Pl. 5, figs. 5-11; Figs. 8-10 (given as *Pithocrinus* sp. in plate explanation, Pl. 4, fig. 1).
  Devonian, upper Emsian, Santa Lucia Formation: Spain

Pithocrinus spinosus Breimer, 1962a, p. 54, Pl. 5, figs. 1-4.
  Devonian, upper Emsian, Santa Lucia Formation: Spain

Pithocrinus waliszewskii (Oehlert), 1896. Kirk, 1945a, p. 344. Breimer, 1962a, p. 57.
  Megistocrinus waliszewskii Oehlert, 1896. Kirk, 1945a, p. 34.
  Devonian, lower Couvinian, Santa Lucia Formation: Spain

PLAGIOCRINUS Wanner, 1924. *P. torynocrinoides* Wanner, 1924. Moore and Laudon, 1943b, p. 76. Ubaghs, 1953, p. 756. Arendt and Hecker, 1964, p. 94.
  Permian

Plagiocrinus jaekeli Wanner, 1924. Ubaghs, 1952, p. 213, figs. 3D, E.
Permian: Timor

PLATYCRINITES Miller, 1821. *P. laevis* Miller, 1821. Moore and Laudon, 1943b, p. 101, fig. 18; 1944, p. 203, Pl. 71, fig. 18. Moore, 1950, p. 42, fig. 11. Ubaghs, 1953, p. 743, figs. 99, 139h. Xu, 1963, p. 108, fig. 1b. Arendt and Hecker, 1964, p. 98.
Platycrinus Wanner, 1942b, p. 209, fig. 5a. Cuénot, 1948, p. 68.
Devonian-Permian

Platycrinites allophylus (Miller), 1891. Bassler and Moodey, 1943, p. 614 (misspelled as *P. allophyllus*). Moore and Laudon, 1944, p. 203, Pl. 77, fig. 16.
Mississippian, Kinderhookian, Chouteau Group: United States: Missouri

Platycrinites americanus (Owen and Shumard), 1852. Moore and Laudon, 1944, p. 203, Pl. 77, fig. 17.
Mississippian, Osagean, Burlington Limestone: United States: Iowa, Missouri

Platycrinites asiatica (Minato), 1951 nom. correct.
Platycrinus asiatica Minato, 1951, p. 357, Pl. 1, fig. 8; Pl. 4, figs. 2a, b (*non* Pl. 5, figs. 2a, b, as given on p. 357).
Lower Carboniferous, Jumonji Stage: Japan

Platycrinites balladoolensis (Wright), 1938. Moore, 1948, p. 390, fig. 11 (misspelled as *P. balladooensis*). Wright, 1956b, p. 280, Pl. 72, figs. 23, 27.
Carboniferous, Visean (D_2): England

Platycrinites basicraniatus Laudon and Severson, 1953, p. 533, Pl. 54, fig. 1; Pl. 55, fig. 24.
Mississippian, Kinderhookian, Lodgepole Limestone: United States: Montana

Platycrinites bellmanensis (Wright), 1942. Wright, 1948, p. 51, Pl. 5, figs. 7-10; 1956b, p. 279, Pl. 71, figs. 9-12, 14-16, 18, 19, 21, 22.
Platycrinus bellmanensis Wright, 1942, p. 270, Pl. 11, figs. 9-16. Wright, 1948, p. 51.
Carboniferous, Visean (C_2): England

Platycrinites bollandensis (Wright), 1938. Wright, 1946b, p. 126, Pl. 7, fig. 1; 1955b, p. 272, Pl. 66, fig. 1; Pl. 71, figs. 24, 25.
Carboniferous, Tournaisian, Mountain Limestone: England; Scotland

Platycrinites bozemanensis (Miller and Gurley), 1897. Laudon and Severson, 1953, p. 532, Pl. 54, figs. 2-7. Laudon, 1967, p. 1492, Pls. 193, 194; Figs. 1-6.
Mississippian, Kinderhookian, Lodgepole Limestone: United States: Montana

Platycrinites burlingtonensis (Owen and Shumard), 1850. Moore and Laudon, 1944, p. 203, Pl. 78, fig. 24.
Platycrinus burlingtonensis (Owen and Shumard), 1850. Breimer, 1962a, p. 139, fig. 32.
Mississippian, Osagean, Burlington Limestone: United States: Iowa, Missouri

Platycrinites canadensis Laudon and Severson, 1953, p. 532, Pl. 54, fig. 8; Pl. 55, fig. 25.
Mississippian, Kinderhookian, Lodgepole Limestone: United States: Montana

Platycrinites conglobatus (Wright), 1937. Wright, 1956b, p. 277, Pl. 66, fig. 8; Pl. 71, figs. 8, 13, 28, 33.
Carboniferous, Visean, Lower Limestone Group: Scotland

Platycrinites contractus (Phillips), 1836. Wright, 1956b, p. 294, Pl. 80, figs. 14, 15.
Carboniferous, Tournaisian: England

Platycrinites crassiconus (Wright), 1937. Wright, 1956b, p. 278, Pl. 71, figs. 20, 23, 35; Pl. 74, figs. 1-5.
Carboniferous, Visean, Lower Limestone Group: Scotland

Platycrinites decadactylus Laudon, Parks and Spreng, 1952, p. 573, Pl. 68, figs. 17, 18; Pl. 69, fig. 25.
Mississippian, Kinderhookian, Banff formation: Canada: Alberta

Platycrinites diadema (McCoy), 1849. Wright, 1956b, p. 273, Pl. 70, figs. 1-4, 8, 9; Pl. 71, fig. 17; Pl. 72, figs. 24-26, 28-32.
Breimer, 1962a, p. 140, fig. 32 (typographical error? on p. 140, as *Pleurocrinus diadema*).
Carboniferous, Tournaisian?, Visean (C_2): Ireland; Scotland; England

Platycrinites directus (Wright), 1938. Wright, 1956b, p. 280, Pl. 72, figs. 21, 22.
Carboniferous, Visean (D_2): England

Platycrinites discoideus (Owen and Shumard), 1850. Moore and Laudon, 1943b, p. 143, Pl. 14, fig. 11; 1944, p. 203, Pl. 77, fig. 26; Pl. 78, fig. 22.
Mississippian, Osagean, Burlington Limestone: United States: Iowa, Missouri

Platycrinites dodgei (Rowley), 1908 nom. correct.
Platycrinus dodgei Rowley, 1908. Williams, 1943, p. 64, Pl. 6, figs. 16-18.
Mississippian, Kinderhookian, Louisiana Limestone: United States: Louisiana, Missouri

Platycrinites expansus (McCoy), 1844. Wright, 1956b, p. 284, Pl. 69, figs. 17-20; Pl. 70, figs. 7, 10, 11.
Carboniferous, Tournaisian?: Ireland

Platycrinites cf. *expansus* (McCoy), 1844 nom. correct.
Platycrinus cf. *expansus* McCoy, 1844. Termier and Termier, 1950, p. 86, Pl. 212, figs. 1-6.
Carboniferous, Westphalian: Morocco

Platycrinites externus Wright, 1956b, p. 282, Pl. 71, figs. 1-5.
Carboniferous, Tournaisian: Scotland

Platycrinites fermanaghensis Wright, 1956b, p. 286, Pl. 69, figs. 13-16; Pl. 70, figs. 12-15.
Carboniferous, Tournaisian?: Ireland

Platycrinites gigas (Gilbertson, *in* Phillips), 1836. Wright, 1946b, p. 126, Pl. 8, figs. 2, 3, 5-7; 1948, p. 51, Pl. 5, fig. 4. Moore, 1948, p. 390, fig. 11. Wright, 1955b, p. 270, Pl. 64, figs. 4, 5; Pl. 66, fig. 5; Pl. 71, figs. 6, 7, 26, 27, 30-32, 34.
Carboniferous, Tournaisian, Mountain Limestone: Scotland

?*Platycrinites gigas* (Gilbertson, *in* Phillips), 1836. Wright, 1946b, p. 126, Pl. 8, fig. 4 (one ray of arms).
Carboniferous, Tournaisian, Mountain Limestone: Scotland

Platycrinites granulatus Miller, 1821. Wright, 1956b, p. 287, Pl. 67, figs. 8, 11.
Carboniferous, Tournaisian: England

Platycrinites hemisphaericus (Meek and Worthen), 1865. Moore and Laudon, 1943b, p. 143, Pl. 14, fig. 9 (misspelled as *P. hemisphericus*); 1944, p. 203, Pl. 78, fig. 21 (misspelled as *P. hemisphericus*). Van Sant, *in* Van Sant and Lane, 1964, p. 125, Pl. 8, figs. 9, 17; Figs. 17,

no. 4; 19, nos. 1, 2 (misspelled as *P. hemisphericus*).
Mississippian, Osagean, Keokuk Limestone: United States: Indiana, Iowa

Platycrinites humilis (Wright), 1938. Wright, 1956b, p. 279, Pl. 72,
figs. 9-12.
Carboniferous, Visean (C_2): England

Platycrinites huntsvillae Troost, 1849. Ubaghs, 1953, p. 743, fig. 55.
Platycrinus penicillus Meek and Worthen, 1860. Butts, 1941, p. 251,
Pl. 132, figs. 45-53; 1948, p. 46, Pl. 6, figs. 11-18.
Mississippian, Meramecian, Ste. Genevieve Limestone: United States:
Virginia, Georgia

Platycrinites incomptus (White), 1863. Laudon and Severson, 1953, p. 532,
Pl. 54, fig. 9.
Mississippian, Kinderhookian, Lodgepole Limestone: United States:
Montana

Platycrinites insulsus Wright, 1956b, p. 282, Pl. 72, figs. 13-16.
Carboniferous, Tournaisian: Scotland

Platycrinites? invertielensis (Wright), 1942. Wright, 1956b, p. 293, Pl. 74,
figs. 14, 15.
Platycrinus invertielensis Wright, 1942, p. 270, Pl. 9, figs. 4, 5.
Wright, 1956b, p. 293.
Carboniferous, Visean, Lower Limestone Group: England

Platycrinites? laciniatus (Phillips), 1836. Wright, 1956b, p. 293, Pl. 80,
figs. 13, 19.
Carboniferous, Tournaisian: Ireland

Platycrinites laevis Miller, 1821. Wright, 1955b, p. 266, Pl. 65, fig. 5;
Pl. 66, figs. 6, 7; Pl. 67, fig. 1; Pl. 68, figs. 1, 3, 5; Pl. 69,
fig. 28.
Carboniferous, Tournaisian: Ireland; England

Platycrinites megastylus (McCoy), 1849. Wright, 1956b, p. 281, Pl. 69,
figs. 1-4.
Carboniferous, Tournaisian?: England

Platycrinites muirkirkensis Wright, 1956b, p. 291, Pl. 74, figs. 23-25.
Carboniferous, Namurian (E_2), Upper Limestone Group: Scotland; England

Platycrinites ornigranulus (McChesney), 1860. Meyer, 1965, fig. II.
Mississippian, Osagean, Burlington Limestone: United States: Iowa

Platycrinites patulus Wright, 1956b, p. 283, Pl. 74, figs. 20-22.
Platycrinus gigas Austin and Austin, 1843 (part, p. 38, Pl. 4, figs. 1a-c).
Wright, 1956b, p. 283.
Carboniferous, Tournaisian?: England

Platycrinites pileatus (Goldfuss), 1838. Wright, 1956b, p. 275, Pl. 69,
figs. 21-24; Pl. 74, figs. 8, 9; Fig. 127.
Carboniferous, Tournaisian-Visean: Scotland; England

Platycrinites pyramidalis Wright, 1956b, p. 285, Pl. 70, figs. 5, 6.
Carboniferous, Tournaisian?: Ireland

Platycrinites quinquenodus (White), 1862. Moore and Laudon, 1944, p. 203,
Pl. 78, fig. 19.
Mississippian, Osagean, Burlington Limestone: United States: Iowa,
Missouri

Platycrinites saffordi (Hall), 1858. Moore and Laudon, 1943b, p. 143,
Pl. 14, fig. 10; 1944, p. 205, Pl. 78, fig. 27.
Mississippian, Osagean, Keokuk Limestone, upper Burlington Limestone:
United States: Indiana, Illinois, Iowa, Missouri

Platycrinites sculptus (Hall), 1858. Moore and Laudon, 1944, p. 203, Pl. 78, fig. 25.
 Mississippian, Osagean, Lake Valley Limestone: United States: New Mexico

Platycrinites selwoodensis Wright, 1955b, explanation of Pl. 64, fig. 9 and footnote; 1956b, p. 289, Pl. 64, fig. 9.
 Carboniferous, Tournaisian: England

Platycrinites smythi Wright, 1955b, explanation of Pl. 66, fig. 3 and footnote; 1956b, p. 291, Pl. 66, fig. 3.
 Carboniferous, Tournaisian: Ireland

Platycrinites spinifer elongatus (Wachsmuth and Springer), 1897. Moore and Laudon, 1944, p. 203, Pl. 78, fig. 23. Termier and Termier, 1950, p. 86, Pl. 211, figs. 1-6; Pl. 226, figs. 36-38 (typographical error, given as figs. 35-37 on p. 86).
 Mississippian, Osagean, upper Visean, Burlington Limestone: United States: Iowa; Morocco

Platycrinites? spiniger (Wright), 1937. Wright, 1956b, p. 292, Pl. 71, fig. 29.
 Carboniferous, Visean, Lower Limestone Group: Scotland

Platycrinites spinosus Austin and Austin, 1842. Wright, 1955b, p. 269, Pl. 65, fig. 6; Pl. 67, figs. 2, 3, 10; Pl. 68, fig. 6; Pl. 69, fig. 29; Pl. 75, fig. 13. Termier and Termier, 1950, p. 86, Pl. 226, figs. 42-44 (*non* fig. 41 as given on p. 86).
 Carboniferous, Tournaisian, lower Visean: Ireland; England; Morocco

Platycrinites striatus Miller, 1821. Wright, 1956b, p. 289, Pl. 66, fig. 2.
 Carboniferous, Tournaisian: Ireland; England

Platycrinites subspinulosus (Hall), 1860. Moore and Laudon, 1944, p. 203, Pl. 78, fig. 20.
 Mississippian, Osagean, Burlington Limestone: United States: Iowa

Platycrinites sunwaptaensis Laudon, Parks and Spreng, 1952, p. 573, Pl. 68, figs. 12-14; Pl. 69, figs. 23, 24.
 Mississippian, Kinderhookian, Banff formation: Canada: Alberta

Platycrinites symmetricus (Wachsmuth and Springer), 1888. Moore and Laudon, 1944, p. 203, Pl. 78, fig. 26. Meyer, 1965, fig. 1B. Arendt and Hecker, 1964, p. 98, fig. 136.
 Platycrinus symmetricus Wachsmuth and Springer, 1888. Cuénot, 1948, p. 68, fig. 90. Breimer, 1962a, p. 139, fig. 32.
 Mississippian, Kinderhookian, Hampton Formation: United States: Iowa

Platycrinites trigintidactylus Austin and Austin, 1842. Wright, 1955b, p. 267, Pl. 66, fig. 4; Pl. 67, figs. 4-6, 9; Pl. 68, figs. 2, 4, 7; Pl. 69, figs. 25-27; Pl. 75, fig. 2.
 Carboniferous, Tournaisian: Ireland; England

Platycrinites verrucosus (White), 1865. Moore and Laudon, 1944, p. 203, Pl. 78, fig. 18.
 Mississippian, Osagean, Burlington Limestone: United States: Iowa, Missouri

Platycrinites? vesiculosus (McCoy), 1849. Wright, 1956b, p. 294.
 Carboniferous, Visean?: England

Platycrinites westheadi (Wright), 1942. Wright, 1955b, p. 272, Pl. 72, figs. 17-20.
 Platycrinus westheadi Wright, 1942, p. 269, Pl. 10, figs. 5-8; 1955b, p. 272.
 Carboniferous, Tournaisian (C_1): Scotland

Platycrinites yandelli (Owen and Shumard), 1850.
   Platycrinus yandelli (Owen and Shumard), 1850. Breimer, 1962a, p. 139,
      fig. 32.
   Mississippian, Osagean, lower Burlington Limestone: United States: Iowa
Platycrinites spp.
   Laudon and others, 1952, p. 574, Pl. 68, fig. 19.
      Mississippian, Kinderhookian, Banff formation: Canada: Alberta
   Laudon and others, 1952, p. 574, Pl. 68, fig. 15.
      Mississippian, Kinderhookian, Banff formation: Canada: Alberta
   Laudon and others, 1952, p. 574, Pl. 68, figs. 16, 20.
      Mississippian, Kinderhookian, Banff formation: Canada: Alberta
   Wright, 1956b, p. 286, Pl. 67, fig. 7; Pl. 69, figs. 5-8; Pl. 75, fig. 1.
      Carboniferous: England, locality unknown
   Wright, 1960, p. 334, Appendix, Pl. A, fig. 2.
      Carboniferous, Tournaisian (Z): England
   Meyer, 1965, p. 1207, fig. 1A.
      Carboniferous, Tournaisian, Carboniferous Limestone: Belgium
   Meyer, 1965, p. 1207, fig. 1C.
      Mississippian, Osagean, Burlington Limestone: United States: Iowa?
   Meyer, 1965, p. 1207, fig. 1D.
      Carboniferous, Tournaisian, Carboniferous Limestone: Belgium
   Meyer, 1965, p. 1207, fig. 1E.
      Mississippian, Osagean, Burlington Limestone: United States: Illinois
   Webster and Lane, 1967, p. 7, Pl. 1, figs. 1-10.
      Permian, Wolfcampian, Bird Spring Formation: United States: Nevada
Platycrinus spp.
   Termier and Termier, 1947, p. 255, Pl. 19, fig. 49.
      Carboniferous: Morocco
   Termier and Termier, 1950, p. 87, Pl. 211, figs. 42, 43; Pl. 15,
      figs. 10, 11.
      Carboniferous, upper Visean: Morocco
   Termier and Termier, 1950, p. 236, Pl. 226, figs. 39-41.
      Carboniferous, upper Visean: Morocco
Platycrinites sp. (ex gr. *bollandensis* Wright, 1938).
   Platycrinus sp. (ex gr. *bollandensis* Wright, 1938). Breimer, 1962a,
      p. 131, Pl. 13, figs. 6-9, 12-14.
   Carboniferous, middle or upper Namurian: Spain
Platycrinites sp. undet.
   Platycrinus sp. undet. Branson, 1944, p. 165, Pl. 30, figs. 11, 13.
   Mississippian, Kinderhookian, Northview Formation: United States:
      Missouri
Platycrinites? sp. indet. Schmidt, 1941, p. 45, fig. 7.
   Devonian, lower Eifelian: Germany

Platycrinus auct. = *Platycrinites*

Platycrinus asiatica Minato, 1951 = *Platycrinites asiatica*

Platycrinus bellmanensis Wright, 1942 = *Platycrinites bellmanensis*

Platycrinus cavus Hall, 1858 = *Plemnocrinus cavus*

Platycrinus diadema McCoy, 1849 = *Pleurocrinus diadema*

Platycrinus gigas Austin and Austin, 1843 = *Platycrinites patulus*

Platycrinus gorbyi Miller, 1891 = *Plemnocrinus gorbyi*

Platycrinus inurbanus Breimer, 1962 = *Pleurocrinus inurbanus*

Platycrinus invertielensis Wright, 1942 = *Platycrinites? invertielensis*

Platycrinus occidentalis Miller, 1891 = *Plemnocrinus occidentalis*

Platycrinus penicillus Meek and Worthen, 1860 = *Platycrinites huntsvillae*

Platycrinus subspinosus Hall, 1858 = *Plemnocrinus subspinosus*

Platycrinus tuberculatus Miller, 1821 = *Pleurocrinus tuberculatus*

Platycrinus tuberosus Hall, 1858 = *Plemnocrinus tuberosus*

Platycrinus westheadi Wright, 1942 = *Platycrinites westheadi*

PLATYHEXACRINUS Schmidt, 1913. *P. inornatus* Schmidt, 1913. Ubaghs, 1953, p. 740. Arendt and Hecker, 1964, p. 96.
Lower-Middle Devonian

Platyhexacrinus grandis Schmidt, 1941, p. 48, fig. 8A.
Middle Devonian, *Cultrijugatus* Zone: Germany

Platyhexacrinus gurievskiensis Dubatolova, 1964, p. 43, Pl. 5, fig. 6.
Lower Devonian: Russia: Kuznetz Basin

**Platyhexacrinus inornatus* Schmidt, 1913. Schmidt, 1941, p. 47, fig. 8B.
Devonian, lower Eifelian: Germany

Platyhexacrinus kegeli Schmidt, 1931. Schmidt, 1952, p. 159, Pl. 9, figs. 2, 3. Breimer, 1962a, p. 84.
Lower Devonian: Spain

Platyhexacrinus? pisum Schmidt, 1941, p. 49, Pl. 16, fig. 8.
Lower Devonian: Germany

PLAXOCRINUS Moore and Plummer, 1938. *Hydreionocrinus crassidiscus* Miller and Gurley, 1894. Moore and Laudon, 1943b, p. 58; 1944, p. 163. Yakovlev, 1947b, p. 749, fig. 2; 1947d, p. 45, fig. 12; 1956a, p. 82, fig. 32; 1958, p. 67, fig. 32; 1964a, p. 76, fig. 34.
Pennsylvanian, Morrowan-Permian, Wolfcampian

Plaxocrinus aplatus Moore and Plummer, 1940. Strimple, 1962e, p. 32, Pl. 1, figs. 5-8; Pl. 2, figs. 13-16.
Pennsylvanian, Desmoinesian, Oologah Limestone: United States: Oklahoma

Plaxocrinus beggsi Strimple, 1961g, p. 52, Pl. 12, figs. 1-3; Figs. 13g, h.
Pennsylvanian, Desmoinesian, Holdenville Shale: United States: Oklahoma

Plaxocrinus? brasilensis Lane, 1964a, p. 364, Pl. 57, figs. 11, 12.
Middle Pennsylvanian, Itaituba Series: Brazil: Pará

**Plaxocrinus crassidiscus* (Miller and Gurley), 1894. Moore and Laudon, 1944, p. 163, Pl. 62, fig. 7; Pl. 63, fig. 7.
Pennsylvanian, Missourian: United States: Missouri

Plaxocrinus discus (Meek and Worthen), 1860. Moore and Laudon, 1944, p. 163, Pl. 63, fig. 6.
Pennsylvanian, Missourian: United States: Illinois

Plaxocrinus dornickensis Strimple, 1949d, p. 15, Pl. 2, figs. 7, 8, 10; Fig. 2.
Pennsylvanian, Atokan, Pumpkin Creek Limestone: United States: Oklahoma

Plaxocrinus aff. *P. dornickensis* Strimple, 1949. Strimple, 1961g, p. 51, Pl. 14, figs. 5, 6.
Pennsylvanian, Desmoinesian, Holdenville Shale: United States: Oklahoma

Plaxocrinus gloukosensis Strimple, 1951b, p. 374, Pl. 57, figs. 1-6.
Pennsylvanian, Virgilian, Haskell Limestone: United States: Kansas

Plaxocrinus kansasensis (Weller), 1898. Strimple, 1961g, p. 46.
Hydreionocrinus kansasensis Weller, 1898.
Upper Pennsylvanian: United States: Kansas

Plaxocrinus laxus Strimple, 1951a, p. 205, Pl. 38, figs. 1, 2; Pl. 39, figs. 1, 2.
Pennsylvanian, Missourian, Lake Bridgeport Shale: United States: Texas

Plaxocrinus lobatus Moore and Plummer, 1940. Plummer, 1950, p. 97, Pl. 21, fig. 17. Heuer, 1958, p. 47, Pl. 5, fig. 17.
  Pennsylvanian, Missourian: United States: Texas

Plaxocrinus mooresi (Whitfield), 1882. Burke, 1967b, p. 298, 3 figs.
  Eupachycrinus mooresi Weller, 1898. La Rocque and Marple, 1955, p. 119, fig. 348.
  Pennsylvanian, Atokan-Desmoinesian, McArthur Shale: United States: Ohio

Plaxocrinus normalis Strimple, 1961g, p. 48, Pl. 11, figs. 5-7; Pl. 15, fig. 5; Pl. 18, figs. 3, 4; Pl. 19, figs. 1, 3; Fig. 13e.
  Pennsylvanian, Desmoinesian, Holdenville Shale: United States: Oklahoma

Plaxocrinus obesus Moore and Plummer, 1940. Moore and Laudon, 1944, p. 163, Pl. 63, fig. 8.
  Pennsylvanian, Desmoinesian, Millsap Lake Formation: United States: Texas

Plaxocrinus aff. *P. obesus* Moore and Plummer, 1940. Strimple, 1961g, p. 47, Pl. 16, fig. 3; Pl. 19, fig. 4; Fig. 13d.
  Pennsylvanian, Desmoinesian, Holdenville Shale: United States: Oklahoma

Plaxocrinus? obesus Moore and Plummer, 1940. Strimple, 1962e, p. 32, Pl. 7, fig. 19.
  Pennsylvanian, Desmoinesian, Oologah Limestone: United States: Oklahoma

Plaxocrinus octarius Strimple, 1961g, p. 54, Pl. 4, figs. 1, 2; Fig. 8.
  Pennsylvanian, Desmoinesian, Holdenville Shale: United States: Oklahoma

Plaxocrinus oeconomicus Strimple, 1951a, p. 206, Pl. 38, figs. 7, 8.
  Pennsylvanian, Missourian, Lake Bridgeport Shale: United States: Texas

Plaxocrinus omphaloides Moore and Plummer, 1940. Plummer, 1950, p. 97, Pl. 21, fig. 15. Heuer, 1958, p. 47, Pl. 5, fig. 15.
  Pennsylvanian, Missourian: United States: Texas

Plaxocrinus parillis Moore and Plummer, 1940. Moore and Laudon, 1944, p. 163, Pl. 63, fig. 5.
  Pennsylvanian, Missourian, Graford Formation: United States: Texas

Plaxocrinus perundatus Moore and Plummer, 1940. Moore and Laudon, 1944, p. 163, Pl. 63, fig. 9.
  Pennsylvanian, Desmoinesian, Millsap Lake Formation: United States: Texas

Plaxocrinus piutae Lane and Webster, 1966, p. 49, Pl. 8, figs. 10-13. Webster and Lane, 1967, p. 29.
  Permian, Wolfcampian, Bird Spring Formation: United States: Nevada

Plaxocrinus politus Moore, 1939 = *Stenopecrinus politus*

Plaxocrinus praevalens Moore, 1939. Moore and Laudon, 1944, p. 163, Pl. 62, fig. 5; Pl. 63, fig. 16.
  Pennsylvanian, Virgilian, Brownville formation: United States: Oklahoma

Plaxocrinus puteus Strimple, 1949d, p. 17, Pl. 3, figs. 1, 2, 5, 7.
  Pennsylvanian, Morrowan, Otterville Limestone: United States: Oklahoma

Plaxocrinus tumulosus Strimple, 1949d, p. 16, Pl. 1, figs. 12, 14, 16, 18 (*non* fig. 19 as given on p. 16).
  Pennsylvanian, Atokan, Pumpkin Creek Limestone: United States: Oklahoma

Plaxocrinus uddeni (Weller), 1909 = *Neozeacrinus uddeni*

Plaxocrinus virginarius Moore, 1939. Moore and Laudon, 1944, p. 163, Pl. 63, fig. 2.
  Pennsylvanian, Virgilian, Brownville formation: United States: Texas

Plaxocrinus sp. Webster and Lane, 1967, p. 29, Pl. 8, fig. 7.
  Permian, Wolfcampian, Bird Spring Formation: United States: Nevada

PLEMNOCRINUS Kirk, 1946. **P. beebei* Kirk, 1946. Kirk, 1946d, p. 435.
Ubaghs, 1953, p. 743. Arendt and Hecker, 1964, p. 98.
Mississippian, Osagean

*Plemnocrinus beebei Kirk, 1946d, p. 436, Pl. 65, figs. 2-5; Pl. 66, figs. 7-9.
Mississippian, Osagean, Burlington Limestone: United States: Iowa

Plemnocrinus bullatus Kirk, 1946d, p. 437, Pl. 65, fig. 1.
Mississippian, Osagean, Burlington Limestone: United States: Iowa

Plemnocrinus cavus (Hall), 1858. Kirk, 1946d, p. 438.
Platycrinus cavus Hall, 1858. Kirk, 1946d, p. 438.
Mississippian, Osagean, Burlington Limestone: United States: Iowa

Plemnocrinus gorbyi (Miller), 1891. Kirk, 1946d, p. 439.
Platycrinus gorbyi Miller, 1891. Kirk, 1946d, p. 439.
Mississippian, Osagean, Burlington Limestone: United States: Iowa

Plemnocrinus homalus Kirk, 1946d, p. 439, Pl. 66, figs. 1-6.
Mississippian, Osagean, Burlington Limestone: United States: Iowa, Missouri

Plemnocrinus occidentalis (Miller), 1891. Kirk, 1946d, p. 440.
Platycrinus occidentalis Miller, 1891. Kirk, 1946d, p. 440.
Mississippian, Osagean, Burlington Limestone: United States: Missouri

Plemnocrinus subspinosus (Hall), 1858. Kirk, 1946d, p. 440.
Platycrinus subspinosus Hall, 1858. Kirk, 1946d, p. 440.
Mississippian, Osagean, Burlington Limestone: United States: Iowa

Plemnocrinus tuberosus(Hall), 1858. Kirk, 1946d, p. 440.
Platycrinus tuberosus Hall, 1858. Kirk, 1946d, p. 440.
Mississippian, Osagean, Burlington Limestone: United States: Iowa

PLESIOCRINUS Wanner, 1937. **P. piriformis* Wanner, 1937. Moore and Laudon, 1943b, p. 101. Ubaghs, 1953, p. 743. Arendt and Hecker, 1964, p. 98.
Permian

PLEUROCRINUS Austin and Austin, 1843. **Platycrinites coronatus* Goldfuss, 1839. Moore and Laudon, 1943b, p. 101. Ubaghs, 1953, p. 743. Arendt and Hecker, 1964, p. 98.
Lower Carboniferous: Permian

Pleurocrinus concavus Wright, 1956b, p. 299, Pl. 80, figs. 2-5.
Carboniferous, Tournaisian: Scotland

Pleurocrinus coplowensis Wright, 1938. Wright, 1956b, p. 300, Pl. 69, figs. 9-12; Pl. 73, figs. 8-11.
Carboniferous, Tournaisian (C_1): Scotland

*Pleurocrinus coronatus (Goldfuss), 1839. Wright, 1956b, p. 296, Pl. 72, figs. 5-8; Fig. 128.
Carboniferous, Tournaisian?-Visean: England; Scotland

Pleurocrinus ellipticus (Phillips), 1836. Wright, 1956b, p. 302, Pl. 73, figs. 17-20; Pl. 80, figs. 9-12.
Carboniferous, Tournaisian: England; Scotland

Pleurocrinus grandis Wright, 1938. Wright, 1956b, p. 297, Pl. 73, figs. 1-7.
Carboniferous, Visean (D_2): England

Pleurocrinus inurbanus Wright, 1938. Moore, 1948, p. 390, fig. 11. Wright, 1956b, p. 297, Pl. 73, figs. 13-16. Breimer, 1962a, p. 140, fig. 32 (mislabeled as *Platycrinus inurbanus*).
Carboniferous, Tournaisian (C_1): Scotland

Pleurocrinus mucronatus (Austin and Austin), 1842. Wright, 1956b, p. 295, Pl. 74, figs. 10-12.
Carboniferous, Tournaisian?-Visean: England

Pleurocrinus rugosus (Miller), 1821. Wright, 1942, p. 271, Pl. 11, figs. 1-8; 1956b, p. 298, Pl. 73, figs. 21-28.
  Carboniferous, Tournaisian?, Mountain Limestone: Ireland; England; Scotland

Pleurocrinus spinosus Breimer, 1962 = *Culicocrinus spinosus*

Pleurocrinus tuberculatus (Miller), 1821. Wright, 1956b, p. 301, Pl. 72, figs. 1-4; Pl. 74, figs. 13, 16-19.
  Platycrinus tuberculatus Miller, 1821. Termier and Termier, 1950, p. 86, Pl. 212, figs. 10, 11.
  Carboniferous, Tournaisian?-Visean: England; Scotland

Pleurocrinus wanneri Wright, 1938. Wright, 1956b, p. 301, Pl. 73, figs. 12, 29, 30.
  Carboniferous, Visean (C_2): England

Pleurocrinus sp. (ex gr. *coplowensis* Wright, 1938). Breimer, 1962a, p. 133, Pl. 14, figs. 1-4.
  Devonian or Carboniferous: Spain, locality unknown

PLUMMERICRINUS Moore and Laudon, 1943. *Pachylocrinus mcguirei* Moore, 1939. Moore and Laudon, 1943b, p. 56; 1944, p. 161. Arendt and Hecker, 1964, p. 88.
  Pennsylvanian, Missourian-Permian, Wolfcampian

Plummericrinus bellirugosus (Moore), 1939. Moore and Laudon, 1943b, p. 58.
  Pachylocrinus bellirugosus Moore, 1939. Moore and Laudon, 1943b, p. 58.
  Pennsylvanian, Virgilian, Wabaunsee Group: United States: Kansas

Plummericrinus braggsi Strimple, 1951 = *Anobasicrinus braggsi*

Plummericrinus colubrosus (Moore), 1939. Moore and Laudon, 1943b, p. 58.
  Pachylocrinus colubrosus Moore, 1939. Moore and Laudon, 1943b, p. 58.
  Permian, Wolfcampian, Ft. Riley Limestone?: United States: Kansas

Plummericrinus emilyae Burke, 1968, p. 10, fig. 3.
  Pennsylvanian, Missourian, Ames Limestone: United States: Ohio

Plummericrinus erectus Strimple, 1954 = *Anobasicrinus perplexus*

Plummericrinus granulosus Strimple, 1954 = *Anobasicrinus granulosus*

Plummericrinus lutanus (Boos), 1939. Moore and Laudon, 1943b, p. 58.
  Pachylocrinus lutanus Boos, 1939. Moore and Laudon, 1943b, p. 58.
  Permian, Wolfcampian, Winfield Limestone: United States: Kansas

Plummericrinus mcguirei (Moore), 1939. Moore and Laudon, 1943b, p. 58, Pl. 5, fig. 1; 1944, p. 161, Pl. 62, fig. 29; Pl. 63, fig. 20.
  Pachylocrinus mcguirei Moore, 1939. Moore and Laudon, 1943b, p. 58.
  Pennsylvanian, Virgilian, Brownville Limestone: United States: Oklahoma

Plummericrinus monongaliensis Burke, 1968, p. 2, fig. 1.
  Pennsylvanian, Missourian, Brush Creek Limestone?: United States: West Virginia

Plummericrinus nettingi Burke, 1968, p. 7, fig. 2.
  Pennsylvanian, Missourian, Cambridge Limestone: United States: Pennsylvania

Plummericrinus ogarai (Moore and Plummer), 1940. Moore and Laudon, 1943b, p. 58. Moore and Laudon, 1944, p. 161, Pl. 62, fig. 35; Pl. 63, fig. 19.
  Pachylocrinus ogarai Moore and Plummer, 1940. Moore and Laudon, 1943b, p. 58.
  Pennsylvanian, Missourian, Graford Formation: United States: Texas

Plummericrinus pittsburghensis Burke, 1968, p. 15, fig. 4.
  Pennsylvanian, Missourian, Ames Limestone: United States: Pennsylvania

Plummericrinus striatus Strimple, 1954a, p. 204, Pl. 23, figs. 1, 2, 8.
  Pennsylvanian, Missourian, Francis Formation: United States: Oklahoma

Plummericrinus twenhofeli (Moore), 1939. Moore and Laudon, 1943b, p. 58.
  Pachylocrinus twenhofeli Moore, 1939. Moore and Laudon, 1943b, p. 58.
  Permian, Wolfcampian, Ft. Riley Limestone: United States: Kansas

Plummericrinus sp. Strimple, 1954a, p. 207, Pl. 23, figs. 3-5.
  Pennsylvanian, Missourian, Wann Formation: United States: Oklahoma

POLUSOCRINUS Strimple, 1951. *P. avanti* Strimple, 1951. Strimple, 1951c,
  p. 24. Arendt and Hecker, 1964, p. 89.
  Pennsylvanian, Morrowan-Permian, Wolfcampian

Polusocrinus amplus (Lane and Webster), 1966. Webster and Lane, 1967,
  p. 31, Pl. 5, fig. 11.
  Aesiocrinus amplus Lane and Webster, 1966, p. 55, Pl. 5, figs. 2, 4, 5.
  Webster and Lane, 1967, p. 31.
  Permian, Wolfcampian, Bird Spring Formation: United States: Nevada

Polusocrinus avanti Strimple, 1951c, p. 24, Pl. 5, figs. 4-6; 1961g,
  p. 102, Pl. 16, fig. 1.
  Pennsylvanian, Desmoinesian-Missourian, Holdenville Shale, Avant Limestone: United States: Oklahoma

Polusocrinus calyculoides Lane, 1964b, p. 682, Pl. 112, figs. 1-3.
  Pennsylvanian, Morrowan, Callville Limestone: United States: Nevada

Polusocrinus ochelataensis Strimple, 1952a, p. 12, figs. 1-8.
  Pennsylvanian, Missourian, Wann Formation: United States: Oklahoma

Polusocrinus pachyplax Lane, 1964b, p. 632, Pl. 112, figs. 4-10; Fig. 1.
  Pennsylvanian, Morrowan, Callville Limestone: United States: Nevada

Polusocrinus rosa Strimple, 1951c, p. 25, Pl. 5, fig. 11.
  Pennsylvanian, Virgilian, Nelagoney Formation: United States: Oklahoma

Polusocrinus tactilis (Strimple), 1951. Strimple, 1961g, p. 102.
  Aesiocrinus tactilis Strimple, 1951b, p. 375, Pl. 56, figs. 5-7; 1961g,
  p. 102.
  Pennsylvanian, Missourian, Raytown Limestone: United States: Kansas

POLYCRINUS Jaekel, 1918. *P. ramulatus* Jaekel, 1918. Moore and Laudon,
  1943b, p. 54.
  Ordovician

POLYGONOCRINUS Strimple, 1961. *P. multiextensus* Strimple, 1961. Strimple,
  1961g, p. 59.
  Pennsylvanian, Desmoinesian

Polygonocrinus multiextensus Strimple, 1961g, p. 61, Pl. 2, fig. 6;
  Pl. 11, figs. 1, 2.
  Pennsylvanian, Desmoinesian, Holdenville Shale: United States: Oklahoma

POLYPELTES Angelin, 1878. *P. granulatus* Angelin, 1878. Ubaghs, 1953,
  p. 742.
  Silurian-Lower Devonian

Polypeltes granulatus Angelin, 1878. Ubaghs, 1956, p. 554, Pls. 1-4;
  Figs. 1, 2, 5, 6.
  Silurian-Lower Devonian: Sweden

POROCRINUS Billings, 1857. *P. conicus* Billings, 1857. Moore and Laudon,
  1943b, p. 50; 1944, p. 153, Pl. 52, fig. 22. Cuénot, 1948, p. 60.
  Ubaghs, 1953, p. 750, fig. 33b. Ramsbottom, 1961, p. 16, fig. 6.
  Xu, 1962, p. 45, fig. 1.
  Ordovician-Early Silurian

Porocrinus conicus Billings, 1857. Moore and Laudon, 1943b, p. 131,
  Pl. 2, fig. 7; 1944, p. 153, Pl. 53, fig. 13. Wilson, 1946, p. 39.

Kesling and Paul, 1968, p. 26, Pl. 6, figs. 1-5; Pl. 7, figs. 1-5; Figs. 11, 12.
Ordovician, Champlainian, Trenton Limestone, Cobourg Limestone, Hull Formation: Canada: Ontario

Porocrinus elegans Kesling and Paul, 1968, p. 25, Pl. 3, figs. 1-8; Pl. 4, figs. 7-9; Fig. 9.
Ordovician, Champlainian, Trenton Group: United States: Minnesota

Porocrinus fayettensis Slocum, 1924. Kesling and Paul, 1968, p. 31, Pl. 4, figs. 5, 6; Pl. 8, figs. 1-8; Figs. 2, 13.
Ordovician, Cincinnatian, Maquoketa Shale: United States: Iowa

Porocrinus pyramidatus Kesling and Paul, 1968, p. 25, Pl. 5, figs. 1-6; Pl. 6, figs. 6, 7; Fig. 10.
Ordovician, Cincinnatian, Clermont Shale: United States: Iowa

Porocrinus scoticus Ramsbottom, 1961, p. 17, Pl. 5, fig. 8.
Ordovician, Ashgillian, Upper Drummuck Group: England

Porocrinus shawi Schuchert, 1900. Kesling and Paul, 1968, p. 31, Pl. 4, figs. 1-4; Fig. 14.
Ordovician, Champlainian: Baffin Land

Porocrinus smithi Grant, 1881. Cuénot, 1948, p. 56, fig. 71.
Ordovician: Canada

POTERIOCRINITES Miller, 1821. *P. crassus* Miller, 1821. Moore and Laudon, 1943b, p. 54, fig. 5; 1944, p. 155, Pl. 52, fig. 18. Moore, 1962b, p. 15, fig. 10, no. 7.
Poteriocrinus Miller, 1821. Termier and Termier, 1947, p. 256, Pl. 19, figs. 44, 56. Philip, 1963, fig. 2b.
Devonian-Permian

Poteriocrinites circumtextus Miller and Gurley, 1894 = *Parisocrinus crawfordsvillensis*

Poteriocrinites (Graphiocrinus; Scaphiocrinus) coreyi Meek and Worthen, 1869 = *Abrotocrinus coreyi*

Poteriocrinites crassus Miller, 1821. Wright, 1944, p. 269, Pl. 11, figs. 4, 6; 1950, p. 2, Pl. 2, figs. 2, 4-8, 12-17; Pl. 3, fig. 7; Pl. 5, fig. 4.
Poteriocrinus crassus Miller, 1821. Kostic-Podgorska, 1958, p. 107, Pl. 33, figs. 3, 5-7.
Rhabdocrinus crassus (Miller), 1821. Wright, 1944, p. 269.
Carboniferous, Tournaisian-lower Visean: Ireland; England; Scotland; Poland

Poteriocrinites (Graphiocrinus; Scaphiocrinus) delicatus Meek and Worthen, 1869 = *Aphelecrinus delicatus*

Poteriocrinites isacobus (Austin and Austin), 1846. Wright, 1950, p. 9, Pl. 5, fig. 6.
Poteriocrinus isacobus Austin and Austin, 1846. Wright, 1950, p. 9.
Carboniferous, Tournaisian: Ireland

Poteriocrinites jeffriesi (Rowley), 1908.
Poteriocrinus jeffriesi Rowley, 1908. Williams, 1943, p. 66, Pl. 6, figs. 19, 20.
Mississippian, Kinderhookian, Louisiana Limestone: United States: Louisiana, Missouri

Poteriocrinites kopfi (Goldring), 1935. Goldring, 1945, p. 64, Pl. 1, fig. 6.
Devonian, Erian, Moscow Shale: United States: New York

Poteriocrinites macoupinensis Worthen, 1873 = *Hypselocrinus? macoupinensis*

Poteriocrinites macropleurus (Hall), 1861. Moore and Laudon, 1944, p. 155, Pl. 58, fig. 3.
    Mississippian, Osagean, Burlington Limestone, Lake Valley Limestone: United States: Iowa, New Mexico

Poteriocrinites? magnus (Wright), 1938. Wright, 1950, p. 9, Pl. 3, fig. 4.
    Poteriocrinus magnus Wright, 1937. Wright, 1950, p. 9.
    Carboniferous, Visean, Lower Limestone Group: Scotland; England

Poteriocrinites multicosta Goldring, 1954, p. 27, Pl. 4, figs. 1-5.
    Devonian, Erian, Kashong Shale, Ludlowville Shale: United States: New York

Poteriocrinites nassa (Hall), 1862. Goldring, 1954, p. 30, Pl. 5, fig. 1.
    Devonian, Erian, Moscow Shale: United States: New York

Poteriocrinites plicatus (Austin and Austin), 1842. Wright, 1950, p. 7, Pl. 2, figs. 10, 11; Pl. 4.
    Carboniferous, Tournaisian: England

Poteriocrinites quinquangularis (Austin and Austin), 1842. Wright, 1950, p. 11, Pl. 5, figs. 7, 8, 10.
    Carboniferous, Tournaisian?: Ireland; England

Poteriocrinites (Scaphiocrinus) randolphensis Worthen, 1873 = *Aphelecrinus randolphensis*

Poteriocrinites robustus Austin, *in* Wright, 1950, p. 10, Pl. 5, figs. 2, 3.
    Carboniferous, Tournaisian?: England; Scotland

Poteriocrinites (Scaphiocrinus) rudis Meek and Worthen, 1869 = *Tropiocrinus rudis*

Poteriocrinites (Scaphiocrinus) striatus Meek and Worthen, 1869 = *Acylocrinus striatus*

Poteriocrinites verus (Miller and Gurley), 1890. Van Sant, *in* Van Sant and Lane, 1964, p. 90.
    Mississippian, Osagean, Indian Creek Beds: United States: Indiana

Poteriocrinites spp.
    Lakeman, 1950, p. 104, fig. 3.
        Carboniferous
    Laudon and others, 1952, p. 547, Pl. 65, figs. 9, 10.
        Mississippian, Kinderhookian, Banff formation: Canada: Alberta
    Poteriocrinus sp. Wright, 1942, p. 273, Pl. 10, fig. 11.
        Lower Carboniferous, Mountain Limestone: Scotland

Poteriocrinites? sp. Lane and Webster, 1966, p. 23, Pl. 3, figs. 2-4.
    Permian, Wolfcampian, Bird Spring Formation: United States: Nevada

Poteriocrinus auct. = *Poteriocrinites*

Poteriocrinus albersi Miller and Gurley, 1896 = *Ramulocrinus albersi*

Poteriocrinus anomalos Wetherby, 1880 = *Cymbiocrinus anomalos*

Poteriocrinus (Scaphiocrinus) bayensis Meek and Worthen, 1865 = *Aphelecrinus bayensis*

Poteriocrinus bockschii Geinitz, 1846 = *Ureocrinus bockschii*

Poteriocrinus bozemanensis Miller and Gurley, 1898 = *Ramulocrinus bozemanensis*

Poteriocrinus conicus Phillips, 1836 = *Bollandocrinus conicus*

Poteriocrinus conicus Phillips, 1843 (part) = *Bollandocrinus erectus*

Poteriocrinus crassimanus McCoy, 1849 = *Pachylocrinus? crassimanus*

Poteriocrinus crawfordsvillensis Miller and Gurley, 1890 = *Hypselocrinus indianensis*

Poteriocrinus crineus Hall, 1864 = *Aphelecrinus crineus*

Poteriocrinus dactyloides Austin and Austin, 1842 = *Ophiurocrinus dactyloides*

Poteriocrinus? enormis Meek and Worthen, 1861 = *Zygotocrinus enormis*

Poteriocrinus granulosus McCoy, *in* Sedgwick and McCoy, 1851 = *Hydreionocrinus* cf. *amplus*

Poteriocrinus impressus Phillips, 1836 = *Blothrocrinus impressus*

Poteriocrinus indianensis Meek and Worthen, 1865 = *Hypselocrinus indianensis*

Poteriocrinus isacobus Austin and Austin, 1846 = *Poteriocrinites isacobus*

Poteriocrinus jeffriesi Rowley, 1908 = *Poteriocrinites jeffriesi*

Poteriocrinus lasallensis Worthen, 1875 = *Hypselocrinus? lasallensis*

Poteriocrinus longidactylus Austin and Austin, 1847 = *Blothrocrinus longidactylus*

?*Poteriocrinus longidactylus* (de Koninck) Wright, 1925 = *Woodocrinus longidactylus*

Poteriocrinus magnus Wright, 1937 = *Poteriocrinites? magnus*

Poteriocrinus (*Moscovicrinus?*) *quenstedti* Golowkinsky, 1868. Yakovlev, 1939b, p. 61, Pl. 10, fig. 6.
  Upper Permian: Russia: Kazan Mountains

Poteriocrinus nanus Roemer, 1863 = *Iteacrinus nanus*

Poteriocrinus nuciformis McCoy, 1849 = *Ureocrinus bockschii*

Poteriocrinus okawensis Worthen, 1882 = *Aphelecrinus okawensis*

Poteriocrinus peculiaris Worthen, 1883 = *Aphelecrinus peculiaris*

Poteriocrinus pentonensis Wright, 1937 = *Rhabdocrinus pentonensis*

Poteriocrinus rhenanus Müller, *in* Zeiler and Wirtgen, 1855 (part) = *Orthocrinus simplex*

Poteriocrinus rhenanus Müller, *in* Zeiler and Wirtgen, 1855 (part) = *Antihomocrinus zeileri*

Poteriocrinus rhenanus Müller, *in* Zeiler and Wirtgen, 1855 (part) = indeterminable form belonging questionably to *Eifelocrinus*, *Botryocrinus*, or *Propoteriocrinus* sp. Schmidt, 1941, p. 220.

Poteriocrinus ribblesdalensis Wright, 1942 = *Barycrinus ribblesdalensis*

Poteriocrinus rostratus Austin and Austin, 1842 = *Phacelocrinus rostratus*

Poteriocrinus rugosus Grenfell, 1876 = *Blothrocrinus rugosus*

Poteriocrinus subramosus Miller and Gurley, 1890 = *Parisocrinus crawfordsvillensis*

Poteriocrinus ulrichi Worthen, 1889 = *Ramulocrinus ulrichi*

Poteriocrinus venustus Worthen, 1882 = *Aphelecrinus venustus*

Poteriocrinus (*Scaphiocrinus*) *wachsmuthi* Meek and Worthen, 1861 = *Holcocrinus wachsmuthi*

Poteriocrinus sp. Wright, 1942 = *Poteriocrinites* sp.

POTTSICRINUS Jillson, 1960. *P. quinstilus* Jillson, 1960. Jillson, 1960, p. 37.
  Mississippian, Meramecian

Pottsicrinus quinstilus Jillson, 1960, p. 37, illustrated on title page; 1965, p. 52, illustrated (a basal circlet?).
  Mississippian, Meramecian, Warsaw Limestone: United States: Kentucky

PRADOCRINUS de Verneuil, 1850. *P. baylii* de Verneuil, 1850. Breimer, 1962a, p. 26.
Devonian

Pradocrinus baylii de Verneuil, 1850. Breimer, 1962a, p. 27, Pl. 3, figs. 2-9.
Devonian, lower Emsian, La Vid Formation, numerous unnamed strata: Spain

Prininocrinus Goldring, 1938 = *Scytalocrinus* Wachsmuth and Springer, 1879.

PROAPSIDOCRINUS Wanner, 1924. *P. permicus* Wanner, 1924. Moore and Laudon, 1943b, p. 74. Ubaghs, 1953, p. 756. Arendt and Hecker, 1964, p. 94.
Permian

PROCHOIDOCRINUS Wanner, 1937. *P. nodosus* Wanner, 1937. Moore and Laudon, 1943b, p. 50. Arendt and Hecker, 1964, p. 85. Lane, 1967b, p. 11, fig. 5.
Permian

PROCTOTHYLACOCRINUS Kier, 1952. *P. longus* Kier, 1952. Kier, 1952, p. 72.
Devonian, Erian

Proctothylacocrinus esseri Kesling, 1965c, p. 77, Pl. 1, figs. 1-5; Pl. 2, figs. 1, 2. Kesling, 1968b, p. 133, Pls. 1, 2; Figs. 1-4.
Devonian, Erian, Silica Formation: United States: Ohio

Proctothylacocrinus longus Kier, 1952, p. 72, Pl. 2, figs. 1-4; Pl. 4, fig. 1. Kesling, 1965c, p. 78, Pl. 3, figs. 1-7; Pl. 4, figs. 1-6; Pl. 5, figs. 1-12; Fig. 1.
Devonian, Erian, Silica Formation: United States: Ohio

PROHEXACRINUS Yakovlev, 1946. *P. arcticus* Yakovlev, 1946. Yakovlev, 1946a, p. 154. Arendt and Hecker, 1964, p. 96.
Upper Silurian

Prohexacrinus arcticus Yakovlev, 1946a, p. 154, illustrated. Arendt and Hecker, 1964, p. 96, fig. 135.
Upper Silurian: Russia: northern Ural Mountains

Proindocrinus Yakovlev, 1939 = *Indocrinus*

Proindocrinus piszowi (Yakovlev), 1926 = *Indocrinus piszowi*

PROLOBOCRINUS Wanner, 1937. *P. permicus* Wanner, 1937. Moore and Laudon, 1943b, p. 59.
Permian

Prolobocrinus permicus Wanner, 1937. Wanner, 1949b, p. 83, figs. 1, 2.
Permian: Timor

PROMELOCRINUS Jaekel, 1902. *P. anglicus* Jaekel, 1902. Moore and Laudon, 1943b, p. 96. Ubaghs, 1953, p. 741.
Silurian

Promelocrinus anglicus Jaekel, 1902. Ubaghs, 1953, p. 741, fig. 82c; 1958b, p. 299, figs. 16, 18.
Silurian, Wenlockian: England

Promelocrinus fulminatus (Angelin), 1878. Ubaghs, 1953, p. 741, fig. 82b; 1958b, p. 268, Pl. 1, figs. 2-4; Figs. 3, 4, 16.
Silurian, Wenlockian, Slite Group: Sweden

Promelocrinus radiatus (Angelin), 1878. Ubaghs, 1958b, p. 272, Pl. 2; Figs. 5-7.
Periechocrinus radiatus Angelin, 1878. Ubaghs, 1958b, p. 272.
Silurian, Wenlockian, Slite Group: Sweden

Promelocrinus sp. Ubaghs, 1958b, p. 299, fig. 17A.
Silurian, Wenlockian: England

PROPHYLLOCRINUS Wanner, 1916. *P. dentatus* Wanner, 1916. Moore and Laudon, 1943b, p. 74. Ubaghs, 1953, p. 756. Arendt and Hecker, 1964, p. 94.
Permian

Prophyllocrinus dentatus Wanner, 1916. Yakovlev, 1949a, p. 265, fig. 1g.
Permian: Timor

PROPOTERIOCRINUS Schmidt, 1934. *P. follmanni* Schmidt, 1934. Moore and Laudon, 1943b, p. 54.
Devonian

Propoteriocrinus follmanni Schmidt, 1934. Schmidt, 1941, p. 155, Pl. 9, figs. 1, 2; Fig. 43.
Devonian, Sandstone in upper Coblentzian: Germany

Propoteriocrinus follmanni brevior Schmidt, 1941, p. 157, Pl. 9, figs. 4, 5.
Devonian, upper Coblentzian, Hohenrhein Schichten: Germany

Propoteriocrinus galerus Schmidt, 1941, p. 159, Pl. 19, fig. 3.
Devonian, Coblentzian, quartzite: Germany

Propoteriocrinus(?) *nothus* Schmidt, 1941, p. 162, Pl. 9, fig. 3.
Devonian, Coblentzian, quartzite: Germany

Propoteriocrinus papillaxialis Schmidt, 1941, p. 158, Pl. 13, fig. 2; Pl. 19, fig. 4.
Devonian, upper Coblentzian: Germany

Propoteriocrinus scopae Schmidt, 1934. Schmidt, 1941, p. 30, Pl. 1, fig. 3; Pl. 3, fig. 2; Fig. 3.
Devonian, Coblentzian, Hunsrück Schiefer: Germany

Propoteriocrinus turgidus Schmidt, 1941, p. 160, Pl. 10, figs. 5, 6; Fig. 44.
Devonian, upper Coblentzian: Germany

Propoteriocrinus(?) sp. indet. Schmidt, 1941, p. 163.
Devonian, upper Coblentzian: Germany

PROTAXOCRINUS Springer, 1906. *Taxocrinus ovalis* Angelin, 1878. Moore and Laudon, 1943b, p. 69, fig. 10; 1944, p. 177, Pl. 67, fig. 11. Cuénot, 1948, p. 66. Moore, 1950, p. 37, fig. 40. Ubaghs, 1953, p. 755. Philip, 1963, fig. 1a.
Ordovician-Devonian

Protaxocrinus aberrans Bouška, 1956b, p. 325, Pl. 1, figs. 6-8.
Silurian, lower Ludlovian, Kopanina Beds: Bohemia

Protaxocrinus elegans (Billings), 1857. Moore and Laudon, 1943b, p. 136, Pl. 7, fig. 11; 1944, p. 177, Pl. 68, fig. 14. Wilson, 1946, p. 30. Ubaghs, 1953, p. 755, fig. 150.
Ordovician, Champlainian, Cobourg Limestone, Trenton Limestone: Canada: Ontario

Protaxocrinus girvanensis Ramsbottom, 1961, p. 19, Pl. 6, figs. 14, 15.
Ordovician, Ashgillian, Upper Drummuck Group: England

Protaxocrinus laevis (Billings), 1857. Moore and Laudon, 1943b, p. 136, Pl. 7, fig. 10; 1944, p. 177, Pl. 69, fig. 7. Wilson, 1946, p. 30. Ubaghs, 1953, p. 755, fig. 151.
Ordovician, Champlainian, Cobourg Limestone, Trenton Limestone: Canada: Ontario

Protaxocrinus robustus Springer, 1920. Moore and Laudon, 1944, p. 177, Pl. 54, fig. 13.
Silurian, Niagaran, Beech River Formation: United States: Tennessee

Protaxocrinus svobodai Bouška, 1956b, p. 323, Pl. 1, figs. 1-5.
  Silurian, lower Ludlovian, Kopanina Beds: Bohemia

PROTENCRINUS Jaekel, 1918. *P. moscoviensis* Jaekel, 1918. Moore and
  Laudon, 1943b, p. 59.
  Carboniferous-Permian

Protencrinus lobatus Yakovlev, 1948, p. 121, figs. 5, 6. Yakovlev, *in*
  Yakovlev and Ivanov, 1956, p. 74, Pl. 13, fig. 6.
  Permian, Artinskian: Russia: Timan

*Protencrinus moscoviensis Jaekel, 1918. Yakovlev and Ivanov, 1956, p. 31,
  fig. 6.
  Carboniferous: Russia: Moscow Basin

PTERINOCRINUS Goldring, 1923. *P. quinquenodus* Goldring, 1923. Moore and
  Laudon, 1943b, p. 84; 1944, p. 187. Ubaghs, 1953, p. 737.
  Devonian

Pterinocrinus decembrachiatus Breimer, 1962a, p. 12, Pl. 1, fig. 3.
  Devonian, Emsian, red shale in Ferroñes Limestone: Spain

Pterinocrinus ehrlicheri Lehmann, 1955, p. 135, Pl. 14, figs. 1-3; Pl. 15,
  figs. 4, 5.
  Devonian, Coblentzian, Hunsrück Schiefer: Germany

*Pterinocrinus quinquenodus Goldring, 1923. Moore and Laudon, 1943b, p. 138,
  Pl. 9, fig. 12; 1944, p. 187, Pl. 72, fig. 22.
  Devonian, Chautauquan, Chemung Formation: United States: New York

PTEROTOCRINUS Lyon and Casseday, 1859. **Asterocrinus capitalis* Lyon, 1857.
  Moore and Laudon, 1943b, p. 96; 1944, p. 199. Cuénot, 1948, p. 69.
  Ubaghs, 1953, p. 741, figs. 114, 115. Arendt and Hecker, 1964, p. 97.
  Gutschick, 1965, p. 639, fig. 5.
  Mississippian, Chesterian

Pterotocrinus acutus Wetherby, 1879. Horowitz, 1965, p. 42, Pl. 5,
  figs. 7-9.
  Mississippian, Chesterian, Glen Dean Formation: United States: Indiana,
  Kentucky

Pterotocrinus bifurcatus Wetherby, 1879. Moore and Laudon, 1944, p. 199,
  Pl. 78, fig. 17. Cuénot, 1948, p. 38, fig. 48. Ubaghs, 1953, p. 741,
  fig. 115. Horowitz, 1965, p. 41, Pl. 4, figs. 5-10.
  Mississippian, Chesterian, Glen Dean Formation: United States: Indiana,
  Kentucky

*Pterotocrinus capitalis (Lyon), 1857. Moore and Laudon, 1943b, p. 142,
  Pl. 13, fig. 2; 1944, p. 199, Pl. 78, fig. 16. Weller, 1947, p. 581,
  figs. 1-4.
  Mississippian, Chesterian, Golconda Formation: United States: Kentucky,
  Illinois

Pterotocrinus coronarius (Lyon), 1857. Ubaghs, 1953, p. 741, fig. 114.
  Gutschick, 1965, p. 644, fig. 6.
  Mississippian, Chesterian: United States: Kentucky

Pterotocrinus depressus Lyon and Casseday, 1860. Horowitz, 1965, p. 45,
  Pl. 5, figs. 16-19.
  Mississippian, Chesterian, Glen Dean Formation: United States: Indiana,
  Kentucky

Pterotocrinus edestus Gutschick, 1965, p. 642, Pl. 79, figs. 13-28.
  Mississippian, Chesterian, Kinkaid Limestone: United States: Kentucky,
  Illinois

Pterotocrinus pegasus Gutschick, 1965, p. 643, Pl. 79, figs. 34-39; Figs. 6A-F.
  Mississippian, Chesterian, Kinkaid Limestone: United States: Illinois

Pterotocrinus serratus Weller, 1920. Butts, 1941, p. 251, Pl. 132, figs. 54-60.
Mississippian, Chesterian, Gasper limestone: United States: West Virginia

Pterotocrinus spatulatus Wetherby, 1879. Butts, 1941, p. 251, Pl. 132, fig. 61. Horowitz, 1965, p. 44, Pl. 5, figs. 14, 15.
Mississippian, Chesterian, Glen Dean Formation: United States: Tennessee, Indiana, Kentucky

Pterotocrinus? springerensis Elias, 1957, p. 386, Pl. 41, fig. 11.
Mississippian, Chesterian, Redoak Hollow Formation: United States: Oklahoma

Pterotocrinus tridecibrachiatus Gutschick, 1965, p. 641, Pl. 79, figs. 1-12; Pl. 80, figs. 1-12; Figs. 4, 5B,C.
Mississippian, Chesterian, Kinkaid Limestone: United States: Kentucky, Illinois

Pterotocrinus vannus Sutton, 1934. Horowitz, 1965, p. 43, Pl. 5, figs. 10-13.
Mississippian, Chesterian, Glen Dean Formation: United States: Indiana

Pterotocrinus wanlessi Gutschick, 1965, p. 644, Pl. 80, figs. 49-66.
Mississippian, Chesterian, Kinkaid Limestone: United States: Illinois, Kentucky

Pterotocrinus spp. A.
Horowitz, 1965, p. 46, Pl. 5, figs. 1-3.
Mississippian, Chesterian, Glen Dean Formation: United States: Indiana
Gutschick, 1965, p. 644, Pl. 80, figs. 13-23.
Mississippian, Chesterian, Kinkaid Limestone: United States: Illinois, Kentucky

Pterotocrinus spp. B.
Horowitz, 1965, p. 48, Pl. 5, figs. 4-6.
Mississippian, Chesterian, Glen Dean Formation: United States: Indiana
Gutschick, 1965, p. 645, Pl. 80, figs. 24-37.
Mississippian, Chesterian, Kinkaid Limestone: United States: Illinois, Kentucky

Pterotocrinus spp. C.
Horowitz, 1965, p. 49, Pl. 5, fig. 20.
Mississippian, Chesterian, Glen Dean Formation: United States: Kentucky
Gutschick, 1965, p. 645, Pl. 80, figs. 38-48.
Mississippian, Chesterian, Kinkaid Limestone: United States: Illinois, Kentucky

PTYCHOCRINUS Wachsmuth and Springer, 1885. *Gaurocrinus splendens* Miller, 1883. Moore and Laudon, 1943b, p. 83, fig. 13; 1944, p. 187, Pl. 70, fig. 9. Moore, 1950, p. 42, fig. 11. Ubaghs, 1953, p. 737.
Ordovician-Early Silurian

**Ptychocrinus splendens* (Miller), 1883. Moore and Laudon, 1944, p. 187, Pl. 72, fig. 19.
Silurian, Albian, Girardeau Formation: United States: Missouri

PYCNOCRINUS Miller, 1883. *Glyptocrinus shafferi* Miller, 1873. Moore and Laudon, 1943b, p. 96. Ubaghs, 1953, p. 741.
Ordovician

Pycnocrinus ornatus (Billings), 1857. Wilson, 1946, p. 29.
Ordovician, Champlainian, Cobourg Limestone, Hull Limestone: Canada: Ontario

PYCNOSACCUS Angelin, 1878. *Cyathocrinites scrobiculatus Hisinger, 1840.
   Moore and Laudon, 1943b, p. 74, fig. 11; 1944, p. 181, Pl. 66,
   fig. 1. Ubaghs, 1953, p. 756. Arendt and Hecker, 1964, p. 94.
   Silurian-Devonian

Pycnosaccus bucephallus bohemicus Bouška, 1943, p. 254, Pl. 1, figs. 1-6;
   Fig. 6.
   Silurian: Bohemia

Pycnosaccus bucephallus bohemicus? Bouška, 1943, p. 254, Pl. 1, fig. 7.
   Silurian: Bohemia

Pycnosaccus patei Springer, 1920. Moore and Laudon, 1944, p. 181, Pl. 54,
   fig. 4.
   Silurian, Niagaran, Beech River Formation: United States: Tennessee

Pycnosaccus semilukensis Yakovlev, 1956b, p. 92, Pl. 1, figs. 4, 5. Arendt
   and Hecker, 1964, p. 94, Pl. 14, fig. 4.
   Devonian, Frasnian: Russia: Voronezh Region

Pycnosaccus tenuibrachiatus Springer, 1920. Moore and Laudon, 1943b, p. 137,
   Pl. 8, fig. 3; 1944, p. 181, Pl. 69, fig. 11.
   Silurian, Cayugan, Keyser Limestone: United States: West Virginia

PYGMAEOCRINUS Bouška, 1946. *P. kettneri Bouška, 1946. Bouška, 1946, p. 1.
   Ubaghs, 1953, p. 747. Strimple, 1963d, p. 36, fig. 8b.
   Lower Devonian

*Pygmaeocrinus kettneri Bouška, 1946, p. 2, Pl. 1, figs. 1-4. Strimple,
   1963d, p. 37, figs. 7a-d.
   Lower Devonian: Bohemia

PYXIDOCRINUS Müller, 1855. *Actinocrinus prumiensis Wirtgen and Zeiler,
   1855. Breimer, 1962a, p. 34.
   Lower-Middle Devonian

Pyxidocrinus? bifrons (Schmidt), 1931. Breimer, 1962a, p. 44.
   Pithocrinus bifrons (Schmidt), 1931. Kirk, 1945a, p. 346.
   Megistocrinus? bifrons Schmidt, 1931. Schmidt, 1952, p. 152, Pl. 10,
   figs. 1a-c; Fig. 10.
   Devonian, lower Couvinian, Aranao Limestone: Spain

Pyxidocrinus collensis Breimer, 1962a, p. 35, Pl. 6, figs. 1-8.
   Devonian, Emsian, La Vid Formation: Spain

Pyxidocrinus latus Breimer, 1962a, p. 40, Pl. 6, figs. 9-14.
   Devonian, lower Emsian, La Vid Formation: Spain

*Pyxidocrinus prumiensis Wirtgen and Zeiler, 1855. Breimer, 1962a, p. 35.
   Middle Devonian: Germany

Pyxidocrinus sanmigueli (Astre), 1925. Breimer, 1962a, p. 43.
   Periechocrinus sanmigueli Astre, 1925. Breimer, 1962a, p. 43.
   Age and locality unknown in Spain

QUANTOXOCRINUS Webby, 1965. *Q. ussheri Webby, 1965. Webby, 1965, p. 12.
   Devonian, upper Givetian or lower Frasnian

*Quantoxocrinus ussheri Webby, 1965, p. 13, Pl. 4; Fig. 1.
   Devonian, upper Givetian or lower Frasnian, Ilfracombe Beds: England

QUINIOCRINUS Schmidt, 1941. *Q. erectus Schmidt, 1941. Schmidt, 1941,
   p. 33. Moore, 1962b, p. 14, fig. 6, no. 5. Strimple, 1963d, p. 24,
   fig. 2b.
   Devonian, Eifelian

*Quiniocrinus erectus Schmidt, 1941, p. 35, fig. 4 (typographical error,
   given as fig. 5). Strimple, 1963d, p. 24, fig. 2a.
   Devonian, lowest Eifelian, Orthocrinus Schichten: Germany

RAMSEYOCRINUS Bates, 1968. **Dendrocrinus cambriensis* Hicks, 1873. Bates, 1968, p. 406.
Ordovician, Arenigian

*Ramseyocrinus cambriensis (Hicks), 1873. Bates, 1968, p. 407, Pl. 76, figs. 1-5.
Dendrocrinus cambriensis Hicks, 1873. Bates, 1968, p. 407.
Iocrinus? cambriensis (Hicks), 1873. Ramsbottom, 1961, p. 5, Pl. 3, figs. 9-11. Bates, 1968, p. 407.
Ordovician, Arenigian, Porth Gain Beds: England

RAMULOCRINUS Laudon, Parks and Spreng, 1952. **R. nigelensis* Laudon, Parks and Spreng, 1952. Laudon and others, 1952, p. 557.
Mississippian

Ramulocrinus albersi (Miller and Gurley), 1896. Laudon and others, 1952, p. 558 (misspelled as *R. albertsi*).
Poteriocrinus albersi Miller and Gurley, 1896. Laudon and others, 1952, p. 558.
Mississippian, Chesterian, Renault Formation, Gasper Formation: United States: Illinois, Alabama

Ramulocrinus bozemanensis (Miller and Gurley), 1898. Laudon and others, 1952, p. 558.
Poteriocrinus bozemanensis Miller and Gurley, 1898. Laudon and others, 1952, p. 558.
Lower Mississippian, Madison Limestone: United States: Montana

Ramulocrinus granuliferus (Miller and Gurley), Laudon and Severson, 1953 = *Decadocrinus depressus*

Ramulocrinus halli (Hall), 1861. Laudon and Severson, 1953, p. 520.
Scaphiocrinus halli Hall, 1861. Laudon and Severson, 1953, p. 520.
Mississippian, Osagean, upper Burlington Limestone: United States: Iowa

Ramulocrinus livingstonensis Laudon and Severson, 1953, p. 519, Pl. 51, figs. 16-19; Pl. 55, figs. 8, 9.
Mississippian, Kinderhookian, Lodgepole Limestone: United States: Montana

*Ramulocrinus nigelensis Laudon, Parks and Spreng, 1952, p. 558, Pl. 65, figs. 39, 40; Pl. 69, fig. 8.
Mississippian, Kinderhookian, Banff formation: Canada: Alberta

Ramulocrinus repertus (Miller and Gurley), 1890. Laudon and others, 1952, p. 558.
Decadocrinus repertus (Miller and Gurley), 1890. Moore and Laudon, 1944, p. 169, Pl. 64, fig. 26.
Scaphiocrinus repertus Miller and Gurley, 1890. Laudon and others, 1952, p. 558.
Mississippian, Osagean, Keokuk Limestone: United States: Indiana

Ramulocrinus rudis (Meek and Worthen), 1873. Laudon and Severson, 1953, p. 520.
Scaphiocrinus rudis Meek and Worthen, 1873. Laudon and Severson, 1953, p. 520.
Mississippian, Osagean, upper Burlington Limestone: United States: Iowa

Ramulocrinus ulrichi (Worthen), 1889. Laudon and Severson, 1953, p. 520.
Poteriocrinus ulrichi Worthen, 1889. Laudon and Severson, 1953, p. 520.
Mississippian, Osagean, Keokuk Limestone: United States: Iowa

Raphanocrinus? wachsmuthi Oehlert, 1887 = *Griphocrinus wachsmuthi*

RETEOCRINUS Billings, 1859. **R. stellaris* Billings, 1859. Moore and Laudon, 1943b, p. 82, fig. 13; 1944, p. 183, Pl. 70, fig. 1. Moore,

1950, p. 42, fig. 11. Ubaghs, 1953, p. 735.
Ordovician-Early Silurian

Reteocrinus nealli (Hall), 1866. Moore and Laudon, 1943b, p. 138, Pl. 9, fig. 1 (misspelled as *R. onealli*); 1944, p. 183, Pl. 72, figs. 1, 8.
Ordovician, Champlainian, Richmond Formation: United States: Ohio

*Reteocrinus stellaris Billings, 1859. Wilson, 1946, p. 25.
Ordovician, Champlainian, Cobourg Limestone, Hull Limestone: Canada: Ontario

REVALOCRINUS Jaekel, 1918. **R. costatus* Jaekel, 1918. Moore and Laudon, 1943b, p. 32. Ubaghs, 1953, p. 749. Arendt and Hecker, 1964, p. 82.
Ordovician

*Revalocrinus costatus Jaekel, 1918. Arendt and Hecker, 1964, p. 82, fig. 117.
Middle Ordovician: Russia: Estonia

RHABDOCRINUS Wright, 1944. **Poteriocrinus scotocarbonarius* Wright, 1937. Wright, 1944, p. 266.
Carboniferous, Visean-Namurian

Rhabdocrinus pentonensis (Wright), 1937. Wright, 1944, p. 269; 1950, p. 16, Pl. 2, fig. 3.
Poteriocrinus pentonensis Wright, 1937. Wright, 1944, p. 269.
Carboniferous, Visean, Lower Limestone Group: Scotland; England

*Rhabdocrinus scotocarbonarius (Wright), 1937. Wright, 1944, p. 267, Pl. 10, figs. 1-15; Pl. 11, figs. 1-3, 5, 8-14; 1950, p. 13, Pl. 2, figs. 1, 9, 18; Pl. 3, figs. 1-3, 5, 6, 8-21; Figs. 2-4.
Poteriocrinus scotocarbonarius Wright, 1937. Wright, 1944, p. 269.
Carboniferous, Visean, Lower Limestone Group, Seafield Tower Limestone: Scotland; England

Rhabdocrinus swaledalensis Wright, 1950, p. 16, Pl. 1, figs. 1-3.
Carboniferous, Namurian, *Woodocrinus* Limestone: England

Rhabdocrinus vatagini Arendt, 1962, p. 118, figs. 1, 2. Arendt, 1963b, p. 1675, figs. 1, 2.
Carboniferous, Visean, Steshev Stratum: Russia: Tarusa

RHADINOCRINUS Jaekel, 1895. **R. rhenanus* Jaekel, 1895. Moore and Laudon, 1943b, p. 54. Ubaghs, 1953, p. 751. Arendt and Hecker, 1964, p. 86.
Devonian

Rhadinocrinus dactylus Schmidt, 1934 = *Iteacrinus dactylus*

*Rhadinocrinus rhenanus Jaekel, 1895. Schmidt, 1941, p. 152, fig. 41.
Devonian, Coblentzian, Koblenz Quartzite: Germany

Rhadinocrinus trifidus Schmidt, 1941, p. 152, Pl. 12, figs. 1-3.
Devonian, upper Coblentzian, Stadtfeld Schichten: Germany

Rhadinocrinus sp. indet. Schmidt, 1941, p. 154, fig. 42.
Devonian, Coblentzian, Herdorf Schichten: Germany

RHAPHANOCRINUS Wachsmuth and Springer, 1885. **Glyptocrinus subnodosus* Walcott, 1883. Moore and Laudon, 1943b, p. 82, fig. 13; 1944, p. 183. Ubaghs, 1953, p. 736. Arendt and Hecker, 1964, p. 95.
Ordovician-Devonian

Rhaphanocrinus esthoniae Jaekel (unable to find original date of designation). Arendt and Hecker, 1964, p. 95, Pl. 14, fig. 11.
Middle Ordovician: Estonia

Rhaphanocrinus sculptus (Miller), 1882. Moore and Laudon, 1944, p. 183, Pl. 72, fig. 23.
Ordovician, Champlainian, Richmond Formation: United States: Ohio

RHENOCRINUS Jaekel, 1906. **R. ramosissimus* Jaekel, 1906. Moore and Laudon, 1943b, p. 54.
Lower Devonian

*Rhenocrinus ramosissimus Jaekel, 1906. Cuénot, 1948, p. 47, fig. 61 (incorrectly given as *R. ramosus*). Moore and Jeffords, 1968, p. 11, fig. 1, no. 8.
Lower Devonian, Hunsrück Slate: Germany

RHIPIDOCRINUS Beyrich, *in* Zittel, 1879. **Rhodocrinites crenatus* Goldfuss, 1831. Moore and Laudon, 1943b, p. 83. Ubaghs, 1953, p. 737. Arendt and Hecker, 1964, p. 95.
Devonian

*Rhipidocrinus crenatus (Goldfuss), 1831. Breimer, 1960, p. 256, fig. 5.
Devonian, Eifelian: Germany

Rhipidocrinus spec. cf. *R. perloricatus* Schmidt, 1905. Breimer, 1960, p. 248, Pl. 1, figs. 1, 2; Figs. 1-4.
Devonian, Givetian, Fleringer Schichten: Germany

RHODOCRINITES Miller, 1821. **R. verus* Miller, 1821. Moore and Laudon, 1943b, p. 83; 1944, p. 185, Pl. 70, fig. 8. Ubaghs, 1953, p. 736, fig. 139b. Philip, 1963, fig. 5c. Arendt and Hecker, 1964, p. 95. *Rhodocrinus* Miller, 1821. Termier and Termier, 1947, p. 255, Pl. 19, fig. 48.
Silurian-Lower Carboniferous

Rhodocrinites baccatus (Wright), 1937. Wright, 1958, p. 313, Pl. 76, figs. 1-6; Fig. 131.
Carboniferous, Visean, Ardross Limestone, shales above White Limestone, Lower Limestone Group: Scotland

Rhodocrinites barrisi (Hall), 1861. Moore and Laudon, 1943b, p. 138, Pl. 9, fig. 8; 1944, p. 185, Pl. 73, fig. 7.
Mississippian, Osagean, Burlington Limestone, Reeds Spring Formation, Lake Valley Limestone: United States: Iowa, Missouri, Illinois, New Mexico

Rhodocrinites beanei Strimple and Boyt, 1965, p. 223, figs. 1, 2.
Mississippian, Kinderhookian, Hampton Formation: United States: Iowa

Rhodocrinites brewsteri Laudon, Parks and Spreng, 1952, p. 561, Pl. 67, figs. 1-6; Pl. 69, fig. 12.
Mississippian, Kinderhookian, Banff formation: Canada: Alberta

Rhodocrinites cavanaughi (Laudon), 1933. Moore and Laudon, 1944, p. 185, Pl. 73, fig. 10.
Mississippian, Kinderhookian, Gilmore City Limestone: United States: Iowa

Rhodocrinites cirrusi Laudon, Parks and Spreng, 1952, p. 563, Pl. 67, figs. 8, 9; Pl. 69, fig. 14. Laudon and Severson, 1953, p. 524, Pl. 52, figs. 3-5.
Mississippian, Kinderhookian, Banff formation, Lodgepole Limestone: Canada: Alberta; United States: Montana

Rhodocrinites douglassi (Miller and Gurley), 1897. Laudon and Severson, 1953, p. 524, Pl. 52, figs. 13-19.
Mississippian, Kinderhookian, Lodgepole Limestone: United States: Montana

Rhodocrinites douglassi serpens (Laudon), 1933. Moore and Laudon, 1944, p. 185, Pl. 73, fig. 9.
Mississippian, Kinderhookian, Gilmore City Formation: United States: Iowa

Rhodocrinites? granulatus Austin and Austin, *in* Wright, 1958, p. 313, Pl. 76, fig. 11; Appendix, Pl. B, fig. 9.
Carboniferous, Tournaisian: Ireland

Rhodocrinites kirbyi (Wachsmuth and Springer), 1890. Moore and Laudon, 1944, p. 185, Pl. 73, fig. 12.
Mississippian, Kinderhookian, Hampton Formation: United States: Iowa

Rhodocrinites macrotumidus Laudon and Severson, 1953, p. 524, Pl. 52, figs. 9-12; Pl. 55, figs. 17, 18.
Mississippian, Kinderhookian, Lodgepole Limestone: United States: Montana

Rhodocrinites nodulosus (Hall), Le Maître, 1952 = *Dimerocrinites nodulosus*

Rhodocrinites nortoni (Goldring), 1933. Goldring, 1954, p. 5.
Devonian, Ulsterian, Oriskany Sandstone: United States: Maine

Rhodocrinites ornatus Dubatolova, 1964, p. 31, Pl. 3, figs. 5-7.
Devonian, Eifelian: Russia: Kuznetz Basin

Rhodocrinites pantheri Laudon, Parks and Spreng, 1952, p. 562, Pl. 67, fig. 7; Pl. 69, fig. 13.
Mississippian, Kinderhookian, Banff formation: Canada: Alberta

Rhodocrinites platyacron (Yakovlev, *in* Yakovlev and Ivanov), 1956. Arendt and Hecker, 1964, p. 95, Pl. 14, fig. 12.
Rhodocrinus platyacron Yakovlev, *in* Yakovlev and Ivanov, 1956, p. 46, Pl. 16, fig. 2.
Lower Carboniferous: Asiatic Russia

Rhodocrinites priminodosus Laudon and Severson, 1953, p. 525, Pl. 52, figs. 6-8; Pl. 55, fig. 16.
Mississippian, Kinderhookian, Lodgepole Limestone: United States: Montana

Rhodocrinites aff. *quinquangularis* Miller, 1821. Larrauri and Mercadillo, 1944, p. 50, Pl. 6, fig. 2 (misspelled as *Rodocrinites*).
Devonian: Spanish Sahara

Rhodocrinites rubiformis (Yakovlev, *in* Yakovlev and Ivanov), 1956, new comb.
Rhodocrinus rubiformis Yakovlev, *in* Yakovlev and Ivanov, 1956, p. 46, Pl. 16, fig. 3.
Lower Carboniferous: Asiatic Russia

Rhodocrinites skourensis (Delpey), 1939, nom. correct.
Rhodocrinus skourensis Delpey, 1939. Termier and Termier, 1950, p. 83, Pl. 208, figs. 1-4 (*non* Pl. 208, fig. 16, as given on p. 83).
Carboniferous, upper Visean: Morocco

Rhodocrinites tuberculatus (Wachsmuth and Springer), 1897. Moore and Laudon, 1944, p. 185, Pl. 73, fig. 6.
Mississippian, Osagean, Lake Valley Limestone: United States: New Mexico

Rhodocrinites verisimilis (Grenfell), 1876. Wright, 1958, p. 311, Pl. 76, figs. 13-15, 19, 20, 26-28; Fig. 130.
Carboniferous, Tournaisian (K_2): England

*Rhodocrinites verus Miller, 1821. Wright, 1958, p. 309, fig. 129.
Carboniferous, Tournaisian: England

Rhodocrinus Termier and Termier, 1947 = *Rhodocrinites*

Rhodocrinus benedicti Miller, 1892 = *Cribanocrinus benedicti*

Rhodocrinus bridgerensis Miller and Gurley, 1897 = *Cribanocrinus bridgerensis*

Rhodocrinus coxanus Worthen, 1882 = *Cribanocrinus coxanus*

Rhodocrinus halli Lyon, 1862 = *Griphocrinus halli*

Rhodocrinus insculptus Goldring, 1935 = *Griphocrinus insculptus*

Rhodocrinus (*Acanthocrinus*) *nodulosus* Hall, 1862 = *Griphocrinus nodulosus*

Rhodocrinus parvus Miller, 1891 = *Cribanocrinus parvus*

Rhodocrinus punctatus Weller, 1909 = *Cribanocrinus punctatus*

Rhodocrinus skourensis Delpey, 1939 = *Rhodocrinites skourensis*

Rhodocrinus wachsmuthi Hall, 1861 = *Cribanocrinus wachsmuthi*

Rhodocrinus watersianus Wachsmuth and Springer, 1889 = *Cribanocrinus watersianus*

Rhodocrinus whitei Hall, 1861 = *Cribanocrinus whitei*

Rhodocrinus wortheni Hall, 1858 = *Cribanocrinus wortheni*

Rhodocrinus wortheni var. *urceolatus* Wachsmuth and Springer, 1897 = *Cribanocrinus urceolatus*

RHOPALOCRINUS Wachsmuth and Springer, 1879. *Taxocrinus gracilis* Schultze, 1866. Ubaghs, 1953, p. 751.
Devonian

Rhopalocrinus gracilis (Schultze), 1867. Sieverts-Doreck, 1953, p. 86, Pl. 4, fig. 12.
Middle Devonian: Germany

RHOPOCRINUS Kirk, 1942. *R. spinosus* Kirk, 1942. Moore and Laudon, 1943b, p. 56. Arendt and Hecker, 1964, p. 88.
Mississippian

RISTNACRINUS Öpik, 1934. *R. marinus* Öpik, 1934. Moore and Laudon, 1943b, p. 31. Ubaghs, 1953, p. 747. Yakovlev, 1956c, p. 155, figs. 1, 2. Moore, 1962b, p. 35, fig. 2, no. 1. Arendt and Hecker, 1964, p. 80.
Middle Ordovician

Ristnacrinus marinus Öpik, 1934. Ubaghs, 1953, p. 747, fig. 7. Moore, 1962b, p. 35, Pl. 1, fig. 6. Arendt and Hecker, 1964, p. 80, Pl. 8, fig. 1.
Middle Ordovician, Jewe Formation: Russia: Estonia

ROEMEROCRINUS Wanner, 1916. *R. gracilis* Wanner, 1916. Moore and Laudon, 1943b, p. 55. Arendt and Hecker, 1964, p. 92.
Permian

Roemerocrinus turbinatus Wanner, 1937. Wanner, 1949a, p. 4.
Roemerocrinus gracilis forma *turginata* Wanner, 1937. Wanner, 1949a, p. 4.
Permian: Timor

RUMPHIOCRINUS Wanner, 1924. *R. singularis* Wanner, 1924. Moore and Laudon, 1943b, p. 74. Ubaghs, 1953, p. 755. Arendt and Hecker, 1964, p. 94.
Permian

SACCOCRINUS Hall, 1852. *S. speciosus* Hall, 1852. Moore and Laudon, 1943b, p. 92; 1944, p. 191, Pl. 71, fig. 1.
Silurian-Devonian

Saccocrinus benedicti Miller, 1892. Moore and Laudon, 1943b, p. 139, Pl. 10, fig. 5; 1944, p. 191, Pl. 75, fig. 17. Strimple, 1963d, p. 109, Pl. 10, figs. 8, 9.
Habrocrinus benedicti Slocum, 1908. La Rocque and Marple, 1955, p. 73, fig. 145.
Silurian, Niagaran, Laurel Limestone, Henryhouse Formation, Cedarville Limestone, Guelph formation: United States: Indiana, Oklahoma, Ohio

Saccocrinus benedicti Springer, 1926 (part) = *Stiptocrinus spinosus*

Saccocrinus howardi Miller, 1892 = *Stiptocrinus howardi*

Saccocrinus? intrastigmatus Schmidt, 1931 = *Stamnocrinus intrastigmatus*

SAGENOCRINITES Austin and Austin, 1842. *Actinocrinites? expansus* Phillips, 1839. Moore and Laudon, 1943b, p. 73, fig. 10;   1944, p. 179, Pl. 67, fig. 7. Ubaghs, 1953, p. 755, fig. 15b.
  Periechocrinites Austin and Austin, 1843. Ramsbottom, 1951b, p. 1041; 1954, p. 687.
  Sagenocrinus Austin and Austin, 1842. Cuénot, 1948, p. 66.
  Silurian

Sagenocrinites americanus (Springer), 1902. Moore and Laudon, 1943b, p. 137, Pl. 8, fig. 12;   1944, p. 179, Pl. 54, fig. 11.
  Silurian, Niagaran, Waldron Shale: United States: Indiana

Sagenocrinites clarki (Springer), 1920. Moore and Laudon, 1943b, p. 137, Pl. 8, fig. 11;   1944, p. 179, Pl. 54, fig. 10.
  Silurian, Niagaran, Beech River Formation: United States: Tennessee

*Sagenocrinites expansus (Phillips), 1839. Ubaghs, 1953, p. 755, fig. 15b.
  Periechocriniteo articulosus Austin and Austin, 1843. Ramsbottom, 1951b, p. 1042.
  Silurian, Wenlockian, Wenlock Limestone: England

SAMPSONOCRINUS Miller and Gurley, 1895. *S. hemisphericus* Miller and Gurley, 1895. Moore and Laudon, 1943b, p. 93.
  Lower Carboniferous

*Sampsonocrinus hemisphericus Miller and Gurley, 1895. Brower, 1967, p. 678, Pl. 75, fig. 10; Figs. 2A, B.
  Mississippian, Kinderhookian, Compton Limestone: United States: Missouri

Sampsonocrinus? globosus (Wachsmuth and Springer), 1897. Brower, 1965, p. 789.
  Steganocrinus? globosus Wachsmuth and Springer, 1897. Kirk, 1943a, p. 261. Brower, 1965, p. 789.
  Mississippian, Kinderhookian: United States: Iowa

Sampsonocrinus loricatus (Schlotheim), 1820. Wright, 1955a, p. 227, Pl. 52, figs. 12-14; Figs. 113, 114.
  Encrinites loricatus Schlotheim, 1820. Wright, 1955a, p. 227.
  Carboniferous, Tournaisian (C_1): England

Sampsonocrinus westheadi (Wright), 1947. Wright, 1955a, p. 230, Pl. 54, figs. 18-20, 22; Pl. 57, figs. 17-20; Pl. 63, figs. 1-4, 8-12, 17-20; Fig. 115. Brower, 1965, p. 789.
  Steganocrinus westheadi Wright, 1947, p. 101, Pl. 3, figs. 4-7; 1955a, p. 230.
  Carboniferous, Tournaisian: Scotland

SAROCRINUS Kirk, 1942. *S. nitidus* Kirk, 1942. Moore and Laudon, 1943b, p. 58. Arendt and Hecker, 1964, p. 89.
  Mississippian, Osagean

Sarocrinus granilineus (Miller and Gurley), 1890. Van Sant, *in* Van Sant and Lane, 1964, p. 98, Pl. 5, fig. 1.
  Mississippian, Osagean, Borden Group: United States: Indiana

Sarocrinus plenus Kirk, 1942. Van Sant, *in* Van Sant and Lane, 1964, p. 99.
  Mississippian, Osagean: United States: Indiana

Scaphiocrinus bellus Miller and Gurley, 1890 = *Decadocrinus bellus*

Scaphiocrinus disparilis Miller and Gurley, 1890 = *Scytalocrinus disparilis*

Scaphiocrinus doris Hall, 1861 = *Springericrinus doris*

Scaphiocrinus elegantulus Wachsmuth and Springer, *in* Miller, 1889 = *Aphelecrinus elegantulus*

Scaphiocrinus granuliferus Miller and Gurley, 1890 = *Decadocrinus depressus*

Scaphiocrinus gurleyi White, 1878 = *Pachylocrinus aequalis*

Scaphiocrinus halli Hall, 1861 = *Ramulocrinus halli*

Scaphiocrinus nodobrachiatus Hall, 1861 = *Holcocrinus? nodobrachiatus*

Scaphiocrinus notatus Miller and Gurley, 1896 = *Nactocrinus notatus*

Scaphiocrinus porrectus Miller, 1892 = *Scytalocrinus robustus*

Scaphiocrinus repertus Miller and Gurley, 1890 = *Ramulocrinus repertus*

Scaphiocrinus rudis Meek and Worthen, 1873 = *Ramulocrinus rudis*

Scaphiocrinus scoparius Hall, 1858 = *Aphelecrinus scoparius*

Scaphiocrinus spinobrachiatus Hall, 1861 = *Holcocrinus spinobrachiatus*

Scaphiocrinus tortuosus Hall, 1861 = *Acylocrinus tortuosus*

Scaphiocrinus washburni Beede, 1899 = *Haeretocrinus washburni*

SCHEDEXOCRINUS Strimple, 1961. **S. gibberellus* Strimple, 1961. Strimple, 1961g, p. 27.
   Pennsylvanian, Desmoinesian-Permian, Wolfcampian

*Schedexocrinus gibberellus Strimple, 1961g, p. 29, Pl. 2, figs. 1-3; Pl. 9, figs. 3-5; Pl. 11, figs. 3, 4; Pl. 17, figs. 4-6; Pl. 19, fig. 6; Figs. 13i, j; 14a, b; 15. Branson, 1962, p. 162, fig. 1.
   Pennsylvanian, Desmoinesian, Holdenville Shale: United States: Oklahoma

Schedexocrinus sp. Webster and Lane, 1967, p. 30, Pl. 4, fig. 8; Pl. 8, fig. 8.
   Permian, Wolfcampian, Bird Spring Formation: United States: Nevada

SCHISTOCRINUS Moore and Plummer, 1940. **S. torquatus* Moore and Plummer, 1940. Moore and Laudon, 1943b, p. 58; 1944, p. 165.
   Pennsylvanian

Schistocrinus azygous (Strimple), 1949. Strimple, 1961g, p. 100, Pl. 9, figs. 1, 2.
   Malaiocrinus azygous Strimple, 1949d, p. 19, Pl. 3, figs. 4, 8, 9, 12; 1961g, p. 100.
   Pennsylvanian, Desmoinesian, Holdenville Shale: United States: Oklahoma

Schistocrinus confertus Moore and Plummer, 1940 = *Sciadiocrinus confertus*

Schistocrinus ovalis Strimple, 1951d, p. 194, figs. 2-5. Tischler, 1963, p. 1067, fig. 6C.
   Pennsylvanian, Desmoinesian, Oologah Limestone, Madera Formation: United States: Oklahoma, Colorado

Schistocrinus parvus Moore and Plummer, 1940 = *Sciadiocrinus parvus*

Schistocrinus planulatus Moore and Plummer, 1940 = *Sciadiocrinus planulatus*

*Schistocrinus torquatus Moore and Plummer, 1940. Moore and Laudon, 1944, p. 165, Pl. 62, fig. 12; Pl. 64, fig. 8.
   Pennsylvanian, Missourian, Winterset formation: United States: Missouri

SCHIZOCRINUS Hall, 1847. **S. nodosus* Hall, 1847. Ubaghs, 1953, p. 741.
   Ordovician

*Schizocrinus nodosus Hall, 1847. Wilson, 1946, p. 29. Welby, 1962, p. 37, fig. 19.
   Ordovician, Champlainian, Glen Falls Limestone: United States: New York, Vermont

SCHMIDTOCRINUS Haarmann, 1921. **Rhenocrinus winterfeldi* Jaekel, 1906. Moore and Laudon, 1943b, p. 54.
   Middle Devonian

SCHULTZICRINUS Springer, 1911. *S. typus Springer, 1911. Moore and
    Laudon, 1943b, p. 50; 1944, p. 154. Ubaghs, 1953, p. 750.
    Devonian

*Schultzicrinus typus Springer, 1911. Moore and Laudon, 1943b, p. 131,
    Pl. 2, fig. 12; 1944, p. 154, Pl. 57, figs. 5, 6.
    Devonian, Erian, Onondaga Limestone: United States: New York

SCIADIOCRINUS Moore and Plummer, 1938. *Zeacrinus (Hydreionocrinus?)
    acanthophorus Meek and Worthen, 1870. Moore and Laudon, 1943b,
    p. 58; 1944, p. 165.
    Pennsylvanian, Desmoinesian-Virgilian

*Sciadiocrinus acanthophorus (Meek and Worthen), 1870. Moore and Laudon,
    1944, p. 165, Pl. 62, fig. 11; Pl. 64, fig. 1.
    Pennsylvanian, Desmoinesian: United States: Illinois

Sciadiocrinus confertus (Moore and Plummer), 1940. Strimple, 1961g, p. 100.
    Schistocrinus confertus Moore and Plummer, 1940. Moore and Laudon, 1944,
    p. 165, Pl. 64, fig. 32. Strimple, 1961g, p. 100.
    Pennsylvanian, Desmoinesian, Millsap Lake Formation: United States:
    Texas

Sciadiocrinus disculus Moore and Plummer, 1940. Moore and Laudon, 1944,
    p. 165, Pl. 64, fig. 7.
    Pennsylvanian, Missourian, Mineral Wells Formation, Graford Formation:
    United States: Texas

Sciadiocrinus harrisae Moore and Plummer, 1940. Moore and Laudon, 1944,
    p. 165, Pl. 62, fig. 17; Pl. 64, fig. 6. Strimple, 1962e, p. 30,
    Pl. 9, figs. 5, 6.
    Pennsylvanian, Desmoinesian, Oologah Limestone, Millsap Lake Formation:
    United States: Oklahoma, Texas

Sciadiocrinus humilis Strimple, 1951b, p. 373, Pl. 56, figs. 11-14 (misspelled
    as Sciadocrinus).
    Pennsylvanian, Virgilian, Stull Shale: United States: Kansas

Sciadiocrinus parvus (Moore and Plummer), 1940. Strimple, 1961g, p. 100.
    Schistocrinus parvus Moore and Plummer, 1940. Moore and Laudon, 1944,
    p. 165, Pl. 64, fig. 27. Strimple, 1961g, p. 100.
    Pennsylvanian, Desmoinesian, Millsap Lake Formation: United States:
    Texas

Sciadiocrinus planulatus (Moore and Plummer), 1940. Strimple, 1961g, p. 100.
    Schistocrinus planulatus Moore and Plummer, 1940. Strimple, 1961g,
    p. 100.
    Pennsylvanian, Desmoinesian, Millsap Lake Formation: United States:
    Texas

SCOLIOCRINUS Jaekel, 1895. *S. eremita Jaekel, 1895. Moore and Laudon,
    1943b, p. 50. Ubaghs, 1953, p. 750.
    Devonian

SCOTIACRINUS Wright, 1945. *Pachylocrinus tyriensis Wright, 1937. Wright,
    1945, p. 119. Arendt and Hecker, 1964, p. 88.
    Carboniferous, Visean-Namurian

Scotiacrinus ardrossensis Wright, 1945, p. 120, Pl. 4, figs. 4-6, 10;
    1951b, p. 76, Pl. 35, figs. 13-16.
    Carboniferous, Visean, Calciferous Sandstone series: Scotland

*Scotiacrinus tyriensis (Wright), 1937. Wright, 1945, p. 120, Pl. 4,
    figs. 1-3, 7; 1951b, p. 75, Pl. 35, figs. 6, 17-19; Figs. 38, 39.
    Pachylocrinus tyriensis Wright, 1937. Wright, 1945, p. 120.
    Carboniferous, Visean, Lower Limestone Group: Scotland

Scotiacrinus yoredalensis Wright, 1945, p. 121, Pl. 4, fig. 8; 1951b, p. 77, Pl. 35, fig. 20.
Carboniferous, Namurian (E_1), red beds above Main Limestone: England
SCYPHOCRINITES Zenker, 1833. *S. elegans* Zenker, 1833. Moore and Laudon, 1943b, p. 96. Ubaghs, 1953, p. 741. Van Sant, *in* Van Sant and Lane, 1964, p. 53, fig. 16. Yakovlev, 1964b, p. 70, fig. 108. Arendt and Hecker, 1964, p. 97. Moore and Jeffords, 1968, p. 11, fig. 3, no. 1.
Camarocrinus Hall, 1879. Ubaghs, 1953, p. 741, fig. 121. Yeltyschewa, 1968, Pl. 1, fig. 14.
Lobolithus Waagen and Jahn, 1899. Ubaghs, 1953, p. 741.
Scyphocrinus Zenker, 1833. Cuénot, 1948, p. 70.
Silurian-Devonian

Scyphocrinites cinctus Strimple, 1963d, p. 99, Pl. 7, figs. 4, 5; Pl. 8, fig. 11; Pl. 10, fig. 7; Pl. 12, fig. 3.
Silurian, Niagaran, Henryhouse Formation: United States: Oklahoma

Scyphocrinites cinctus? Strimple, 1963d, p. 101, Pl. 11, figs. 2, 3; Pl. 12, fig. 5.
Silurian, Niagaran, Henryhouse Formation: United States: Oklahoma

Scyphocrinites aff. *cinctus* Strimple, 1963. Yeltyschewa, 1968, p. 37, Pl. 5, fig. 1.
Silurian, Borschovskian horizon: Russia: Podolsky

Scyphocrinites decoratus (Waagen and Jahn), 1899. Yeltyschewa, 1968, p. 36, Pl. 5, fig. 2.
Silurian, Borschovskian horizon: Russia: Podolsky

*Scyphocrinites elegans Zenker, 1833. Ubaghs, 1953, p. 741, figs. 58, 90. Le Maître and Heddebaut, 1963, p. 273. Yakovlev, 1964b, p. 65, fig. 87.
Scyphocrinus elegans Zenker, 1833. Termier and Termier, 1949, p. 50, Pl. 2, figs. 13-15; 1950, p. 85, Pl. 209; Pl. 210, figs. 1-6, 10 (*non* figs. 7-9 as given on p. 85); Pl. 216, figs. 1-4.
Silurian: United States; Morocco

Scyphocrinites elegans? Zenker, 1833. Yeltyschewa, 1968, p. 34, Pl. 2, figs. 1, 3-5 (*non* fig. 6 as given on p. 34).
Silurian, Borschovskian horizon: Russia: Podolsky

Scyphocrinites excavatus (Schlotheim), 1820. Svobodo and others, 1968, p. 310, Pl. 49, fig. 2.
Silurian, Pridoli Formation: Bohemia

Scyphocrinites excavatus schlotheimi (Waagen and Jahn), 1899. Yeltyschewa, 1968, p. 35, Pl. 1, figs. 1-13; Pl. 2, fig. 2.
Scyphocrinus excavatus var. *schlotheimi* Waagen and Jahn, 1899. Yakovlev, 1950, p. 93, fig. 1,I.
Silurian, Borschovskian horizon: Russia: Podolsky

Scyphocrinites excavatus var. *schluteri* (Waagen and Jahn), 1899.
Scyphocrinus excavatus var. *schluteri* Waagen and Jahn, 1899. Yakovlev, 1950, p. 93, fig. 1,II.
No age or locality given

Scyphocrinites gibbosus (Springer), 1917. Strimple, 1963d, p. 102, Pl. 7, fig. 6.
Devonian, Ulsterian, Haragan Formation: United States: Oklahoma

Scyphocrinites mariannae (Yakovlev), 1953. Arendt and Hecker, 1964, p. 97, Pl. 16, figs. 1-3.
Camarocrinus mariannae Yakovlev, 1953, p. 27, Pl. 1, figs. 1-3; 1956a, p. 110, fig. 42, nos. 1, 2; 1958, p. 97, fig. 42, nos. 1, 2; 1964a, p. 97, fig. 42, nos. 1, 2.
Scyphocrinus mariannae Yakovlev, 1953. Yakovlev, 1950, p. 93, fig. 1,III

(nomen nudum); 1953, p. 29, Pl. 3, figs. 1-5; 1964a, p. 101, fig. 41, no. 3.
Silurian, Karagandinian: Russia: Kazakhstan

Scyphocrinites saffordi (Hall), 1879.
Camarocrinus saffordi Hall, 1879. Yakovlev, 1953, p. 19, figs. 1, 4v; 1956a, p. 104, fig. 39, nos. 1, 2; 1958, p. 91, fig. 39, nos. 1, 2; 1964a, p. 95, fig. 41, nos. 1, 2.
Devonian: United States

Scyphocrinites stellatus-mutabilis (Hall-Springer) group. Le Maître and Heddebaut, 1963, p. 273.
Silurian, Wenlockian: France

Scyphocrinites ulrichi (Schuchert), 1903.
Camarocrinus ulrichi Schuchert, 1903. Moore and Laudon, 1944, p. 209, Pl. 79, fig. 34.
Devonian, Ulsterian, Haragan Marl: United States: Oklahoma

Scyphocrinites spp.
Yakovlev, 1953, p. 20, fig. 2.
  Silurian: Russia
Yakovlev, 1964b, p. 67, fig. 98.
  Silurian: North America
Horný, 1964, p. 91, fig. 3.
  Silurian-Devonian, Pridoli Beds: Bohemia

Scyphocrinus spp.
Branson, 1944, p. 113, Pl. 17, fig. 17.
  Devonian, Ulsterian, Bailey Limestone: United States: Missouri
Termier and Termier, 1947, p. 255, Pl. 19, fig. 8.
  Lower Devonian: Morocco
Termier and Termier, 1950, p. 85, Pl. 210, fig. 12.
  Lower Devonian: Morocco
Korejwo and Teller, 1964, p. 236, Pl. 2, fig. 1.
  Upper Silurian, *Pristiograptus bugensius* Zone: Poland

SCYTALOCRINUS Wachsmuth and Springer, 1879. *Scaphiocrinus robustus* Hall, 1861. Moore and Laudon, 1943b, p. 59, fig. 7; 1944, p. 169. Moore, 1950, p. 34, fig. 4h. Arendt and Hecker, 1964, p. 89.
Prininocrinus Goldring, 1938. Moore and Laudon, 1943b, p. 59.
Devonian-Pennsylvanian

Scytalocrinus aftonensis Strimple, 1951g, p. 294, figs. 1-4.
Mississippian, Chesterian, Fayetteville Formation: United States: Oklahoma

Scytalocrinus arbiglandensis Wright, 1934 = *Ophiurocrinus arbiglandensis*

Scytalocrinus beggi Wright, 1938 = *Ophiurocrinus beggi*

Scytalocrinus decadactylus (Meek and Worthen), 1860. Van Sant, *in* Van Sant and Lane, 1964, p. 96.
Mississippian, Osagean, Keokuk Limestone: United States: Indiana?

Scytalocrinus disparilis (Miller and Gurley), 1890. Van Sant, *in* Van Sant and Lane, 1964, p. 96, Pl. 5, figs. 4, 6.
Scaphiocrinus disparilis Miller and Gurley, 1890. Van Sant, *in* Van Sant and Lane, 1964, p. 96.
Mississippian, Osagean, Borden Group: United States: Indiana

Scytalocrinus kalmiusi Yakovlev, *in* Yakovlev and Ivanov, 1956, p. 45, Pl. 12, fig. 6. Arendt and Hecker, 1964, p. 89, Pl. 12, fig. 1.
Lower Carboniferous: Russia: Donetz Basin

Scytalocrinus loreus (Sladen), 1821. Wright, 1951a, p. 35.
Lower Carboniferous: England

Scytalocrinus robustus (Hall), 1861. Van Sant, *in* Van Sant and Lane, 1964, p. 97, Pl. 5, figs. 10, 11, 13, 15.
  Scaphiocrinus porrectus Miller, 1892. Van Sant, *in* Van Sant and Lane, 1964, p. 97.
  Mississippian, Osagean, Borden Group, Burlington-Keokuk transition beds: United States: Indiana, Iowa, Illinois

Scytalocrinus sansabensis Moore and Plummer, 1940 = *Hypselocrinus sansabensis*

Scytalocrinus seafieldensis Wright, 1948, p. 48, Pl. 5, fig. 6; 1951a, p. 34, Pl. 10, fig. 2.
  Carboniferous, Visean, Lower Limestone Group: Scotland

Scytalocrinus validus Wachsmuth and Springer, 1897. Moore and Laudon, 1943b, p. 132, Pl. 3, fig. 2; 1944, p. 169, Pl. 59, fig. 11.
  Mississippian, Osagean, Borden Group: United States: Indiana

SELLARDSICRINUS Moore and Plummer, 1940. *S. marrsae* Moore and Plummer, 1940. Moore and Laudon, 1943b, p. 62; 1944, p. 175.
  Pennsylvanian, Desmoinesian

Sellardsicrinus marrsae Moore and Plummer, 1940. Moore and Laudon, 1944, p. 175, Pl. 64, fig. 29.
  Pennsylvanian, Desmoinesian, Millsap Lake Formation: United States: Texas

SENARIOCRINUS Schmidt, 1934. *S. maucheri* Schmidt, 1934. Moore and Laudon, 1943b, p. 30. Ubaghs, 1953, p. 746. Moore, 1962a, p. 30, fig. 5, no. 6 (*non* fig. 5, no. 7, as given on p. 31). Arendt and Hecker, 1964, p. 81.
  Lower Devonian

Senariocrinus maucheri Schmidt, 1934. Moore, 1962a, p. 30, Pl. 3, fig. 1; Fig. 19. Strimple, 1963d, p. 54, figs. 12b, c.
  Lower Devonian, Hunsrück Limestone: Germany

SIDEROCRINUS Wood, 1909. *S. ornatus* Troost, *in* Wood, 1909. Ubaghs, 1953, p. 737.
  Silurian

SIGAMBROCRINUS Schmidt, 1941. *S. laevis* Schmidt, 1941. Schmidt, 1941, p. 178.
  Devonian, Eifelian

Sigambrocrinus laevis Schmidt, 1941, p. 178, Pl. 11, fig. 7; Pl. 19, fig. 2; Fig. 48a.
  Cyathocrinus loganensis Müller, 1859. Schmidt, 1941, p. 178.
  Devonian, Eifelian, *Orthocrinus* Schichten: Germany

SINOCRINUS Tien, 1926. *S. microgranulosus* Tien, 1926. Moore and Laudon, 1943b, p. 60. Moore, 1962b, p. 10, fig. 1, no. 10. Arendt and Hecker, 1964, p. 93.
  Carboniferous

SIPHONOCRINUS Miller, 1888. *Glyptocrinus nobilis* Hall, 1861. Moore and Laudon, 1943b, p. 84; 1944, p. 187. Ubaghs, 1953, p. 737, fig. 27. Strimple, 1963d, p. 72, fig. 26 (part).
  Silurian

Siphonocrinus armosus (McChesney), 1861. Ubaghs, 1953, p. 737, fig. 27.
  Silurian, Niagaran: United States

Siphonocrinus dignus Strimple, 1963d, p. 73, Pl. 3, figs. 6, 7; Fig. 20a.
  Silurian, Niagaran, Henryhouse Formation: United States: Oklahoma

Siphonocrinus nobilis (Hall), 1865. Moore and Laudon, 1944, p. 187, Pl. 72, fig. 21.

Silurian, Niagaran, Racine Dolomite: United States: Indiana, Wisconsin, Illinois, Iowa

Siphonocrinus pentagonus Wachsmuth and Springer, 1897 = *Ochlerocrinus pentagonus*

SITULACRINUS Breimer, 1962. *S. costatus* Breimer, 1962. Breimer, 1962a, p. 153.
Devonian, Emsian

Situlacrinus costatus Breimer, 1962a, p. 154, Pl. 15, figs. 1-6; Figs. 36, 37.
Devonian, Emsian, La Vid Shale: Spain

SPANIOCRINUS Wanner, 1924. *S. validus* Wanner, 1924. Moore and Laudon, 1943b, p. 60, fig. 8. Arendt and Hecker, 1964, p. 92.
Permian

Spaniocrinus transcaucasicus Yakovlev, 1933. Yakovlev, *in* Yakovlev and Ivanov, 1956, p. 83, Pl. 19, fig. 3. Arendt and Hecker, 1964, p. 92, Pl. 13, fig. 19.
Upper Permian: Russia: Transcaucasian area

Spaniocrinus validus Wanner, 1924. Arendt and Hecker, 1964, p. 92, Pl. 13, fig. 18.
Permian: Timor

SPHAEROCRINUS Roemer, 1851. *Cyathocrinites geometricus* Goldfuss, 1831. Moore and Laudon, 1943b, p. 50. Ubaghs, 1953, p. 750. Xu, 1962, p. 45, fig. 1.
Devonian

Sphaerocrinus paucisculptus Wanner, 1942a, p. 31, Pl. 1, figs. 3a-c.
Devonian, Eifelian: Germany

Sphaerocrinus sp. indet. Schmidt, 1952, p. 160, Pl. 9, figs. 4a-c.
Lower Devonian: Spain

SPHAEROTOCRINUS Goldring, 1923. *S. ornatus* Goldring, 1923. Moore and Laudon, 1943b, p. 82; 1944, p. 185. Ubaghs, 1953, p. 737. Arendt and Hecker, 1964, p. 95.
Devonian

Sphaerotocrinus ornatus Goldring, 1923. Moore and Laudon, 1943b, p. 138, Pl. 9, fig. 3; 1944, p. 185, Pl. 72, fig. 24.
Devonian, Ulsterian, Lewistown Limestone: United States: Pennsylvania

SPHENISOCRINUS Wanner, 1937. *S. spinosus* Wanner, 1937. Moore and Laudon, 1943b, p. 56. Arendt and Hecker, 1964, p. 88.
Permian

SPRINGERICRINUS Jaekel, 1918. *Poteriocrinus magniventrus* Springer, 1911. Moore and Laudon, 1943b, p. 54; 1944, p. 157.
Mississippian, Osagean

Springericrinus doris (Hall), 1861. Moore and Laudon, 1943b, p. 133, Pl. 4, fig. 6; 1944, p. 157, Pl. 58, fig. 12.
Scaphiocrinus doris Hall, 1861. Moore and Laudon, 1943b, p. 133.
Mississippian, Osagean, Burlington Limestone: United States: Iowa

Springericrinus magniventrus (Springer), 1911. Moore and Laudon, 1944, p. 157, Pl. 58, fig. 9. Van Sant, *in* Van Sant and Lane, 1964, p. 90, Pl. 4, figs. 13, 17.
Mississippian, Osagean, Keokuk Limestone: United States: Indiana

SPYRIDIOCRINUS Oehlert, 1889. *S. cheuxi* Oehlert, 1889. Ubaghs, 1950, p. 108, fig. 8; 1953, p. 735.
Lahuseniocrinus Tschernyschew, 1892. Moore and Laudon, 1943b, p. 82.

Ubaghs, 1950, p. 108.
Lower-Middle Devonian

Spyridiocrinus cheuxi Oehlert, 1889. Ubaghs, 1950, p. 109, Pl. 1, figs. 1-6; Figs. 2-6.
Lower-Middle Devonian, Calcaire d'Angers: Belgium

Spyridiocrinus tirlensis (Tschernyschew), 1892. Ubaghs, 1950, p. 109, fig. 1.
Lahuseniocrinus tirlensis Tschernyschew, 1892. Ubaghs, 1950, p. 109.
Lower Devonian: Russia: Ural Mountains

STACHYOCRINUS Wanner, 1916. *S. zea* Wanner, 1916. Moore and Laudon, 1943b, p. 60. Wanner, 1949a, p. 42. Arendt and Hecker, 1964, p. 90.
Permian

Stachyocrinus stefaninii (Yakovlev), 1934. Wanner, 1949a, p. 42.
Erisocrinus stefaninii Yakovlev, 1934. Wanner, 1949a, p. 42.
Permian: Sicily

Stachyocrinus timanicus Yakovlev, 1941b, p. 103, fig. 2. Yakovlev, *in* Yakovlev and Ivanov, 1956, p. 75, Pl. 13, fig. 3. Arendt and Hecker, 1964, p. 90, Pl. 12, fig. 5.
Lower Permian (P_1): Russia: Timan

STAMNOCRINUS Breimer, 1962. *Saccocrinus(?) intrastigmatus* Schmidt, 1931. Breimer, 1962a, p. 59.
Lower-Middle Devonian

Stamnocrinus devonicus (Springer), 1911. Breimer, 1962a, p. 64.
Dorycrinus devonicus Springer, 1911. Breimer, 1962a, p. 64.
Devonian, Erian, Hamilton Group: United States: Indiana

Stamnocrinus intrastigmatus (Schmidt), 1931. Breimer, 1962a, p. 59, Pl. 4, fig. 5; Pl. 7, figs. 1-7.
Pithocrinus intrastigmatus (Schmidt), 1931. Kirk, 1945a, p. 345. Breimer, 1962a, p. 59.
Saccocrinus(?) intrastigmatus Schmidt, 1931. Schmidt, 1952, p. 149, Pl. 9, figs. 5-7. Breimer, 1962a, p. 59.
Devonian, lower Emsian, red calcareous shale: Spain

Stamnocrinus sp. 1. Breimer, 1962a, p. 65, Pl. 4, fig. 4.
Devonian, lower Emsian, red calcareous shale: Spain

Stamnocrinus sp. 2. Breimer, 1962a, p. 65, fig. 12.
Devonian: locality unknown

STEGANOCRINUS Meek and Worthen, 1866. *Actinocrinus pentagonus* Hall, 1858. Moore and Laudon, 1943b, p. 93, fig. 15; 1944, p. 193, Pl. 71, fig. 12. Ubaghs, 1953, p. 739. Strimple, 1963d, fig. 27 (part). Arendt and Hecker, 1964, p. 96.
Mississippian, Osagean

Steganocrinus albersi Miller and Gurley, 1897 = *Actinocrinites multiradiatus*

Steganocrinus altus Brower, 1965, p. 780, Pl. 93, figs. 7, 8, 13; Fig. 3.
Mississippian, Osagean, Nunn Member: United States: New Mexico

Steganocrinus araneolus (Meek and Worthen), 1860 = *Steganocrinus pentagonus*

Steganocrinus benedicti Miller, 1892 = *Actinocrinites benedicti*

Steganocrinus burlingtonensis Brower, 1965, p. 788, Pl. 93, figs. 9, 10, 16; Fig. 7.
Mississippian, Osagean, Burlington Limestone: United States: Iowa

Steganocrinus concinnus (Shumard), 1855. Kirk, 1943a, p. 260. Brower, 1965, p. 784, Pl. 93, figs. 1-4.
Mississippian, Osagean, Burlington Limestone: United States: Missouri, Iowa, Illinois

Steganocrinus concinnus Wachsmuth and Springer, 1897 = *Steganocrinus elongatus*

Steganocrinus elongatus Kirk, 1943a, p. 261. Brower, 1965, p. 783, Pl. 93, figs. 11, 12.
   Steganocrinus concinnus Wachsmuth and Springer, 1897. Kirk, 1943a, p. 261.
   Mississippian, Osagean, Burlington Limestone: United States: Iowa, Missouri

Steganocrinus? globosus Wachsmuth and Springer, 1897 = *Sampsonocrinus? globosus*

Steganocrinus longus Brower, 1965, p. 781, Pl. 94, figs. 1-4; Fig. 4.
   Mississippian, Osagean, Nunn Member: United States: New Mexico

Steganocrinus multistriatus Brower, 1965, p. 782, Pl. 93, figs. 14, 15, 17.
   Mississippian, Osagean, Burlington Limestone: United States: Missouri

Steganocrinus pentagonus (Hall), 1858. Kirk, 1943a, p. 261, figs. 1, 2. Moore and Laudon, 1943b, p. 139, Pl. 10, fig. 12; 1944, p. 193, Pl. 77, fig. 6. Laudon, 1948, Pl. 3. Brower, 1965, p. 776, Pl. 91, figs. 1-15; Pl. 92, figs. 1-37; Fig. 1; 1967, p. 683, Pl. 76, fig. 9.
   Steganocrinus araneolus (Meek and Worthen), 1860. Kirk, 1943a, p. 260, fig. 4. Brower, 1965, p. 776.
   Mississippian, Osagean, Burlington Limestone, Fern Glen Limestone, Nunn Member: United States: Iowa, Missouri, New Mexico

Steganocrinus planus Brower, 1965, p. 787, Pl. 94, figs. 10-12; Fig. 6.
   Mississippian, Osagean, Burlington Limestone: United States: Missouri

Steganocrinus robustus Brower, 1965, p. 786, Pl. 94, figs. 5-8.
   Mississippian, Osagean, Burlington Limestone: United States: Missouri, Iowa

Steganocrinus sculptus (Hall), 1858. Moore and Laudon, 1943b, p. 140, Pl. 11, fig. 3; 1944, p. 193, Pl. 77, fig. 5.
   Mississippian, Osagean, Burlington Limestone: United States: Iowa, Missouri

Steganocrinus validus (Meek and Worthen), 1860. Kirk, 1943a, p. 261. Brower, 1965, p. 785, Pl. 93, figs. 5, 6; Fig. 5.
   Mississippian, Osagean, Burlington Limestone: United States: Iowa

Steganocrinus westheadi Wright, 1947 = *Sampsonocrinus westheadi*

Steganocrinus sp. Brower, 1965, p. 783, Pl. 94, fig. 9.
   Mississippian, Osagean, Burlington Limestone: United States: Missouri

STEGOCRINUS Sieverts-Doreck, 1962. *S. dohmi* Sieverts-Doreck, 1962. Sieverts-Doreck, 1962, p. 107.
Devonian

Stegocrinus bloosi Sieverts-Doreck, 1962, p. 115, Pl. 14, fig. 6; Fig. 7.
   Middle Devonian, Fleringer Schichten: Germany

Stegocrinus dohmi Sieverts-Doreck, 1962, p. 112, Pl. 14, figs. 1-3; Figs. 3-5.
Devonian, upper *Calceola* Beds: Germany

Stegocrinus cf. *dohmi* Sieverts-Doreck, 1962, p. 114, Pl. 14, fig. 4; Fig. 6.
Devonian, upper *Calceola* Beds: Germany

Stegocrinus sp. Sieverts-Doreck, 1962, p. 114, Pl. 14, fig. 5.
Devonian, upper *Calceola* Beds: Germany

STELIDIOCRINUS Angelin, 1878. **S*. capitulum* Angelin, 1878. Moore and Laudon, 1943b, p. 98. Ubaghs, 1953, p. 742.
?Ordovician-Silurian

STELLAROCRINUS Strimple, 1940. **Cyathocrinus stillativus* White, 1879. Moore and Laudon, 1943b, p. 55; 1944, p. 157, Pl. 52, fig. 14.
Pennsylvanian, Desmoinesian-Permian, Wolfcampian

Stellarocrinus angulatus (Miller and Gurley), 1894. Strimple, 1962e, p. 41, Pl. 3, figs. 2-4.
Pennsylvanian, Desmoinesian, Oologah Limestone: United States: Oklahoma

Stellarocrinus comptus Webster and Lane, 1967, p. 13, Pl. 3, fig. 3.
Permian, Wolfcampian, Bird Spring Formation: United States: Nevada

Stellarocrinus cuneatus Lane and Webster, 1966, p. 24, Pl. 3, figs. 5, 7; Pl. 13, fig. 21; Fig. 7. Webster and Lane, 1967, p. 12.
Permian, Wolfcampian, Bird Spring Formation: United States: Nevada

Stellarocrinus florealis (Moore and Plummer), 1940. Moore and Laudon, 1944, p. 157, Pl. 60, fig. 14.
Pennsylvanian, Desmoinesian, Millsap Lake Formation: United States: Texas

Stellarocrinus geometricus (Moore and Plummer), 1940. Moore and Laudon, 1943b, p. 133, Pl. 4, fig. 1; 1944, p. 157, Pl. 58, fig. 5.
Pennsylvanian, Missourian, Iola Beds: United States: Kansas

Stellarocrinus minimus Strimple, 1962e, p. 42, Pl. 3, figs. 5-7.
Pennsylvanian, Desmoinesian, Oologah Limestone: United States: Oklahoma

Stellarocrinus petalosus Strimple, 1961g, p. 110, Pl. 14, figs. 3, 4; Pl. 15, fig. 4; Pl. 16, fig. 2; Pl. 19, fig. 7.
Pennsylvanian, Desmoinesian, Holdenville Shale: United States: Oklahoma

**Stellarocrinus stillativus* (White), 1879. Moore and Laudon, 1943b, p. 133, Pl. 4, fig. 2; 1944, p. 157, Pl. 57, fig. 13.
Pennsylvanian, Virgilian, Oread Limestone: United States: Kansas

Stellarocrinus texani Strimple, 1951a, p. 202, Pl. 37, figs. 13-16.
Pennsylvanian, Missourian, Lake Bridgeport Shale: United States: Texas

Stellarocrinus virgilensis Strimple, 1951f, p. 675, Pl. 98, figs. 1, 2, 5, 6.
Pennsylvanian, Virgilian, Nelagony Formation: United States: Oklahoma

Stellarocrinus sp. Webster and Lane, 1967, p. 13, Pl. 3, figs. 4, 6.
Permian, Wolfcampian, Bird Spring Formation: United States: Nevada

STEMMATOCRINUS = *Erisocrinus*

Stemmatocrinus cernuus Trautschold, 1867 = *Erisocrinus cernus*

Stemmatocrinus trautscholdi Wachsmuth and Springer, 1885 = *Gaulocrinus trautscholdi*

Stemmatocrinus? veryi Rowley, 1903 = *Gaulocrinus veryi*

STENOPECRINUS Strimple, 1961. **Perimestocrinus planus* Strimple, 1952. Strimple, 1961g, p. 39.
Pennsylvanian, Morrowan-Permian, Wolfcampian

Stenopecrinus hexagonus (Strimple), 1952. Strimple, 1961g, p. 40.
Perimestocrinus hexagonus Strimple, 1952e, p. 785, Pl. 112, figs. 1-3; Pl. 113, figs. 1-7. Strimple, 1961g, p. 40.
Pennsylvanian, Missourian, Wann Formation: United States: Oklahoma

Stenopecrinus moseleyi (Strimple), 1951. Strimple, 1961g, p. 40.
Perimestocrinus moseleyi Strimple, 1951a, p. 205, Pl. 39, figs. 5-10;

1961g, p. 40.
Pennsylvanian, Missourian, Lake Bridgeport Shale: United States: Texas

Stenopecrinus planus (Strimple), 1952. Strimple, 1961g, p. 40, Pl. 1, figs. 4, 5; Figs. 5a, b.
Perimestocrinus planus Strimple, 1952e, p. 787, Pl. 112, figs. 4-9; 1961g, p. 40.
Pennsylvanian, Missourian, shale 30 ft above Torpedo Sandstone: United States: Oklahoma

Stenopecrinus politus (Moore), 1939. Strimple, 1961g, p. 40.
Plaxocrinus politus Moore, 1939. Moore and Laudon, 1944, p. 163, Pl. 63, fig. 1. Strimple, 1961g, p. 40.
Pennsylvanian, Virgilian, Nelagoney Formation: United States: Oklahoma

Stenopecrinus rugosus Strimple, 1961g, p. 42, Pl. 2, figs. 4, 5; Pl. 14, fig. 7; Figs. 7a, b.
Pennsylvanian, Morrowan, Wapanucka Limestone: United States: Oklahoma

Stenopecrinus? xerophilus Lane and Webster, 1966, p. 50, Pl. 12, figs. 3, 4, 6.
Permian, Wolfcampian, Bird Spring Formation: United States: Nevada

STEPHANOCRINUS Conrad, 1842. *S. angulatus* Conrad, 1842.
Silurian

Stephanocrinus angulatus Conrad, 1842. Fay, 1961, p. 236, Pl. 1, figs. 1-4; 1962a, p. 206, Pl. 35, figs. 1-9; Fig. 1.
Silurian, Niagaran, Rochester Shale: United States: New York

Stephanocrinus gemmiformis Hall, 1852. Wilson, 1949, p. 251, Pl. 25, figs. 16, 17. Fay, 1960, p. 256, Pl. 1; Pl. 2, figs. 1-8.
Silurian, Niagaran, Waldron Shale: United States: Tennessee

STEREOBRACHICRINUS Mather, 1915. *S. pustullosus* Mather, 1915. Moore and Laudon, 1943b, p. 30. Arendt and Hecker, 1964, p. 81.
Pennsylvanian

Stereocrinus Barris, 1878 = *Dolatocrinus*

Stereocrinus barrisi Wachsmuth and Springer, 1897 = *Dolatocrinus barrisi*

Stereocrinus helderbergensis Springer, 1921 = *Dolatocrinus helderbergensis*

Stereocrinus triangulatus Barris, 1878 = *Dolatocrinus triangulatus*

Stereocrinus triangulatus liratus Barris, 1878 = *Dolatocrinus triangulatus*

Stereocrinus vandiveri Branson and Wilson, 1922 = *Dolatocrinus vandiveri*

STINOCRINUS Kirk, 1941. *S. granulosus* Kirk, 1941. Moore and Laudon, 1943b, p. 55; 1944, p. 158. Arendt and Hecker, 1964, p. 88.
Mississippian, Osagean

Stinocrinus nanus (Miller and Gurley), 1896. Moore and Laudon, 1944, p. 159, Pl. 60, fig. 15 (misspelled as *S. manus*).
Batocrinus nanus Miller and Gurley, 1896.
Mississippian, Osagean, Keokuk Limestone: United States: Indiana

STIPTOCRINUS Kirk, 1946. *S. spinosus* Kirk, 1946. Kirk, 1946a, p. 33. Ubaghs, 1953, p. 739.
Silurian, Niagaran

Stiptocrinus benedicti (Miller), 1892. Kirk, 1946a, p. 34.
Saccocrinus benedicti Miller, 1892. Kirk, 1946a, p. 34.
Silurian, Niagaran, Laurel Limestone: United States: Indiana

Stiptocrinus carinatus Kirk, 1946a, p. 34.
Saccocrinus benedicti Springer, 1926 (part, Pl. 10, fig. 14). Kirk, 1946a, p. 34.
Silurian, Niagaran, Laurel Limestone: United States: Indiana

Stiptocrinus chicagoensis (Weller), 1900. Kirk, 1946a, p. 35.
  Periechocrinus chicagoensis Weller, 1900. Kirk, 1946a, p. 35.
  Silurian, Niagaran, Racine Dolomite: United States: Illinois

Stiptocrinus farringtoni (Slocom), 1908. Kirk, 1946a, p. 35.
  Habrocrinus farringtoni Slocom, 1908. Kirk, 1946a, p. 35.
  Silurian, Niagaran, Racine Dolomite: United States: Illinois

Stiptocrinus howardi (Miller), 1892. Kirk, 1946a, p. 35.
  Saccocrinus howardi Miller, 1892. Kirk, 1946a, p. 35.
  Silurian, Niagaran, Laurel Limestone: United States: Indiana

Stiptocrinus nodosus (Springer), 1926. Kirk, 1946a, p. 35.
  Aorocrinus nodosus Springer, 1926. Kirk, 1946a, p. 35.
  Silurian, Niagaran, Decatur Limestone: United States: Tennessee

*Stiptocrinus spinosus Kirk, 1946a, p. 35.
  Saccocrinus benedicti Springer, 1926 (part, Pl. 10, figs. 11, 12).
    Kirk, 1946a, p. 35.
  Silurian, Niagaran, Laurel Limestone: United States: Indiana

Stiptocrinus sp. Kirk, 1946a, p. 36.
  Habrocrinus sp. Foerste, 1917. Kirk, 1946a, p. 36.
  Silurian, Niagaran, Cedarville Dolomite: United States: Ohio

STOMIOCRINUS Wanner, 1937. *S. subglobosus* Wanner, 1937. Moore and Laudon,
  1943b, p. 96. Ubaghs, 1953, p. 741. Arendt and Hecker, 1964, p. 96.
  Permian

Stomiocrinus permiensis (Yakovlev), 1927. Yakovlev, *in* Yakovlev and Ivanov,
  1956, p. 67, Pl. 10, figs. 21-26. Arendt and Hecker, 1964, p. 96,
  Pl. 15, figs. 4, 5.
  Permian, Artinskian: Russia: central Ural Mountains

STORTHINGOCRINUS Schultze, 1866. **Platycrinus fritillus* Müller, 1855.
  Moore and Laudon, 1943b, p. 31. Ubaghs, 1953, p. 746.
  Devonian

*Storthingocrinus fritillus (Müller, *in* Zeiler and Wirtgen), 1855. Kesling
  and Smith, 1963, p. 188.
  Symbathocrinus celechovicensis Remes, 1929. Kesling and Smith, 1963,
    p. 188.
  Middle Devonian: Czechoslovakia

Storthingocrinus haugi Oehlert, 1896. Breimer, 1962a, p. 145; 1962b,
  p. 326. ICZN, China, 1965, p. 45.
  Devonian: Spain

Storthingocrinus cf. *haugi* Oehlert, 1897. Schmidt, 1952, p. 155, Pl. 10,
  figs. 2a, b.
  Lower Devonian: Spain

Storthingocrinus labiatus Schmidt, 1931. Schmidt, 1952, p. 156, Pl. 10,
  figs. 3a, b. Breimer, 1962a, p. 145; 1962b, p. 326. ICZN, China,
  1965, p. 45.
  Lower Devonian: Spain

STREBLOCRINUS Koenig and Meyer, 1965. **S. brachiatus* Koenig and Meyer,
  1965. Koenig and Meyer, 1965, p. 393. Lane, 1967b, p. 13, fig. 2.
  Devonian, Erian

*Streblocrinus brachiatus Koenig and Meyer, 1965, p. 394, figs. 2, 3B, 4h-n.
  Devonian, Erian, Windom Member: United States: New York

STREPTOCRINUS Wachsmuth and Springer, 1886. **Ophiocrinus crotalurus*
  Angelin, 1878. Moore and Laudon, 1943b, p. 54.
  Silurian

STRONGYLOCRINUS Wanner, 1916. *S. molengraffi Wanner, 1916. Moore and Laudon, 1943b, p. 55. Arendt and Hecker, 1964, p. 92.
Permian

Strongylocrinus uralicus Yakovlev, 1937. Yakovlev, in Yakovlev and Ivanov, 1956, p. 63, Pl. 11, fig. 16. Arendt and Hecker, 1964, p. 92, Pl. 13, figs. 12, 13.
Permian, Artinskian: Russia: central Ural Mountains

STROPHOCRINUS Sardeson, 1899. *S. dicyclicus Sardeson, 1899. Moore and Laudon, 1943b, p. 49. Ubaghs, 1953, p. 750.
Ordovician

STROTOCRINUS Meek and Worthen, 1866. *Actinocrinus regalis Hall, 1860. Moore and Laudon, 1943b, p. 93, fig. 15; 1944, p. 193, Pl. 71, fig. 14. Ubaghs, 1953, p. 739. Arendt and Hecker, 1964, p. 96.
Mississippian

Strotocrinus? asperrimus Meek and Worthen, 1869 = Cusacrinus asperrimus

Strotocrinus ectypus Meek and Worthen, 1869 = Cusacrinus ectypus

Strotocrinus glyptus (Hall), 1860. Moore and Laudon, 1944, p. 193, Pl. 77, fig. 9. Laudon, 1948, Pl. 3.
Mississippian, Osagean, Burlington Limestone: United States: Iowa, Missouri

*Strotocrinus regalis (Hall), 1860. Melendez, 1947, p. 335, fig. 178.
Lower Carboniferous: Spain

Strotocrinus sp. Ubaghs, 1953, p. 739, fig. 138a. Yakovlev, 1964b, p. 70, fig. 112.
Mississippian

STUARTWELLERCRINUS Moore and Plummer, 1938. *Cibolocrinus turbinatus Weller, 1909. Moore and Laudon, 1943b, p. 60, fig. 8; 1944, p. 175. Arendt and Hecker, 1964 (misspelled as Stuartwellecrinus).
Pennsylvanian, Morrowan-Permian, Wolfcampian

Stuartwellercrinus argentinei Strimple, 1949b, p. 125, Pl. 4, figs. 1-4; Fig. 1b; 1960b, p. 155, fig. 2E.
Pennsylvanian, Missourian, Argentine Limestone: United States: Missouri

Stuartwellercrinus corbatoi Lane and Webster, 1966, p. 37, Pl. 8, figs. 1-9; Figs. 11-14. Webster and Lane, 1967, p. 20, Pl. 2, fig. 12.
Permian, Wolfcampian, Bird Spring Formation: United States: Nevada

Stuartwellercrinus praedecta Strimple, 1961c, p. 188, Pl. 1, figs. 7-9.
Pennsylvanian, Morrowan, Wapanucka Limestone: United States: Oklahoma

Stuartwellercrinus pusillus (Wanner), 1937. Wanner, 1949a, p. 48, Pl. 3, fig. 22.
Cibolocrinus pusillus Wanner, 1937. Wanner, 1949a, p. 48.
Permian: Timor

Stuartwellercrinus symmetricus (Weller), 1909. Moore and Laudon, 1943b, p. 135, Pl. 6, fig. 6; 1944, p. 175, Pl. 60, fig. 19.
Permian, Wolfcampian, Cibolo Formation: United States: Texas

Stuartwellercrinus sp. Termier and Termier, 1958, p. 53, Pl. 2, figs. a-c.
Permian: Tunisia

Stuartwellercrinus? sp. Termier and Termier, 1950, p. 91, Pl. 222, figs. 9-13.
Carboniferous, Westphalian: Morocco

STYLOCRINUS Sandberger and Sandberger, 1856. *Platycrinites scaber Goldfuss, in Sandberger and Sandberger, 1856. Moore and Laudon, 1943b,

p. 31. Ubaghs, 1953, p. 746.
Silurian-Devonian

Stylocrinus? canandaigua Goldring, 1923 = *Synbathocrinus canandaigus*

Stylocrinus elimatus Strimple, 1963d, p. 34, Pl. 1, figs. 6-8.
Silurian, Niagaran, Henryhouse Formation: United States: Oklahoma

SUNDACRINUS Wanner, 1916. *S. granulatus* Wanner, 1916. Moore and Laudon, 1943b, p. 64, fig. 8. Cuénot, 1948, p. 64. Arendt and Hecker, 1964, p. 92.
Permian

Sundacrinus granulatus Wanner, 1916. Arendt and Hecker, 1964, p. 92, Pl. 13, fig. 15; Fig. 133.
Permian: Timor

Sundacrinus septentrionalis Yakovlev, 1937. Yakovlev, *in* Yakovlev and Ivanov, 1956, p. 62, Pl. 11, fig. 18. Arendt and Hecker, 1964, p. 92, Pl. 13, fig. 14.
Permian, Artinskian: Russia: central Ural Mountains

SUNWAPTACRINUS Laudon, Parks and Spreng, 1952. *S. brazeauensis* Laudon, Parks and Spreng, 1952. Laudon and others, 1952, p. 569.
Mississippian, Kinderhookian

Sunwaptacrinus brazeauensis Laudon, Parks and Spreng, 1952, p. 570, Pl. 67, figs. 21-23; Pl. 69, figs. 20, 21.
Mississippian, Kinderhookian, Banff formation: Canada: Alberta

SYCOCRINITES Austin and Austin, 1842. *S. anapeptamenus* Austin and Austin, 1842. Moore and Laudon, 1943b, p. 50. Arendt and Hecker, 1964, p. 85. Lane, 1967b, p. 12, fig. 6.
Carboniferous

Sycocrinites anapeptamenus Austin and Austin, 1842. Wright, 1952a, p. 133, Pl. 36, figs. 7-9.
Carboniferous, Visean: England

Sycocrinites jacksoni Austin and Austin, 1942 = *Lageniocrinus jacksoni*

SYGCAULOCRINUS Ulrich, 1924. *S. typus* Ulrich, 1924. Moore and Laudon, 1943b, p. 29; 1944, p. 145, Pl. 52, fig. 7. Ubaghs, 1953, p. 746. Moore, 1962b, p. 12, fig. 3, no. 7.
Ordovician-Early Silurian

Sygcaulocrinus typus Ulrich, 1924. Moore and Laudon, 1944, p. 145, Pl. 52, fig. 7.
Ordovician, Cincinnatian, Wykoff Member: United States: Iowa

SYNAPTOCRINUS Springer, 1920. *Forbesiocrinus nuntius* Hall, 1862. Moore and Laudon, 1943b, p. 76, fig. 11; 1944, p. 183, Pl. 66, fig. 10. Ubaghs, 1953, p. 756.
Devonian, Erian

Synaptocrinus nuntius (Hall), 1862. Moore and Laudon, 1943b, p. 136, Pl. 7, fig. 5; 1944, p. 183, Pl. 68, fig. 7.
Devonian, Erian, Hamilton Group: United States: New York

SYNARMOCRINUS Lane, 1964. *S. brachiatus* Lane, 1964. Lane, 1964b, p. 678. Strimple, 1966a, p. 8.
Pennsylvanian, Morrowan-Atokan

Synarmocrinus brachiatus Lane, 1964b, p. 678, Pl. 112, figs. 14, 15.
Pennsylvanian, Atokan, Bird Spring Formation: United States: Nevada

Synarmocrinus depressus Washburn, 1968, p. 123, Pl. 2, figs. 7-9.
Pennsylvanian, Morrowan, Oquirrh Formation: United States: Utah

Synarmocrinus fundundus Strimple, 1966a, p. 8, Pl. 2, figs. 1-4.
  Pennsylvanian, Atokan, Atoka Formation: United States: Oklahoma

SYNBATHOCRINUS Phillips, 1836. *S. conicus* Phillips, 1836. Moore and
  Laudon, 1943b, p. 31, figs. 1, 3; 1944, p. 151, Pl. 52, fig. 10.
  Moore, 1950, figs. 1a, b. Ubaghs, 1953, p. 746. Moore, 1962b,
  p. 16, fig. 11, no. 3. Philip, 1963, p. 266, fig. 4d. Van Sant, *in*
  Van Sant and Lane, 1964, p. 70, fig. 26, no. 2.
  Devonian, Ulsterian-Mississippian, Meramecian

Synbathocrinus anglicus Wright, 1946b, p. 124, Pl. 7, fig. 10; Pl. 8,
  fig. 1. Moore, 1948, p. 390, fig. 11. Wright, 1952a, p. 136,
  Pl. 36, figs. 18, 19. Kesling and Smith, 1963, p. 190.
  Carboniferous, Tournaisian: Scotland

Synbathocrinus angularis Miller and Gurley, 1894. Kesling and Smith, 1963,
  p. 190.
  Mississippian, Osagean, New Providence Shale: United States: Kentucky

Synbathocrinus antiquus Strimple, 1952 = *Abyssocrinus antiquus*

Synbathocrinus blari Miller, 1891. Kesling and Smith, 1963, p. 190.
  Mississippian, Meramecian, Warsaw Formation: United States: Missouri

Synbathocrinus brevis Meek and Worthen, 1869. Kesling and Smith, 1963,
  p. 191.
  Mississippian, Osagean, lower Burlington Limestone: United States:
  Iowa

Synbathocrinus campanulatus (Wanner), 1916 = *Taidocrinus campanulatus*

Synbathocrinus campanulatus elongatus (Wanner), 1916 = *Taidocrinus campanulatus elongatus*

Synbathocrinus campanulatus inflatus (Wanner), 1916 = *Taidocrinus campanulatus inflatus*

Synbathocrinus canandaigua (Goldring), 1923. Koenig, 1965, p. 401,
  figs. 1A, 2a-j.
  Stylocrinus? canandaigua Goldring, 1923. Koenig, 1965, p. 401.
  Devonian, Erian, Windom Shale: United States: New York

Synbathocrinus celechovicensis (Remes), 1929 = *Storthingocrinus fritillus*

*Synbathocrinus conicus Phillips, 1836. Wright, 1946b, p. 124, Pl. 7,
  figs. 2-7. Moore, 1948, p. 390, fig. 11. Wright, 1952a, p. 134,
  Pl. 36, figs. 10, 21, 22. Kesling and Smith, 1963, p. 191.
  Carboniferous, Tournaisian: Scotland; England

Synbathocrinus constrictus (Wanner), 1916 = *Taidocrinus constrictus*

Synbathocrinus constrictus sinuosus (Wanner), 1916 = *Taidocrinus constrictus sinuosus*

Synbathocrinus dentatus Owen and Shumard, 1852. Moore and Laudon, 1943b,
  p. 130, Pl. 1, fig. 9; 1944, p. 151, Pl. 55, fig. 16. Ubaghs,
  1953, p. 746, fig. 18e. Kesling and Smith, 1963, p. 191.
  Mississippian, Osagean, Burlington Limestone, Fern Glen Limestone:
  United States: Iowa, Missouri

Synbathocrinus expansus Goldring, 1935. Kesling and Smith, 1963, p. 191.
  Devonian, Erian, West Brook Member: United States: New York

Synbathocrinus granulatus (Troost), 1849. Kesling and Smith, 1963, p. 192.
  Mississippian, Osagean, Fort Payne Chert: United States: Tennessee

Synbathocrinus granuliferus Wetherby, 1880. Kesling and Smith, 1963, p. 192.
  Mississippian, Osagean, New Providence Shale: United States: Kentucky

Synbathocrinus hamiltonensis Springer, 1923. Kesling and Smith, 1963, p. 192.
  Devonian, Erian, Moscow Shale: United States: New York

Synbathocrinus illinoisensis Miller and Gurley, 1896. Kesling and Smith, 1963, p. 192.
Mississippian, Osagean, Burlington Limestone: United States: Illinois

Synbathocrinus matutinus Hall, 1858. Moore and Laudon, 1944, p. 151, Pl. 56, fig. 22. Kesling and Smith, 1963, p. 192, Pl. 1, figs. 22, 23.
Devonian, Erian, Hamilton Formation: United States: Iowa

Synbathocrinus melba Strimple, 1938 = *Taidocrinus? melba*

Synbathocrinus michiganensis Kesling and Smith, 1963, p. 189, Pl. 1, figs. 1-16, 20, 21 (*non* figs. 1-21 as given on p. 189).
Devonian, Erian, Dock Street Clay, Thunder Bay Limestone: United States: Michigan

Synbathocrinus sp. cf. *S. michiganensis* Kesling and Smith, 1963, p. 192, Pl. 1, figs. 17-19.
Devonian, Erian, Thunder Bay Limestone: United States: Michigan

Synbathocrinus onondaga Springer, 1923. Kesling and Smith, 1963, p. 193. Koenig, 1965, p. 402, fig. 2B.
Devonian, Ulsterian, Onondaga Limestone: United States: Kentucky

Synbathocrinus oweni Hall, 1860. Kesling and Smith, 1963, p. 193.
Mississippian, Kinderhookian, Rockford Limestone: United States: Indiana

Synbathocrinus robustus Shumard, 1866. Moore and Laudon, 1943b, p. 130, Pl. 1, fig. 10; 1944, p. 151, Pl. 55, fig. 11; Pl. 56, fig. 23. Kesling and Smith, 1963, p. 194.
Mississippian, Osagean, New Providence Shale, Keokuk Limestone: United States: Kentucky, Illinois, Tennessee

Synbathocrinus sulcatus (Goldring), 1923. Moore and Laudon, 1944, p. 151, Pl. 56, fig. 17. Kesling and Smith, 1963, p. 194.
Devonian, Ulsterian, Onondaga Limestone: United States: New York

Synbathocrinus swallovi Hall, 1858. Kesling and Smith, 1963, p. 194. Van Sant, *in* Van Sant and Lane, 1964, p. 71, Pl. 1, fig. 6; Figs. 23, no. 4; 26, no. 1.
Mississippian, Osagean, Edwardsville Formation: United States: Indiana

Synbathocrinus tennesseensis Roemer, 1860 = *Pisocrinus (Parapisocrinus) tennesseensis*

Synbathocrinus texasensis Moore and Ewers, 1942, p. 94, figs. 1-28. Kesling and Smith, 1963, p. 194 (misspelled as *S. texanensis*).
Mississippian, Osagean, Chappel Limestone: United States: Texas

Synbathocrinus troosti (Wood), 1909. Kesling and Smith, 1963, p. 195.
Mississippian, Osagean, Fort Payne Chert: United States: Tennessee

Synbathocrinus wachsmuthi Meek and Worthen, 1869. Moore and Laudon, 1944, p. 151, Pl. 55, fig. 10. Kesling and Smith, 1963, p. 195.
Mississippian, Osagean, Burlington Limestone: United States: Iowa

Synbathocrinus wortheni Hall, 1858. Kesling and Smith, 1963, p. 195.
Mississippian, Osagean, Burlington Limestone: United States: Iowa

Synbathocrinus sp. Termier and Termier, 1950, p. 87, Pl. 218, fig. 39.
Carboniferous, upper Visean: Morocco

SYNCHIROCRINUS Jaekel, 1918. *S. nitidus* Bather, 1893. Moore, 1962a, p. 27, figs. 3, no. 3; 5, no. 7 (*non* fig. 5, no. 4, as given on p. 27). Strimple, 1963d, p. 51, fig. 13 (part).
Cheirocrinus Slater, 1873. Moore, 1962a, p. 26.
Lower Silurian-Lower Devonian

Synchirocrinus barrisi (Worthen), 1875. Moore, 1962a, p. 27.
   Calceocrinus barrisi Worthen, 1875. Moore, 1962a, p. 27.
   Devonian, Erian, Cedar Valley Formation, Alpena Limestone: United
      States: Iowa, Michigan

Synchirocrinus bassleri (Springer), 1926. Moore, 1962a, p. 27.
   Calceocrinus bassleri Springer, 1926. Moore and Laudon, 1943b, p. 130,
      Pl. 1, fig. 1; 1944, p. 145, Pl. 55, fig. 15. Moore, 1962a, p. 27.
   Silurian, Niagaran, Beech River Formation: United States: Tennessee

Synchirocrinus bifurcatus (Springer), 1926. Moore, 1962a, p. 27, Pl. 2,
      fig. 7.
   Calceocrinus bifurcatus Springer, 1926. Moore, 1962a, p. 27.
   Silurian, Niagaran, Beech River Formation: United States: Tennessee

Synchirocrinus centervillensis (Foerste), 1936. Moore, 1962a, p. 27.
   Calceocrinus centervillensis Foerste, 1936. Moore, 1962a, p. 27.
   Silurian, Medinan, Brassfield Limestone: United States: Ohio

Synchirocrinus divisus Strimple, 1963d, p. 54, Pl. 2, figs. 1-4; Fig. 16d.
   Silurian, Niagaran, Henryhouse Formation: United States: Oklahoma

Synchirocrinus foerstei (Springer), 1926. Moore, 1962a, p. 27, Pl. 3,
      fig. 8; Fig. 15.
   Calceocrinus foerstei Springer, 1926. Moore, 1962a, p. 27.
   Silurian, Niagaran, Beech River Formation: United States: Tennessee

Synchirocrinus gradatus (Salter), 1873. Moore, 1962a, p. 28, Pl. 3,
      fig. 3.
   Cheirocrinus gradatus Salter, 1873. Moore, 1962a, p. 28.
   Silurian, Wenlockian, Wenlock Limestone: England

Synchirocrinus? *halli* (Ringueberg), 1889. Moore, 1962a, p. 27, Pl. 2,
      fig. 5.
   Calceocrinus halli Ringueberg, 1889. Moore, 1962a, p. 27.
   Silurian, Niagaran, Rochester Shale: United States: New York

Synchirocrinus inclinus (Ramsbottom), 1952. Moore, 1962a, p. 28.
   Calceocrinus inclinus Ramsbottom, 1952, p. 38, Pl. 4, figs. 4-7;
      Fig. 2E. Moore, 1962a, p. 28.
   Silurian, Wenlockian, Wenlock Limestone: England

Synchirocrinus interpres (Bather), 1893. Moore, 1962a, p. 27.
   Calceocrinus interpres Bather, 1893. Moore, 1962a, p. 27.
   Silurian, Gotlandian: Sweden

Synchirocrinus keyserensis (Springer), 1926. Moore, 1962a, p. 27.
   Halysiocrinus keyserensis Springer, 1926. Moore, 1962a, p. 27.
   Devonian, Ulsterian, Keyser Limestone: United States: West Virginia

Synchirocrinus marylandensis (Springer), 1926. Moore, 1962a, p. 27.
   Halysiocrinus marylandensis Springer, 1926. Moore, 1962a, p. 27.
   Devonian, Ulsterian, Oriskany Sandstone: United States: West Virginia

Synchirocrinus nitidus (Bather), 1893. Moore, 1962a, p. 27, Pl. 2,
      fig. 4; Pl. 3, fig. 4; Fig. 2.
   Calceocrinus nitidus Bather, 1893. Ramsbottom, 1952, p. 41, fig. 2A.
      Moore, 1962a, p. 27.
   Synchirocrinus anglicus Jaekel, 1918. Moore, 1962a, p. 27.
   Silurian, Wenlockian, Wenlock Limestone: England; Sweden

Synchirocrinus pugil (Bather), 1893. Moore, 1962a, p. 27, fig. 16.
   Calceocrinus pugil Bather, 1893. Ramsbottom, 1952, p. 38, Pl. 4,
      figs. 2, 3; Fig. 2C. Moore, 1962a, p. 27.
   Silurian, Wenlockian, Gotlandian, Wenlock Limestone: England; Sweden

Synchirocrinus quadratus Strimple, 1963d, p. 57, Pl. 2, figs. 5, 6; Figs. 15a-c.
   Silurian, Niagaran, Henryhouse Formation: United States: Oklahoma

Synchirocrinus serialis (Salter), 1873. Moore, 1962a, p. 28, Pl. 3, fig. 2.
   Cheirocrinus serialis Salter, 1873. Moore, 1962a, p. 26.
   Silurian, Wenlockian, Wenlock Limestone: England
Synchirocrinus tenax (Bather), 1893. Moore, 1962a, p. 27, Pl. 1, fig. 6.
   Calceocrinus tenax Bather, 1893. Moore, 1962a, p. 27.
   Silurian, Gotlandian: Sweden
Synchirocrinus tucanus (Bather), 1893. Moore, 1962a, p. 27, Pl. 2, fig. 6.
   Calceocrinus tucanus Bather, 1893. Moore, 1962a, p. 27.
   Silurian, Gotlandian: Sweden
Synchirocrinus typus (Ringueberg), 1889. Moore, 1962a, p. 27.
   Calceocrinus typus Ringueberg, 1889. Moore, 1962a, p. 27.
   Silurian, Niagaran, Rochester Shale: United States: New York
SYNDETOCRINUS Kirk, 1933. *S. dartae* Kirk, 1933. Moore and Laudon, 1943b, p. 50. Bouška, 1950, p. 14, fig. 2. Ubaghs, 1953, p. 750. Arendt and Hecker, 1964, p. 84.
   Silurian
Syndetocrinus bohemicus Bouška, 1950, p. 15, Pl. 2, figs. 1-11; Fig. 5. Arendt and Hecker, 1964, p. 84, Pl. 8, figs. 10-12.
   Silurian, Wenlockian-Ludlovian boundary, Budňany Limestone: Bohemia
Syndetocrinus depressus Bouška, 1950, p. 18, Pl. 2, figs. 12, 13a.
   Silurian, Wenlockian-Ludlovian boundary, Budňany Limestone: Bohemia
Syndetocrinus uralicus Yakovlev, 1949c, p. 18, Pl. 2, figs. 3-5.
   Silurian, Ludlovian: Russia: Ural Mountains
Syndetocrinus sp. Yakovlev, 1949c, p. 18, Pl. 2, fig. 6.
   Silurian, Ludlovian: Russia: Tian-Shan
SYNEROCRINUS Jaekel, 1897. *Forbesiocrinus incurvus* Trautschold, 1867. Moore and Laudon, 1943b, p. 74, fig. 10; 1944, p. 179, Pl. 67, fig. 3. Ubaghs, 1953, p. 755. Arendt and Hecker, 1964, p. 93.
   Carboniferous
Synerocrinus formosus Moore and Plummer, 1940. Moore and Laudon, 1943b, p. 137, Pl. 8, fig. 9; 1944, p. 179, Pl. 56, fig. 20. Strimple, 1962e, p. 51, Pl. 3, fig. 1; Pl. 8, figs. 2, 3.
   Pennsylvanian, Desmoinesian, Millsap Lake Formation, Oologah Limestone: United States: Texas, Oklahoma
Synerocrinus incurvus (Trautschold), 1867. Yakovlev, 1939a, p. 67, Pl. 12, figs. 2, 3. Yakovlev and Ivanov, 1956, p. 36, Pl. 4, fig. 9; Figs. 9-12. Yakovlev, *in* Yakovlev and Ivanov, 1956, p. 45. Ivanova, 1958, p. 134, Pl. 11, fig. 3. Arendt and Hecker, 1964, p. 93, Pl. 14, fig. 3.
   Middle Carboniferous (C_2m): Russia: Moscow Basin, Donetz Basin
SYNTOMOCRINUS Wanner, 1916. *S. sundaicus* Wanner, 1916. Moore and Laudon, 1943b, p. 76. Ubaghs, 1953, p. 756. Arendt and Hecker, 1964, p. 94.
   Permian
SYNYPHOCRINUS Trautschold, 1881. *S. cornutus* Trautschold, 1881. Arendt and Hecker, 1964, p. 88.
   Carboniferous, Namurian-Permian, Wolfcampian
Synyphocrinus cornutus Trautschold, 1881. Yakovlev and Ivanov, 1956, p. 14, Pl. 2, figs. 1-4; Fig. 2. Arendt and Hecker, 1964, p. 88, Pl. 10, figs. 5, 6.
   Middle Carboniferous (C_2m): Russia: Moscow Basin
Synyphocrinus magnus Yakovlev and Ivanov, 1956, p. 14, Pl. 1, figs. 3-5. Ivanova, 1958, p. 132, Pl. 11, fig. 1. Arendt and Hecker, 1964, p. 88, Pl. 10, fig. 4.
   Middle Carboniferous (C_2m): Russia: Moscow Basin

Synyphocrinus permicus Lane and Webster, 1966, p. 27, Pl. 4, figs. 1-7; Fig. 8 (misspelled as *Synyphocrius*, p. 27). Webster and Lane, 1967, p. 15, Pl. 2, figs. 13, 14.
Permian, Wolfcampian, Bird Spring Formation: United States: Nevada

TAIDOCRINUS Tolmatchoff, 1924. *T. poljenowi Tolmatchoff, 1924. Moore and Laudon, 1943b, p. 31.
Pennsylvanian, Missourian-Permian, Wolfcampian

Taidocrinus campanulatus (Wanner), 1916. Moore and Ewers, 1942, p. 105.
Synbathocrinus campanulatus (Wanner), 1916. Moore and Ewers, 1942, p. 105.
Permian: Timor

Taidocrinus campanulatus elongatus (Wanner), 1916. Moore and Ewers, 1942, p. 105.
Synbathocrinus campanulatus elongatus (Wanner), 1916. Moore and Ewers, 1942, p. 105.
Permian: Timor

Taidocrinus campanulatus inflatus (Wanner), 1916. Moore and Ewers, 1942, p. 105.
Synbathocrinus campanulatus inflatus (Wanner), 1916. Moore and Ewers, 1942, p. 105.
Permian: Timor

Taidocrinus constrictus (Wanner), 1916. Moore and Ewers, 1942, p. 105.
Synbathocrinus constrictus (Wanner), 1916. Moore and Ewers, 1942, p. 105.
Permian: Timor

Taidocrinus constrictus sinuosus (Wanner), 1916. Moore and Ewers, 1942, p. 105.
Synbathocrinus constrictus sinuosus (Wanner), 1916. Moore and Ewers, 1942, p. 105.
Permian: Timor

Taidocrinus? melba (Strimple), 1938. Moore and Ewers, 1942, p. 105.
Synbathocrinus melba Strimple, 1938. Moore and Ewers, 1942, p. 105. Strimple, 1959a, p. 121, Pl. 1, figs. 12, 13; Figs. 1, 2.
Pennsylvanian, Missourian, Wann Formation: United States: Oklahoma

TALANTEROCRINUS Moore and Plummer, 1940. *T. jaekeli Moore and Plummer, 1940. Moore and Laudon, 1943b, p. 74, fig. 10; 1944, p. 181, Pl. 67, fig. 6. Moore, 1950, p. 37, fig. 40. Ubaghs, 1953, p. 755. Arendt and Hecker, 1964, p. 94.
Carboniferous

Talanterocrinus jaekeli Moore and Plummer, 1940. Wright, 1946a, p. 37, Pl. 3, fig. 6; 1954a, p. 153, Pl. 43, fig. 9; Pl. 44, figs. 1, 8.
Carboniferous, Visean, Lower Limestone Group: Scotland

Talanterocrinus mcguirei (Moore), 1939. Moore and Laudon, 1944, p. 181, Pl. 56, fig. 30.
Pennsylvanian, Virgilian, Brownville Limestone: United States: Oklahoma

Talanterocrinus redesdalensis Wright, 1952b, p. 320, Pl. 13, figs. 1-3, 7; Figs. 1, 2; 1954a, p. 154, Pl. 42, figs. 3, 6, 7, 11; Figs. 85-89.
Carboniferous, Visean (D_1), Calciferous Sandstone series: Scotland

Talanterocrinus strimplei Wright, 1946a, p. 36, Pl. 3, figs. 8, 9, 13, 15, 16; 1954a, p. 156, Pl. 43, figs. 5-7, 14, 15.
Carboniferous, Visean, Lower Limestone Group: Scotland

Talanterocrinus swaledalensis Wright, 1954a, p. 157, Pl. 42, fig. 4.
Carboniferous, Namurian (E_1), red beds above Main Limestone: England

TALAROCRINUS Wachsmuth and Springer, 1881. *Dichocrinus cornigerus
Shumard, 1857. Moore and Laudon, 1943b, p. 96; 1944, p. 199.
Cuénot, 1948, p. 69. Ubaghs, 1953, p. 741. Arendt and Hecker, 1964,
p. 97.
Mississippian

*Talarocrinus cornigerus (Shumard), 1857. Butts, 1941, p. 250, Pl. 132,
figs. 30-34. Moore and Laudon, 1944, p. 199, Pl. 78, fig. 12.
Mississippian, Chesterian, Ohara formation, Gasper limestone: United
States: Alabama, Virginia

Talarocrinus ovatus Worthen, 1882. Moore and Laudon, 1943b, p. 141,
Pl. 12, fig. 15; 1944, p. 199, Pl. 78, fig. 11.
Mississippian, Chesterian, Renault Formation: United States: Illinois

Talarocrinus sp. Butts, 1941, p. 250, Pl. 132, figs. 35-42.
Mississippian, Chesterian, Gasper Formation: United States: Virginia

TANAOCRINUS Wachsmuth and Springer, 1885. *T. typus Wachsmuth and Springer,
1897. Moore and Laudon, 1943b, p. 90, fig. 13; 1944, p. 189. Moore,
1950, p. 42, fig. 11. Strimple, 1963d, p. 90, fig. 27 (part).
Ordovician, Cincinnatian

*Tanaocrinus typus Wachsmuth and Springer, 1897. Moore and Laudon, 1943b,
p. 139, Pl. 10, fig. 3; 1944, p. 189, Pl. 72, fig. 3.
Ordovician, Cincinnatian, Waynesville Limestone: United States: Ohio

TARACHIOCRINUS Strimple, 1962. *Ataxiacrinus multiramus Strimple, 1961.
Strimple, 1962d, p. 135.
Ataxiacrinus Strimple, 1961g (non Lyon, 1869), p. 89.
Pennsylvanian, Morrowan-Missourian

Tarachiocrinus meadowensis (Strimple), 1949. Strimple, 1962d, p. 136.
Ataxiacrinus meadowensis (Strimple), 1949. Strimple, 1961g, p. 90.
Mooreocrinus meadowensis Strimple, 1949d, p. 26, Pl. 4, figs. 12-14.
Pennsylvanian, Missourian, Meadow Limestone: United States: Nebraska

*Tarachiocrinus multiramus (Strimple), 1961. Strimple, 1962d, p. 136.
Ataxiacrinus multiramus Strimple, 1961g, p. 90, Pl. 5, figs. 1-3;
Pl. 10, figs. 1-3.
Pennsylvanian, Desmoinesian, Holdenville Shale: United States: Oklahoma

Tarachiocrinus optimus (Strimple), 1951. Strimple, 1962d, p. 136.
Ataxiacrinus optimus (Strimple), 1951. Strimple, 1961g, p. 90.
Dicromyocrinus optimus Strimple, 1951c, p. 15, Pl. 3, figs. 1-4.
Pennsylvanian, Morrowan, Brentwood Limestone: United States: Oklahoma

Tarachiocrinus periodus (Strimple), 1951. Strimple, 1962d, p. 136.
Ataxiacrinus periodus (Strimple), 1951. Strimple, 1961g, p. 90.
Dicromyocrinus periodus Strimple, 1951b, p. 375, Pl. 56, figs. 1-4.
Pennsylvanian, Missourian, Winterset Limestone: United States: Kansas

Tarachiocrinus tapajosi (Strimple), 1960. Strimple, 1962 = Dicromyocrinus
tapajosi

TAXOCRINUS Phillips, 1843. *Cyathocrinus? macrodactylus Phillips, 1841.
Moore and Laudon, 1943b, p. 69, fig. 10; 1944, p. 177, Pl. 67,
fig. 10. Moore, 1950, p. 37, fig. 40. Ubaghs, 1953, p. 755.
Devonian-Carboniferous

Taxocrinus bellmanensis Wright, 1954a, p. 169, Pl. 42, figs. 12, 14, 15.
Carboniferous, Visean (C_2 or S_1): Scotland

Taxocrinus colletti White, 1880. Moore and Laudon, 1943b, p. 136, Pl. 7,
fig. 12; 1944, p. 177, Pl. 69, fig. 5. Van Sant, in Van Sant and
Lane, 1964, p. 103, Pl. 6, figs. 1, 3-5, 10; Figs. 12, nos. 5-8; 13.
Moore and others, 1968a, p. 20, Pl. 4, fig. 4.
Mississippian, Osagean, Borden Group: United States: Indiana

Taxocrinus communis (Hall), 1864. Moore and others, 1968a, p. 20, Pl. 4, fig. 5.
Mississippian, lower Waverly Group: United States: Ohio

Taxocrinus coplowensis Wright, 1946b, p. 122, Pl. 7, figs. 8, 9. Wright, 1954a, p. 167, Pl. 42, figs. 1, 2.
Carboniferous, Tournaisian: Scotland

Taxocrinus giddingsei (Hall), 1858. Moore and Laudon, 1944, p. 177, Pl. 69, fig. 13.
Mississippian, Osagean: United States: Missouri

Taxocrinus hookheadensis Wright, 1954a, p. 169, Pl. 45, figs. 1, 4; Pl. 46, figs. 1, 2.
Carboniferous, Tournaisian: Ireland

Taxocrinus huntsvillae Springer, 1920. Moore and Laudon, 1944, p. 177, Pl. 69, fig. 16.
Mississippian, Chesterian: United States: Alabama

Taxocrinus intermedius Wachsmuth and Springer, 1888. Moore and Laudon, 1944, p. 177, Pl. 69, fig. 1. Ubaghs, 1953, p. 683, fig. 21h (misspelled as *Taxocrinites intermedius*). Yakovlev, 1964b, p. 58, fig. 62.
Mississippian, Kinderhookian, Hampton Formation: United States: Iowa

Taxocrinus lobatus (Hall), 1862. Ubaghs, 1953, p. 755, fig. 138b.
Devonian, Erian, Moscow Shale: United States: New York

Taxocrinus nobilis (Phillips), 1836. Wright, 1954a, p. 166, Pl. 44, fig. 7.
Carboniferous, Tournaisian?: England

Taxocrinus ornatus Springer, 1920. Moore and Laudon, 1944, p. 177, Pl. 69, fig. 9.
Mississippian, Osagean, Burlington Limestone: United States: Iowa

Taxocrinus telleri Springer, 1920. Moore and Laudon, 1944, p. 177, Pl. 68, fig. 5.
Devonian, Erian, Hamilton Group: United States: Wisconsin

Taxocrinus ungula Miller and Gurley, 1896. Ubaghs, 1953, p. 755, figs. 15f, 106. Van Sant, *in* Van Sant and Lane, 1964, p. 104.
Mississippian, Osagean, Keokuk Limestone: United States: Indiana

Taxocrinus whitfieldi (Hall), 1858. Moore and Laudon, 1944, p. 177, Pl. 69, fig. 6.
Mississippian, Chesterian, Glen Dean Limestone: United States: Kentucky, Illinois

Taxocrinus spp.
Termier and Termier, 1950, p. 92, Pl. 217, figs. 1, 2.
  Carboniferous, Westphalian: Morocco
Termier and Termier, 1950, p. 220, Pl. 218, figs. 40-45.
  Carboniferous, Westphalian: Morocco
Termier and Termier, 1950, p. 230, Pl. 223, figs. 19-21.
  Carboniferous, Visean: Morocco
Ubaghs, 1953, p. 755, fig. 15g.
  No age given

Taxocrinus sp. indet. Schmidt, 1941, p. 189, Pl. 25, fig. 6.
Devonian, Coblentzian: Germany

TEBAGACRINUS Termier and Termier, 1958. *Coenocrinus solignaci* Valette, 1934. Termier and Termier, 1958, p. 51.
  Coenocrinus Valette, 1934 (*non* Forbes, 1852). Termier and Termier, 1958, p. 51.
  Permian

Tebagacrinus solignaci (Valette), 1934. Termier and Termier, 1958, p. 52, Pl. 1, figs. a-c, h, i (figs. h, i columnals).
  Coenocrinus solignaci Valette, 1934. Termier and Termier, 1958, p. 52.
  Coenocrinus elegans Valette, 1934. Termier and Termier, 1958, p. 53.
  Permian: Tunisia

TECHNOCRINUS Hall, 1859. *Mariacrinus (Technocrinus) andrewsi* Hall, 1859. Moore and Laudon, 1943b, p. 96; 1944, p. 200. Ubaghs, 1953, p. 742.
  Silurian-Devonian

Technocrinus andrewsi (Hall), 1859. Moore and Laudon, 1944, p. 200, Pl. 74, fig. 8.
  Devonian, Ulsterian, Oriskany Sandstone: United States: Maryland

TELEIOCRINUS Wachsmuth and Springer, 1881. *Actinocrinus umbrosus* Hall, 1858. Moore and Laudon, 1943b, p. 93; 1944, p. 193, Pl. 71, fig. 13. Ubaghs, 1953, p. 739, fig. 31. Arendt and Hecker, 1964, p. 96.
  Mississippian, Osagean

Teleiocrinus rudis (Hall), 1860. Moore and Laudon, 1943b, p. 140, Pl. 11, fig. 6; 1944, p. 193, Pl. 77, fig. 11. Laudon, 1948, Pl. 3 (part).
  Mississippian, Osagean, upper Burlington Limestone: United States: Iowa, Missouri

Teleiocrinus? sibiricus Yakovlev, *in* Yakovlev and Ivanov, 1956, p. 48, Pl. 18, fig. 3. Arendt and Hecker, 1964, p. 96, Pl. 14, fig. 13.
  Carboniferous, Turonian (C_1): Russia: Kuznetz Basin

Teleiocrinus umbrosus (Hall), 1858. Moore and Laudon, 1943b, p. 140, Pl. 11, fig. 5; 1944, p. 193, Pl. 77, fig. 1. Laudon, 1948, Pl. 3 (part).
  Mississippian, Osagean, upper Burlington Limestone: United States: Iowa, Missouri

Teleiocrinus sp. Ubaghs, 1953, p. 739, fig. 31.
  Mississippian, Osagean, Burlington Limestone: United States: Iowa

TELIKOSOCRINUS Strimple, 1951. *T. caespes* Strimple, 1951. Strimple, 1951e, p. 260.
  Mississippian, Chesterian

Telikosocrinus caespes Strimple, 1951e, p. 260, figs. 1-4.
  Mississippian, Chesterian, Pitkin Limestone: United States: Oklahoma

Telikosocrinus residuus Strimple, 1951e, p. 262, fig. 13.
  Mississippian, Chesterian, Pitkin Limestone: United States: Oklahoma

TEMNOCRINUS Springer, 1902. *Cyathocrinites tuberculatus* Miller, 1821. Moore and Laudon, 1943b, p. 73. Ubaghs, 1953, p. 755.
  Silurian

TENAGOCRINUS Wanner, 1929. *Embryocrinus sulcatus* Wanner, 1916. Moore and Laudon, 1943b, p. 50. Yakovlev, 1947d, p. 44, fig. 7; 1956a, p. 80, fig. 27; 1958, p. 65, fig. 27; 1964a, p. 75, fig. 29. Arendt and Hecker, 1964, p. 85. Lane, 1967b, p. 11, fig. 5.
  Permian

Tenagocrinus sulcatus (Wanner), 1916. Ubaghs, 1953, p. 672, figs. 10e, f. Yakovlev, 1964b, p. 56, fig. 55.
  Permian: Timor

TERATOCRINUS Wanner, 1924. *T. spathulifer* Wanner, 1924.
  Permian

Teratocrinus spathulifer Wanner, 1924. Sieverts-Doreck, 1962, p. 109, fig. 2.
Permian: Timor

TETRABRACHIOCRINUS Yakovlev, 1934. *T. fabianii* Yakovlev, 1934. Moore and Laudon, 1943b, p. 55.
Permian

TETRAMEROCRINITES Austin and Austin, 1843. *T. formosus* Austin and Austin, 1843. Wright, 1955b, p. 258.
Carboniferous, Tournaisian

Tetramerocrinites formosus Austin and Austin, 1843. Wright, 1955b, p. 258, Pl. 80, fig. 17; Fig. 126.
Carboniferous, Tournaisian: Ireland

TETRAPLEUROCRINUS Wanner, 1942. *T. eifelensis* Wanner, 1942. Wanner, 1942a, p. 28.
Devonian, Eifelian

Tetrapleurocrinus eifelensis Wanner, 1942a, p. 28, Pl. 1, figs. 1a-c, 2a-e; Figs. 2, 3.
Devonian, Eifelian: Germany

TEXACRINUS Moore and Plummer, 1940. *T. gracilis* Moore and Plummer, 1940. Moore and Laudon, 1943b, p. 56. Arendt and Hecker, 1964, p. 88.
Pennsylvanian, Desmoinesian-Permian, Wolfcampian

Texacrinus associatus Strimple, 1952c, p. 219, figs. 1-4.
Pennsylvanian, Desmoinesian, Oologah Limestone: United States: Oklahoma

Texacrinus compactus Strimple, 1952c, p. 219, figs. 10, 11, 16.
Pennsylvanian, Missourian, Francis Formation: United States: Oklahoma

Texacrinus coniformis Strimple, 1961g, p. 95, Pl. 1, figs. 1-3; Pl. 16, fig. 6.
Pennsylvanian, Desmoinesian, Holdenville Shale: United States: Oklahoma

Texacrinus distortus Webster and Lane, 1967, p. 19, Pl. 4, figs. 7, 9; Pl. 7, figs. 1, 5.
Permian, Wolfcampian, Bird Spring Formation: United States: Nevada

Texacrinus interruptus Strimple, 1952c, p. 216, fig. 8.
Pennsylvanian, Missourian, Francis Formation: United States: Oklahoma

Texacrinus irradiatus Strimple, 1952c, p. 218, figs. 9, 12-15.
Pennsylvanian, Missourian, Francis Formation: United States: Oklahoma

Texacrinus progressus Strimple, 1952c, p. 219, figs. 5-7.
Pennsylvanian, Virgilian, Nelagony Formation: United States: Oklahoma

Texacrinus uddeni (Moore and Plummer), 1940. Strimple, 1961g, p. 94. Plummericrinus uddeni (Moore and Plummer), 1940. Moore and Laudon, 1943b, p. 58; 1944, p. 161, Pl. 63, fig. 18.
Pennsylvanian, Missourian, Mineral Wells Formation: United States: Texas

THALAMOCRINUS Miller and Gurley, 1895. *T. ovatus* Miller and Gurley, 1895. Moore and Laudon, 1943b, p. 50; 1944, p. 153. Ubaghs, 1953, p. 750. Xu, 1962, p. 45, fig. 1.
Silurian

Thalamocrinus elongatus Springer, 1926. Strimple, 1963d, p. 66, Pl. 3, fig. 5.
Silurian, Niagaran, Henryhouse Formation: United States: Oklahoma

Thalamocrinus ovatus Miller and Gurley, 1895. Moore and Laudon, 1943b, p. 131, Pl. 2, fig. 11; 1944, p. 153, Pl. 54, fig. 24. Amsden,

1949, p. 80, Pl. 12, figs. 8, 9.
Silurian, Niagaran, Beech River Formation, Brownsport Formation: United States: Tennessee

THALLOCRINUS Jaekel, 1895. *T. hauchecornei Jaekel, 1895. Moore and Laudon, 1943b, p. 101, fig. 18. Ubaghs, 1953, p. 743.
Devonian

Thallocrinus acifer Schmidt, 1941, p. 23, Pl. 4, fig. 1; Fig. 2.
Devonian, Hunsrück Schiefer: Germany

THAMNOCRINUS Goldring, 1923. *T. springeri Goldring, 1923. Moore and Laudon, 1943b, p. 92; 1944, p. 191, Pl. 70, fig. 13. Ubaghs, 1953, p. 739.
Devonian, Erian

*Thamnocrinus springeri Goldring, 1923. Moore and Laudon, 1943b, p. 139, Pl. 10, fig. 6; 1944, p. 191, Pl. 73, fig. 17. Goldring, 1945, p. 57, Pl. 1, figs. 1-3.
Devonian, Erian, Moscow Shale: United States: New York

THELOREUS Moore, 1962. *Phimocrinus jouberti Oehlert, 1882. Moore, 1962b, p. 44.
Lower Devonian

*Theloreus jouberti (Oehlert), 1882. Moore, 1962b, p. 44, fig. 17. Strimple, 1963d, p. 29, fig. 4b (given as Phimocrinus jouberti in text).
Phimocrinus jouberti Oehlert, 1882. Moore, 1962b, p. 44.
Lower Devonian, sandstone: France

THENAROCRINUS Bather, 1890. *T. callipygus Bather, 1890. Moore and Laudon, 1943b, p. 54, figs. 1, 6. Moore, 1950, p. 34, fig. 4c. Ubaghs, 1953, p. 751. Moore, 1962b, p. 15, fig. 9, no. 3. Philip, 1963, fig. 4b. Yakovlev, 1964b, p. 65, fig. 90b.
Silurian, Wenlockian

Thenarocrinus gracilis Bather, 1891. Ubaghs, 1953, p. 751, fig. 14c.
Silurian, Wenlockian, Wenlock Limestone: England

THETIDICRINUS Wanner, 1916. *T. piriformis Wanner, 1916. Moore and Laudon, 1943b, p. 50. Arendt and Hecker, 1964, p. 85. Lane, 1967b, p. 11, fig. 5.
Permian

Tholiacrinus Strimple, 1962 = Endelocrinus

Tholiacrinus bifidus (Moore and Plummer), Strimple, 1962 = Endelocrinus bifidus

Tholiacrinus parinodosarius (Strimple), Strimple, 1962 = Endelocrinus parinodosarius

Tholiacrinus rectus (Moore and Plummer), Strimple, 1962 = Endelocrinus rectus

Tholiacrinus rimulatus Strimple, 1962 = Endelocrinus rimulatus

Tholiacrinus undulatus (Strimple), Strimple, 1962 = Endelocrinus undulatus

THOLOCRINUS Kirk, 1939. *Hydreionocrinus spinosus Wood, 1909. Moore and Laudon, 1943b, p. 58; 1944, p. 167. Arendt and Hecker, 1964, p. 89.
Mississippian

Tholocrinus foveatus Strimple, 1951 = Zeusocrinus foveatus

Tholocrinus spinosus (Wood), 1909. Moore and Laudon, 1944, p. 167, Pl. 59, fig. 2. Ubaghs, 1953, p. 726, figs. 118, 119. Horowitz, 1965, p. 22, Pl. 2, figs. 5-10.

Hydreionocrinus depressus Cuénot, 1948, p. 61, fig. 77.
Mississippian, Chesterian, Glen Dean Formation: United States:
Kentucky, Indiana

Tholocrinus wetherbyi (Wachsmuth and Springer), 1886. Moore and Laudon,
1943b, p. 134, Pl. 5, fig. 6; 1944, p. 167, Pl. 59, fig. 4.
Mississippian, Chesterian, Glen Dean Formation: United States: Kentucky

THYLACOCRINUS Oehlert, 1878. *T. vannioti* Oehlert, 1878. Moore and Laudon,
1943b, p. 82; 1944, p. 185. Ubaghs, 1953, p. 737. Arendt and
Hecker, 1964, p. 95.
Devonian

Thylacocrinus clarkei Wachsmuth and Springer, 1897. Moore and Laudon,
1944, p. 185, Pl. 72, fig. 27. Le Maître, 1958b, p. 128, Pl. 3,
fig. 10.
Devonian, Erian, Moscow Shale: United States: New York

Thylacocrinus? ignotus Philip, 1961, p. 145, Pl. 8, figs. 1, 2, 8; Figs. 1-3.
Lower Devonian: Australia: Victoria

Thylacocrinus vannioti Oehlert, 1878. Le Maître, 1958b, p. 128, Pl. 3,
fig. 9.
Devonian?, locality unknown: France

Thylacocrinus aff. *vannioti* Oehlert, 1878. Le Maître, 1958b, p. 129,
Pl. 2, fig. 6; Pl. 3, fig. 6.
Devonian, lower Couvinian: Algeria

THYRIDOCRINUS Kirk, 1944. *Lecythiocrinus? problematicus* Springer, 1926.
Kirk, 1944e, p. 388.
Silurian, Niagaran

Thyridocrinus patulus (Slocum), 1908. Kirk, 1944e, p. 389.
Achradocrinus patulus Slocum, 1908. Kirk, 1944e, p. 389.
Silurian, Niagaran, Niagara Limestone: United States: Illinois

Thyridocrinus problematicus (Springer), 1926. Kirk, 1944e, p. 389.
Lecythiocrinus? problematicus Springer, 1926. Kirk, 1944e, p. 389.
Silurian, Niagaran, Laurel Limestone: United States: Indiana

Thysanocrinus arborescens Talbot, 1905 = *Ambicocrinus arborescens*

TIARACRINUS Schultze, 1866. *T. quadrifrons* Schultze, 1866. Moore and
Laudon, 1943b, p. 30. Ubaghs, 1953, p. 747.
Devonian

Tiaracrinus moravicus Ubaghs and Bouček, 1962, p. 42, Pls. 1, 2; Figs. 1-3.
Devonian, Emsian: Moravia

Tiaracrinus oehlerti Schlüter, 1881. Le Maître, 1958c, p. 1069.
Devonian, Coblentzian: France

Tiaracrinus quadrifrons Schultze, 1866. Le Maître, 1958b, p. 135, Pl. 1,
figs. 1-6; 1958c, p. 1068, fig. 1.
Devonian, Coblentzian (F_2): Algeria; Bohemia

Tiaracrinus soyei Oehlert, 1882. Le Maître, 1958b, p. 139, Pl. 1,
figs. 8-10; 1958c, p. 1069.
Devonian, Coblentzian: Algeria; France

Tiaracrinus tetraedra Jaekel, 1901. Le Maître, 1958b, p. 135, Pl. 1,
fig. 14; 1958c, p. 1069.
Devonian, Couvinian: Germany

TIMORECHINUS Wanner, 1910. *T. mirabilis* Wanner, 1910. Moore and Laudon,
1943b, p. 59, fig. 7. Moore, 1950, p. 37, fig. 4q.
Timorocrinus Wanner, 1912. Cuénot, 1948, p. 62.
Permian

Timorechinus mirabilis Wanner, 1910. Yakovlev, 1964b, p. 62, fig. 80 (incorrectly given as *Timorocrinus*, p. 62).
Timorocrinus mirabilis Wanner, 1910. Cuénot, 1948, p. 62, fig. 79.
Permian: Timor

TIMOROCIDARIS Wanner, 1920. *T. sphaeracantha* Wanner, 1920.
Permian

Timorocidaris cf. *fusiformis* Wanner, 1940. Wanner, 1950, p. 360, fig. 1.
Permian: Timor

Timorocidaris sphaeracantha Wanner, 1920. Lakeman, 1950, p. 100, fig. 1. Wanner, 1950, p. 366, figs. 2, 3.
Permian: Timor

Timorocidaris sp. Lakeman, 1950, p. 100, figs. 2, 4, 7, 8.
Permian: Timor

TRAMPIDOCRINUS Lane and Webster, 1966. *T. phiala* Lane and Webster, 1966. Lane and Webster, 1966, p. 17.
Permian, Wolfcampian

Trampidocrinus bellicus Lane and Webster, 1966, p. 21, Pl. 2, figs. 3, 5-7; Pl. 3, fig. 1. Webster and Lane, 1967, p. 10, Pl. 1, figs. 14, 15; Fig. 3.
Permian, Wolfcampian, Bird Spring Formation: United States: Nevada

Trampidocrinus phiala Lane and Webster, 1966, p. 20, Pl. 2, figs. 1, 2, 4; Fig. 6. Webster and Lane, 1967, p. 10, Pl. 1, figs. 11, 12.
Permian, Wolfcampian, Bird Spring Formation: United States: Nevada

Trampidocrinus sp. Lane and Webster, 1967, p. 14, fig. 4.
Permian, Wolfcampian, Bird Spring Formation: United States: Nevada

TRAUTSCHOLDICRINUS Yakovlev and Ivanov, *in* Yakovlev, 1939. *T. miloradowitschi* Yakovlev, 1939. Yakovlev and Ivanov, *in* Yakovlev, 1939a, p. 66.
Ivanovicrinus Yakovlev, *in* Arendt and Hecker, 1964, p. 91.
Middle-upper Carboniferous

Trautscholdicrinus miloradowitschi Yakovlev, 1939a, p. 66, Pl. 12, fig. 4. Yakovlev and Ivanov, 1956, p. 23, Pl. 6, figs. 1, 2.
Ivanovicrinus miloradovitschi (Yakovlev), 1939. Yakovlev, *in* Arendt and Hecker, 1964, p. 91, Pl. 13, fig. 1.
Middle-upper Carboniferous: Russia: Moscow Basin

Trautscholdicrinus sp. Yakovlev and Ivanov, 1956, p. 22, Pl. 6, fig. 3.
Upper Carboniferous (C_3ks): Russia: Moscow Basin

Trematocrinus robustus Hall, 1860 = *Gilbertsocrinus robustus*

TRIACRINUS Münster, 1839. *T. pyriformis* Münster, 1839. Moore and Laudon, 1943b, p. 30. Cuénot, 1948, p. 60. Ubaghs, 1953, p. 746, fig. 11b. Bouška, 1956a, p. 37, fig. 10. Moore, 1962b, p. 14, fig. 6, no. 4. Strimple, 1963d, p. 37, fig. 8a. Arendt and Hecker, 1964, p. 81. Weyer, 1965, p. 969, unnumbered fig.
Silurian-Devonian

Triacrinus Termier and Termier, 1950, *non* Münster, 1839 = *Becharocrinus*

Triacrinus altus (Müller), 1856. Weyer, 1965, p. 970.
Devonian, Eifelian: Germany

Triacrinus depressus (Müller), 1856. Weyer, 1965, p. 970.
Devonian, Eifelian: Germany

Triacrinus elongatus Follman, 1887. Strimple, 1963d, figs. 9a, 10j. Weyer, 1965, p. 970.
Devonian, Hunsrück Slate: Germany

Triacrinus granulatus Münster, 1839. Weyer, 1965, p. 974, Pl. 1, fig. 2; Pl. 2, figs. la-e.
Upper Devonian, Clymenien Limestone: Thüringia

Triacrinus koenigswaldi Schmidt, 1934. Weyer, 1965, p. 970.
Devonian, Hunsrück Slate: Germany

Triacrinus kutscheri Schmidt, 1934. Schmidt, 1941, p. 17, Pl. 1, fig. 1. Weyer, 1965, p. 970.
Devonian, Hunsrück Slate: Germany

Triacrinus lutulentus Dubatolova, 1964, p. 16, Pl. 1, fig. 2. Weyer, 1965, p. 970.
Devonian, Eifelian: Russia: Kuznetz Basin

Triacrinus paradoxus Termier and Termier, 1950 = *Becharocrinus paradoxus*

Triacrinus prantli Bouška, 1956a, p. 38, Pl. 6, figs. 2-5. Weyer, 1965, p. 970.
Lower Devonian, Prokop Limestone: Bohemia

**Triacrinus pyriformis* Münster, 1839. Weyer, 1965, p. 971, Pl. 1, figs. la-e.
Upper Devonian, *Wocklumeria* Stage, Clymenien Limestone: Thüringia

Triacrinus regnelli Bouška, 1956a, p. 39, Pl. 6, figs. 6, 7. Strimple, 1963d, figs. 10e-g. Weyer, 1965, p. 970.
Silurian, Ludlovian, Kopanina Limestone: Bohemia

TRIBOLOPORUS Kesling and Paul, 1968. **T. cryptoplicatus* Kesling and Paul, 1968. Kesling and Paul, 1968, p. 13.
Ordovician, Champlainian

**Triboloporus cryptoplicatus* Kesling and Paul, 1968, p. 13, Pl. 1, figs. 1-8; Figs. 3a, 7.
Ordovician, Champlainian, Benbolt Limestone: United States: Virginia

Triboloporus xystrotus Kesling and Paul, 1968, p. 21, Pl. 2, figs. 1-7; Figs. 3b, 8.
Ordovician, Champlainian, Chambersburg Limestone: United States: Pennsylvania

TRIBRACHIOCRINUS McCoy, 1847. **T. clarkii* McCoy, 1847. Moore and Laudon, 1943b, p. 64. Arendt and Hecker, 1964, p. 92.
Carboniferous-Permian

Tribrachiocrinus caledonicus Wright, 1936 = *Hosieocrinus caledonicus*

**Tribrachiocrinus clarkei* McCoy, 1847. Philip, 1964, p. 199, Pl. 3; Fig. 1.
Permian, Darlington Shale: Australia: New South Wales

TRICERACRINUS Bramlette, 1943. **T. moorei* Bramlette, 1943. Bramlette, 1943, p. 550.
Permian, Wolfcampian

**Triceracrinus moorei* Bramlette, 1943, p. 551, Pl. 96, figs. 1-7.
Perimestocrinus (*Triceracrinus*) *moorei* (Bramlette), 1943. Strimple, 1961g, p. 19, figs. 3, 4.
Permian, Wolfcampian, Drake formation: United States: Texas

Triceracrinus sp. Bramlette, 1943, p. 552, Pl. 96, figs. 8-15.
Permian, Wolfcampian, Drake formation: United States: Texas

TRICHINOCRINUS Moore and Laudon, 1943. **T. terranovicus* Moore and Laudon, 1943. Moore and Laudon, 1943a, p. 265. Ubaghs, 1953, p. 737.
Ordovician, Canadian

**Trichinocrinus terranovicus* Moore and Laudon, 1943a, p. 265, Pls. 1, 2; Fig. 1.
Ordovician, Canadian, Table Head Limestone: Canada: Newfoundland

Trichotocrinus Olsson, 1912 = *Melocrinites*

Trichotocrinus harrisi Olsson, 1912 = *Melocrinites harrisi*

TRIMEROCRINUS Wanner, 1916. **T. pumilus* Wanner, 1916. Moore and Laudon, 1943b, p. 62, fig. 8. Arendt and Hecker, 1964, p. 91.
Permian

Trimerocrinus platypleura Yakovlev, 1930. Yakovlev, *in* Yakovlev and Ivanov, 1956, p. 60, Pl. 10, fig. 31. Arendt and Hecker, 1964, p. 91, Pl. 13, figs. 8, 9.
Permian, Artinskian: Russia: Krasnoufimsk

**Trimerocrinus pumilus* Wanner, 1916. Arendt and Hecker, 1964, p. 91, Pl. 13, fig. 10; Fig. 132.
Permian: Timor

TROCHOCRINITES Portlock, 1843. **T. laevis* Portlock, 1843. Ramsbottom, 1961, p. 29.
Ordovician, Caradocian

Trochocrinites gottlandicus Pander, *in* Helmerson, 1858 = *Melocrinites gottlandicus*

**Trochocrinites laevis* Portlock, 1843. Ramsbottom, 1961, p. 29, Pl. 3, fig. 12.
Ordovician, Caradocian: Ireland

TROPHOCRINUS Kirk, 1930. **T. tumidus* Kirk, 1930. Moore and Laudon, 1943b, p. 30; 1944, p. 149. Ubaghs, 1953, p. 747. Arendt and Hecker, 1964, p. 81.
Mississippian, Kinderhookian-Osagean

Trophocrinus bicornis Strimple and Koenig, 1956, p. 1238, fig. 3, nos. 5-12.
Mississippian, Osagean, St. Joe Formation: United States: Oklahoma

Trophocrinus brevis Strimple and Koenig, 1956, p. 1238, fig. 3, nos. 13-16.
Mississippian, Osagean, St. Joe Formation: United States: Oklahoma

**Trophocrinus tumidus* Kirk, 1930. Moore and Laudon, 1944, p. 149, Pl. 56, fig. 11. Strimple and Koenig, 1956, p. 1238, fig. 3, nos. 17-20.
Mississippian, Osagean, Welden Limestone: United States: Oklahoma

Trophocrinus variabilis Strimple and Koenig, 1956, p. 1237, fig. 3, nos. 1-4.
Mississippian, Kinderhookian, Welden Limestone: United States: Oklahoma

TROPIOCRINUS Kirk, 1947. **Poteriocrinites (Scaphiocrinus) rudis* Meek and Worthen, 1869. Kirk, 1947, p. 299.
Mississippian, Osagean

Tropiocrinus carinatus Kirk, 1947, p. 300, Pl. 1, fig. 7.
Mississippian, Osagean, Burlington-Keokuk transition: United States: Iowa

Tropiocrinus planatus Kirk, 1947, p. 301, Pl. 1, fig. 8.
Mississippian, Osagean, Burlington Limestone: United States: Iowa

**Tropiocrinus rudis* (Meek and Worthen), 1869. Kirk, 1947, p. 303.
Poteriocrinites (Scaphiocrinus) rudis Meek and Worthen, 1869. Kirk, 1947, p. 303.
Mississippian, Osagean, Burlington Limestone: United States: Iowa

TRYBLIOCRINUS Geinitz, 1867. *Glyptocrinus (in error for *Trybliocrinus*) *flatheanus* Geinitz, 1867. Ubaghs, 1953, p. 742. Breimer, 1962a, p. 85.
Devonian

**Trybliocrinus flatheanus* (Geinitz), 1867. Ubaghs, 1953, p. 742, fig. 33h; 1956b, p. 557, fig. 3. Breimer, 1960, p. 257, figs. 6, 7; 1962a, p. 86, Pl. 8, figs. 12, 13; Pl. 9, figs. 1-4; Pl. 10, figs. 1-7;

Figs. 16-24, 26. Author unknown, 1962, p. 89, unnumbered fig.
Hadrocrinus hispaniae Schmidt, 1931. Schmidt, 1952, p. 128, Pls. 1-8;
  Figs. 3-9. Ubaghs, 1956b, p. 558. Almela and Sanch, 1958, Pl. 9,
  figs. 4, 4a. Breimer, 1962a, p. 86.
Devonian, Emsian, Calcaire de Ferrones, La Vid Shale: Spain

Tundracrinus Yakovlev, 1928 = *Zeacrinites*

Tundracrinus polaris Yakovlev, 1928 = *Zeacrinites polaris*

TUNGUSKOCRINUS Arendt, 1963. *T. ivanovae* Arendt, 1963. Arendt, 1963a,
  p. 132; 1965b, p. 1116.
Middle Ordovician

*Tunguskocrinus ivanovae Arendt, 1963a, p. 133, figs. 1, 2; 1965b, p. 1117,
  figs. 1, 2.
Middle Ordovician, Mangazeyan Stage, Stolbovaya Suite: Russia:
  Stolbovaya River mouth

Turbinicrinites Troost, in Wood, 1909 = *Melocrinites*

Turbinicrinites verneuili Troost, in Wood, 1909 = *Melocrinites verneuili*

Turbinocrinus Wachsmuth and Springer, 1881 = *Melocrinites*

Tylacocrinus sp. Le Maître, 1952, p. 97, Pl. 22, fig. 13 (uncertain what
  this specimen is, may be a *Megistocrinus*; the generic name *Tylacocrinus*
  is not described, and may be a misspelling of *Tylocrinus* = *Megistocrinus*).
Devonian, Eifelian: Algeria

TYRIEOCRINUS Wright, 1945. *T. laxus* Wright, 1945. Wright, 1945, p. 114.
  Arendt and Hecker, 1964, p. 90.
Carboniferous, Visean

*Tyrieocrinus laxus Wright, 1945, p. 114, Pl. 3, figs. 1, 6-15, 17, 19, 21,
  23; Figs. 1-4; 1952a, p. 117, Pl. 19, figs. 5-7, 15, 16, 18-22.
Carboniferous, Visean, Lower Limestone Group: Scotland

TYTTHOCRINUS Weller, 1930. *T. comptus* Weller, 1930. Moore and Laudon,
  1943b, p. 50. Arendt and Hecker, 1964, p. 85. Lane, 1967b, p. 13,
  fig. 2.
Octocrinus Peck, 1936. Koenig and Meyer, 1965, p. 392.
Devonian-Pennsylvanian

Tytthocrinus alamogordoensis Strimple and Koenig, 1956, p. 1241, fig. 4,
  nos. 1-9. Koenig and Meyer, 1965, p. 392, fig. 1B.
Mississippian, Osagean, Nunn Shale: United States: New Mexico

Tytthocrinus arcarius Koenig and Meyer, 1965, p. 392, figs. 1a, 4a-g.
Devonian, Erian, Windom Shale: United States: New York

*Tytthocrinus comptus Weller, 1930. Koenig and Meyer, 1965, p. 393,
  fig. 1D.
Amphipsalidocrinus comptus (Weller), 1930. Peck and Connelly, 1951,
  p. 414, figs. 7, 8.
Pennsylvanian, Morrowan: United States: Indiana

Tytthocrinus inconsuetus (Peck), 1936. Koenig and Meyer, 1965, p. 392,
  fig. 1C.
Amphipsalidocrinus inconsuetus (Peck), 1936. Peck and Connelly, 1951,
  p. 414, figs. 1-4.
Octocrinus inconsuetus Peck, 1936. Peck and Connelly, 1951, p. 414.
Mississippian, Osagean, Fern Glen Limestone: United States: Missouri

ULOCRINUS Miller and Gurley, 1890. *U. buttsi* Miller and Gurley, 1890.
  Moore and Laudon, 1943b, p. 60; 1944, p. 175. Wright and Strimple,

1945, p. 222. Cuénot, 1948, p. 62. Arendt and Hecker, 1964, p. 90. Carboniferous-Permian

Ulocrinus americanus Weller, 1909 = *Parulocrinus americanus*

Ulocrinus blairi Miller and Gurley, 1894 = *Parulocrinus blairi*

*Ulocrinus buttsi Miller and Gurley, 1890. Moore and Laudon, 1944, p. 175, Pl. 56, fig. 27. Wright and Strimple, 1945, p. 222, Pl. 9, figs. 8, 9; Figs. 4, 5. Ubaghs, 1953, p. 677, fig. 16e. Cronoble, 1960, p. 97, figs. 1-3. Yakovlev, 1964b, p. 66, figs. 91zh, z, i.
Pennsylvanian, Missourian, Argentine Limestone, Hogshooter Limestone: United States: Missouri, Oklahoma

Ulocrinus caverna Strimple, 1949 = *Parulocrinus caverna*

Ulocrinus doliolus Wright, 1936 = *Ureocrinus doliolus*

Ulocrinus elongatus Strimple, 1961g, p. 75, Pl. 6, figs. 5, 6.
Pennsylvanian, Desmoinesian, Holdenville Shale: United States: Oklahoma

Ulocrinus globulus de Koninck, 1858, Yakovlev, 1949 = *Ureocrinus globularis*

Ulocrinus kansasensis Miller and Gurley, 1890. Moore and Laudon, 1944, p. 175, Pl. 56, fig. 26. Wright and Strimple, 1945, p. 222, fig. 1. Ubaghs, 1953, p. 677, fig. 16f (misspelled as *U. kansassensis*).
Pennsylvanian, Missourian, Argentine Limestone: United States: Missouri

Ulocrinus nuciformis (McCoy), 1849 = *Ureocrinus bockschii*

Ulocrinus sangamonensis (Meek and Worthen), 1860. Moore and Laudon, 1943b, p. 135, Pl. 6, fig. 7; 1944, p. 175, Pl. 56, fig. 28. Wright and Strimple, 1945, p. 222, fig. 3. Ubaghs, 1953, p. 677, fig. 16d. Yakovlev, 1964b, p. 66, figs. 91g, d, e.
Pennsylvanian, Missourian: United States: Illinois

Ulocrinus uralensis Yakovlev, 1930. Yakovlev, *in* Yakovlev and Ivanov, 1956, p. 63, Pl. 11, fig. 14. Arendt and Hecker, 1964, p. 90, Pl. 12, fig. 9.
Permian, Artinskian: Russia: central Ural Mountains

Ulocrinus sp. Tischler, 1963, p. 1067, figs. 6D, E.
Pennsylvanian, Missourian?, Madera Formation: United States: Colorado

ULRICHICRINUS Springer, 1926. **U. oklahomae* Springer, 1926. Moore and Laudon, 1943b, p. 55; 1944, p. 158. Arendt and Hecker, 1964, p. 88.
Mississippian, Osagean-Pennsylvanian, Morrowan

Ulrichicrinus chesterensis Strimple, 1949d, p. 29, Pl. 5, figs. 8-11.
Mississippian, Chesterian, Fayetteville Formation: United States: Oklahoma

Ulrichicrinus coryphaeus (Miller), 1891. Moore and Laudon, 1943b, p. 133, Pl. 4, fig. 5; 1944, p. 158, Pl. 59, fig. 13.
Mississippian, Osagean, Borden Group: United States: Indiana

**Ulrichicrinus oklahomae* Springer, 1926. Moore and Laudon, 1944, p. 158, Pl. 59, fig. 14.
Pennsylvanian, Morrowan, Wapanucka Limestone: United States: Oklahoma

UPEROCRINUS Meek and Worthen, 1865. **Actinocrinus pyriformis* Shumard, 1855. Moore and Laudon, 1943b, p. 95, fig. 16; 1944, p. 195. Ubaghs, 1953, p. 740, fig. 23. Lane, 1963b, p. 921.
Mississippian, Osagean

Uperocrinus aequibrachiatus (McChesney), 1860. Moore and Laudon, 1943b, p. 141, Pl. 12, fig. 8; 1944, p. 197, Pl. 76, fig. 16.
Mississippian, Osagean, Burlington Limestone: United States: Iowa, Missouri

Uperocrinus longirostris (Hall), 1858. Moore and Laudon, 1943b, p. 141, Pl. 12, fig. 7; 1944, p. 197, Pl. 76, fig. 23. Laudon, 1948, Pl. 2 (part). Ubaghs, 1953, p. 740, fig. 23. Yakovlev, 1964b, p. 60, fig. 74.
Mississippian, Osagean, Burlington Limestone: United States: Iowa, Missouri, Illinois

Uperocrinus marinus (Miller and Gurley), 1890. Van Sant, *in* Van Sant and Lane, 1964, p. 118, Pl. 8, fig. 3.
Batocrinus marinus Miller and Gurley, 1890. Van Sant, *in* Van Sant and Lane, 1964, p. 118.
Lobocrinus spiniferus Wachsmuth and Springer, 1897. Van Sant, *in* Van Sant and Lane, 1964, p. 118.
Mississippian, Osagean, Borden Group: United States: Indiana

Uperocrinus nashvillae (Troost), 1849. Moore and Laudon, 1943b, p. 141, Pl. 12, fig. 9; 1944, p. 195, Pl. 76, fig. 25. Laudon, 1948, Pl. 2 (part).
Mississippian, Osagean, Keokuk Limestone: United States: Tennessee, Kentucky, Indiana, Illinois, Missouri

*Uperocrinus pyriformis (Shumard), 1855. Moore and Laudon, 1944, p. 195, Pl. 76, fig. 24. Laudon, 1948, Pl. 2 (part). Lane, 1963b, p. 928, fig. 5.
Mississippian, Osagean, Burlington Limestone, Fern Glen Limestone, St. Joe Formation, Lake Valley Limestone: United States: Iowa, Missouri, Arkansas, Oklahoma, New Mexico

UREOCRINUS Wright and Strimple, 1945. **Poteriocrinus bockschii* Geinitz, 1846. Wright and Strimple, 1945, p. 225. Arendt and Hecker, 1964, p. 90.
Lower Carboniferous

*Ureocrinus bockschii (Geinitz), 1846. Wright and Strimple, 1945, p. 225, Pl. 9, figs. 1-5, 7, 10. Wright, 1948, p. 50, Pl. 5, fig. 5; 1952a, p. 112, Pl. 19, figs. 1-4, 8, 9; Pl. 34, figs. 1-43; Figs. 49-60; 1952b, p. 325, Pl. 13, fig. 6. Ubaghs, 1953, p. 677, figs. 16a-c. Yakovlev, 1957, p. 12, figs. 2a, b; 1964b, p. 66, figs. 91a, b, v.
Poteriocrinus bockschii Geinitz, 1846. Wright and Strimple, 1945, p. 225.
Ulocrinus nuciformis (McCoy), 1849. Wright and Strimple, 1945, p. 227.
Cromyocrinus nuciformis (McCoy), 1849. Strimple, 1961g, p. 66.
Carboniferous, Visean, Lower Limestone Group, Calciferous Sandstone series: Scotland; Germany

Ureocrinus doliolus (Wright), 1936. Wright and Strimple, 1945, p. 227, Pl. 9, fig. 6. Wright, 1952a, p. 116, Pl. 19, figs. 10-13; Figs. 61-65; 1952b, p. 325, Pl. 13, figs. 4, 5.
Ulocrinus doliolus Wright, 1936. Wright and Strimple, 1945, p. 227.
Carboniferous, Visean, Calciferous Sandstone series: Scotland

Ureocrinus globularis (de Koninck), 1858. Yakovlev, 1956a, p. 93, fig. 37, nos. 1, 5; 1958, p. 79, fig. 37, nos. 1, 5; 1961, p. 418, Pl. 3, figs. a-f; 1964a, p. 86, fig. 39, nos. 1, 5.
Ulocrinus globulus de Koninck, 1858. Yakovlev, 1949b, p. 897, figs. 1, I; 2.
Lower Carboniferous: Russia: Donetz Basin

Ureocrinus? sp. Wright, 1952a, p. 117, Pl. 12, figs. 7, 8.
Lower Carboniferous: Scotland

UTHAROCRINUS Moore and Plummer, 1938. **Delocrinus pentanodus* Mather, 1915. Moore and Laudon, 1943b, p. 58; 1944, p. 163.
Pennsylvanian, Morrowan-Permian, Wolfcampian

Utharocrinus fabulosus Strimple, 1950 = *Perimestocrinus fabulosus*

Utharocrinus facilis Strimple, 1950 = *Perimestocrinus facilis*
Utharocrinus granulosus Strimple, 1939 = *Perimestocrinus granulosus*
Utharocrinus habitus Strimple, 1950 = *Perimestocrinus habitus*
Utharocrinus oreadensis Moore, 1939 = *Perimestocrinus oreadensis*
**Utharocrinus pentanodus* (Mather), 1915. Moore and Laudon, 1944, p. 163, Pl. 62, fig. 13; Pl. 64, fig. 18.
  Pennsylvanian, Morrowan, Brentwood Limestone: United States: Oklahoma
Utharocrinus quinquacutus Moore, 1939 = *Perimestocrinus quinquacutus*
Utharocrinus spinosus Strimple, 1950 = *Perimestocrinus spinosus*
Utharocrinus topekensis Moore, 1939 = *Perimestocrinus topekensis*
Utharocrinus sp. Lane and Webster, 1966, p. 52, Pl. 3, fig. 6.
  Permian, Wolfcampian, Bird Spring Formation: United States: Nevada

VASOCRINUS Lyon, 1857. **V. valens* Lyon, 1857. Moore and Laudon, 1943b, p. 51; 1944, p. 154, Pl. 52, fig. 23. Ubaghs, 1953, p. 750. Arendt and Hecker, 1964, p. 85.
  Devonian

Vasocrinus sculptus Lyon, 1857. Moore and Laudon, 1944, p. 154, Pl. 57, fig. 3.
  Devonian, Erian, Hamilton Group, Sellersburg Limestone: United States: Ohio, Indiana, Kentucky

Vasocrinus sp. cf. *V. sculptus* Lyon, 1857. Breimer, 1962a, p. 151, Pl. 14, figs. 15-17; Fig. 35.
  Devonian: Spain

Vasocrinus stellaris (Schultze), 1867. Breimer, 1962a, p. 149, Pl. 14, figs. 11-14; Fig. 34.
  Devonian: Spain

Vasocrinus? sulcosutura Wanner, 1942a, p. 33, Pl. 1, figs. 4a-c.
  Devonian, Eifelian: Germany

Vasocrinus turbinatus Kirk, 1929. Breimer, 1962a, p. 148, Pl. 14, figs. 8-10.
  Devonian: Spain

**Vasocrinus valens* Lyon, 1857. Moore and Laudon, 1943b, p. 131, Pl. 2, fig. 13; 1944, p. 154, Pl. 57, fig. 2. Breimer, 1962a, p. 147, Pl. 14, figs. 5-7.
  Middle Devonian, Onondaga Formation, Santa Lucia Formation, La Vid Shale: United States: Kentucky; Spain

VOSEKOCRINUS Jaekel, 1918. **V. granulatus* Jaekel, 1918. Moore and Laudon, 1943b, p. 52.
  Ordovician

WACHSMUTHICRINUS Springer, 1902. **Forbesiocrinus thiemi* Hall, 1861. Moore and Laudon, 1943b, p. 74, fig. 10; 1944, p. 181, Pl. 67, fig. 4. Moore, 1950, p. 37, fig. 40. Ubaghs, 1953, p. 755. Arendt and Hecker, 1964, p. 93.
  Devonian- Mississippian

Wachsmuthicrinus dubjanskii Yakovlev, 1956b, p. 93, Pl. 1, figs. 1, 2. Arendt and Hecker, 1964, p. 93, Pl. 14, fig. 2.
  Wachsmuthicrinus sp. Yakovlev, 1946d, p. 263, unnumbered fig; 1946e, p. 263, unnumbered fig.
  Devonian, Semiluki Beds: Russia: Voronezh region

Wachsmuthicrinus ponderosus Springer, 1920. Wright, 1954a, p. 166, Pl. 44, fig. 5.
Carboniferous, Visean, Lower Limestone Group: Scotland

Wachsmuthicrinus spinosulus (Miller and Gurley), 1894. Moore and Laudon, 1944, p. 181, Pl. 68, fig. 8 (misspelled as *W. spinulosus*).
Mississippian, Osagean, Knobstone formation: United States: Indiana, Kentucky

WANNEROCRINUS Marez Oyens, 1940. *W. glans* Marez Oyens, 1940. Moore and Laudon, 1943b, p. 92. Sieverts-Doreck, 1951b, p. 105, figs. 1, 2b. Ubaghs, 1953, p. 739.
Permian

Wannerocrinus glans Marez Oyens, 1940. Moore and Laudon, 1942, p. 75, figs. 5D-F.
Permian: Timor

WHITEOCRINUS Jaekel, 1918. *Belemnocrinus florifer* Wachsmuth and Springer, 1877. Moore and Laudon, 1943b, p. 31. Ubaghs, 1953, p. 747.
Mississippian

Whiteocrinus exsculptus Strimple, 1939 = *Stellarocrinus exsculptus*

Whiteocrinus stillativus (White), 1879 = *Stellarocrinus stillativus*

Whiteocrinus sp. indet. Strimple, 1939 = *Stellarocrinus* sp. indet.

WILSONICRINUS Springer, 1926. *W. discoideus* Springer, 1926. Moore and Laudon, 1943b, p. 82; 1944, p. 185. Ubaghs, 1953, p. 737. Arendt and Hecker, 1964, p. 95.
Silurian

Wilsonicrinus discoideus Springer, 1926. Moore and Laudon, 1944, p. 185, Pl. 72, fig. 26.
Silurian, Niagaran, Laurel formation: United States: Indiana

WOODOCRINUS de Koninck, 1854. *W. macrodactylus* de Koninck, 1854. Moore and Laudon, 1943b, p. 55. Arendt and Hecker, 1964, p. 88.
Lower-Middle Carboniferous

Woodocrinus decodactylus de Koninck and Wood, 1858. Wright, 1951b, p. 58 (nomen nudum).
Carboniferous: England

Woodocrinus expansus de Koninck and Wood, 1858. Wright, 1951b, p. 55, Pl. 21, fig. 2; Pl. 27, figs. 1, 3, 4; Pl. 28, figs. 2-4; Pl. 29, figs. 1, 3.
Carboniferous, Namurian (E_1), red beds above Main Limestone: England

Woodocrinus cf. *expansus* Wright, 1924 = *Woodocrinus pentonensis*

Woodocrinus fimbriatus de Koninck, *in* Wright, 1951b, p. 58, Pl. 34, fig. 2.
Carboniferous, Namurian (E_1), red beds above Main Limestone: England

Woodocrinus goniodactylus Wright, 1936 = *Hydreionocrinus amplus*

Woodocrinus gravis Wright, 1936. Wright, 1951b, p. 53, Pl. 25, figs. 2-9; Figs. 16-20.
Carboniferous, Visean, Lower Limestone Group: Scotland

Woodocrinus liddesdalensis Wright, 1936. Wright, 1951b, p. 52, Pl. 22; Pl. 23, figs. 1, 2; Pl. 24, fig. 1; Pl. 25, fig. 1; Pl. 26, figs. 1, 2; Pl. 30, fig. 10.
Carboniferous, Visean, Lower Limestone Group: Scotland

Woodocrinus? longidactylus (de Koninck, *in* Wright), 1925. Wright, 1951b, p. 59, Pl. 29, fig. 2.
Poteriocrinus? longidactylus (de Koninck, *in* Wright), 1925. Wright,

1951b, p. 39.
Carboniferous, Namurian (E_1), red beds above Main Limestone: England

*Woodocrinus macrodactylus de Koninck, 1854. Wright, 1951b, p. 48,
Pls. 20, 21; Pl. 23, figs. 3-5; Pl. 24, figs. 2, 3, 5; Pl. 27, fig. 1;
Pl. 28, figs. 3, 4; Pl. 29, fig. 4; Pl. 30, figs. 2, 5, 9; Fig. 15.
Carboniferous, Namurian (E_1), red beds above Main Limestone: England

Woodocrinus pentonensis Wright, 1951b, p. 56, Pl. 26, figs. 3, 4.
Woodocrinus cf. expansus de Koninck and Wood, 1858, Wright, 1924. Wright, 1951b, p. 56.
Carboniferous, Visean, Lower Limestone Group: Scotland

Woodocrinus whytei Wright, 1936. Wright, 1951b, p. 57, Pl. 24, fig. 4; Pl. 26, fig. 5; Pl. 31, fig. 15.
Carboniferous, Visean, Lower Limestone Group: Scotland

Woodocrinus? sp. indet. Schmidt, 1934 (part) = Orocrinus hercynicus

Woodocrinus? sp. indet. Schmidt, 1934 (part) = Orocrinus grundensis

Woodocrinus? sp. indet. Schmidt, 1934 (part) = Orocrinus winterbergensis

WRIGHTOCRINUS Moore, 1940. *Allagecrinus yakovlevi Wanner, 1929. Moore and Laudon, 1943b, p. 31. Ubaghs, 1953, p. 747. Moore, 1962b, p. 16, fig. 12, no. 5. Arendt and Hecker, 1964, p. 81.
Lower Carboniferous

Wrightocrinus biplex (Wright), 1932. Wright, 1952a, p. 142, Pl. 39, fig. 16; Pl. 40, figs. 12-17, 26.
Carboniferous, Visean, Lower Limestone Group: Scotland

*Wrightocrinus yakovlevi (Wanner), 1929. Ubaghs, 1953, p. 747, fig. 21b.
Permian: Timor

XENOCATILLOCRINUS Wanner, 1937. *X. wrighti Wanner, 1937. Moore and Laudon, 1943b, p. 31. Ubaghs, 1953, p. 747. Moore, 1962b, p. 16, fig. 12, no. 8. Arendt and Hecker, 1964, p. 81.
Permian

XENOCRINUS Miller, 1881. *X. penicillus Miller, 1881. Moore and Laudon, 1943b, p. 90, fig. 13; 1944, p. 189. Moore, 1950, p. 42, fig. 11. Ubaghs, 1953, p. 738. Ramsbottom, 1961, p. 21, fig. 9. Strimple, 1963d, p. 89, fig. 27 (part).
Ordovician-Silurian

Xenocrinus Jahn, 1892, non Miller, 1881 = Melocrinites

Xenocrinus baeri (Meek), 1872. Moore and Laudon, 1943b, p. 139, Pl. 10, fig. 1; 1944, p. 189, Pl. 72, fig. 2.
Ordovician, Cincinnatian: United States: Ohio, Indiana

Xenocrinus multiramus Ramsbottom, 1961, p. 21, Pl. 6, figs. 1-4.
Ordovician, Ashgillian, upper Drummuck Group: England

*Xenocrinus penicillus Miller, 1881. Ubaghs, 1953, p. 738, fig. 20a.
Silurian, Albian, Liberty Limestone: United States: Ohio

Xenocrinus sp. Ramsbottom, 1961, p. 22.
Ordovician, Ashgillian, upper Drummuck Group: England

ZEACRINITES Troost, in Hall, 1858. *Z. magnoliaeformis Troost, in Hall, 1858. Moore and Laudon, 1943b, p. 58, figs. 1, 7; 1944, p. 165. Moore, 1950, p. 34, fig. 4k. Arendt and Hecker, 1964, p. 88.
Tundracrinus Yakovlev, 1928. Moore and Laudon, 1943b, p. 55.
Mississippian-Permian

Zeacrinites acicularis Moore and Plummer, 1938.
　　Zeacrinus acicularis Moore and Plummer, 1938. Sieverts-Doreck, 1962,
　　　　p. 112, fig. 1.
　　Pennsylvanian, Morrowan, Brentwood Limestone: United States: Oklahoma

Zeacrinites doverensis (Miller and Gurley), 1896. Horowitz, 1965, p. 19,
　　Pl. 1, figs. 16-18; Fig. 2A.
　　Mississippian, Chesterian, Glen Dean Formation: United States: Indiana

Zeacrinites impressus (McCoy), 1851. Wright, 1952a, p. 109, Pl. 34,
　　figs. 50, 56, 63, 69; Fig. 48.
　　Cupressocrinus impressus McCoy, 1851. Wright, 1952a, p. 109.
　　Carboniferous, Visean: England

Zeacrinites konincki (Bather), 1912. Wright, 1952a, p. 107, Pl. 32,
　　figs. 1-6, 8-10, 12-17; Pl. 33, figs. 1, 2, 13, 15; Pl. 34,
　　figs. 44-49, 51-55, 57-62, 64-68, 70; Figs. 43-47. Ubaghs, 1953,
　　p. 704, fig. 72. Moore, 1962b, p. 24, fig. 14, no. 3.
　　Carboniferous, Visean, Lower Limestone Group, Mountain Limestone, Visean
　　Limestone: Scotland; England; Belgium

*ized*Zeacrinites magnoliaeformis* Troost, *in* Hall, 1858. Moore and Laudon, 1943b,
　　p. 134, Pl. 5, fig. 8; 1944, p. 165, Pl. 61, fig. 1. Moore, 1962b,
　　p. 24, fig. 14, no. 1.
　　Mississippian, Chesterian, Gasper Formation: United States: Alabama

Zeacrinites peculiaris (Miller and Gurley), 1896. Strimple, 1962j, p. 310,
　　Pl. 1, figs. 1-5; Figs. 1-7.
　　Mississippian, Chesterian, Fayetteville Formation: United States:
　　Oklahoma, Kentucky

Zeacrinites polaris (Yakovlev), 1928 (nom. correct.).
　　Zeacrinus polaris (Yakovlev), 1928. Yakovlev, 1939b, p. 60, Pl. 10,
　　　　fig. 5. Yakovlev, *in* Yakovlev and Ivanov, 1956, p. 78, Pl. 14;
　　　　Pl. 15, fig. 1. Arendt and Hecker, 1964, p. 88, Pl. 11, fig. 5.
　　Tundracrinus polaris Yakovlev, 1928. Yakovlev, *in* Yakovlev and Ivanov,
　　　　1956, p. 78.
　　Lower Permian: Russia: Arctic Urals

Zeacrinites schmitowi (Yakovlev and Ivanov), 1956 (nom. correct.).
　　Zeacrinus schmitowi Yakovlev and Ivanov, 1956, p. 26, Pl. 7, fig. 2;
　　　　Fig. 4. Arendt and Hecker, 1964, p. 88, Pl. 11, fig. 6.
　　Upper Carboniferous (C_3ks): Russia: Moscow Basin

Zeacrinites? sellardsi Moore and Plummer, 1940. Moore and Laudon, 1944,
　　p. 165, Pl. 63, fig. 17.
　　Pennsylvanian, Missourian, Graford Formation: United States: Texas

Zeacrinites trapeziatus (Sutton and Hagan), 1939. Horowitz, 1965, p. 16,
　　Pl. 1, figs. 3-5; Fig. 2D.
　　Mississippian, Chesterian, Glen Dean Formation: United States: Indiana

Zeacrinites wortheni (Hall), 1858. Moore and Laudon, 1944, p. 165, Pl. 61,
　　fig. 2. Moore, 1962b, p. 24, fig. 14, no. 2. Horowitz, 1965, p. 17,
　　Pl. 1, figs. 6-8, 15, 19; Figs. 2B, C.
　　Mississippian, Chesterian, Glen Dean Formation: United States: Indiana,
　　Kentucky

Zeacrinites sp. A. Horowitz, 1965, p. 19, Pl. 1, figs. 1, 2; Fig. 2F.
　　Mississippian, Chesterian, Glen Dean Formation: United States: Indiana

Zeacrinites sp. B. Horowitz, 1965, p. 20, Pl. 1, figs. 9-11; Fig. 2E.
　　Mississippian, Chesterian, Glen Dean Formation: United States: Indiana

Zeacrinites sp. C. Horowitz, 1965, p. 20, Pl. 1, figs. 12, 13; Fig. 2G.
　　Mississippian, Chesterian, Glen Dean Formation: United States: Indiana

Zeacrinites sp. D. Horowitz, 1965, p. 21, Pl. 1, fig. 14; Fig. 2H.
    Mississippian, Chesterian, Glen Dean Formation: United States: Indiana
Zeacrinites sp. E. Horowitz, 1965, p. 21, Pl. 2, figs. 1, 2.
    Mississippian, Chesterian, Glen Dean Formation: United States: Indiana
Zeacrinus mooresi Whitfield, 1882 = *Aatocrinus mooresi*
Zeacrinus scoparius Hall, 1861 = *Ascetocrinus scoparius*
Zenkericrinus Waagen and Jahn, 1899 = *Melocrinites*
Zenkericrinus melocrinoides Waagen and Jahn, 1899 = *Melocrinites melocrinoides*
ZEUSOCRINUS Strimple, 1961. *Tholocrinus foveatus* Strimple, 1951.
    Strimple, 1961g, p. 21.
    Mississippian, Chesterian

Zeusocrinus foveatus (Strimple), 1961g, p. 21.
    Tholocrinus foveatus Strimple, 1951f, p. 674, Pl. 99, figs. 1, 2.
    Mississippian, Chesterian, Pitkin Limestone: United States: Oklahoma
ZOPHOCRINUS Miller, 1891. *Z. howardi* Miller, 1891. Moore and Laudon,
    1943b, p. 30; 1944, p. 145. Ubaghs, 1953, p. 747.
    Silurian

Zophocrinus angulatus Strimple, 1952b, p. 76, figs. 6, 7.
    Silurian, Niagaran, Henryhouse Formation: United States: Oklahoma
Zophocrinus howardi Miller, 1891. Moore and Laudon, 1943b, p. 130, Pl. 1,
    fig. 16; 1944, p. 147, Pl. 54, fig. 20. Le Maître, 1958b, p. 134,
    Pl. 1, fig. 7.
    Silurian, Niagaran, Laurel Limestone, Beech River Formation: United
    States: Indiana, Tennessee
ZOSTOCRINUS Kirk, 1948. *Z. ornatus* Kirk, 1948. Kirk, 1948, p. 708.
    Devonian, Erian

Zostocrinus ornatus Kirk, 1948, p. 709, Pl. 1, figs. 12-16. Goldring,
    1954, p. 34, Pl. 5, figs. 7, 8; Pl. 6, figs. 1-9.
    Devonian, Erian, Ludlowville Shale, Kashong Member: United States:
    New York
ZYGOTOCRINUS Kirk, 1943. *Z. fragilis* Kirk, 1943. Kirk, 1943b, p. 640.
    Mississippian, Osagean

Zygotocrinus enormis (Meek and Worthen), 1861. Kirk, 1943b, p. 642.
    Poteriocrinus? enormis Meek and Worthen, 1861. Kirk, 1943b, p. 642.
    Mississippian, Osagean, lower Burlington Limestone: United States: Iowa
Zygotocrinus fragilis Kirk, 1943b, p. 642, Pl. 1, figs. 1-4.
    Mississippian, Osagean, lower Burlington Limestone: United States: Iowa

SECTION 2. UNIDENTIFIED CROWNS AND PARTS OF CROWNS

Abrotocrinus sp. Termier and Termier, 1950, p. 220, Pl. 218, figs. 29, 30 (infrabasal circlet).
Carboniferous, Westphalian: Morocco

Actinocrinitidae. Moore, 1954, p. 144, fig. 8C (plate diagram).

Adunata indéterminé. Termier and Termier, 1950, p. 204, Pl. 210, fig. 11 (theca).
Gotlandian: Morocco

Allegacrinidae gen. et sp. nov. Hill and Woods, 1964, p. 24, Pl. P 12, fig. 20 (theca).
Lower Permian, Oxtrack Formation: Australia: Queensland

Amphicrinus sp. Yakovlev, 1961, p. 417, Pl. 2, fig. 11 (brachial).
Lower Carboniferous: Russia: Donetz Basin

Antedon sp. Lue and Hy, 1965, p. 373, Pl. 1, figs. 1-4 (mould of set of arms).
Upper Permian, Lungtan Coal Series: China: Kiangsu Province

Apical glyptus (Moore), 1939. Moore and Laudon, 1944, p. 209, Pl. 79, fig. 24 (infrabasal? circlet).
Pennsylvanian, Desmoinesian, Mineral Coal cap: United States: Kansas

cf. Apical glyptus Moore, 1939. Termier and Termier, 1950, p. 208, Pl. 212, fig. 9 (infrabasal? circlet).
Carboniferous, Westphalian: Morocco

Apical regularis (Moore), 1939. Moore and Laudon, 1944, p. 209, Pl. 79, fig. 25 (infrabasal? circlet).
Pennsylvanian, Missourian, Stanton Limestone: United States: Oklahoma

Apical urna (Moore), 1939. Moore and Laudon, 1944, p. 209, Pl. 79, fig. 21 (infrabasal? circlet).
Mississippian, Osagean, Burlington Limestone: United States: Missouri

Archaeocrinidae and Rhodocrinitidae. Moore, 1954, p. 144, fig. 7A (plate diagram).

Arm fragment. Yeltyschewa and Schewtschenko, 1960, Pl. 1, fig. 11.
Lower Carboniferous: Russia: Tien-Shan Mountains

Basal circlets, Poteriocrinida. Yakovlev, 1961, Pl. 2, figs. 5, 6.
Lower Carboniferous: Russia: Donetz Basin

Base de Flexibilia. Termier and Termier, 1950, p. 218, Pl. 217, figs. 3, 4 (infrabasal? circlet).
Carboniferous, upper Visean: Morocco

Base indéterminé. Termier and Termier, 1950, p. 208, Pl. 212, figs. 12, 13 (infrabasal? circlet).
Carboniferous, Westphalian: Morocco

Base indéterminé. Termier and Termier, 1950, p. 236, Pl. 226, figs. 11, 12 (infrabasal? circlet).
Carboniferous, Westphalian: Morocco

Base indéterminé. Termier and Termier, 1950, p. 230, Pl. 223, figs. 22-24 (infrabasal? circlet).
    Carboniferous, Visean: Morocco

Brachial acutus (Moore), 1939. Moore and Laudon, 1944, p. 209, Pl. 79, fig. 19.
    Pennsylvanian, Desmoinesian, Mineral Coal cap: United States: Kansas

Brachial sinus (Moore and Plummer), 1939. Moore and Laudon, 1944, p. 209, Pl. 79, fig. 11.
    Pennsylvanian, Morrowan, Brentwood Limestone: United States: Arkansas

Brachial gen. et sp. indet. Sieverts-Doreck, 1951, p. 141.
    Lower Carboniferous: Germany

Brachials gen. et sp. indet. Sieverts-Doreck, 1951, p. 138, figs. 4, 5.
    Lower Carboniferous: Germany

Brachials, Poteriocrinida. Yakovlev, 1961, Pl. 2, fig. 7.
    Lower Carboniferous: Russia: Donetz Basin

Bras. Termier and Termier, 1950, p. 226, Pl. 221, figs. 16-18 (partial arm).
    Carboniferous, upper Visean: Morocco

Bras de Crinoide pinnule Articulata. Termier and Termier, 1950, p. 222, Pl. 219, fig. 32 (pinnules).
    Carboniferous, Westphalian: Morocco

Bras d'un Crinoide indéterminé. Termier and Termier, 1950, p. 216, Pl. 216, figs. 12-16 (brachials).
    Carboniferous, Namurian: Morocco

Bras pinulifere. Termier and Termier, 1950, p. 226, Pl. 221, fig. 15 (arm segment).
    Carboniferous, upper Visean: Morocco

Calceocrinidae. Sieverts-Doreck, 1953, pl. 78, fig. 8 (crown).
    Devonian: Germany

Calices indéterminé. Termier and Termier, 1950, p. 200, Pl. 208, figs. 29-31, 33-35 (theca).
    Carboniferous, Westphalian: Morocco

Carabocrinus sp. Roy, 1941, p. 81, fig. 42 (basal).
    Ordovician, Richmondian: Baffin Land

Codiacrinidae? Le Maître, 1958b, p. 141, Pl. 2, fig. 4, fig. 3 (theca).
    Devonian, Coblentzian, or Couvinian: Morocco

Crinoid cup. Lane, 1962, p. 909, Pl. 128, fig. 24.
    Pennsylvanian, Atokan?, Ely Group: United States: Nevada

Crinoid gen. et sp. undet. Branson, 1944, p. 165, Pl. 30, fig. 14 (theca).
    Mississippian, Kinderhookian, Northview Formation: United States: Missouri

Crinoid gen. et sp. undet. Branson, 1944, p. 165, Pl. 30, figs. 16, 17 (theca).
    Mississippian, Kinderhookian, Northview Formation: United States: Missouri

Crinoid plate. Walter, 1953, Pl. 73, fig. 8 (primibrach).
    Permian, Ochoan, Rustler Formation: United States: Texas

Crinoid plate. Easton, 1962, p. 37, Pl. 4, fig. 9 (tegmen plate).
    Mississippian, Chesterian, Heath Formation: United States: Montana

Crinoid plates. Heuer, 1958, p. 39, Pl. 1, fig. 1 (tegmen spine), fig. 2 (brachial), fig. 3 (basal), fig. 4 (infrabasal? circlet), fig. 5 (brachial), figs. 6, 7 (radial).

Pennsylvanian, Desmoinesian, East Mountain Shale: United States: Texas

Crinoidal remains. Newell and others, 1953, Pl. 23, fig. 46 (infrabasal? circlet), fig. 48 (radial or primibrach).
Permian, Guadalupian, Lamar Limestone: United States: Texas

Crinoidea A. Lochman and Hu, 1960, p. 826, Pl. 98, figs. 53-55 (indeterminate echinoderm plate).
Upper Cambrian, Du Noir Limestone: United States: Wyoming

Crinoidea B. Lochman and Hu, 1960, p. 827, Pl. 99, fig. 55 (indeterminate echinoderm plate).
Upper Cambrian, Du Noir Limestone: United States: Wyoming

Crinoidea II. Laursen, 1943, p. 13, Pl. 2, fig. 1 (crown).
Middle Silurian: Denmark

Crinoideorus gen. et sp. indet. Schmidt, 1941, p. 218, fig. 62 (arm).
Lower Devonian, highest Koblenz Schichten: Germany

Crinoid stems and plates. Newell and others, 1953, p. 165, Pl. 35, fig. 7 (tegmen spine or spined brachial), fig. 8 (tegmen plate or theca plate?), fig. 9 (tegmen spine or spined primibrach), fig. 10 primibrach), fig. 11 (basal).
Lower Permian, Copacabana Group: Peru

Cupressocrinites schlotheimi rectangularis Schmidt, 1941, p. 103, fig. 27 (brachials and a radial).
Devonian, Eifelian: Germany

Cyathocrinoidea indéterminé. Termier and Termier, 1950, p. 220, Pl. 218, figs. 31, 32 (theca).
Devonian: Morocco

Delocrinus sp. Heuer, 1958, p. 41, Pl. 2 (incorrect explanation for plate), fig. 7 (brachial), fig. 8 (brachial), fig. 9 (radial) fig. 12 (brachial).
Permian, Wolfcampian, Harpersville Formation: United States: Texas

Dimerocrinitidae, Lampterocrinidae, and Ptychocrinidae. Moore, 1954, p. 144, fig. 7B (plate diagram).

Empreinte des bras et du pourtour d'un calice de Crinoide indéterminé. Termier and Termier, 1950, p. 224, Pl. 220, fig. 1 (crown).
Ordovician: Morocco

Ethelocrinus? *kirki* Elias, 1957, p. 385, Pl. 41, figs. 4, 5 (theca plates).
Mississippian, Chesterian, Redoak Hollow Formation: United States: Oklahoma

Facetal canalis (Moore), 1939. Moore and Laudon, 1944, p. 209, Pl. 79, fig. 23 (radial).
Pennsylvanian, Desmoinesian, Mineral Coal cap: United States: Kansas

Facetal circularia Byrne and Seeberger, 1942, p. 223, Pl. 1, fig. 11 (radial).
Permian, Wolfcampian, Grant Shale: United States: Kansas

Facetal corrugatus (Moore), 1939. Moore and Laudon, 1944, p. 209, Pl. 79, fig. 16 (radial).
Pennsylvanian, Missourian, Coffeyville Formation: United States: Kansas

Facetal lineatus (Moore), 1939. Moore and Laudon, 1944, p. 209, Pl. 79, fig. 18 (radial).
Pennsylvanian, Desmoinesian, Oologah Limestone: United States: Oklahoma

Facetal triangulata Byrne and Seeberger, 1942, p. 224, Pl. 1, figs. 7-9 (radial).
Lower Permian, Wolfcampian, Hughes Creek Shale to Grant Shale: United States: Kansas

Facetal tuberculata Byrne and Seeberger, 1942, p. 224, Pl. 1, fig. 10 (radial).
Permian, Wolfcampian, Hughes Creek Shale: United States: Kansas

Facetal tuberculatus (Moore), 1939. Moore and Laudon, 1944, p. 209, Pl. 79, fig. 15 (radial).
Pennsylvanian, Desmoinesian, Mineral Coal cap: United States: Kansas

Family, genus, and species unknown, Species D. Webster and Lane, 1967, p. 31, Pl. 2, figs. 3,4 (radial), fig. 8 (brachial).
Permian, Wolfcampian, Bird Spring Formation: United States: Nevada

Family Poteriocrinidae. Yakovlev, 1949a, p. 265, fig. 1v (radial); 1956a, p. 85, fig. 35, no. 3; 1964a, p. 79, fig. 37, no. 3.
No age given, probably Permian: Russia

Fragments de Crinoides indéterminés. Termier and Termier, 1950, p. 236, Pl. 226, figs. 13, 21 (brachials).
Devonian, Siegenian: Morocco

Fragments de Crinoides indéterminés. Termier and Termier, 1950, p. 236, Pl. 226, fig. 16 (radial).
Devonian, Siegenian: Morocco

Genus and species unknown, Species A. Lane and Webster, 1966, p. 56, Pl. 9, fig. 7 (crown).
Permian, Wolfcampian, Bird Spring Formation: United States: Nevada

Genus and species unknown, Species A. Webster and Lane, 1967, p. 12, fig. 4, no. 2 (part of one ray, basals to brachials).
Permian, Wolfcampian, Bird Spring Formation: United States: Nevada

Genus and species unknown, Species B. Lane and Webster, 1966, p. 57, Pl. 9, fig. 8 (set of arms).
Permian, Wolfcampian, Bird Spring Formation: United States: Nevada

Genus and species unknown, Species B. Webster and Lane, 1967, p. 20, Pl. 8, fig. 5; Fig. 4, no. 1 (crown).
Permian, Wolfcampian, Bird Spring Formation: United States: Nevada

Gen. indet. et sp. nov. indet. 1. Wanner, 1916 = *Calceolispongia* sp. nov.

Gen. nov. et sp. nov. indet. 4. Wanner, 1924 = *Calceolispongia* sp. nov.

Genus unknown, Species C. Webster and Lane, 1967, p. 22, Pl. 4, figs. 10, 11 (crown).
Permian, Wolfcampian, Bird Spring Formation: United States: Nevada

Glyptocrinina. Moore, 1954, p. 144, fig. 8A (plate diagram).

Hydreionocrinus sp. Heuer, 1958, p. 41 (incorrect plate for explanation), Pl. 2, figs. 10, 11 (tegmen spines).
Permian, Wolfcampian, Harpersville Formation: United States: Texas

Indéterminé. Termier and Termier, 1950, p. 218, Pl. 217, fig. 16 (partial theca).
Carboniferous, lower Westphalian: Morocco

Indéterminés. Termier and Termier, 1950, p. 222, Pl. 219, figs. 24, 25 (brachials).
Lower Devonian: Morocco

Isocrinus sp. skelettelemente van crinoiden. Zankl, 1965, p. 551, Pl. 2, figs. 1a, b (pinnules).
No age given

Jonkerocrinus Wanner, 1924. Moore and Laudon, 1943b, p. 104 (anal sac).
   Permian

Lecanocrinidae? Le Maître, 1958b, p. 145, Pl. 1, figs. 11-13 (theca).
   Devonian, Couvinian: Algeria

Plaque de Flexibilia indéterminée. Termier and Termier, 1950, p. 222,
   Pl. 219, figs. 35-39 (basal).
   Carboniferous, upper Visean: Morocco

Plate diagrams of cups. Yakovlev, 1947d, p. 40, fig. 2; 1956a, p. 76,
   fig. 22, nos. 1, 2; 1958, p. 61, fig. 22; 1964a, p. 72, fig. 24,
   nos. 1, 2.

Platycrinus permiensis Yakovlev, *in* Yakovlev and Ivanov, 1956, p. 66,
   Pl. 10, fig. 14 (radial).
   Permian: Russia: Ural Mountains

Platycrinus sp. Termier and Termier, 1950, p. 86, Pl. 227, figs. 1-10
   (brachials).
   Carboniferous, Visean: Morocco

Platycrinus sp. Termier and Termier, 1950, p. 86, Pl. 227, figs. 19-25
   (radials).
   Carboniferous, Visean: Morocco

Platycrinus sp. Termier and Termier, 1950, p. 87, Pl. 211, figs. 7-10
   (radials).
   Carboniferous, Westphalian: Morocco

Polycerus Fischer, 1811. Moore and Laudon, 1943b, p. 104 (plates).
   Carboniferous

Polygonal hexagona Byrne and Seeberger, 1942, p. 225, Pl. 1, figs. 16, 17
   (basal).
   Permian, Wolfcampian, Florena Shale to Grant Shale: United States:
   Kansas

Polygonal pentagona Byrne and Seeberger, 1942, p. 224, Pl. 1, figs. 12, 13
   (basal).
   Permian, Wolfcampian, Hughes Creek Shale to Grant Shale: United
   States: Kansas

Polygonal tenuirugosus (Moore), 1939. Moore and Laudon, 1944, p. 209,
   Pl. 79, fig. 22 (basal).
   Pennsylvanian, Missourian, Missouri Formation: United States: Iowa

Poteriocrinidae Yakovlev, 1949a, p. 265, fig. 1v; 1956a, p. 86, fig. 35,
   no. 3; 1958, p. 72, fig. 35, no. 3; 1964a, p. 79, fig. 37, no. 3;
   1964b, p. 58, fig. 66 (labeled *Moscovicrinus*).
   Carboniferous

Poteriocrinidae gen. indet. et sp. nov. indet. Wanner, 1949a, p. 54, Pl. 3,
   fig. 31 (set of arms).
   Permian: Timor

Poteriocrinus crassus Miller. Termier and Termier, 1950, p. 89,
   Pl. 218, fig. 1 (set of arms).
   Carboniferous, upper Visean: Morocco

Prophyllocrinus sp. Yakovlev, 1949a, p. 265, fig. 1d; 1956a, p. 87,
   fig. 35, no. 4; 1958, p. 72, fig. 35, no. 4; 1964a, p. 79, fig. 37,
   no. 4; 1964b, p. 58, fig. 67 (radial).
   Permian: Timor

Radiale d'un Dendrocrinoidea indéterminé. Termier and Termier, 1950,
   p. 228, Pl. 222, figs. 23, 24 (radial).
   Carboniferous, Westphalian: Morocco

Radial gen. et sp. indet. Sieverts-Doreck, 1951a, p. 141, fig. 8 (radial).
   Lower Carboniferous: Germany

Scytalocrinus sp. indet. Sieverts-Doreck, 1951a, p. 140, fig. 6 (brachial).
  Lower Carboniferous: Germany

Skelettelemente von Crinoiden. Zankl, 1965, p. 551, Pl. 1, fig. 1a
   (pinnule), fig. 1b (brachial).
  No age given

Tanaocrinidae. Moore, 1954, p. 144, fig. 8B (plate diagram).

Tanaocrinina. Moore, 1954, p. 144, fig. 8D (plate diagram).

Tegminal expansus (Moore), 1939. Moore and Laudon, 1944, p. 209, Pl. 79,
   fig. 17 (tegmen plate).
  Pennsylvanian, Desmoinesian, Mineral Coal cap: United States: Kansas

Tegminale? de *Platycrinus* sp. Termier and Termier, 1950, p. 238, Pl. 227,
   figs. 25, 26 (tegmen plate).
  Carboniferous, Visean: Morocco

Ulocrinus occidentalis. Heuer, 1958, p. 41 (incorrect explanation for
   plate), Pl. 2, fig. 13 (basal).
  Permian, Wolfcampian, Harpersville Formation: United States: Texas

Unidentified crinoid crown. Trunko, 1966, p. 93, Pls. 3-6.
  Devonian, Givetian, Ton Schiefer: Germany

Unidentified crinoid cups. Wilson, 1949, p. 243, Pl. 21, figs. 1-3.
  Silurian, Albian, Brassfield Limestone: United States: Tennessee

Part II. Columnals

SECTION 1. IDENTIFIED COLUMNALS

Acanthocrinus gracilior vermicularis Schmidt, 1941, p. 211, fig. 58a.
  Devonian, Coblentzian, upper Coblentzian beds: Germany

Actinocrinites cingulatus Goldfuss, 1866 = *Pentagonocyclicus cingulatus*

Adelocrinus Phillips, 1841, Moore and Laudon, 1943b, p. 103.
  Devonian

Agathocrinus acanthaceus Schewtschenko, 1967, p. 87, Pl. 10, figs. 7, 8, fig. 9; 1968b, p. 83, Pl. 10, figs. 7, 8; Fig. 9.
  Lower Devonian, Kshtutskian horizon: Russia: Zeravshan Mountains

AMPHOLENIUM Moore and Jeffords, 1968. *A. apolegma* Moore and Jeffords, 1968. Moore and Jeffords, 1968, p. 75.
  Mississippian, Osagean.

Ampholenium apolegma Moore and Jeffords, 1968, p. 75, Pl. 23, fig. 2.
  Mississippian, Osagean, Floyds Knob Formation: United States: Kentucky

AMSDENANTERIS Moore and Jeffords, 1968 *A. tennesseensis* Moore and Jeffords, 1968. Moore and Jeffords, 1968, p. 60.
  Silurian, Niagaran

Amsdenanteris tennesseensis Moore and Jeffords, 1968, p. 60, Pl. 11, figs. 7, 8.
  Silurian, Niagaran, Brownsport Formation: United States: Tennessee

Ancyrocrinus Hall, 1862. Moore and Laudon, 1943b, p. 103.
  Devonian

Ancyrocrinus spinosus Hall, 1862. Moore and Laudon, 1944, p. 209, Pl. 79, fig. 30.
  Devonian: United States: New York

ANTHINOCRINUS Yeltyschewa and Sizova, *in* Stukalina, 1961. *A. ludlowicus* Stukalina, 1961. Stukalina, 1961, p. 34.
  Silurian-lower Carboniferous

Anthinocrinus abditus Schewtschenko, 1966, p. 151, Pl. 4, fig. 1; Fig. 20.
  Lower Devonian, Pandzhrutshian horizon: Russia: Zeravshan Mountains

Anthinocrinus acutulus Dubatolova, 1964, p. 57, Pl. 7, figs. 9-11.
  Lower Devonian: Russia: Kuznetz Basin

Anthinocrinus arenosus Yeltyschewa and Dubatolova, 1967, p. 45, Pl. 5, figs. 13-15.
  Lower Carboniferous: Russia: Upper Priamur

Anthinocrinus carbonicus Yeltyschewa, *in* Yeltyschewa and Dubatolova, 1967, p. 44, Pl. 5, figs. 16-18.
  Carboniferous, upper Tournaisian-lower Visean: Russia: Kazakhstan

Anthinocrinus cognatus Dubatolova, 1964, p. 56, Pl. 7, figs. 7, 8.
  Lower Devonian: Russia: Kuznetz Basin

Anthinocrinus conspicuus Dubatolova, 1964 = *Facetocrinus conspicuus*

Anthinocrinus eugeniae Yeltyschewa and Dubatolova, 1967, p. 40, Pl. 5, figs. 3-6 (*non* figs. 3-7 as given in text), fig. 17. Modzalevskaya, 1967, Pl. 4, fig. 17.
  Lower-Middle Devonian, Jmatchinskian suite: Russia: Upper Priamur

Anthinocrinus floreus Yeltyschewa, *in* Yeltyschewa and Dubatolova, 1967,
  p. 37, Pl. 4, figs. 7, 8; Fig. 16. Modzalevskaya, 1967, p. 546,
  Pl. 4, figs. 18, 19. Sizova, 1960, p. 53 (nomen nudem).
  Lower-Middle Devonian, Jmatchinskian suite: Russia: Kazakhstan,
  Kuznetz Basin, Urals, Upper Priamur, Far East

Anthinocrinus aff. *floreus* Yeltyschewa, *in* Dubatolova, 1964, p. 59, Pl. 7,
  figs. 13-15.
  Devonian, Eifelian: Russia: Kuznetz Basin, Pribalkhash

Athinocrinus floreus gracilis (Yeltyschewa and Dubatolova), 1961.
  Yeltyschewa and Dubatolova, 1967, p. 38, Pl. 4, fig. 15.
  Pentagonopentagonalis floreus var. *gracila* Yeltyschewa and Dubatolova,
  1961, p. 558, Pl. D-87, figs. 10, 11. Yeltyschewa and Dubatolova,
  1967, p. 38.
  Devonian, Givetian: Russia: Altay Mountains, Upper Priamur

Anthinocrinus floreus var. *magna* Sizova, 1960, p. 53 (nomen nudem).
  Devonian, Eifelian-Frasnian: Russia: central Kazakhstan

Anthinocrinus incisus Yeltyschewa, *in* Dubatolova, 1964, p. 60, Pl. 8,
  fig. 1.
  Devonian, Frasnian: Russia: Kuznetz Basin, main Devonian field

Anthinocrinus luchi Yeltyschewa, 1968, p. 40, Pl. 5, figs. 23-25.
  Silurian, Skalskian horizon: Russia: Podolsk

*Anthinocrinus ludlowicus Stukalina, 1961, p. 34, Pl. 2, fig. 6; 1965b,
  p. 135, Pl. 2, figs. 12-14. Dubatolova, 1968, p. 147, Pl. 15,
  fig. 7; Fig. 5.
  Upper Silurian-Devonian, Karaespinian horizon, lower Rekovian Beds:
  Russia: Aksarli Mountains, central Kazakhstan, Salair

Anthinocrinus minimus Yeltyschewa and Dubatolova, 1967, p. 42, Pl. 5,
  figs. 7-9 (*non* figs. 7, 8, as given in text); Fig. 18.
  Modzalevskaya, 1967, Pl. 4, fig. 16.
  Middle Devonian, Jmatchinskian suite: Russia: Upper Priamur

Anthinocrinus petalatus Yeltyschewa and Dubatolova, 1967, p. 43, Pl. 4,
  fig. 10; Fig. 19. Modzalevskaya, 1967, Pl. 4, fig. 20.
  Middle Devonian, Jmatchinskian suite: Russia: Upper Priamur

Anthinocrinus podolicus Yeltyschewa, 1968, p. 39, Pl. 5, figs. 26-29.
  Silurian, Skalskian horizon: Russia: Podolsk

Anthinocrinus primaevus Sizova, *in* Yeltyschewa and Dubatolova, 1967,
  p. 39, Pl. 4, figs. 16-18 (misspelled as *A. primevus*).
  Lower-Middle Devonian: Russia: Upper Priamur

Anthinocrinus cf. *primaevus* Sizova, *in* Dubatolova, 1967, p. 35, Pl. 6,
  figs. 8, 9; Fig. 2.
  Lower Devonian, Eifelian: Russia: Tas-Hayahtah Mountain, Pribalkhash

Anthinocrinus quinqueangularis Dubatolova, 1964 = *Facetocrinus
  quinqueangularis*

Anthinocrinus quinquefidus Dubatolova, 1968, p. 148, Pl. 15, figs. 8, 9;
  Fig. 6.
  Devonian, Tom-Chumshian horizon, lower Rekovian Beds: Russia: Salair

Anthinocrinus raricostatus Yeltyschewa and Dubatolova, 1967, p. 41, Pl. 5,
  figs. 10-12. Modzalevskaya, 1967, p. 545, Pl. 3, fig. 7.
  Lower-Middle Devonian, Bolsheneverskian suite: Russia: Kazakhstan,
  Altay, Far East, Upper Priamur

Anthinocrinus sangulus Schewtschenko, 1966, p. 152, Pl. 4, fig. 16;
  Fig. 21.
  Lower Devonian, Pandzhrutskian horizon: Russia: Zeravshan Mountains,
  southern Tien-Shan Mountains

Anthinocrinus terminalis Schewtschenko, 1966, p. 153, Pl. 4, fig. 2; Fig. 22.
   Lower Devonian, Pandzhrutskian horizon: Russia: Zeravshan Mountains, southern Tien-Shan Mountains

Anthinocrinus urkaensis Yeltyschewa and Dubatolova, 1967, p. 46, Pl. 5, figs. 1, 2.
   Lower Carboniferous: Russia: Upper Priamur

APERTOCRINUS Stukalina, 1968. *A. apertus* Yeltyschewa and Stukalina, *in* Stukalina, 1968. Stukalina, 1968c, p. 83.
   Middle-Upper Ordovician

Apertocrinus apertus Yeltyschewa and Stukalina, *in* Stukalina, 1968c, p. 84, figs. 2a, b.
   Middle-Upper Ordovician: Russia: central Kazakhstan, Siberian Platform

APIASTRUM Moore and Jeffords, 1968. *A. candidum* Moore and Jeffords, 1968. Moore and Jeffords, 1968, p. 58.
   Devonian, Helderbergian

Apiastrum candidum Moore and Jeffords, 1968, p. 58, Pl. 10, fig. 9; Pl. 11, figs. 2, 3.
   Devonian, Helderbergian, Birdsong Shale: United States: Tennessee

Arthroacantha? spp.
   Termier and Termier, 1950, p. 85, Pl. 213, figs. 16-21; Pl. 224, figs. 48, 49.
      Lower-Middle Devonian: Morocco
   Termier and Termier, 1950, p. 214, Pl. 215, figs. 21, 22.
      Devonian, Emsian: Morocco

Cf. *Arthroacantha*? sp. Termier and Termier, 1950, p. 232, Pl. 224, figs. 33, 34.
   Devonian, Siegenian: Morocco

Ascarum Gregorio, 1930. Moore and Laudon, 1943, p. 103.
   Permian

Aspidocrinus Hall, 1858. Moore and Laudon, 1943b, p. 103.
   Devonian

Asterocrinus munsteri Eichwald, 1860. Yeltyschewa, 1964a, p. 60, Pl. 1, figs. 23-28; Pl. 4, fig. 20.
   Lower Ordovician: Russia: Kazakhstan

ASTEROMISCHUS Moore and Jeffords, 1968. *A. stellatus* Moore and Jeffords, 1968. Moore and Jeffords, 1968, p. 54.
   Mississippian, Chesterian

Asteromischus stellatus Moore and Jeffords, 1968, p. 54, Pl. 9, figs. 1-3.
   Mississippian, Chesterian, Paint Creek Formation: United States: Illinois

Asteromischus stellatus? Moore and Jeffords, 1968, p. 54, Pl. 8, fig. 11.
   Mississippian, Chesterian, Paint Creek Formation: United States: Illinois

Astroporites Lambe, 1896. Moore and Laudon, 1943b, p. 103.
   Ordovician

Atactocrinus? species undetermined. Moore and Jeffords, 1968, p. 37, Pl. 2, fig. 13.
   Ordovician, Cincinnatian, Fernvale Limestone: United States: Missouri, Minnesota, Iowa, Illinois

AVICANTUS Moore and Jeffords, 1968. *A. dunbari* Moore and Jeffords, 1968. Moore and Jeffords, 1968, p. 70.
   Devonian, Helderbergian

Avicantus dunbari Moore and Jeffords, 1968, p. 70, Pl. 20, figs. 4-6.
  Devonian, Helderbergian, Birdsong Shale: United States: Tennessee

Barycrinus? aff. *B. herculeus* Meek and Worthen, 1868. Elias, 1957,
  p. 384, Pl. 41, figs. 1, 2.
  Mississippian, Chesterian, Redoak Hollow Formation: United States:
  Oklahoma

Barycrinus sp. Sangil, 1955, p. 251, fig. 2A.
  No age given

BARYSCHYR Moore and Jeffords, 1968. *B. anosus* Moore and Jeffords, 1968.
  Moore and Jeffords, 1968, p. 64.
  Pennsylvanian, Atokan

Baryschyr anosus Moore and Jeffords, 1968, p. 64, Pl. 14, figs. 6, 7;
  Pl. 15, figs. 7-11; Pl. 16, figs. 1-11. Moore and others, 1968a,
  p. 11, Pl. 2, figs. 6, 7.
  Pennsylvanian, Atokan, Pumpkin Creek Limestone: United States:
  Oklahoma

BAZARICRINUS Stukalina, 1968. *B. bazarensis* Stukalina, 1968. Stukalina,
  1968c, p. 90.
  Silurian, Wenlockian

Bazaricrinus bazarensis Stukalina, 1968c, p. 90, fig. 2z.
  Silurian, Wenlockian: Russia: central Kazakhstan

BLOTHRONAGMA Moore and Jeffords, 1968. *B. cinctutum* Moore and Jeffords,
  1968. Moore and Jeffords, 1968, p. 63.
  Pennsylvanian, Atokan

Blothronagma cinctutum Moore and Jeffords, 1968, p. 63, Pl. 15,
  figs. 1-6; Pl. 16, figs. 12-15. Moore and others, 1968a, p. 11,
  Pl. 2, fig. 5.
  Pennsylvanian, Atokan, Pumpkin Creek Limestone: United States:
  Oklahoma

Bornium Gregorio, 1930. Moore and Laudon, 1943b, p. 103.
  Permian

Botryocrinus sp.
  Termier and Termier, 1950, p. 88, Pl. 226, figs. 23, 24.
    Ordovician, Llandeilian: Morocco
  Ubaghs, 1953, p. 751, fig. 86.

BYSTROWICRINUS Yeltyschewa, *in* Yeltyschewa and Stukalina, 1963.
  Pentagonopentagonalis quinquelobatus Yeltyschewa, 1955.
  Yeltyschewa and Stukalina, 1963, p. 27.
  Middle Ordovician-Lower Silurian

Bystrowicrinus angustilobatus Yeltyschewa, *in* Yeltyschewa and Stukalina,
  1963, p. 29, Pl. 2, fig. 6; Fig. 2 (misspelled as *B. angustilobatuc*,
  p. 29).
  Upper Ordovician-Lower Silurian, Vezenbergskian horizon-Llandoverian:
  Russia: Estonia, Nova Zemlya, central Taymyr

Bystrowicrinus compositus (Yeltyschewa), 1955. Yeltyschewa and Stukalina,
  1963, p. 31, Pl. 2, figs. 4, 5; Fig. 3.
  Pentagonopentagonalis compositus Yeltyschewa, 1955a, p. 42, Pl. 23,
    fig. 1; 1960, p. 10, Pl. 1, figs. 1, 2. Yeltyschewa and Stukalina,
    1963, p. 31. Schewtschenko, 1964, p. 11, Pl. 2, figs. 6, 7; Pl. 3,
    fig. 3.
  Middle Ordovician-Lower Silurian: Russia: Siberian Platform, Taymyr,
  Nova Zemlya, central Tadzhikstan

Bystrowicrinus quinquelobatus (Yeltyschewa), 1955. Yeltyschewa and
  Stukalina, 1963, p. 28, Pl. 2, figs. 7, 8; Fig. 1.
  Pentagonopentagonalis quinquelobatus Yeltyschewa, 1955a, p. 43, Pl. 23,
    figs. 4, 5; Pl. 37, figs. 5, 6; 1960, p. 11, Pl. 2, figs. 1-6.
  Yeltyschewa and Stukalina, 1963, p. 28. Schewtschenko, 1964, p. 11,

Pl. 1, figs. 3-5; Pl. 2, fig. 12; Pl. 3, fig. 7.
Middle Ordovician-Lower Silurian: Russia: Siberian Platform, Taymyr, Nova Zemlya, Pamir Mountains, central Tadzhikstan

Camptocrinus beaveri Moore and Jeffords, 1968, p. 48, Pl. 5, fig. 10; Pl. 6, fig. 4.
Mississippian, Chesterian, Paint Creek Formation: United States: Illinois

Camptocrinus? tasmaniensis Sieverts-Doreck, 1942, p. 225, figs. 1-4.
Permian: Australia; Tasmania

CATAGRAPHIOCRINUS Stukalina, 1968. *Pentagonopentagonalis quindecemlobatus* Yeltyschewa and Stukalina, 1963. Stukalina, 1968c, p. 88.
Upper Ordovician-Lower Silurian

Catagraphiocrinus quindecemlobatus (Yeltyschewa and Stukalina), 1963. Stukalina, 1968c, p. 88.
Pentagonopentagonalis quindecemlobatus Yeltyschewa and Stukalina, 1963, p. 41, Pl. 1, figs. 19, 20; Fig. 10. Stukalina, 1968c, p. 88.
Upper Ordovician-Lower Silurian: Russia: central Taymyr

CATHOLICORHACHIS Moore and Jeffords, 1968. *C. multifaria* Moore and Jeffords, 1968. Moore and Jeffords, 1968, p. 63.
Pennsylvanian, Desmoinesian

Catholicorhachis multifaria Moore and Jeffords, 1968, p. 63, Pl. 14, fig. 8.
Pennsylvanian, Desmoinesian, Mingus Shale: United States: Texas

CIONERISMA Moore and Jeffords, 1968. *C. exile* Moore and Jeffords, 1968. Moore and Jeffords, 1968, p. 76.
Silurian, Niagaran

Cionerisma exile Moore and Jeffords, 1968, p. 76, Pl. 23, fig. 5.
Silurian, Niagaran, Waldron Shale: United States: Indiana

Cleiocrinus grandis Billings, 1859. Wilson, 1946, p. 24.
Ordovician, Champlainian, Cobourg Limestone: Canada: Ontario

Cleiocrinus magnificus Billings, 1859. Wilson, 1946, p. 24.
Ordovician, Champlainian, Cobourg Limestone, Hull Formation: Canada: Ontario, Quebec

CLEMATIDISCUS Moore and Jeffords, 1968. *C. denotatus* Moore and Jeffords, 1968. Moore and Jeffords, 1968, p. 74.
Devonian, Erian

Clematidiscus denotatus Moore and Jeffords, 1968, p. 74, Pl. 22, figs. 9-11.
Devonian, Erian, Wanakah Shale: United States: New York

Coenocrinus Valette, 1934. Moore and Laudon, 1943b, p. 104.
Permian

Coenocrinus solignaci Valette, 1934 = *Tebagacrinus solignaci*

Cophinus Murchison, 1839. Moore and Laudon, 1943b, p. 104.
Silurian

CRENATAMES Moore and Jeffords, 1968. *C. amicabilis* Moore and Jeffords, 1968. Moore and Jeffords, 1968, p. 71.
Devonian, Erian

Crenatames amicabilis Moore and Jeffords, 1968, p. 71, Pl. 21, figs. 8, 9.
Devonian, Erian, Wanakah Shale: United States: New York

Crisantum Gregorio, 1930. Moore and Laudon, 1943b, p. 104.
Permian

Cromyocrinus sp. Yakovlev, 1956a, p. 118, fig. 44, nos. 1-3; 1958, p. 102, fig. 44, nos. 1-3; 1964a, p. 105, fig. 46.
   Middle Carboniferous: Russia: Moscow Basin

Crotalocrinites rugosus Miller, 1821. Stukalina, 1965b, p. 137, pl. 1, figs. 10-13; Fig. 3.
   Upper Silurian: Russia: central Kazakhstan

Crotalocrinites sp. Yeltyschewa, 1960, p. 7, Pl. 5, figs. 11-13.
   Silurian: Russia: Siberian Platform

Crotalocrinites spp. nomen correctum.
   Crotalocrinus sp. Yeltyschewa, 1955a, p. 41, Pl. 54, figs. 4, 5.
      Silurian, Llandovery: Russia: Siberian Platform
   Crotalocrinus sp. Sangil, 1955, p. 251, fig. 2F.
      No age given
   Crotalocrinus sp. indet. Schmidt, 1952, Pl. 8, figs. 5a, b.
      Devonian: Germany

Ctenocrinus sp. (aff. *pinnatus* Goldfuss). Kettner and Prantl, 1942, p. 13, Pl. 1, figs. 4-6.
   Devonian: Czechoslovakia

?*Ctenocrinus portezuelensis* Rusconi, 1952, p. 35, fig. 6. Sieverts-Doreck, 1957b, p. 154.
   Ordovician or younger (Sieverts-Doreck, 1957b), Salagastanse horizon: Argentina

Culmicrinus n. sp. Sieverts-Doreck, 1964, p. 4, Pl. 1, figs. 1-9; Pl. 3.
   Carboniferous, Namurian?: Spain

Culmicrinus n. sp.? Sieverts-Doreck, 1964, p. 9.
   Carboniferous, Namurian?, Middle Limestone Series: Spain

Cupressocrinites crassus Yeltyschewa and Dubatolova, 1961 = *C. scaber*

Cupressocrinites elegans (Schewtschenko), 1959, nomen correctum.
   Cupressocrinus elegans Schewtschenko, 1959, p. 8, Pl. 1, figs. 8, 9.
   Devonian, Eifelian: Russia: Zeravshan Range

Cupressocrinites gracilis Goldfuss, 1826. Yeltyschewa and Dubatolova, 1961, p. 552, Pl. D-86, fig. 4. Dubatolova, 1964, p. 29, Pl. 3, figs. 3, 4; Figs. 3, 4.
   Devonian, Eifelian, Shandinsky layer: Russia: Altay Mountains, Kuznetz Basin

Cupressocrinites minor Yeltyschewa, *in* Yeltyschewa and Dubatolova, 1961, p. 553, Pl. D-86, fig. 5.
   Devonian, Eifelian, Shandinsky layer: Russia: Altay Mountains

Cupressocrinites ovatus (Schewtschenko), 1959, nomen correctum.
   Cupressocrinus ovatus Schewtschenko, 1959, p. 9, Pl. 1, figs. 5-7.
   Devonian, Eifelian: Russia: Zeravshan Range

Cupressocrinites pentaporus Eichwald, 1860 = *Dianthiocoeloma pentaporus*

Cupressocrinites planus (Schewtschenko), 1959, nomen correctum.
   Cupressocrinus planus Schewtschenko, 1959, p. 7, Pl. 1, figs. 1-4.
   Devonian, Eifelian: Russia: Zeravshan Range

Cupressocrinites rossicus Antropov, 1954, nomen correctum.
   Devonian

Cupressocrinites rossicus gr. Cyclotetradigonopae (Antropov), 1954, nomen correctum.
   Cupressocrinus rossicus gr. Cyclotetradigonopae Antropov, 1954, p. 15, Pl. 1, figs. 1-3.
   Cupressocrinus rossicus gr. Cyclotetradigonopae Antropov, 1954, p. 15, Pl. 1, fig. 6.
   Devonian: Russia

Cupressocrinites rossicus gr. Ellipsisodigonopae (Antropov), 1954, nomen correctum.

Cupressocrinus rossicus gr. Ellipsisodigonopae Antropov, 1954, p. 15, Pl. 1, figs. 4, 5.
Devonian: Russia

Cupressocrinites? salagastianus (Rusconi), 1952, nomen correctum.
?*Cupressocrinus salagastianus* Rusconi, 1952, p. 37, figs. 7, 8. Sieverts-Doreck, 1957b, p. 154.
Ordovician or younger (Sieverts-Doreck, 1957b), Salagastanse horizon: Argentina

Cupressocrinites scaber Schultze, 1867. Dubatolova, 1964, p. 28, Pl. 2, fig. 9; Pl. 3, figs. 1, 2.
Cupressocrinites crassus Goldfuss, 1826. Yeltyschewa and Dubatolova, 1961, p. 552, Pl. D-87, fig. 1. Dubatolova, 1964, p. 28.
Devonian, Eifelian, Shandinsky layer: Russia: Altay Mountains, Kuznetz Basin

Cupressocrinites tripartitus Schewtschenko, 1968, p. 284, Pl. 66, fig. 8; Fig. 36.
Devonian, Eifelian: Russia: Tien-Shan, Zeravshan Range

Cupressocrinites spp. nomen correctum.
Cupressocrinus sp. Sangil, 1955, p. 251, fig. 2B.
No age given
Cupressocrinus sp. Schewtschenko, 1959, p. 9, Pl. 1, figs. 10, 11.
Devonian, Eifelian: Russia: Zeravshan Range

Cyathocrinites goliathus (Waagen), 1885.
"*Cyathocrinus*" *goliathus* Waagen, 1885. Termier and Termier, 1958, p. 53, Pl. 2, figs. d-n.
Permian: Tunisia

Cyathocrinites pinnatus Goldfuss, 1831. Termier and Termier, 1950, p. 93, Pl. 213, figs. 22-29.
Lower Devonian: Morocco

Cyathocrinites virgalensis (Waagen), 1885.
"*Cyathocrinus*" *virgalensis* Waagen, 1885. Termier and Termier, 1958, p. 55, Pl. 3, figs. a-f.
Permian: Tunisia

CYCLOCAUDEX Moore and Jeffords, 1968. *C. typicus* Moore and Jeffords, 1968. Moore and Jeffords, 1968, p. 65.
Mississippian, Osagean-Pennsylvanian, Virgilian

Cyclocaudex aptus Moore and Jeffords, 1968, p. 66, Pl. 17, fig. 9.
Mississippian, Osagean, New Providence Shale: United States: Kentucky

Cyclocaudex congregalis Moore and Jeffords, 1968, p. 65, Pl. 17, fig. 10.
Mississippian, Osagean, New Providence Shale: United States: Kentucky

Cyclocaudex costatus Moore and Jeffords, 1968, p. 66, Pl. 17, fig. 8.
Pennsylvanian, Virgilian, South Bend Shale: United States: Texas

Cyclocaudex insaturatus Moore and Jeffords, 1968, p. 66, Pl. 17, fig. 11.
Pennsylvanian, Desmoinesian, Cabaniss Formation: United States: Kansas

Cyclocaudex jucundus Moore and Jeffords, 1968, p. 66, Pl. 17, figs. 6, 7.
Pennsylvanian, Virgilian, Chaffin Limestone, Belknap Limestone: United States: Texas

Cyclocaudex plenus Moore and Jeffords, 1968, p. 66, Pl. 17, fig. 12; Pl. 18, figs. 1-5. Moore and others, 1968a, p. 19, Pl. 2, figs. 1-4; Pl. 3, fig. 9. Jeffords and Miller, 1968, p. 11, Pl. 1, figs. 1-13; Figs. 1A, 2A, 3, 4A, B, 5A, B.
Pennsylvanian, Virgilian, Chaffin Limestone, Black Ranch Limestone,

Belknap Limestone, Gunsight Limestone, Wayland Shale: United States: Texas

Cyclocaudex typicus Moore and Jeffords, 1968, p. 65, Pl. 17, figs. 1-5.
Pennyslvanian, Virgilian, Belknap Limestone: United States: Texas

CYCLOCAUDICULUS Moore and Jeffords, 1968. *C. regularis* Moore and Jeffords, 1968. Moore and Jeffords, 1968, p. 83.
Pennyslvanian, Virgilian

Cyclocaudiculus regularis Moore and Jeffords, 1968, p. 83, Pl. 27, fig. 12.
Pennsylvanian, Virgilian, Chaffin Limestone: United States: Texas

CYCLOCHARAX Moore and Jeffords, 1968. *C. fasciatus* Moore and Jeffords, 1968. Moore and Jeffords, 1968, p. 76.
Silurian, Niagaran

Cyclocharax fasciatus Moore and Jeffords, 1968, p. 76, Pl. 23, figs. 7, 8.
Silurian, Niagaran, Waldron Shale: United States: Indiana

Cyclocharax modestus Moore and Jeffords, 1968, p. 76, Pl. 23, fig. 1.
Silurian, Niagaran, Waldron Shale: United States: Indiana

CYCLOCION Moore and Jeffords, 1968. *C. distinctus* Moore and Jeffords, 1968. Moore and Jeffords, 1968, p. 78.
Mississippian, Chesterian

Cyclocion distinctus Moore and Jeffords, 1968, p. 78, Pl. 24, figs. 15-61; Pl. 25, figs. 1-7.
Mississippian, Chesterian, Paint Creek Formation: United States: Illinois

CYCLOCRISTA Moore and Jeffords, 1968. *C. lineolata* Moore and Jeffords, 1968. Moore and Jeffords, 1968, p. 79.
Pennsylvanian, Virgilian-Permian, Wolfcampian

Cyclocrista cheneyi Moore and Jeffords, 1968, p. 80, Pl. 26, figs. 7, 8.
Permian, Wolfcampian, Waldrip Shale: United States: Texas

Cyclocrista lineolata Moore and Jeffords, 1968, p. 80, Pl. 25, figs. 13, 14.
Pennsylvanian, Virgilian, Chaffin Limestone: United States: Texas

Cyclocrista martini Miller, *in* Moore and Jeffords, 1968, p. 80, Pl. 26, figs. 1-6. Jeffords and Miller, 1968, p. 13, Pl. 4, figs. 1-9; Figs. 1D, 2D, 3, 4E, F, 5D.
Pennsylvanian, Virgilian, Wayland Shale: United States: Texas

CYCLOCYCLICUS Yeltyschewa, 1955, nomen vetitum.
Yeltyschewa, 1955a, p. 46; 1956, p. 42, fig. 1-5E; 1959, p. 233, fig. 1-5E; 1964b, p. 75, fig. 114-5E. Dubatolova, 1964, p. 9, fig. 1-5E.
Ordovician-Permian

Cyclocyclicus aequiplicatus Yeltyschewa and Dubatolova, 1960a, p. 372, Pl. 70, fig. 5; Fig. 45; 1967, p. 58, Pl. 6, fig. 9; Fig. 23 (*non* Fig. 24 as listed on p. 58).
Devonian, upper Givetian: Russia: Far East, Priamur

Cyclocyclicus ampliatus Yeltyschewa, 1968, p. 47, Pl. 5, figs. 21, 22.
Silurian, Borschovskian horizon: Russia: Podolsk

Cyclocylicus arenarius Yeltyschewa and Schewtschenko, 1960, p. 119, Pl. 1, figs. 10, 12-15.
Carboniferous, Visean: Russia: Tien-Shan Mountains

Cyclocyclicus arenarius var. *cingulata* Yeltyschewa and Schewtschenko, 1960, p. 120, Pl. 1, figs. 6, 7.
  Carboniferous, Visean: Russia: Tien-Shan Mountains

Cyclocyclicus arenarius var. *granulosa* Yeltyschewa and Schewtschenko, 1960, p. 121, Pl. 2, figs. 1, 2, 5-8, 11 (*non* fig. 9 as given in text).
  Carboniferous, Visean: Russia: Tien-Shan Mountains

Cyclocyclicus arenarius var. *laevis* Sizova, 1960, p. 59, nomen nudum.
  Carboniferous, upper Visean-Namurian: Russia, Kazakhstan

Cyclocyclicus brevitodentatus Dubatolova and Shao, 1959, p. 47, Pl. 1, fig. 14.
  Lower-middle Carboniferous: China: Kwangtung Province

Cyclocyclicus chaneensis Dubatolova and Shao, 1959, p. 53, Pl. 2, fig. 6.
  Upper Permian: China: Hunan Province

Cyclocyclicus circumvalatus Yeltyschewa, *in* Dubatolova and Shao, 1959, p. 48, Pl. 2, fig. 1.
  Carboniferous, Tournaisian: China: Hunan and Kwangtung Provinces

Cyclocyclicus crassiformis Yeltyschewa, 1964a, p. 76, Pl. 2, fig. 10.
  Lower Ordovician: Russia: Estonia

Cyclocyclicus crista Yeltyschewa, Sizova, 1960, p. 58, nomen nudum.
  Carboniferous, Visean: Russia: central Kazakhstan

Cyclocyclicus crystalliferus Yeltyschewa, 1964a, p. 78, Pl. 3, figs. 5, 6.
  Lower Ordovician: Russia: Estonia

Cyclocyclicus aff. *dignus* Yeltyschewa, *in* Dubatolova and Shao, 1959, p. 53, Pl. 2, fig. 7.
  Carboniferous, Visean: China: Hunan Province

Cyclocyclicus disparies Dubatolova and Shao, 1959, p. 56, Pl. 2, fig. 10.
  Upper Permian: China: Hunan Province

Cyclocyclicus gyratus Yeltyschewa, *in* Yeltyschewa and Dubatolova, 1961, p. 559, Pl. D-87, fig. 18.
  Devonian, Eifelian, Shandinsky layer: Russia: Salair

Cyclocyclicus hissariensis Yeltyschewa and Schewtschenko, 1960, p. 122, Pl. 1, figs. 8, 9.
  Lower Carboniferous: Russia: Tien-Shan Mountains

Cyclocyclicus kuangtungensis Dubatolova and Shao, 1959, p. 52, Pl. 2, fig. 5.
  Lower Carboniferous, Tournaisian: China: Hunan and Kwangtung Provinces

Cyclocyclicus echinatus Yeltyschewa, *in* Stukalina, 1965b, p. 135, Pl. 2, figs. 3, 4.
  Upper Silurian, Aynasuyskian horizon: Russia: central Kazakhstan

Cyclocyclicus infacetus Stukalina, 1960, p. 108, fig. 7.
  Silurian, Llandoverian: Russia: Chingiz Mountains

Cyclocyclicus mirandus Yeltyschewa (unpub.), Dubatolova and Shao, 1959, p. 46, Pl. 1, fig. 12: unnumbered fig.
  Carboniferous, Visean: China: Hunan and Kwangtung Provinces

Cyclocyclicus modestus Yeltyschewa and Dubatolova, 1961, p. 559, Pl. D-86, fig. 19.
  Devonian, Salairkinsky layer: Russia: Salair

Cyclocyclicus mui Dubatolova and Shao, 1959, p. 54, Pl. 2, fig. 9.
  Upper Permian: China: Hunan, Kwangtung, and Sychuan Provinces

Cyclocyclicus nuraensis Sizova, 1960, p. 55, nomen nudum.
  Devonian, Famennian: Russia: central Kazakhstan

Cyclocyclicus orbitus Dubatolova, 1964, p. 98, Pl. 14, fig. 15.
  Devonian, Givetian: Russia: Kuznetz Basin

Cyclocyclicus orenarius var. *ordinata* Yeltyschewa (unpub.), Dubatolova and Shao, 1959, p. 54, Pl. 2, fig. 8.
  Carboniferous, Visean: China: Hunan Province

Cyclocyclicus paludatus Dubatolova, 1964, p. 97, Pl. 14, figs. 10-12.
  Lower Devonian: Russia: Kuznetz Basin

Cyclocyclicus perpusillus Dubatolova and Shao, 1959, p. 47, Pl. 1, fig. 13.
  Carboniferous, Visean: China: Kwangtung Province

Cyclocyclicus proximus Dubatolova, 1964, p. 97, Pl. 14, figs. 13, 14.
  Lower Devonian: Russia: Kuznetz Basin

Cyclocyclicus pulcher Yeltyschewa (unpub.), Dubatolova and Shao, 1959, p. 50, Pl. 2, fig. 3.
  Carboniferous, Tournaisian: China: Kwangtung Province

Cyclocyclicus rarus Schewtschenko, 1966, p. 169, Pl. 5, fig. 13; Fig. 41.
  Upper Silurian, Shishkatian horizon: Russia: Zeravshan Mountains

Cyclocyclicus rugosus Yeltyschewa and Schewtschenko, 1960, p. 121, Pl. 1, figs. 1-5.
  Carboniferous, Visean: Russia: Tien-Shan Mountains

Cyclocyclicus scalariformis Yeltyschewa, 1960, p. 24, Pl. 6, figs. 5-7.
  Silurian, Wenlockian: Russia: Siberian Platform

Cyclocyclicus sibiricus Yeltyschewa, 1960, p. 24, Pl. 4, figs. 6, 7.
  Silurian, Llandoverian?: Russia: Siberian Platform

Cyclocyclicus strigiliferus Yeltyschewa and Dubatolova, 1967, p. 59, Pl. 6, fig. 20; Fig. 24.
  Lower Carboniferous: Russia: Upper Priamur

Cyclocyclicus subincrustatus Schewtschenko, 1968a, p. 281, Pl. 66, fig. 6; Fig. 33.
  Middle Carboniferous: Russia: central Pamir

Cyclocyclicus subsulcatus Yeltyschewa and Schewtschenko, 1960, p. 122, Pl. 2, figs. 3, 4.
  Lower Carboniferous: Russia: Darvus

Cyclocyclicus suni Dubatolova and Shao, 1959, p. 51, Pl. 2, fig. 4.
  Carboniferous, Tournaisian: China: Hunan and Kwangtung Provinces

*Cyclocyclicus tenuis Yeltyschewa, 1955a, p. 46, Pl. 54, fig. 6; 1960, p. 23, Pl. 6, figs. 1-4.
  Silurian, Llandoverian-Wenlockian: Russia: Siberian Platform

Cyclocyclicus tieni Dubatolova and Shao, 1959, p. 49, Pl. 2, fig. 2.
  Carboniferous, Tournaisian: China: Hunan and Kwangtung Provinces

Cyclocyclicus tortousus Sizova, 1960, p. 59, nomen nudum.
  Carboniferous, Visean: Russia: central Kazakhstan

Cyclocyclicus ungulatus Yeltyschewa, 1966, p. 65, Pl. 3, fig. 6.
  Middle Ordovician, Idavereskian horizon: Russia: Estonia

Cyclocyclicus variabilis Yeltyschewa, 1964a, p. 77, Pl. 3, figs. 1-4.
  Lower Ordovician: Russia: Estonia

CYCLOELLIPTICUS Yeltyschewa, 1956, nomen vetitum.
  Yeltyschewa, 1956, p. 42, fig. 1-5F; 1959, p. 233, fig. 1-5F; 1964b, p. 75, fig. 114-5F. Dubatolova, 1964, p. 9, fig. 1-5F.
  Theoretical genus, no known species.

CYCLOHEXAGONALIS Yeltyschewa, 1956, nomen vetitum.
  Yeltyschewa, 1956, p. 42, fig. 1-5D; 1959, p. 233, fig. 1-5D; 1946b, p. 75, fig. 114-5D. Dubatolova, 1964, p. 9, fig. 1-5D.
  Theoretical genus, no known species.

CYCLOMISCHUS Moore and Jeffords, 1968. *C. shelbyensis* Moore and Jeffords, 1968. Moore and Jeffords, 1968, p. 58.
  Silurian, Niagaran

Cyclomischus alternatus Moore and Jeffords, 1968, p. 59, Pl. 11, fig. 1.
  Silurian, Niagaran, Waldron Shale: United States: Indiana

**Cyclomischus shelbyensis* Moore and Jeffords, 1968, p. 59, Pl. 11, fig. 4.
  Silurian, Niagaran, Waldron Shale: United States: Indiana

Cyclomischus tennesseensis Moore and Jeffords, 1968, p. 59, Pl. 12, fig. 1.
  Silurian, Niagaran, Waldron Shale: United States: Tennessee

CYCLOMONILE Moore and Jeffords, 1968. *C. monile* Moore and Jeffords, 1968. Moore and Jeffords, 1968, p. 58.
  Ordovician, Trentonian

**Cyclomonile monile* Moore and Jeffords, 1968, p. 58, Pl. 10, figs. 3, 4.
  Ordovician, Trentonian, Catheys Limestone: United States: Tennessee

CYCLOPAGODA Moore and Jeffords, 1968. *C. alternata* Moore and Jeffords, 1968. Moore and Jeffords, 1968, p. 57.
  Ordovician, Cincinnatian

**Cyclopagoda alternata* Moore and Jeffords, 1968, p. 57, Pl. 10, figs. 5, 6.
  Ordovician, Cincinnatian: Ohio

Cyclopagoda costata Moore and Jeffords, 1968, p. 57, Pl. 10, figs. 1, 2.
  Ordovician, Cincinnatian: Ohio

CYCLOPENTAGONALIS Yeltyschewa, 1956, nomen vetitum.
  Yeltyschewa, 1956, p. 42, fig. 1-5C; 1959, p. 233, fig. 1-5C; 1964a, p. 76; 1964b, p. 75, fig. 114-5C. Dubatolova, 1964, p. 9, fig. 1-5C.
  Lower-Middle Ordovician

Cyclopentagonalis balticus Yeltyschewa, 1964a, p. 76, Pl. 3, figs. 25, 26.
  Lower Ordovician: Russia: Estonia

Cyclopentagonalis guttaeformis Yeltyschewa, 1966, p. 63, Pl. 3, figs. 1, 2, 7, 8.
  Middle Ordovician: Russia: Estonia, Leningrad region

Cyclopentagonalis hrevicaenesis Yeltyschewa, 1966, p. 63, Pl. 2, figs. 11-14.
  Middle Ordovician: Russia: Estonia, Leningrad region

Cyclopentagonalis pulcher var. *carbonica* Sizova, 1960, p. 60, nomen nudum (species assigned initially to *Cyclocyclicus*).
  Carboniferous, upper Visean-Namurian: Russia: central Kazakhstan

Cyclopentagonalis stella Yeltyschewa, 1966, p. 62, Pl. 1, figs. 1-6.
  Middle Ordovician: Russia: Estonia, Leningrad region

CYCLOSCAPUS Moore and Jeffords, 1968. *C. laevis* Moore and Jeffords, 1968. Moore and Jeffords, 1968, p. 83.
  Pennsylvanian, Desmoinesian

**Cycloscapus laevis* Moore and Jeffords, 1968, p. 83, Pl. 27, fig. 15.
  Pennsylvanian, Desmoinesian, Mingus Shale: United States: Texas

CYCLOSTELECHUS Moore and Jeffords, 1968. *C. turritus* Moore and
    Jeffords, 1968. Moore and Jeffords, 1968, p. 75.
    Mississippian, Osagean

Cyclostelechus turritus Moore and Jeffords, 1968, p. 76, Pl. 23, fig. 3.
    Mississippian, Osagean, Burlington Limestone: United States: Iowa

CYCLOTETRAGONALIS Yeltyschewa, 1956, nomen vetitum.
    Yeltyschewa, 1956, p. 42, fig. 1-5B; 1959, p. 233, fig. 1-5B; 1964b,
    p. 75, fig. 114-5B. Dubatolova, 1964, p. 9, fig. 1-5B.
    Theoretical genus, no known species.

CYCLOTRIGONALIS Yeltyschewa, 1956, nomen vetitum.
    Yeltyschewa, 1956, p. 42, fig. 1-5A; 1959, p. 233, fig. 1-5A; 1964b,
    p. 75, fig. 114-5A. Dubatolova, 1964, p. 9, fig. 1-5A.
    Theoretical genus, no known species.

CYLINDROCAULISCUS Moore and Jeffords, 1968. *C. fiski* Moore and
    Jeffords, 1968. Moore and Jeffords, 1968, p. 62.
    Pennsylvanian, Desmoinesian

Cylindrocauliscus fiski Moore and Jeffords, 1968, p. 62, Pl. 14, fig. 5.
    Pennsylvanian, Desmoinesian, Mingus Shale: United States: Texas

CYPHOSTELECHUS Moore and Jeffords, 1968. *C. claudus* Moore and Jeffords,
    1968. Moore and Jeffords, 1968, p. 62.
    Pennsylvanian, Desmoinesian

Cyphostelechus claudus Moore and Jeffords, 1968, p. 63, Pl. 14, fig. 4.
    Pennsylvanian, Desmoinesian, Mingus Shale: United States: Texas

Cystocrinus Roemer, 1860. Moore and Laudon, 1943b, p. 104.
    Silurian

DECACRINUS Yeltyschewa, 1957. *D. pennatus* Yeltyschewa, 1957.
    Yeltyschewa, 1957, p. 221. Stukalina, 1968b, p. 256.
    Ordovician-Devonian

Decacrinus antiquus Yeltyschewa, 1964a, p. 66, Pl. 3, figs. 7-9.
    Lower Ordovician: Russia: Popovka River

Decacrinus ludlowensis Schewtschenko, 1966, p. 148, Pl. 6, figs. 16-18;
    Fig. 18.
    Upper Silurian, Tiverskian Stage, Shishkatskian horizon: Russia:
    Tien-Shan Mountains, Zeravshan Mountains

Decacrinus orientalis Yeltyschewa, 1957, p. 221, Pl. 1, figs. 5-7, Fig. 2.
    Yeltyschewa and Dubatolova, 1967, p. 31, Pl. 4, figs. 11-14;
    Fig. 14. Modzalevskaya, 1967, p. 545, Pl. 3, fig. 8.
    Lower-Middle Devonian, Bolshenevershaya Suite: Russia: Upper Priamur

Decacrinus ornatus Stukalina, 1968b, p. 260, Pl. 1, figs. 12, 13;
    Figs. 1-4, 2-5. Levitskiy and others, 1968, p. 72, fig. 2, no. 5.
    Lower Devonian, Pribalkhashian: Russia: central Kazakhstan, Far East

Decacrinus ovalis Stukalina, 1968b, p. 257, Pl. 1, figs. 7-11; Figs. 1,
    no. 2; 2, no. 3. Levitskiy and others, 1968, p. 71, fig. 2, no. 3.
    Upper Silurian-Lower Devonian, Karaespinian-Pribalkhashian: Russia:
    central Kazakhstan, northern Pribalkhash

Decacrinus aff. *pennatus* Yeltyschewa, 1957. Modzalevskaya, 1967, p. 545,
    Pl. 3, fig. 6.
    Lower Devonian, Bolsheneverskaya Suite: Russia: Upper Priamur

Decacrinus pennatus Yeltyschewa, 1957, p. 221, Pl. 1, figs. 1-4; Fig. 1;
    1964b, p. 80, Pl. 15, fig. 22. Yeltyschewa and Dubatolova, 1967,
    p. 33, Pl. 5, fig. 19. Moore and Jeffords, 1968, p. 35, Pl. 2,
    fig. 2. Stukalina, 1968b, p. 259, Pl. 1, figs. 1-6; Figs. 1, no. 3;
    2, no. 4; Levitskiy and others, 1968, p. 71, fig. 2, no. 4.

Upper Silurian-Lower Devonian, Karaespinian-Pribalkhashian layer:
Russia: southern Baltic, Kazakhstan, northern Pribalkhash

Decacrinus tenuicrenulatus Moore and Jeffords, 1968, p. 35, Pl. 2, fig. 1.
Devonian, Ulsterian, Birdsong Shale: United States: Tennessee

DENTIFEROCRINUS Stukalina, 1968. *Pentagonopentagonalis dentiferus* Yeltyschewa, 1955. Stukalina, 1968c, p. 85.
Middle Ordovician-Lower Silurian

Dentiferocrinus dentiferus (Yeltyschewa), 1955. Stukalina, 1968c, p. 85.
Pentagonopentagonalis dentiferus Yeltyschewa, 1955a, p. 45, Pl. 54, figs. 1, 2; 1960, p. 16, Pl. 4, figs. 1-5; 1964b, p. 75, Pl. 15, fig. 19. Stukalina, 1968c, p. 85.
Silurian, Llandoverian: Russia: Siberian Platform

Dentiferocrinus dividuus (Yeltyschewa), 1955. Stukalina, 1968c, p. 85.
Pentagonopentagonalis dividuus Yeltyschewa, 1955a, p. 44, Pl. 37, figs. 1, 2; 1960, p. 15, Pl. 3, figs. 10, 11. Stukalina, 1968c, p. 85.
Upper Ordovician, Dolborskian horizon: Russia: Siberian Platform

Dentiferocrinus proximus (Yeltyschewa and Stukalina), 1963. 1968c, p. 85.
Pentagonopentagonalis proximus Yeltyschewa and Stukalina, 1963, p. 48, Pl. 3, figs. 1-7; Pl. 4, figs. 1a, 2; Fig. 16. Stukalina, 1968c, p. 85.
Middle-Upper Ordovician: Russia: Vaigach

DESIDIAMPHIDIA Moore and Jeffords, 1968. *D. frondea* Moore and Jeffords, 1968. Moore and Jeffords, 1968, p. 70.
Devonian, Erian

Desidiamphidia frondea Moore and Jeffords, 1968, p. 70, Pl. 20, figs. 9, 10.
Devonian, Erian, Wanakah Shale: United States: New York

Desmidocrinus macrodactylus Angelin, 1878. Dubatolova, 1968, p. 144, Pl. 15, fig. 1; Fig. 2.
Devonian, Tom-Chumskian horizon, Lower Rekovian Beds: Russia: Salair

Desmidocrinus salagastianus Rusconi, 1951, p. 256, nomen nudum; 1952, p. 36, Pl. 1, fig. 3. Sieverts-Doreck, 1957b, p. 154.
Ordovician or younger (Sieverts-Doreck, 1957b), Salagastense horizon: Argentina

Cf. *Diamenocrinus* sp. Termier and Termier, 1950, p. 222, Pl. 219, fig. 4.
Devonian, Emsian: Morocco

DIANTHICOELOMA Moore and Jeffords, 1968. *D. insuetum* Moore and Jeffords, 1968. Moore and Jeffords, 1968, p. 50.
Ordovician

Dianthicoeloma insuetum Moore and Jeffords, 1968, p. 50, Pl. 7, figs. 1-3.
Ordovician, Cincinnatian, Blanchester Formation: United States: Ohio

Dianthicoeloma monile (Eichwald), 1856. Moore and Jeffords, 1968, p. 51, Pl. 7, fig. 8.
Haplocrinus monile Eichwald, 1856. Moore and Jeffords, 1968, p. 51.
Pentagonocyclicus monile (Eichwald), 1856. Yeltyschewa, 1960, p. 18, Pl. 1, figs. 5, 6; 1964a, p. 70, Pl. 1, figs. 1-7; Pl. 4, figs. 1-13; 1964b, p. 75, Pl. 15, figs. 9, 10. Moore and Jeffords, 1968, p. 51
Lower Ordovician, Ustukutskian Stage: Russia: Leningrad region, Taymyr Peninsula; northwestern Yugoslavia

Dianthicoeloma pentaporus (Eichwald), 1840. Moore and Jeffords, 1968, p. 51, Pl. 7, fig. 4.

Cupressocrinites pentaporus Eichwald, 1840. Moore and Jeffords, 1968, p. 51.
  Pentagonocyclicus pentaporus (Eichwald), 1840. Yeltyschewa, 1960, p. 18, Pl. 1, figs. 3, 4; 1964b, p. 75, Pl. 15, figs. 11, 12. Moore and Jeffords, 1968, p. 51.
  Middle Ordovician, Krivolutskian Stage: Russia: Leningrad region, Siberian Platform

Dianthicoeloma pentaporus var. *tuberculata* (Yeltyschewa), 1964, new comb.
  Pentagonocyclicus pentaporus var. *tuberculata* Yeltyschewa, 1964a, p. 72, Pl. 3, figs. 18-22; Pl. 4, figs. 14-16.
  Lower Ordovician: Russia: Estonia

Dicirrocrinus(?) *dicirrocrinus* (Ehrenberg), 1930. Schmidt, 1941, p. 140, Pl. 15, fig. 3.
  Herpetocrinus (*Myelodactylus*) *dicirrocrinus* Ehrenberg, 1930. Schmidt, 1941, p. 140.
  Devonian, Coblentzian, upper Coblentzian Beds: Germany

Dicirrocrinus(?) *duplex* a Schmidt, 1941, p. 141, Pl. 17, fig. 3.
  Devonian, Coblentzian, lower Coblentzian Beds: Germany

Dicirrocrinus(?) *duplex* b Schmidt, 1941, p. 142, Pl. 17, fig. 4.
  Devonian, Coblentzian, lower Coblentzian Beds: Germany

Dicirrocrinus(?) *margaritatus* Schmidt, 1941, p. 142, Pl. 20, fig. 7.
  Devonian, Siegenian, Rauhflaser Beds: Germany

Dicirrocrinus(?) *ramulosus* Schmidt, 1941, p. 143, fig. 37.
  Devonian, Coblentzian, upper Coblentzian Beds: Germany

DIEROCALIPTER Moore and Jeffords, 1968. *D. doter* Moore and Jeffords, 1968. Moore and Jeffords, 1968, p. 73.
  Mississippian, Osagean

*Dierocalipter doter Moore and Jeffords, 1968, p. 73, Pl. 21, fig. 10.
  Mississippian, Osagean, New Providence Shale: United States: Kentucky

DILANTERIS Moore and Jeffords, 1968. *D. trestes* Moore and Jeffords, 1968. Moore and Jeffords, 1968, p. 72.
  Mississippian, Osagean

*Dilanteris trestes Moore and Jeffords, 1968, p. 73, Pl. 21, fig. 13.
  Mississippian, Osagean, New Providence Shale: United States: Kentucky

Dolatocrinus avis Moore and Jeffords, 1968, p. 47, Pl. 6, fig. 14.
  Devonian, Erian, Wanakah Shale: United States: New York

Dolatocrinus exculptus Moore and Jeffords, 1968, p. 47, Pl. 6, figs. 12, 13.
  Devonian, Erian, Wanakah Shale: United States: New York

DWORTSOWAECRINUS Stukalina, *in* Stukalina and Tuyutyan, 1967.
  *Tetragonotetragonalis quadrihamatus Yeltyschewa, 1963. Stukalina and Tuyutyan, 1967, p. 72.
  Upper Ordovician-Lower Silurian

Dwortsowaecrinus dwortsowae Stukalina, *in* Stukalina and Tuyutyan, 1967, p. 74, figs. 6, 10.
  Upper Ordovician, Dulankarinskian horizon: Russia: Kazakhstan

*Dwortsowaecrinus quadrihamatus (Yeltyschewa), 1963. Stukalina and Tuyutyan, 1967, p. 72.
  Tetragonotetragonalis quadrihamatus Yeltyschewa, *in* Yeltyschewa and Stukalina, 1963, p. 55, Pl. 2, figs. 22, 23; Fig. 21. Stukalina and Tuyutyan, 1967, p. 72.
  Lower Silurian, Llandoverian: Russia: Nova Zemlya, central Taymyr

Dwortsowaecrinus robustus Tuyutyan, *in* Stukalina and Tuyutyan, 1967, p. 74, figs. 1-5, 11.
Upper Ordovician, Anderkenskian horizon: Russia: Kazakhstan

Ectenocrinus grandis (Meek), 1873. Jillson, 1963, p. 16, fig. 13.
Ordovician, Cincinnatian, Eden Formation: United States: Kentucky

Ectenocrinus simplex (Hall), 1847. Jillson, 1963, p. 16, fig. 12.
Ordovician, Cincinnatian, Eden Formation: United States: Kentucky

ELLIPSOCYCLICUS Yeltyschewa, 1956, nomen vetitum.
Yeltyschewa, 1956, p. 42, fig. 1-6E; 1959, p. 233, fig. 1-6E; 1964b, p. 75, fig. 114-6E. Dubatolova, 1964, p. 9, fig. 1-6E.
Theoretical genus, no known species.

ELLIPSOELLIPTICUS Yeltyschewa, 1956, nomen vetitum.
Yeltyschewa, 1956, p. 42, figs. 1-6F, 2h; 1959, p. 233, fig. 1-6F; 1964b, p. 75, fig. 114-6F. Dubatolova, 1964, p. 9, fig. 1-6F.
Theoretical genus, no known species.

ELLIPSOHEXAGONALIS Yeltyschewa, 1956, nomen vetitum.
Yeltyschewa, 1956, p. 42, fig. 1-6D; 1959, p. 233, fig. 1-6D; 1964b, p. 75, fig. 114-6D. Dubatolova, 1964, p. 9, fig. 1-6D.
Theoretical genus, no known species.

ELLIPSOPENTAGONALIS Yeltyschewa, 1956, nomen vetitum.
Yeltyschewa, 1956, p. 42, fig. 1-6C; 1959, p. 233, fig. 1-6C; 1964b, p. 75, fig. 114-6C. Dubatolova, 1964, p. 9, fig. 1-6C.
Theoretical genus, no known species.

ELLIPSOTETRAGONALIS Yeltyschewa, 1956, nomen vetitum.
Yeltyschewa, 1956, p. 42, fig. 1-6B; 1959, p. 233, fig. 1-6B; 1964b, p. 75, fig. 114-6B. Dubatolova, 1964, p. 9, fig. 1-6B.
Theoretical genus, no known species.

ELLIPSOTRIGONALIS Yeltyschewa, 1956, nomen vetitum.
Yeltyschewa, 1956, p. 42, fig. 1-6A; 1959, p. 233, fig. 1-6A; 1964b, p. 75, fig. 114-6A. Dubatolova, 1964, p. 9, fig. 1-6A.
Theoretical genus, no known species.

ELYTROCLON Moore and Jeffords, 1968. *E. elimatus* Moore and Jeffords, 1968. Moore and Jeffords, 1968, p. 39.
Mississippian, Osagean

Elytroclon elimatus Moore and Jeffords, 1968, p. 40, Pl. 3, figs. 2, 3.
Mississippian, Osagean, New Providence Shale: United States: Kentucky

Entrochi biarticulati Quenstedt, 1876 = *Hexacrinites*(?) *humilicarinatus*

Entrochi mamillati Quenstedt, 1876 = *Hexacrinites*(?) *humilicarinatus*

Entrochus cingulatus impares Yeltyschewa, 1961 = *Pentagonocyclicus cingulatus*

Entrochus dentatus Quenstedt, 1874-1876. Modzalevskaya, 1967, p. 547, Pl. 6, fig. 26.
Middle Devonian, Oldoiskaya Suite: Russia: Upper Priamur

Entrochus cf. *dentatus* Quenstedt, 1874-1876. Yeltyschewa and Dubatolova, 1961, p. 557, Pl. D-87, figs. 16, 17.
Middle Devonian: Russia: Altay Mountains

Entrochus dentatus var. *echinata* Yeltyschewa and Dubatolova, 1961 = *Pentagonocyclicus dentatus* var. *echinata*

Entrochus ligatus Quenstedt, 1876 = *Pentagonocyclicus ligatus*

Entrochus primus Barrande, *in* Waagen and Jahn, 1899. Yeltyschewa, 1964a, p. 66, Pl. 4, fig. 17.
Devonian: Bohemia

Entrochus sp. A. Sieverts-Doreck, 1951a, p. 129, Pl. 9, fig. 9.
   Lower Carboniferous, Erdbaker Limestone: Germany

Entrochus sp. B. Sieverts-Doreck, 1951a, p. 130, Pl. 9, fig. 8.
   Lower Carboniferous, Erdbaker Limestone: Germany

Entrochus sp. C. Sieverts-Doreck, 1951a, p. 135.
   Lower Carboniferous: Germany

Entrochus sp. D. Sieverts-Doreck, 1951a, p. 136.
   Lower Carboniferous: Germany

Entrochus sp. E. Sieverts-Doreck, 1951a, p. 136, Pl. 9, fig. 11; Fig. 3.
   Lower Carboniferous: Germany

Entrochus sp. F. Sieverts-Doreck, 1951a, p. 139, fig. 7.
   Lower Carboniferous: Germany

Entrochus sp. G. Sieverts-Doreck, 1951a, p. 142.
   Lower Carboniferous: Germany

Entrochus sp. H. Sieverts-Doreck, 1951a, p. 142, Pl. 8, fig. 11.
   Lower Carboniferous: Germany

Entrochus (*Culmicrinus*?) sp. 1. Sieverts-Doreck, 1964, p. 9, Pl. 1, fig. 11.
   Carboniferous, Namurian?, Hangend Bed: Spain

Entrochus (*Culmicrinus*?) sp. 1? Sieverts-Doreck, 1964, p. 10, Pl. 1, fig. 12.
   Carboniferous, Namurian?, Hangend Bed: Spain

Entrochus sp. 2. Sieverts-Doreck, 1964, p. 10, Pl. 1, figs. 13, 14.
   Carboniferous, Namurian?, Hangend Bed: Spain

cf. *Entrochus* sp. 2. Sieverts-Doreck, 1964, p. 11.
   Carboniferous, Namurian?, Pyritkalkbank: Spain

Entrochus sp. 3. Sieverts-Doreck, 1964, p. 11.
   Carboniferous, Namurian?, Middle Limestone Series: Spain

Entrochus sp. indet. Sieverts-Doreck, 1964, p. 11.
   Carboniferous, Namurian?, Middle Limestone Series: Spain

Epalxyocrinus asymmetricus Sizova, 1960, p. 58, nomen nudum.
   Carboniferous, Visean-Namurian: Russia: central Kazakhstan

Epalxyocrinus asymmetricus var. *spina* Sizova, 1960, p. 58, nomen nudum.
   Carboniferous, Visean-Namurian: Russia: central Kazakhstan

Eucalyptocrinites cf. *rosaceus* Goldfuss, 1826. Yeltyschewa and Dubatolova, 1961, p. 553, Pl. D-87, fig. 2.
   Devonian, Eifelian: Russia: Salair

Eucalyptocrinites? shelbyensis Moore and Jeffords, 1968, p. 46, Pl. 5, fig. 3.
   Silurian, Niagaran, Waldron Shale: United States: Indiana

Eucladocrinus? kentuckiensis Moore and Jeffords, 1968, p. 42, Pl. 4, fig. 7.
   Mississippian, Osagean, Borden Group, Burlington Limestone: United States: Tennessee, Indiana, Iowa

Eucladocrinus? springeri Moore and Jeffords, 1968, p. 42, Pl. 4, fig. 1.
   Mississippian, Osagean, Burlington Limestone: United States: Iowa

Eugeniacrinites sp. nomen correctum.
   Eugeniacrinus sp. Sangil, 1955, p. 253, fig. 3D.
   No age given

EULONCHEROSTIGMA Moore and Jeffords, 1968. *E. impunitum* Moore and Jeffords, 1968. Moore and Jeffords, 1968, p. 58.
Mississippian, Osagean

**Euloncherostigma impunitum* Moore and Jeffords, 1968, p. 58, Pl. 11, fig. 6.
Mississippian, Osagean, New Providence Shale: United States: Kentucky

EURAX Moore and Jeffords, 1968. *E. ethas* Moore and Jeffords, 1968. Moore and Jeffords, 1968, p. 68.
Silurian, Niagaran

**Eurax ethas* Moore and Jeffords, 1968, p. 68, Pl. 19, figs. 7-10.
Silurian, Niagaran, Brownsport Formation, Waldron Shale: United States: Tennessee, Indiana

Eurax eugenes Moore and Jeffords, 1968, p. 68, Pl. 19, figs. 1-6.
Silurian, Niagaran-Devonian, Helderbergian, Brownsport Formation, Birdsong Shale: United States: Tennessee

EXAESIODISCUS Moore and Jeffords, 1968. *E. acutus* Moore and Jeffords, 1968. Moore and Jeffords, 1968, p. 73.
Silurian, Niagaran-Devonian, Erian

**Exaesiodiscus acutus* Moore and Jeffords, 1968, p. 73, Pl. 22, figs. 4-6.
Silurian, Niagaran, Waldron Shale: United States: Indiana

Exaesiodiscus minutus Moore and Jeffords, 1968, p. 74, Pl. 22, figs. 7, 8.
Devonian, Erian, Wanakah Shale: United States: New York

Exaesiodiscus truncatus Moore and Jeffords, 1968, p. 73, Pl. 22, figs. 1-3.
Silurian, Niagaran, Waldron Shale: United States: Indiana

EXEDRODISCUS Moore and Jeffords, 1968. *E. excussus* Moore and Jeffords, 1968. Moore and Jeffords, 1968, p. 74.
Mississippian, Osagean

**Exedrodiscus excussus* Moore and Jeffords, 1968, p. 74, Pl. 23, fig. 9.
Mississippian, Osagean, New Providence Shale: United States: Kentucky

FABALIUM Moore and Jeffords, 1968. *F. fabale* Moore and Jeffords, 1968. Moore and Jeffords, 1968, p. 71.
Silurian, Niagaran

**Fabalium fabale* Moore and Jeffords, 1968, p. 71, Pl. 21, fig. 7.
Silurian, Niagaran, Brownsport Formation: United States: Tennessee

FACETOCRINUS Stukalina, 1968. **Pentagonopentagonalis facetus* Stukalina, 1961. Stukalina, 1968c, p. 91.
Upper Silurian-Middle Devonian

Facetocrinus ajnasuensis (Stukalina), 1961. Stukalina, 1968c, p. 91.
Pentagonopentagonalis ajnasuensis Stukalina, 1961, p. 36, Pl. 1, fig. 3; 1968c, p. 91.
Silurian, upper Ludlovian: Russia: central Kazakhstan

Facetocrinus conspicuus (Dubatolova), 1964. Stukalina, 1968c, p. 91.
Athinocrinus conspicuus Dubatolova, 1964, p. 58, Pl. 7, fig. 12. Stukalina, 1968c, p. 91.
Lower Devonian: Russia: Kuznetz Basin

**Facetocrinus facetus* (Stukalina), 1961. Stukalina, 1968c, p. 91.
Pentagonopentagonalis facetus Stukalina, 1961, p. 35, Pl. 1, fig. 7; 1965b, p. 135, Pl. 2, fig. 11; 1968c, p. 91.
Upper Silurian, Karaespinian and Pribalkhashian horizons: Russia: central Kazakhstan

Facetocrinus quinquespinosus (Stukalina), 1961. Stukalina, 1968c, p. 91.
 Pentagonopentagonalis quinquespinosus Stukalina, 1961, p. 35, Pl. 2,
 fig. 5; 1965b, p. 135, Pl. 2, figs. 17-19; 1968c, p. 91.
 Upper Silurian, Ludlovian, Karaespinian and Aynasuyskian horizons:
 Russia: central Kazakhstan

Facetocrinus quinqueangularis (Dubatolova), 1964. Stukalina, 1968c,
 p. 91 (misspelled as *F. quinqulangularis*).
 Anthinocrinus quinqueangularis Dubatolova, 1964, p. 59, Pl. 7,
 figs. 16-19. Stukalina, 1968c, p. 91.
 Devonian, Givetian: Russia: Kuznetz Basin

FASCICRINUS Stukalina, 1968. **F. flabellatus* Yeltyschewa and Stukalina,
 in Stukalina, 1968. Stukalina, 1968c, p. 87.
 Ordovician, Caradocian

Fascicrinus flabellatus Yeltyschewa and Stukalina, *in* Stukalina, 1968c,
 p. 87, fig. 2zh.
 Ordovician, upper Caradocian: Russia: central Kazakhstan

FLORICYCLUS Moore and Jeffords, 1968. **F. hebes* Moore and Jeffords, 1968.
 Moore and Jeffords, 1968, p. 76.
 Mississippian, Osagean-Pennsylvanian, Virgilian

Floricyclus angustimargo Moore and Jeffords, 1968, p. 77, Pl. 24,
 figs. 6, 7.
 Pennsylvanian, Desmoinesian, Minturn Formation: United States:
 Colorado

Floricyclus granulosus Moore and Jeffords, 1968, p. 77, Pl. 24, figs. 1-3;
 Pl. 26, figs. 9-16. Moore and others, 1968a, p. 16, fig. 3, no. 3.
 Pennsylvanian, Virgilian, Wayland Shale: United States: Texas

*Floricyclus hebes Moore and Jeffords, 1968, p. 77, Pl. 24, fig. 5.
 Pennsylvanian, Desmoinesian, Mingus Shale: United States: Texas

Floricyclus kansasensis Moore and Jeffords, 1968, p. 77, Pl. 24, fig. 9.
 Pennsylvanian, Desmoinesian, Cabaniss formation: United States:
 Kansas

Floricyclus pulcher Moore and Jeffords, 1968, p. 77, Pl. 24, fig. 4.
 Pennsylvanian, Desmoinesian, Cabaniss formation: United States:
 Kansas

Floricyclus welleri Moore and Jeffords, 1968, p. 77, Pl. 24, fig. 8.
 Mississippian, Osagean, Brodhead Formation: United States: Kentucky

FLORIPILA Moore and Jeffords, 1968. **F. florealis* Moore and Jeffords,
 1968. Moore and Jeffords, 1968, p. 56.
 Devonian, Erian

*Floripila florealis Moore and Jeffords, 1968, p. 57, Pl. 10, figs. 7, 8.
 Devonian, Erian, Wanakah Shale: United States: New York

FLUCTICHARAX Moore and Jeffords, 1968. **F. undatus* Moore and Jeffords,
 1968. Moore and Jeffords, 1968, p. 70.
 Mississippian, Osagean

*Flucticharax undatus Moore and Jeffords, 1968, p. 71, Pl. 20, figs. 7, 8.
 Mississippian, Osagean, New Providence Shale: United States: Kentucky

cf. *Gennaeocrinus* Termier and Termier, 1950, p. 222, Pl. 219,
 figs. 2, 3.
 Devonian, Gedinnian: Morocco

Gilbertsocrinus aequalis Moore and Jeffords, 1968, p. 39, Pl. 3, fig. 5.
 Mississippian, Osagean, New Providence Shale: United States: Kentucky

Gilbertsocrinus cassiope Moore and Jeffords, 1968, p. 39, Pl. 3, figs. 6, 7.
  Mississippian, Osagean, New Providence Shale: United States: Kentucky
Gilbertsocrinus concinnus Moore and Jeffords, 1968, p. 39, Pl. 3, fig. 12.
  Mississippian, Osagean, Burlington Limestone: United States: Iowa
Gilbertsocrinus vetulus Moore and Jeffords, 1968, p. 38, Pl. 3, fig. 1.
  Devonian, Erian, Wanakah Shale, Centerfield Limestone: United States: New York
?*Gissocrinus salagastensis* Rusconi, 1951, p. 257, nomen nudum; 1952, p. 37, Pl. 1, fig. 4; Fig. 9. Sieverts-Doreck, 1957b, p. 154.
  Ordovician or younger (Sieverts-Doreck, 1957b): Argentina
Goniaster pygmaeus Eichwald, 1860 = *Tetragonocrinus pygmaeus*
GONIOCION Moore and Jeffords, 1968. **G. gonimus* Moore and Jeffords, 1968. Moore and Jeffords, 1968, p. 75.
  Mississippian, Osagean
*Goniocion gonimus Moore and Jeffords, 1968, p. 75, Pl. 23, fig. 6.
  Mississippian, Osagean, Burlington Limestone: United States: Iowa
Goniocion turgidus Moore and Jeffords, 1968, p. 75, Pl. 23, fig. 4.
  Mississippian, Osagean, Burlington Limestone: United States: Iowa
GONIOSTATHMUS Moore and Jeffords, 1968. **G. annexus* Moore and Jeffords, 1968. Moore and Jeffords, 1968, p. 75.
  Mississippian, Osagean
*Goniostathmus annexus Moore and Jeffords, 1968, p. 75, Pl. 22, fig. 12.
  Mississippian, Osagean, Burlington Limestone: United States: Iowa
Grammocrinus Eichwald, 1860. Moore and Laudon, 1943b, p. 104.
  Ordovician
Grammocrinus lineatus Eichwald, 1860. Yeltyschewa, 1964a, p. 67, Pl. 1, figs. 18, 19.
  Lower Ordovician: Russia: Gorodishche
Grammocrinus lineatus var. *brevia* Yeltyschewa, 1964a, p. 67, Pl. 1, figs. 15-17.
  Lower Ordovician: Russia: Popovka River
Grammocrinus tuberculatus Yeltyschewa, 1964a, p. 68, Pl. 1, fig. 20.
  Lower Ordovician: Russia: Popovka River
GRAPHOSTERIGMA Moore and Jeffords, 1968. **G. scriptum* Moore and Jeffords, 1968. Moore and Jeffords, 1968, p. 61.
  Mississippian, Osagean
Graphosterigma grammodes Moore and Jeffords, 1968, p. 62, Pl. 13, fig. 6; Pl. 14, figs. 1-3.
  Mississippian, Osagean, Edwardsville Formation: United States: Indiana
*Graphosterigma scriptum Moore and Jeffords, 1968, p. 62, Pl. 13, fig. 3.
  Mississippian, Osagean, New Providence Shale: United States: Kentucky
Graphosterigma synthetes Moore and Jeffords, 1968, p. 62, Pl. 13, fig. 1.
  Mississippian, Osagean, New Providence Shale: United States: Kentucky
GREGARIOCRINUS Stukalina, 1968. **Pentagonocyclicus forus* Stukalina, 1961. Stukalina, 1968c, p. 91.
  Upper Silurian
*Gregariocrinus forus (Stukalina), 1961. Stukalina, 1968c, p. 91.
  Pentagonocyclicus forus Stukalina, 1961, p. 39, Pl. 1, fig. 8; 1965b, p. 135, Pl. 2, figs. 1, 2; 1968c, p. 91.
  Upper Silurian, Aynasuyskian horizon: Russia: central Kazakhstan

Haplocrinus monile Eichwald, 1856 = *Dianthicoeloma monile*

HATTINANTHERIS Moore and Jeffords, 1968. *H. indianensis* Moore and Jeffords, 1968. Moore and Jeffords, 1968, p. 59.
Silurian, Niagaran

Hattinanteris indianensis Moore and Jeffords, 1968, p. 59, Pl. 11, fig. 5.
Silurian, Niagaran, Waldron Shale: United States: Indiana

Hattinanteris regularis Moore and Jeffords, 1968, p. 59, Pl. 12, fig. 2.
Silurian, Niagaran, Waldron Shale: United States: Tennessee

?*Herpetocrinus* (*Myelodactylus*) *dicirrocrinus* Ehrenberg, 1930 = *Dicirrocrinus*(?) *dicirrocrinus*

Herpetocrinus sp. Yeltyschewa, 1955a, p. 42, Pl. 54, fig. 7.
Silurian, Llandoverian: Russia: Siberian Platform

HETEROSTAURUS Moore and Jeffords, 1968. *H. belknapensis* Moore and Jeffords, 1968. Moore and Jeffords, 1968, p. 83.
Pennsylvanian, Virgilian

Heterostaurus belknapensis Moore and Jeffords, 1968, p. 83, Pl. 27, fig. 14.
Pennsylvanian, Virgilian, Belknap Limestone: United States: Texas

HETEROSTELECHUS Moore and Jeffords, 1968. *H. texanus* Moore and Jeffords, 1968. Moore and Jeffords, 1968, p. 81.
Pennsylvanian, Virgilian-Permian, Wolfcampian

Heterostelechus jeffordsi Miller, *in* Moore and Jeffords, 1968, p. 82, Pl. 28, figs. 2-6. Jeffords and Miller, 1968, p. 12, Pl. 3, figs. 1-9; Figs. 1C, 2C, 3, 4C, D, 5C.
Pennsylvanian, Virgilian, Wayland Shale: United States: Texas

Heterostelechus keithi Miller, *in* Moore and Jeffords, 1968, p. 82, Pl. 28, figs. 8-10. Moore and others, 1968a, p. 11, Pl. 3, figs. 1-3; Fig. 3, no. 1.
Pennsylvanian, Virgilian, Gunsight Limestone, Wayland Shale: United States: Texas

Heterostelechus texanus Moore and Jeffords, 1968, p. 81, Pl. 27, fig. 16; Pl. 28, figs. 1, 7. Moore and others, 1968a, p. 16, fig. 3, no. 4.
Pennsylvanian, Virgilian, Permian, Wolfcampian, Belknap Limestone, Waldrip Shale: United States: Texas

Hexacrinites(?) *argutus* Yeltyschewa, *in* Dubatolova, 1964, p. 41, Pl. 5, figs. 3-5.
Devonian, Frasnian: Russia: Kuznetz Basin, central Devonian Field

Hexacrinites(?) *biconcavus* (Yeltyschewa and Dubatolova), 1960.
Yeltyschewa and Dubatolova, 1967, p. 23, Pl. 2, figs. 7-10; Fig. 11 (*non* Fig. 12 as given in text). Modzalevskaya, 1967, p. 547, Pl. 6, figs. 19, 20.
Hexacrinus biconcavus Yeltyschewa and Dubatolova, 1960, p. 369, Pl. 70, figs. 7, 8.
Middle Devonian, Givetian-Frasnian, Oldoiskaya suite: Russia: Far East, Altay Mountains, Kazakhstan; China

Hexacrinites(?) *cauliculatus* Dubatolova, 1968, p. 145, Pl. 15, figs. 2, 3; Fig. 3.
Lower Devonian, lower Rekovian beds: Russia: Salair

Hexacrinites(?) *dentatus carinatus* Yeltyschewa and Dubatolova, 1967, p. 25, Pl. 2, figs. 1-6; Fig. 12 (*non* fig. 13 as given in text).
Lower-Middle Devonian: Russia: Upper Priamur

Hexacrinites(?) *dentatus echinatus* Yeltyschewa and Dubatolova, 1967, p. 26, Pl. 3, figs. 1-12; Fig. 13 (*non* Fig. 14 as given in text).
Lower-Middle Devonian: Russia: eastern Zabaykal'sk Mountains, Upper Priamur

Hexacrinites humilicarinatus Yeltyschewa, *in* Yeltyschewa and Dubatolova, 1961, p. 555, Pl. D-87, fig. 6. Dubatolova, 1964, p. 37, Pl. 3, figs. 8-11; Fig. 5; 1967, p. 34, Pl. 6, figs. 3-7; Fig. 1.
Entrochi biarticulati Quenstedt, 1876. Dubatolova, 1964, p. 37.
Entrochi mammillati Quenstedt, 1876. Dubatolova, 1964, p. 37.
Lower-Middle Devonian: Russia: Salair, Kuznetz Basin, Armenia

Hexacrinites(?) *inflatus* Stukalina, 1965b, p. 190, Pl. 1, figs. 5-8.
Upper Silurian-Lower Devonian: Russia: Kazakhstan

Hexacrinites kartzevae Yeltyschewa and Dubatolova, 1961, p. 554, Pl. D-87, figs. 3, 4. Dubatolova, 1964, p. 40, Pl. 4, figs. 7-10; Pl. 5, fig. 1.
Devonian, Givetian-Frasnian: Russia: Armenia, Kuznetz Basin, central Devonian Field

Hexacrinus kartzevae var. *kazachstanica* Sizova, 1960, p. 55, nomen nudum.
Devonian, Givetian-Frasnian: Russia: central Kazakhstan

Hexacrinites(?) *maculosus* Dubatolova, 1964, p. 41, Pl. 5, fig. 2.
Devonian, Givetian-Frasnian: Russia: Kuznetz Basin

Hexacrinites(?) *mamillatus* (Yeltyschewa and Dubatolova), 1960.
Yeltyschewa and Dubatolova, 1967, p. 20, Pl. 1, figs. 1-12; Fig. 10 (*non* Fig. 11 as given in text). Modzalevskaya, 1967, p. 547, Pl. 6, figs. 21, 22.
Hexacrinus mamillatus Yeltyschewa and Dubatolova, 1960, p. 367, Pl. 70, figs. 1, 2. Sizova, 1960, p. 53.
Devonian, Coblentzian-Givetian: Russia: Upper Priamur, Far East, Kazakhstan

Hexacrinites paratuberosus Yeltyschewa, 1968, p. 37, Pl. 5, figs. 14, 15.
Silurian, Skalskian horizon: Russia: Podolsk

Hexacrinites(?) *subbiconcavus* Stukalina, 1965b, p. 189, Pl. 1, figs. 1-4.
Lower Devonian: Russia: Kazakhstan

Hexacrinites(?) *sverbilovi* Stukalina, 1965b, p. 191, Pl. 1, figs. 9-13.
Devonian, Frasnian: Russia: Kazakhstan

Hexacrinites tuberosus Yeltyschewa, *in* Dubatolova and Yeltyschewa, 1961 = *Laudonomphalus tuberosus*

Hexacrinites tumidulus Yeltyschewa, 1968, p. 38, Pl. 5, figs. 10-13.
Silurian, Borschovskian horizon: Russia: Podolsk

Hexacrinites spp. nomen correctum.
Hexacrinus? sp. Termier and Termier, 1950, p. 84, Pl. 211, figs. 39-41.
Devonian, Givetian: Morocco
Hexacrinus sp. Termier and Termier, 1950, p. 206, Pl. 211, figs. 27, 28.
Devonian, lower Eifelian: Morocco
Hexacrinus sp. Termier and Termier, 1950, p. 206, Pl. 211, figs. 33, 34.
Devonian, upper Siegenian: Morocco
Hexacrinus sp. Termier and Termier, 1950, p. 206, Pl. 211, figs. 37, 38.
Devonian, Eifelian: Morocco
Hexacrinus sp. Sangil, 1955, p. 251, fig. 2C.
No age given

HEXAGONOCYCLICUS Yeltyschewa, 1956, nomen vetitum.
Yeltyschewa, 1956, p. 42, fig. 1-4E; 1959, p. 233, fig. 1-4E; 1964b,
p. 75, fig. 114-4E. Dubatolova, 1964, p. 9, fig. 1-4E.
Theoretical genus, no known species.

HEXAGONOELLIPTICUS Yeltyschewa, 1956, nomen vetitum.
Yeltyschewa, 1956, p. 42, fig. 1-4F; 1959, p. 233, fig. 1-4F; 1964b,
p. 75, fig. 114-4F. Dubatolova, 1964, p. 9, fig. 1-4F.
Theoretical genus, no known species.

HEXAGONOHEXAGONALIS Yeltyschewa, 1956, nomen vetitum.
Yeltyschewa, 1956, p. 42, fig. 1-4D; 1959, p. 233, fig. 1-4D; 1964b,
p. 75, fig. 114-4D. Dubatolova, 1964, p. 9, fig. 1-4D.
Theoretical genus, no known species.

HEXAGONOPENTAGONALIS Yeltyschewa, 1956, nomen vetitum.
Yeltyschewa, 1956, p. 42, fig. 1-4C; 1959, p. 233, fig. 1-4C; 1964b,
p. 75, fig. 114-4C. Dubatolova, 1964, p. 9, fig. 1-4C.
Theoretical genus, no known species.

HEXAGONOTETRAGONALIS Yeltyschewa, 1956, nomen vetitum.
Yeltyschewa, 1956, p. 42, fig. 1-4B; 1959, p. 233, fig. 1-4B; 1964b,
p. 75, fig. 114-4B. Dubatolova, 1964, p. 9, fig. 1-4B.
Theoretical genus, no known species.

HEXAGONOTRIGONALIS Yeltyschewa, 1956, nomen vetitum.
Yeltyschewa, 1956, p. 42, fig. 1-4A; 1959, p. 233, fig. 1-4A; 1964b,
p. 75, fig. 114-4A. Dubatolova, 1964, p. 9, fig. 1-4A.
Theoretical genus, no known species.

HYPEREXOCHUS Moore and Jeffords, 1968. *H. immodicus Moore and Jeffords,
1968. Moore and Jeffords, 1968, p. 33.
Devonian, Helderbergian

*Hyperexochus immodicus Moore and Jeffords, 1968, p. 33, Pl. 1, figs. 4-7.
Devonian, Helderbergian, Birdsong Shale: United States: Tennessee

Iberocrinus sp. Hernández-Sampelayo, 1954, p. 25, Pl. 3, fig. 1e.
Carboniferous: Spain

Idromecrinus Gregorio, 1930. Moore and Laudon, 1943b, p. 104.
Permian

ILEMATERISMA Moore and Jeffords, 1968. *I. enamma Moore and Jeffords,
1968. Moore and Jeffords, 1968, p. 40.
Mississippian, Osagean

*Ilematerisma enamma Moore and Jeffords, 1968, p. 40, Pl. 3, fig. 11.
Mississippian, Osagean, New Providence Shale: United States: Kentucky

Imperatoria Gregorio, 1930. Moore and Laudon, 1943b, p. 104.
Permian

Iocrinus subcrassus (Meek and Worthen), 1865. Moore and Jeffords, 1968,
p. 32, Pl. 1, fig. 3.
Ordovician, Cincinnatian: United States: Ohio

KASACHSTANOCRINUS Sizova, in Schewtschenko, 1966. *K. asperum
Schewtschenko, 1966. Sizova, in Schewtschenko, 1966, p. 158.
Lower Devonian-lower Carboniferous

*Kasachstanocrinus asperum Schewtschenko, 1966, p. 159, Pl. 4, fig. 19;
Fig. 29.
Lower Devonian, Pandzhrutskian horizon: Russia: Tien-Shan Mountains

Kasachstanocrinus aff. latus Sizova, in Yeltyschewa and Dubatolova, 1967,
p. 48, Pl. 4, fig. 19 (misspelled as Kasakhstanocrinus).
Lower Carboniferous: Russia: Kazakhstan, Upper Priamur

KSTUTOCRINUS Schewtschenko, 1966. **K. sublatus* Schewtschenko, 1966. Schewtschenko, 1966, p. 160.
Lower Devonian

Kstutocrinus doliaris Schewtschenko, 1966, p. 161, Pl. 6, figs. 13-15; Fig. 31.
Lower Devonian, Pandzhrutskian horizon: Russia: Tien-Shan Mountains

Kstutocrinus exsculptus Schewtschenko, 1966, p. 161, Pl. 6, fig. 10; Fig. 32.
Lower Devonian, Panzhrutskian horizon: Russia: Tien-Shan Mountains

Kstutocrinus rugellosus Schewtschenko, 1966, p. 162, Pl. 6, fig. 19; Fig. 33.
Lower Devonian, Pandzhrutskian horizon: Russia: Tien-Shan Mountains

*Kstutocrinus sublatus Schewtschenko, 1966, p. 160, Pl. 6, figs. 11, 12; Fig. 30.
Lower Devonian, Pandzhrutskian horizon: Russia: Tien-Shan Mountains

Kuzbassocrinus Yeltyschewa, 1957 = *Melocrinites* Goldfuss, 1826

Kuzbassocrinus binidigitatus Yeltyschewa, 1957 = *Melocrinites binidigitatus*

Kuzbassocrinus bystrowi Yeltyschewa, 1957 = *Melocrinites bystrowi*

Kuzbassocrinus decemlobatus Yeltyschewa, 1957 = *Melocrinites decemlobatus*

Kuzbassocrinus impalpabilis Dubatolova, 1968 = *Melocrinites impalpabilis*

Kuzbassocrinus paucicostalus Yeltyschewa, 1957 = *Melocrinites paucicostalus*

Kuzbassocrinus yeltyschewae Dubatolova, 1964 = *Melocrinites yeltyschewae*

LAMPROSTERIGMA Moore and Jeffords, 1968. **L. mirificum* Moore and Jeffords, 1968. Moore and Jeffords, 1968, p. 78.
Pennsylvanian, Desmoinesian

Lamprosterigma erathense Moore and Jeffords, 1968, p. 79, Pl. 25, fig. 15.
Pennsylvanian, Desmoinesian, Mingus Shale: United States: Texas

*Lamprosterigma mirificum Moore and Jeffords, 1968, p. 79, Pl. 25, fig. 10.
Pennsylvanian, Desmoinesian, Cabaniss formation: United States: Kansas

Lampterocrinus tennesseensis Roemer, 1860. Moore and Jeffords, 1968, p. 36, Pl. 2, figs. 8-11.
Silurian, Niagaran, Brownsport Formation: United States: Tennessee

Lampterocrinus sp. undet. (A). Moore and Jeffords, 1968, p. 37, Pl. 2, fig. 7.
Silurian, Niagaran, Brownsport Formation: United States: Tennessee

Lampterocrinus sp. undet. (B). Moore and Jeffords, 1968, p. 37, Pl. 2, fig. 12.
Silurian, Niagaran, Brownsport Formation: United States: Tennessee

LAUDONOMPHALUS Moore and Jeffords, 1968. **L. regularis* Moore and Jeffords, 1968. Moore and Jeffords, 1968, p. 71
Middle Devonian

Laudonomphalus ornatus Moore and Jeffords, 1968, p. 72, Pl. 21, fig. 5.
Devonian, Erian, Rockport Limestone: United States: Michigan

*Laudonomphalus regularis Moore and Jeffords, 1968, p. 72, Pl. 21, figs. 1-4.
Devonian, Erian, Bell Shale, Rockport Limestone, Wanakah Shale: United States: Michigan, New York

Laudonomphalus tuberosus (Yeltyschewa, *in* Yeltyschewa and Dubatolova), 1961. Moore and Jeffords, 1968, p. 72, Pl. 21, fig. 6.
Hexacrinites tuberosus Yeltyschewa, *in* Yeltyschewa and Dubatolova, 1961, p. 554, Pl. D-87, fig. 5. Dubatolova, 1964, p. 38, Pl. 3, figs. 12, 13; Pl. 4, fig. 6. Yeltyschewa, 1964b, p. 80, Pl. 15, fig. 23. Moore and Jeffords, 1968, p. 72.
Devonian, Eifelian-Givetian: Russia: Salair, Kuznetz Basin, Armenia

LEPTOCARPHIUM Moore and Jeffords, 1968. *L. gracile* Moore and Jeffords, 1968. Moore and Jeffords, 1968, p. 79.
Pennsylvanian, Desmoinesian

Leptocarphium gracile Moore and Jeffords, 1968, p. 79, Pl. 25, figs. 11, 12.
Pennsylvanian, Desmoinesian, Mingus Shale: United States: Texas

Leptocarphium regulare Moore and Jeffords, 1968, p. 79, Pl. 25, figs. 8, 9.
Pennsylvanian, Desmoinesian, Mingus Shale: United States: Texas

Leseus Gregorio, 1930. Moore and Laudon, 1943b, p. 104.
Permian

Lichenocrinus Hall, 1866. Moore and Laudon, 1943b, p. 104.
Ordovician-Silurian

LOMALEGNUM Moore and Jeffords, 1968. *L. hormidium* Moore and Jeffords, 1968. Moore and Jeffords, 1968, p. 61.
Mississippian, Osagean

Lomalegnum hormidium Moore and Jeffords, 1968, p. 61, Pl. 13, fig. 2.
Mississippian, Osagean, Burlington Limestone: United States: Iowa

MALOVICRINUS Stukalina, 1968. *M. fragosus* Yeltyschewa and Stukalina, *in* Stukalina, 1968. Stukalina, 1968c, p. 86.
Middle-Upper Ordovician

Malovicrinus fragosus Yeltyschewa and Stukalina, *in* Stukalina, 1968c, p. 86, figs. 2d, e.
Middle-Upper Ordovician: Russia: Kazakhstan

MEDINECRINUS Stukalina, 1965. *Pentagonopentagonalis vitreus* Stukalina, 1961. Stukalina, 1965b, p. 140.
Upper Silurian

Medinecrinus vitreus (Stukalina), 1961. Stukalina, 1965b, p. 141, Pl. 2, figs. 7, 8; Fig. 7.
Pentagonopentagonalis vitreus Stukalina, 1961, p. 35, Pl. 1, fig. 5; 1965b, p. 141.
Upper Silurian, Karaespinian horizon: Russia: central Kazakhstan

MEDIOCRINUS Stukalina, 1965. *Pentagonocyclicus medius* Yeltyschewa, *in* Stukalina, 1965. Stukalina, 1965b, p. 139.
Upper Silurian-Devonian

Mediocrinus medius Yeltyschewa, *in* Stukalina, 1965b, p. 139, Pl. 1, figs. 7-9; Figs. 5, 6. Dubatolova, 1968, p. 149, Pl. 15, fig. 10; Fig. 7.
Upper Silurian-Devonian, Aynasuyskian horizon; Tom-Chumshian horizon, lower Rekovian beds: Russia: Kazakhstan, Salair

Mediocrinus rugatus (Stukalina), 1961. Stukalina, 1965b, p. 140, Pl. 2, fig. 15.
Pentagonocyclicus rugatus Stukalina, 1961, p. 39, Pl. 1, fig. 4; 1965b, p. 140.
Upper Silurian, Karaespinian horizon: Russia: Kazakhstan

Melocrinites Goldfuss, 1826. Schewtschenko, 1966, p. 133.
Kuzbassocrinus Yeltyschewa, 1957, p. 223. Schewtschenko, 1966, p. 133.
Devonian

Melocrinites binidigitatus (Yeltyschewa), 1957, new comb.
  Kuzbassocrinus binidigitatus Yeltyschewa, 1957, p. 224, Pl. 2, fig. 3;
    Fig. 4. Dubatolova, 1964, p. 55, Pl. 7, fig. 6. Yeltyschewa and
    Dubatolova, 1967, p. 35, Pl. 4, figs. 1-6; Fig. 15. Modzalevskaya,
    1967, p. 546, Pl. 4, fig. 21.
  Lower-Middle Devonian, Losishinsky layer, Jmatchinskian suite: Russia:
    Altay Mountains, Kuznetz Basin, Upper Priamur

Melocrinites bisulcus Schewtschenko, 1966, p. 141, Pl. 4, fig. 13; Fig. 11.
  Lower Devonian, Pandzhrutskian horizon: Russia: Tien-Shan Mountains

Melocrinites brevilobatus Schewtschenko, 1966, p. 139, Pl. 2, figs. 11-13;
    Fig. 8.
  Lower Devonian, Pandzhrutskian horizon: Russia: Tien-Shan Mountains

Melocrinites bystrowi (Yeltyschewa), 1957. Schewtschenko, 1966, p. 134,
    Pl. 1, figs. 1-4, 8, 10, 11.
  Kuzbassocrinus bystrowi Yeltyschewa, 1957, p. 223, Pl. 2, figs. 1, 2;
    Fig. 3. Yeltyschewa and Dubatolova, 1961, p. 555, Pl. D-87, fig. 7.
    Dubatolova, 1964, p. 53, Pl. 6, fig. 7; Pl. 7, figs. 2, 3. Moore
    and Jeffords, 1968, p. 36, Pl. 2, fig. 3.
  Lower-Middle Devonian, Shandinskian stage: Russia: Kuznetz Basin,
    Karachumsh, Salair, Tadzhikstan

Melocrinites cingulatus Schewtschenko, 1966, p. 137, Pl. 1, figs. 5-7, 9;
    Fig. 6.
  Lower Devonian, Pandzhrutskian horizon: Russia: Tien-Shan Mountains

Melocrinites cylindricus Schewtschenko, 1966, p. 140, Pl. 2, fig. 1;
    Fig. 9.
  Lower Devonian, Pandzhrutskian horizon: Russia: Tien-Shan Mountains

Melocrinites decemlobatus (Yeltyschewa), 1957. Schewtschenko, 1966,
    p. 138, Pl. 2, fig. 14; Pl. 3, figs. 11-14.
  Kuzbassocrinus decemlobatus Yeltyschewa, 1957, p. 225, Pl. 1,
    figs. 8, 9; Fig. 5. Dubatolova, 1964, p. 52, Pl. 6, fig. 6.'
    Yeltyschewa and Dubatolova, 1967, p. 34, Pl. 4, fig. 9.
    Modzalevskaya, 1967, p. 545, Pl. 3, fig. 5.
  Lower Devonian, Pribalkhash layer, Bolsheneverskian suite: Russia:
    Tien-Shan Mountains, Kuznetz Basin, Far East, Tadzhikstan, Upper
    Priamur, Pribalkhash

Melocrinites impalpabilis (Dubatolova), 1968, new comb.
  Kuzbassocrinus impalpabilis Dubatolova, 1968, p. 146, Pl. 15,
    figs. 4-6; Fig. 4.
  Devonian, Tom-Chumskian horizon: Russia: Salair

Melocrinites lanceolatus Schewtschenko, 1966, p. 143, Pl. 2, figs. 5, 6;
    Fig. 13.
  Lower Devonian, Pandzhrutskian horizon: Russia: Tien-Shan Mountains

Melocrinites mixtus Schewtschenko, 1966, p. 144, Pl. 2, figs. 3, 4;
    Fig. 14.
  Lower Devonian, Pandzhrutskian horizon: Russia: Tien-Shan Mountains

Melocrinites parvus Schewtschenko, 1966, p. 136, Pl. 6, figs. 1, 2, 9;
    Fig. 5.
  Upper Silurian, Tiverskian stage, Shishkatskian horizon: Russia:
    Tien-Shan Mountains

Melocrinites paucicostalus (Yeltyschewa), 1957, new comb.
  Kuzbassocrinus paucicostalus Yeltyschewa, 1957, p. 226, Pl. 2,
    figs. 4, 5; Fig. 6.
  Lower Devonian, Pribalkhash layer: Russia: southern Baltic

Melocrinites perpetuus Schewtschenko, 1966, p. 138, Pl. 6, figs. 21, 22;
    Fig. 7.

Upper Silurian, Tiverskian stage, Shishkatskian horizon: Russia: Tien-Shan Mountains

Melocrinites schyschcatus Schewtschenko, 1966, p. 141, Pl. 6, fig. 20; Fig. 10.
Upper Silurian, Tiverskian stage, Shishkatskian horizon: Russia: Tien-Shan Mountains

Melocrinites simplex Schewtschenko, 1966, p. 145, Pl. 2, figs. 7-10; Fig. 15.
Lower Devonian, Pandzhrutskian horizon, Kshtutskian horizon: Russia: Tien-Shan Mountains

Melocrinites subtilis Schewtschenko, 1966, p. 142, Pl. 2, fig. 2; Fig. 12.
Lower Devonian, Pandzhrutskian horizon: Russia: Tien-Shan Mountains

Melocrinites tuberculatus Schewtschenko, 1966, p. 146, Pl. 3, figs. 1-10; Fig. 16.
Lower Devonian, Pandzhrutskian horizon: Russia: Tien-Shan Mountains

Melocrinites verrucosus Schewtschenko, 1966, p. 147, Pl. 4, fig. 18; Fig. 17.
Lower Devonian, Pandzhrutskian horizon: Russia: Tien-Shan Mountains

Melocrinites yeltyschewae (Dubatolova), 1964, new comb.
Kuzbassocrinus yeltyschewae Dubatolova, 1964, p. 54, Pl. 7, figs. 4, 5.
Lower Devonian: Russia: Kuznetz Basin

Metichthyocrinus? sp. Termier and Termier, 1950, p. 92, Pl. 218, figs. 46-52.
Carboniferous, Westphalian: Morocco

Millericrinus sp. Sangil, 1955, p. 253, fig. 3A (misspelled as *Millercrinus*).
No age given

MOOREANTERIS Miller, *in* Moore and Jeffords, 1968. *M. waylandensis* Miller, *in* Moore and Jeffords, 1968. Miller, *in* Moore and Jeffords, 1968, p. 66.
Pennsylvanian, Virgilian

Mooreanteris perforatus Moore and Jeffords, 1968, p. 67, Pl. 18, fig. 6; Pl. 27, fig. 10.
Pennsylvanian, Virgilian, Belknap Limestone: United States: Texas

*Mooreanteris waylandensis Miller, *in* Moore and Jeffords, 1968, p. 67, Pl. 18, figs. 7-11. Moore and others, 1968a, p. 16, fig. 3, no. 2.
Pennsylvanian, Virgilian, Wayland Shale: United States: Texas

Mooreanteris sp. undet. (A), Miller, *in* Moore and Jeffords, 1968, p. 68, Pl. 19, fig. 13.
Pennsylvanian, Virgilian, Wayland Shale: United States: Texas

Moscovicrinus sp. Ivanova, 1958, Pl. 2, fig. 2.
Carboniferous (C_2^m): Russia: Moscow Syncline

MUSIVOCRINUS Termier and Termier, 1958. *M. arnouldi* Termier and Termier, 1958. Termier and Termier, 1949, p. 36, nomen nudum; 1958, p. 55.
Permian

*Musivocrinus arnouldi Termier and Termier, 1958, p. 55, Pl. 4, figs. a-h; Pl. 5, fig. a.
Permian: Tunisia

Musivocrinus sp. Termier and Termier, 1958, p. 58, Pl. 5, figs. b, c.
Permian: Tunisia

Myelodactylus ammonis (Bather), 1893. Moore and Jeffords, 1968, p. 32, Pl. 1, figs. 1, 2.
Silurian, Niagaran, Waynesville Formation: United States: Tennessee

Myelodactylus canaliculatus (Goldfuss), 1826. Sieverts-Doreck, 1953, p. 76, Pl. 4, figs. 1-3.
  Rhodocrinites canaliculatus Goldfuss, 1826. Sieverts-Doreck, 1953, p. 76.
  Devonian, Eifelian, Übergangs Limestone: Germany
Myelodactylus cf. *convolutus* Hall, 1852. Yeltyschewa, 1960, p. 5, Pl. 5, figs. 9, 10.
  Silurian, Llandoverian-Wenlockian: Russia: Siberian Platform, Ural Mountains
Myelodactylus spp.
  Sieverts-Doreck, 1953, p. 78, Pl. 4, figs. 4-6.
  Devonian: Germany
  Yeltyschewa, 1960, p. 6, Pl. 5, fig. 8.
  Silurian: Russia: Siberian Platform
Myrtillocrinus cf. *elongatus* Sandberger and Sandberger, 1856. Dubatolova, 1967, p. 33, Pl. 6, figs. 1, 2.
  Devonian, upper Eifelian-Givetian: Russia: Tas-Hayahtah Mountain
NOTHROSTERIGMA Moore and Jeffords, 1968. *N. merum* Moore and Jeffords, 1968. Moore and Jeffords, 1968, p. 83.
  Pennsylvanian, Virgilian
Nothrosterigma merum Moore and Jeffords, 1968, p. 84, Pl. 27, fig. 13.
  Pennsylvanian, Virgilian, Chaffin Limestone: United States: Texas
OBUTICRINUS Yeltyschewa and Stukalina, 1963. *Pentagonopentagonalis bilobatus* Yeltyschewa, 1960. Yeltyschewa and Stukalina, 1963, p. 32.
  Silurian
Obuticrinus bilobatus (Yeltyschewa), 1960. Yeltyschewa and Stukalina, 1963, p. 32, Pl. 2, figs. 1-3; Fig. 4.
  Pentagonopentagonalis bilobatus Yeltyschewa, 1960, p. 17, Pl. 5, figs. 1-4; 1964b, p. 75, Pl. 15, fig. 21. Schewtschenko, 1964, p. 13, Pl. 1, fig. 1; Pl. 4, fig. 3. Yeltyschewa and Stukalina, 1963, p. 32.
  Silurian: Russia: Taymyr, Siberian Platform, Tadzhikstan
Obuticrinus bullosus Yeltyschewa and Stukalina, 1963, p. 34, Pl. 2, figs. 9-11; Fig. 5.
  Lower Silurian: Russia: Nova Zemlya, Taymyr

Pachycrinites Eichwald, 1840. Moore and Laudon, 1943b, p. 104.
  Lower Carboniferous
Pachyocrinus Billings, 1859. Moore and Laudon, 1943b, p. 104.
  Ordovician
Palecocrinus? sp. Elias, 1957, p. 385, Pl. 41, figs. 6, 7.
  Mississippian, Chesterian, Redoak Hollow Formation: United States: Oklahoma
Palermocrinus Jaekel, 1918. Moore and Laudon, 1943b, p. 104.
  Permian
Palermocrinus jaeckeli Gislen, 1924. Nicosia, 1954, p. 89, figs. 1, 2. Termier and Termier, 1958, p. 55, Pl. 3, fig. g.
  Permian: Sicily; Tunisia
PANDOCRINUS Stukalina, 1965. *P. pandus* Stukalina, 1965. Stukalina, 1965b, p. 138. Moore and Jeffords, 1968, p. 69.
  Upper Silurian-Lower Devonian
Pandocrinus pandus Stukalina, 1965b, p. 138, Pl. 1, figs. 18-20; Fig. 4.
  Upper Silurian, Karaespinskian horizon: Russia: central Kazakhstan

Pandocrinus aff. *pandus* Stukalina, 1965. Dubatolova, 1968, p. 150,
Pl. 15, figs. 11-13; Fig. 8.
Devonian, Tom-Chumskian horizon, lower Rekovian beds: Russia: Salair

Pandocrinus stoloniferus (Hall), 1859. Moore and Jeffords, 1968, p. 69,
Pl. 19, figs. 11, 12; Pl. 20, figs. 1-3; Pl. 21, figs. 14, 15.
Mariacrinus stoloniferus Hall, 1859. Moore and Jeffords, 1968, p. 69.
Upper Silurian, Devonian, Helderbergian, New Scotland Limestone,
Birdsong Shale: Czechoslovakia; United States: New York, Tennessee

Parahexacrinus ellipticus Schewtschenko, 1967, p. 81, Pl. 10, fig. 9;
Fig. 4; 1968, p. 76, Pl. 10, fig. 9; Fig. 4.
Lower Devonian, Kshtutskian horizon: Russia: Zeravshan Mountains

Parahexacrinus glaber Schewtschenko, 1967, p. 79, Pl. 10, figs. 10-12;
Fig. 3; 1968, p. 76, Pl. 10, figs. 10-12; Fig. 3.
Lower Devonian, Kshtutskian horizon: Russia: Zeravshan Mountains

PARTICRINUS Stukalina, 1968. *Pentagonocyclicus partitus* Yeltyschewa,
1960. Stukalina, 1968c, p. 88 (incorrectly designates type species
as *Pentagonopentagonalis partitus* instead of *Pentagonocyclicus
partitus*).
Middle-Upper Ordovician

Particrinus partitus (Yeltyschewa), 1960. Stukalina, 1968c, p. 88.
Pentagonocyclicus partitus Yeltyschewa, 1960, p. 19, Pl. 2, figs. 7-9.
Yeltyschewa and Stukalina, 1963, p. 38, Pl. 2, figs. 12-14; Fig. 8.
Middle-Upper Ordovician: Russia: Siberian Platform, Taymyr

PENNATOCRINUS Stukalina, 1968. *P. subpennatus* Yeltyschewa, *in* Stukalina,
1968. Stukalina, 1968b, p. 261.
Upper Silurian-Lower Devonian

Pennatocrinus parapennatus Stukalina, 1968b, fig. 2, no. 1, nomen nudum.
Levitskiy, and others, 1968, p. 72, fig. 2, no. 1, nomen nudum
(attribute authorship of species to Yeltyschewa).
Upper Silurian-Lower Devonian, Ainasuiskian-Karaespinian horizon:
Russia: Kazakhstan, Pribalkhash

Pennatocrinus subpennatus Yeltyschewa, *in* Stukalina, 1968b, p. 263,
Pl. 1, figs. 14-16; Figs. 1, no. 1; 2, no. 2. Levitskiy, and
others, 1968, p. 72, fig. 2, no. 2.
Pentagonopentagonalis subpennatus Yeltyschewa. Stukalina, 1965b,
p. 135, Pl. 2, figs. 9, 10 (no description). Stukalina, 1968b,
p. 261 (gives name as "nom. in coll.," then gives a description and
illustrates the species attributing authorship to Yeltyschewa, thus
establishing the species as *Pennatocrinus subpennatus*).
Upper Silurian-Lower Devonian, Ainasuiskian-Pribalkhashian: Russia:
Kazakhstan, Pribalkhash

PENTACAULISCUS Moore and Jeffords, 1968. *P. nodosus* Moore and Jeffords,
1968. Moore and Jeffords, 1968, p. 52.
Devonian, Helderbergian

Pentacauliscus nodosus Moore and Jeffords, 1968, p. 52, Pl. 7, figs. 6, 7.
Devonian, Helderbergian, Birdsong Shale: United States: Tennessee

Pentacauliscus nodosus? Moore and Jeffords, 1968, p. 52, Pl. 7, fig. 5.
Devonian, Helderbergian, Birdsong Shale: United States: Tennessee

Pentacrinus lobatus Eichwald, 1860 = *Pentagonopentagonalis lobatus*

Pentagonites Rafinesque, 1819. Moore and Laudon, 1943b, p. 104.
Age unknown

PENTAGONOCYCLICUS Yeltyschewa, 1955, nomen vetitum.
Yeltyschewa, 1955a, p. 46; 1956, p. 42, figs. 1-3E, 2i, 2k; 1959,
p. 233, fig. 1-3E; 1964b, p. 75, fig. 1-3E. Dubatolova, 1964,

p. 9, fig. 1-3E.
Ordovician-Permian

Pentagonocyclicus acanthaceus Yeltyschewa, 1968, p. 44, Pl. 5, figs. 7-9 (misspelled as *P. acantihaceus* on plate explanation).
Silurian, Skalskian horizon: Russia: Podolsk

Pentagonocyclicus acbastanensis Sizova, 1960, p. 56, nomen nudum.
Devonian, Famennian: Russia: Kazakhstan

Pentagonocyclicus acbastanensis var. *unda* Sizova, 1960, p. 56, nomen nudum.
Devonian, Famennian: Russia: Kazakhstan

Pentagonocyclicus aculeatus Dubatolova, 1964, p. 72, Pl. 9, fig. 14.
Lower Devonian: Russia: Kuznetz Basin

Pentagonocyclicus admotus Dubatolova and Shao, 1959, p. 43, Pl. 1, fig. 5.
Upper Permian: China: Hunan Province

Pentagonocyclicus akbaitalicus Schewtschenko, 1968, p. 281, Pl. 66, fig. 4; Fig. 32.
Middle Carboniferous: Russia: Pamir

Pentagonocyclicus aktschetauensis Stukalina, 1960, p. 102, fig. 3.
Silurian, upper Llandoverian, *Pentamerus longiseptatus* beds: Russia: Chingiz Mountains

Pentagonocyclicus altimarginalis Yeltyschewa, 1955a, p. 46, Pl. 37, fig. 4; 1960, p. 20, Pl. 3, figs. 12, 13. Yeltyschewa and Stukalina, 1963, p. 50, Pl. 1, fig. 9; fig. 17; 1964b, p. 75, Pl. 15, fig. 17.
Middle-Late Ordovician, Dolborskian horizon: Russia: Siberian Platform, Taymyr

Pentagonocyclicus arenarius radialis Yeltyschewa and Dubatolova, 1967, p. 57, Pl. 6, fig. 1; Fig. 22.
Lower Carboniferous: Russia: Priamur

Pentagonocyclicus aseriensis Yeltyschewa, 1966, p. 61, Pl. 1, figs. 7-9.
Middle Ordovician: Russia: Estonia

Pentagonocyclicus astericus Schewtschenko, 1966, p. 166, Pl. 5, figs. 1-3; Fig. 37. Yeltyschewa, 1968, p. 46, Pl. 5, fig. 20. Dubatolova, 1968, p. 153, Pl. 16, figs. 10, 11; Fig. 11.
Silurian-Devonian, Tiverskian stage, Tom-Chumskian horizon, lower Rekovian beds: Russia: Tien-Shan Mountains, Podolsk, Salair

Pentagonocyclicus astericus papulosus Dubatolova, 1968, p. 154, Pl. 16, figs. 12, 13; Fig. 12.
Devonian, Tom-Chumskian horizon: Russia: Salair

Pentagonocyclicus bakanasensis Sizova, 1960, p. 55, nomen nudum.
Devonian, Givetian-Frasnian: Russia: Kazakhstan

Pentagonocyclicus baskanensis Sizova, 1960, p. 53, nomen nudum.
Devonian, Coblentzian-Frasnian: Russia: Kazakhstan

Pentagonocyclicus bicostatus Stukalina, 1961, p. 36, Pl. 2, fig. 1.
Upper Silurian: Russia: Kazakhstan

Pentagonocyclicus bifidus Yeltyschewa, 1964a, p. 74, Pl. 3, figs. 23, 24.
Lower Ordovician: Russia: Izvoz

Pentagonocyclicus biplex (Eichwald), 1860. Yeltyschewa, 1964a, p. 71, Pl. 3, figs. 12-14; Pl. 4, fig. 21.
Poteriocrinus biplex Eichwald, 1860. Yeltyschewa, 1964a, p. 71.
Lower Ordovician: Russia: Estonia

Pentagonocyclicus borealis Yeltyschewa, 1955a, p. 46, Pl. 54, fig. 3; 1960, p. 21, Pl. 4, fig. 8. Yeltyschewa and Stukalina, 1963, p. 53, Pl. 1, fig. 18; Fig. 20.
Silurian, Llandoverian: Russia: Siberian Platform, Taymyr

Pentagonocyclicus cingulatus (Goldfuss) 1866. Dubatolova, 1964, p. 87, Pl. 12, fig. 13; Pl. 13, figs. 3, 4.
Devonian, Eifelian: Russia: Kuznetz Basin, Ural Mountains
Actinocrinites cingulatus Goldfuss, 1866. Dubatolova, 1964, p. 87.
Entrochus cingulatus impares Quenstedt, 1874-1876. Yeltyschewa and Dubatolova, 1961, p. 556, Pl. D-87, fig. 8. Dubatolova, 1964, p. 87.

Pentagonocyclicus cf. *circumvalatus* var. *minor* Yeltyschewa, *in* Dubatolova and Shao, 1959, p. 45, Pl. 1, fig. 9; unnumbered fig.
Carboniferous, Visean: China: Hunan Province

Pentagonocyclicus concentricus Yeltyschewa, 1964a, p. 73, Pl. 2, figs. 22-28.
Lower Ordovician: Russia: Estonia, Leningrad Province

Pentagonocyclicus conserratus Yeltyschewa and Dubatolova, 1967, p. 53, Pl. 5, fig. 15.
Devonian, Eifelian: Russia: Upper Priamur

Pentagonocyclicus constrictus Yeltyschewa, 1964a, p. 75, Pl. 2, figs. 19-21; Pl. 4, fig. 23.
Lower Ordovician: Russia: Estonia

Pentagonocyclicus continatus Dubatolova, 1968, p. 151, Pl. 16, figs. 1-8; Fig. 9.
Devonian, Tom-Chumskian horizon, lower Rekovian beds: Russia: Salair

Pentagonocyclicus costatus Schewtschenko, 1966, p. 165, Pl. 5, figs. 7, 8; Fig. 36. Dubatolova, 1968, p. 152, Pl. 16, fig. 9; Fig. 10.
Upper Silurian-Lower Devonian, Tiverskian stage, Shishkatskian horizon—Tom-Chumskian horizon, lower Rekovian beds: Russia: Tien-Shan Mountains, Salair

Pentagonocyclicus curtus Dubatolova, 1964, p. 77, Pl. 10, figs. 11-13.
Lower Devonian: Russia: Kuznetz Basin

Pentagonocyclicus definitus Stukalina, 1961, p. 38, Pl. 1, fig. 1; 1965b, p. 135, Pl. 2, figs. 5, 6.
Upper Silurian, Karaespinskian horizon: Russia: Kazakhstan

Pentagonocyclicus dentatus var. *echinata* (Yeltyschewa, *in* Yeltyschewa and Dubatolova), 1961. Dubatolova, 1964, p. 88, Pl. 12, fig. 12; Pl. 13, figs. 1, 2.
Entrochus dentatus var. *echinata* Yeltyschewa, *in* Yeltyschewa and Dubatolova, 1961, p. 557, Pl. D-87, fig. 13.
Devonian, Eifelian, Shandinskian layer: Russia: Kuznetz Basin, Salair

Pentagonocyclicus dvinaeboreae Yeltyschewa, *in* Dubatolova and Shao, 1959, p. 44, Pl. 1, fig. 7, nomen correctum (given as *P. dvinae-boreae*).
Lower-middle Carboniferous: China: Kwangtung Province; Russia: North Dvina River Basin

Pentagonocyclicus egiasarowi Yeltyschewa, 1960, p. 22, Pl. 5, figs. 5, 6.
Silurian, Wenlockian: Russia: Siberian Platform

Pentagonocyclicus elegans Yeltyschewa, 1960, p. 20, Pl. 4, figs. 9-12; 1964b, p. 75, Pl. 15, fig. 13.
Silurian, Llandovery: Russia: Siberian Platform, Nova Zemlya

Pentagonocyclicus equitans Yeltyschewa, 1966, p. 62, Pl. 1, figs. 10-12.
Middle Ordovician, Azeriskian horizon: Russia: Estonia

Pentagonocyclicus exculcatus Dubatolova, 1964, p. 79, Pl. 11, figs. 11, 12.
Lower Devonian: Russia: Kuznetz Basin

Pentagonocyclicus expolites Dubatolova, 1964, p. 83, Pl. 11, fig. 8.
  Lower Devonian: Russia: Kuznetz Basin

Pentagonocyclicus expressus Dubatolova and Shao, 1959, p. 43, Pl. 1, fig. 4.
  Lower-Middle Carboniferous: China: Kwangtung Province

Pentagonocyclicus falsus Dubatolova and Shao, 1959, p. 45, Pl. 1, fig. 10.
  Lower Carboniferous, Visean: China: Hunan Province

Pentagonocyclicus festus Schewtschenko, 1966, p. 164, Pl. 5, fig. 12; Fig. 35.
  Upper Silurian, Tiverskian stage, Shishkatskian horizon: Russia: Tien-Shan Mountains

Pentagonocyclicus filigerum Schewtschenko, 1966, p. 168, Pl. 6, figs. 23, 24; Fig. 40.
  Upper Silurian-Lower Devonian: Russia: Tien-Shan Mountains

Pentagonocyclicus firmus Yeltyschewa, *in* Modzalevskaya, 1967, p. 547, Pl. 6, fig. 25 (no description), nomen nudum.
  Middle Devonian, Oldoiskian suite: Russia: Upper Priamur

Pentagonocyclicus forus Stukalina, 1961 = *Gregariocrinus forus*

Pentagonocyclicus frondeus Stukalina, 1961, p. 38, Pl. 2, fig. 2.
  Upper Silurian: Russia: Kazakhstan

Pentagonocyclicus glaber Yeltyschewa, *in* Yeltyschewa and Dubatolova, 1961, p. 559, Pl. D-87, fig. 12. Dubatolova, 1964, p. 76, Pl. 10, figs. 9, 10, 15.
  Lower Devonian, Krekovskian layer: Russia: Salair

Pentagonocyclicus gradatus Yeltyschewa, *in* Dubatolova, 1964, p. 79, Pl. 11, figs. 3, 4.
  Lower Devonian: Russia: Kuznetz Basin, Pribalkhash

Pentagonocyclicus granatus Dubatolova, 1964, p. 78, Pl. 10, fig. 14; Pl. 11, figs. 1, 2.
  Lower Devonian: Russia: Kuznetz Basin

Pentagonocyclicus haldaranensis Yeltyschewa and Schewtschenko, 1960, p. 122, Pl. 2, figs. 9, 10.
  Lower Carboniferous: Russia: Darvas

Pentagonocyclicus hobotschaloensis Dubatolova, 1967, p. 39, Pl. 6, figs. 22-25; Fig. 6.
  Devonian, Eifelian: Russia: Tas-Hayahtah Mountain

Pentagonocyclicus hsianghsiagnensis Dubatolova and Shao, 1959, p. 46, Pl. 1, fig. 16.
  Carboniferous, Visean: China: Hunan Province

Pentagonocyclicus humilicristatus Yeltyschewa, 1955b, p. 36, Pl. 15, figs. 2, 3.
  Devonian: Russia: Minusinsk Trough

Pentagonocyclicus humilis Dubatolova, 1964, p. 84, Pl. 11, fig. 13; Pl. 12, fig. 5; 1967, p. 40, Pl. 6, figs. 19-21.
  Devonian, Eifelian: Russia: Kuznetz Basin, Altay, Tas-Hayahtah Mountain

Pentagonocyclicus imatschensis Yeltyschewa and Dubatolova, 1961, p. 558, Pl. D-87, figs. 14, 15; 1967, p. 49, Pl. 6, figs. 10, 11; Fig. 20 (*non* Fig. 21 as given in text). Modzalevskaya, 1967, p. 546, Pl. 4, fig. 22.
  Lower-Middle Devonian, Jmatchinskian suite: Russia: Altay Mountains, Upper Priamur

Pentagonocyclicus inaequalis Yeltyschewa, 1966, p. 60, Pl. 3, figs. 12, 13.
  Middle Ordovician: Russia: Estonia

Pentagonocyclicus incelebratus Yeltyschewa and Dubatolova, 1967, p. 51,
   Pl. 7, figs. 1-4.
   Lower-Middle Devonian: Russia: Upper Priamur
Pentagonocyclicus inconditus Dubatolova, 1964, p. 85, Pl. 12, figs. 9-11;
   Fig. 12.
   Devonian, Eifelian: Russia: Kuznetz Basin
Pentagonocyclicus indissimilis Schewtschenko, 1964, p. 18, Pl. 2,
   figs. 10, 11; Pl. 3, figs. 4, 5; 1968a, p. 278, Pl. 66, fig. 2;
   Fig. 29.
   Silurian, lower Wenlockian: Russia: Tadzhikstan, Zeravshan Mountains
Pentagonocyclicus infimus Dubatolova, 1964, p. 91, Pl. 13, figs. 11-14
   (given as figs. 11-13 in text).
   Devonian, Frasnian: Russia: Kuznetz Basin
Pentagonocyclicus inflatus Dubatolova, 1964, p. 71, Pl. 9, figs. 11, 15;
   Fig. 9.
   Lower Devonian: Russia: Kuznetz Basin
Pentagonocyclicus insectus Yeltyschewa, *in* Dubatolova, 1964, p. 92,
   Pl. 14, figs. 1, 2.
   Devonian, Frasnian: Russia: Kuznetz Basin, central Devonian Field
Pentagonocyclicus insignis Dubatolova, 1964, p. 80, Pl. 11, fig. 10.
   Lower Devonian: Russia: Kuznetz Basin
Pentagonocyclicus ivanovi Yeltyschewa and Dubatolova, 1967, p. 55, Pl. 5,
   figs. 20-22; Fig. 21.
   Lower-Middle Devonian: Russia: Upper Priamur
Pentagonocyclicus jucundus Dubatolova, 1964, p. 81, Pl. 11, figs. 5-7.
   Devonian, Eifelian: Russia: Kuznetz Basin
Pentagonocyclicus kegelensis Yeltyschewa, 1966, p. 59, Pl. 3, figs. 14-23.
   Middle Ordovician: Russia: Estonia, Leningrad Province
Pentagonocyclicus kounradensis Sizova, 1960, p. 57, nomen nudum.
   Carboniferous, Tournaisian: Russia: Kazakhstan
Pentagonocyclicus kuangsinensis Dubatolova and Shao, 1959, p. 43, Pl. 1,
   fig. 3.
   Carboniferous, Visean: China: Kwangtung Province
Pentagonocyclicus lenitus Stukalina, 1960, p. 101, fig. 2.
   Silurian, upper Llandoverian, *Pentamerus longiseptatus* Zone: Russia:
   Chingiz Mountains
Pentagonocyclicus lenticularis Stukalina, 1961, p. 38, Pl. 2, fig. 4;
   1965b, p. 135, Pl. 2, fig. 16.
   Upper Silurian, Karaespinskian horizon: Russia: Kazakhstan
Pentagonocyclicus lesnikovae Yeltyschewa, 1964 = *Sokolovicrinus
   lesnikovae*
Pentagonocyclicus levidensis Dubatolova, 1964, p. 86, Pl. 12, figs. 6-8.
   Devonian, Eifelian: Russia: Kuznetz Basin
Pentagonocyclicus ligatus (Quenstedt), 1876.   Dubatolova, 1964, p. 74,
   Pl. 10, figs. 4-6.
   Entrochus ligatus Quenstedt, 1876.   Yeltyschewa and Dubatolova, 1961,
   p. 556, Pl. D-87, fig. 9.   Dubatolova, 1964, p. 74.
   Devonian, Eifelian: Russia: Kuznetz Basin, Salair; Germany
Pentagonocyclicus maltus Dubatolova, 1964, p. 86, Pl. 12, figs. 14, 15;
   Fig. 13.
   Devonian, Eifelian: Russia: Kuznetz Basin

Pentagonocyclicus meditatus Yeltyschewa, *in* Modzalevskaya, 1967, Pl. 6, fig. 23, no description, nomen nudum.
   Middle Devonian, Oldoiskian suite: Russia: Upper Priamur

Pentagonocyclicus monile (Eichwald), 1860 =*Dianthicoeloma monile*

Pentagonocyclicus monocostatus Stukalina, 1961, p. 37, Pl. 1, fig. 2; 1965b, p. 135, Pl. 1, figs. 14-16.
   Upper Silurian, Karaespinskian horizon: Russia: Kazakhstan

Pentagonocyclicus multicius Dubatolova, 1964, p. 94, Pl. 14, fig. 3.
   Devonian, Frasnian: Russia: Kuznetz Basin

Pentagonocyclicus mundus Dubatolova, 1964, p. 89, Pl. 13, figs. 7, 8.
   Devonian, Givetian: Russia: Kuznetz Basin

Pentagonocyclicus muschketowi Schewtschenko, 1964, p. 17, Pl. 2, fig. 2; Pl. 4, fig. 2; 1968a, p. 280, Pl. 66, fig. 3; Fig. 31.
   Silurian, lower Wenlockian: Russia: Tien-Shan Mountains, Tadzhikstan

Pentagonocyclicus nieczlawiensis Yeltyschewa, 1968, p. 45, Pl. 4, figs. 13-16.
   Silurian, Borschovskian horizon: Russia: Podolsk

Pentagonocyclicus novellus Sizova, 1960, p. 57, nomen nudum.
   Carboniferous, Tournaisian: Russia: Kazakhstan

Pentagonocyclicus obscurus Dubatolova, 1964, p. 90, Pl. 13, figs. 5-6.
   Devonian, Givetian: Russia: Kuznetz Basin

Pentagonocyclicus observabilis Dubatolova, 1964, p. 94, Pl. 14, figs. 8, 9; Fig. 14.
   Devonian, Famennian: Russia: Kuznetz Basin

Pentagonocyclicus obstupendus Dubatolova, 1964, p. 82, Pl. 12, figs. 1-4; Fig. 11.
   Devonian, Eifelian: Russia: Kuznetz Basin

Pentagonocyclicus occultus Dubatolova, 1964, p. 73, Pl. 9, fig. 12.
   Lower Devonian: Russia: Kuznetz Basin

Pentagonocyclicus oldicus Yeltyschewa and Dubatolova, 1960, p. 371, Pl. 70, fig. 6; 1967, p. 53, Pl. 6, figs. 4-8. Modzalevskaya, 1967, Pl. 6, fig. 24.
   Devonian, Givetian, Oldoiskian suite: Russia: Far East, Upper Priamur

Pentagonocyclicus opertus Dubatolova, 1964, p. 73, Pl. 9, fig. 13; Pl. 10, figs. 1-3; Fig. 10.
   Devonian, Eifelian: Russia: Kuznetz Basin

Pentagonocyclicus paragaudius Dubatolova, 1964, p. 93, Pl. 14, fig. 4.
   Devonian, Frasnian: Russia: Kuznetz Basin

Pentagonocyclicus partitus Yeltyschewa, 1960 = *Particrinus partitus*

Pentagonocyclicus paucus Dubatolova, 1964, p. 95, Pl. 14, figs. 5-7; Fig. 15.
   Devonian, Famennian: Russia: Kuznetz Basin

Pentagonocyclicus pentaporus (Eichwald), 1860 = *Dianthicoeloma pentaporus*

Pentagonocyclicus pentaporus var. *tuberculata* Yeltyschewa, 1964 = *Dianthicoeloma pentaporus* var. *tuberculata*

Pentagonocyclicus persimilis Dubatolova, 1964, p. 83, Pl. 11, fig. 9.
   Lower Devonian: Russia: Kuznetz Basin

Pentagonocyclicus primitivus Schewtschenko, 1964, p. 16, Pl. 2, fig. 8; Pl. 3, fig. 6; 1968a, p. 279, Pl. 66, fig. 1; Fig. 30.

Silurian, Llandovery-Wenlockian: Russia: Tien-Shan Mountains, Tadzhikstan

Pentagonocyclicus pulverlus Dubatolova and Shao, 1959, p. 42, Pl. 1, fig. 2.
Lower-Middle Carboniferous: China: Kwangtung Province

Pentagonocyclicus rimosus Schewtschenko, 1966, p. 167, Pl. 5, figs. 9-11, 17; Fig. 39.
Lower Devonian, Kshtutskian horizon: Russia: Tien-Shan Mountains

Pentagonocyclicus rugatus Stukalina, 1961 = *Mediocrinus rugatus*

Pentagonocyclicus salebrosus Stukalina, 1961, p. 37, Pl. 2, fig. 7.
  Pentagonopentagonalis salebrosus Stukalina, 1965b, p. 135, Pl. 2, figs. 20, 21.
Upper Silurian, Karaespinskian horizon, Pribalkhashskian horizon: Russia: Kazakhstan

Pentagonocyclicus saragaschensis Yeltyschewa, 1955b, p. 36, Pl. 15, fig. 4.
Devonian: Russia: Minusinsk Trough

Pentagonocyclicus scabrum Schewtschenko, 1966, p. 167, Pl. 5, figs. 14-16; Fig. 38.
Upper Silurian, Tiverskian stage-Devonian, Eifelian: Russia: Tien-Shan Mountains

Pentagonocyclicus schansinicus Dubatolova and Shao, 1959, p. 45, Pl. 1, fig. 11.
Carboniferous, Visean: China: Hunan Province

Pentagonocyclicus serratus Yeltyschewa, 1966, p. 64, Pl. 2, figs. 1-4.
Middle Ordovician, Ukhakuskian horizon: Russia: Estonia, Leningrad Province

Pentagonocyclicus simplex Yeltyschewa, 1955b, p. 37, Pl. 15, fig. 5.
Devonian: Russia: Minusinsk Trough

Pentagonocyclicus singularis Dubatolova, 1964, p. 91, Pl. 13, figs. 9, 10.
Devonian, Frasnian: Russia: Kuznetz Basin

Pentagonocyclicus stellatus Dubatolova and Shao, 1959, p. 44, Pl. 1, fig. 6.
Lower-Middle Carboniferous: China: Kwangtung Province

Pentagonocyclicus stellatus (Eichwald), 1860. Yeltyschewa, 1964a, p. 71, Pl. 3, figs. 15-17; Pl. 4, fig. 24.
  Platycrinus stellatus Eichwald, 1860. Yeltyschewa, 1964a, p. 71.
Lower Ordovician: Russia: Estonia

Pentagonocyclicus submersus Dubatolova, 1968, p. 154, Pl. 15, figs. 14, 15; Fig. 13.
Devonian, Tom-Chumskian horizon: Russia: Salair

Pentagonocyclicus superadornatus Schewtschenko, 1964, p. 16, Pl. 3, fig. 1.
Silurian, Llandoverian: Russia: Tadzhikstan

Pentagonocyclicus tainaensis Yeltyschewa, 1968, p. 44, Pl. 5, figs. 3-6.
Silurian, Borschovskian horizon: Russia: Podolsk

Pentagonocyclicus tajmirensis Yeltyschewa and Stukalina, 1963, p. 52, Pl. 1, fig. 6; Fig. 19.
Silurian, Llandoverian: Russia: Taymyr

Pentagonocyclicus tenuitas Yeltyschewa and Stukalina, 1963, p. 51, Pl. 4, figs. 1b, 3, 4; Fig. 18.
Upper Ordovician: Russia: Vaigach

Pentagonocyclicus textus Dubatolova, 1964, p. 75, Pl. 10, figs. 7, 8.
  Lower Devonian: Russia: Kuznetz Basin

Pentagonocyclicus tianschanicus Schewtschenko, 1966, p. 163, Pl. 5, figs. 4-6; Fig. 34.
  Devonian, Pandzhrutskian horizon: Russia: Tien-Shan Mountains

Pentagonocyclicus tipariensis Yeltyschewa and Dubatolova, 1967, p. 54, Pl. 6, figs. 2, 3.
  Devonian, Givetian: Russia: Upper Priamur

Pentagonocyclicus tricostatus Stukalina, 1961, p. 37, Pl. 2, fig. 3; 1965b, p. 135, Pl. 1, fig. 17 (mistakenly listed as *P. bicostatus* in text?).
  Upper Silurian, Karaespinskian horizon: Russia: Kazakhstan

Pentagonocyclicus triformis Dubatolova and Shao, 1959, p. 44, Pl. 1, fig. 8.
  Carboniferous, Visean: China: Hunan Province

Pentagonocyclicus vallatus Stukalina, 1960, p. 100, fig. 1.
  Silurian, Llandoverian, *Pentamerus longiseptatus* Zone: Russia: Chingiz Mountains

Pentagonocyclicus vastus Yeltyschewa and Dubatolova, 1960, p. 370, Pl. 70, figs. 3, 4; Fig. 44; 1967, p. 50, Pl. 7, figs. 5-9. Modzalevskaya, 1967, p. 546, Pl. 4, fig. 23; Pl. 5, fig. 1.
  Lower-Middle Devonian: Russia: Far East, Upper Priamur

Pentagonocyclicus vulgaris Yeltyschewa, 1955b, p. 36, Pl. 15, fig. 1.
  Devonian: Russia: Minusinsk Trough

Pentagonocyclopa cf. *P. dispar* (Moore), 1939. Walter, 1953, p. 701, Pl. 73, figs. 15-17.
  Permian, Rustler Formation: United States: Texas

PENTAGONOELLIPTICUS Yeltyschewa, 1956, nomen vetitum.
  Yeltyschewa, 1956, p. 42, fig. 1-3F; 1959, p. 233, fig. 1-3F; 1964b, p. 75, fig. 114-3F. Dubatolova, 1964, p. 9, fig. 1-3F.
  Theoretical genus, no known species.

PENTAGONOHEXAGONALIS Yeltyschewa, 1956, nomen vetitum.
  Yeltyschewa, 1956, p. 42, fig. 1-3D; 1956, p. 233, fig. 1-3D; 1964b, p. 75, fig. 114-3D. Dubatolova, 1964, p. 9, fig. 1-3D.
  Theoretical genus, no known species.

PENTAGONOMISCHUS Moore and Jeffords, 1968. *P. plebeius* Moore and Jeffords, 1968. Moore and Jeffords, 1968, p. 53.
  Mississippian, Chesterian

Pentagonomischus plebeius Moore and Jeffords, 1968, p. 53, Pl. 8, figs. 1-7.
  Mississippian, Chesterian, Paint Creek Formation: United States: Illinois

Pentagonomischus plebeius? Moore and Jeffords, 1968, p. 54, Pl. 8, fig. 8.
  Mississippian, Chesterian, Paint Creek Formation: United States: Illinois

PENTAGONOPENTAGONALIS Yeltyschewa, 1955, nomen vetitum.
  Yeltyschewa, 1955a, p. 42; 1956, p. 42, fig. 1-3C; 1959, p. 233, fig. 1-3C; 1964b, p. 75, fig. 114-3C. Dubatolova, 1964, p. 9, fig. 1-3C.
  Ordovician-Silurian

Pentagonopentagonalis ajnasuensis Stukalina, 1961 = *Facetocrinus ajnasuensis*

Pentagonopentagonalis angustilobatus Yeltyschewa, *in* Schewtschenko. 1964,
p. 12, Pl. 2, figs. 3-5, 9; Pl. 4, fig. 4.
Middle Ordovician-Silurian: Russia: Siberian Platform, Pamir, Pribaltic, Tadzhikstan

Pentagonopentagonalis antiquus (Eichwald), 1860. Yeltyschewa, 1966, p. 57, Pl. 2, figs. 23-26.
Pentacrinus antiquus Eichwald, 1860. Yeltyschewa, 1966, p. 57.
Middle Ordovician: Russia: Estonia, Leningrad Province

Pentagonopentagonalis artificiosus Yeltyschewa, 1964a, p. 69, Pl. 3, fig. 27; Pl. 4, fig. 22.
Lower Ordovician: Russia: Estonia

Pentagonopentagonalis bilobatus Yeltyschewa, 1960 = *Obuticrinus bilobatus*

Pentagonopentagonalis bondarewi Yeltyschewa and Stukalina, 1963, p. 46, Pl. 1, figs. 13-17; Fig. 14 (misspelled as *P. bondarevi* in text and plate explanations).
Upper Ordovician: Russia: Taymyr

Pentagonopentagonalis catagraphus Stukalina, 1960, p. 107, fig. 6.
Silurian, upper Llandoverian, *Pentamerus longiseptatus* Zone: Russia: Chingiz Mountains

Pentagonopentagonalis collariformis Yeltyschewa, 1964a, p. 68, Pl. 1, figs. 21, 22.
Lower Ordovician: Russia: Estonia

Pentagonopentagonalis compositus Yeltyschewa, 1955 = *Bystrowicrinus compositus*

Pentagonopentagonalis comptus Yeltyschewa, 1955a, p. 45, Pl. 37, fig. 7; 1960, p. 14, Pl. 3, figs. 6-9. Yeltyschewa and Stukalina, 1963, p. 36, Pl. 1, figs. 10-12; Fig. 6; 1964b, p. 75, Pl. 15, fig. 16.
Middle Ordovician-Silurian: Russia: Siberian Platform, Taymyr

Pentagonopentagonalis concinnus Yeltyschewa and Stukalina, 1963, p. 43, Pl. 1, figs. 1-3; Fig. 12.
Upper Ordovician: Russia: Vaigach

Pentagonopentagonalis dauritschensis Schewtschenko, 1964, p. 14, Pl. 2, fig. 1; Pl. 3, fig. 2; 1968a, p. 283, Pl. 66, fig. 7; Fig. 35.
Silurian, upper Llandoverian: Russia: Tadzhikstan, Tien-Shan Mountains

Pentagonopentagonalis dentiferus Yeltyschewa, 1960 = *Dentiferocrinus dentiferus*

Pentagonopentagonalis dividuus Yeltyschewa, 1960 = *Dentiferocrinus dividuus*

Pentagonopentagonalis facetus Stukalina, 1961 = *Facetocrinus facetus*

Pentagonopentagonalis floreus var. *gracila* Yeltyschewa and Dubatolova, 1961 = *Anthinocrinus floreus gracilis*

Pentagonopentagonalis fragilis Yeltyschewa, 1955a, p. 43, Pl. 23, fig. 6; 1960, p. 9, Pl. 2, fig. 14.
Middle Ordovician, Mangazeyskian horizon: Russia: Siberian Platform

Pentagonopentagonalis hrustalnjensis Yeltyschewa and Stukalina, 1963 = *Ramulicrinus hrustalnjensis*

Pentagonopentagonalis klunnikowi Schewtschenko, 1964, p. 15, Pl. 1, fig. 2; Pl. 4, fig. 1; 1968a, p. 282, Pl. 66, fig. 5; Fig. 34.
Silurian, lower Wenlockian: Russia: Tien-Shan Mountains, Tadzhikstan

Pentagonopentagonalis kokajgyrensis Stukalina, 1960, p. 104, fig. 4.
Silurian, Llandoverian, *Holorhynchus cinghizicus* beds: Russia: Chingiz Mountains

Pentagonopentagonalis lobatus (Eichwald), 1860. Yeltyschewa, 1966, p. 58, Pl. 3, figs. 3-5.
Pentacrinus lobatus Eichwald, 1860. Yeltyschewa, 1966, p. 58.
Middle Ordovician: Russia: Estonia, Leningrad Province

Pentagonopentagonalis mirabilis Yeltyschewa, 1955a, p. 44, Pl. 23, fig. 2; 1960, p. 7, Pl. 1, figs. 7, 8. Yeltyschewa and Stukalina, 1963, p. 47, Pl. 1, fig. 7; Fig. 15.
Middle Ordovician, Mangazeyskian horizon: Russia: Siberian Platform, Taymyr

Pentagonopentagonalis morcocaensis Yeltyschewa, 1960, p. 9, Pl. 3, figs. 1-5; 1964b, p. 75, Pl. 15, fig. 14.
Lower-Middle Ordovician: Russia: Siberian Platform

Pentagonopentagonalis multipartitus Yeltyschewa, 1960 = *Ramulicrinus multipartitus*

Pentagonopentagonalis papillaris Yeltyschewa, 1960, p. 13, Pl. 5, fig. 7.
Silurian, Llandoverian: Russia: Siberian Platform

Pentagonopentagonalis privus Yeltyschewa, 1964 = *Squameocrinus privus*

Pentagonopentagonalis probus Stukalina, 1961, p. 36, Pl. 1, fig. 6.
Upper Silurian: Russia: Kazakhstan

Pentagonopentagonalis proximus Yeltyschewa and Stukalina, 1963 = *Dentiferocrinus proximus*

Pentagonopentagonalis quindecemlobatus Yeltyschewa and Stukalina, 1963 = *Catagraphiocrinus quindecemlobatus*

Pentagonopentagonalis quinquelobatus Yeltyschewa, 1955 = *Bystrowicrinus quinquelobatus*

Pentagonopentagonalis quinquespinosus Stukalina, 1961 = *Facetocrinus quinquespinosus*

Pentagonopentagonalis ramosus Yeltyschewa and Stukalina, 1963, p. 42, Pl. 1, figs. 4, 5; Fig. 11.
Silurian, Llandoverian: Russia: Taymyr

Pentagonopentagonalis subpennatus Yeltyschewa, 1968 = *Pennatocrinus subpennatus*

Pentagonopentagonalis tolenensis Stukalina, 1960, p. 106, fig. 5.
Silurian, upper Llandoverian, *Pentamerus longiseptatus* Zone: Russia: Chingiz Mountains

Pentagonopentagonalis tridens Yeltyschewa, 1955a, p. 43, Pl. 23, fig. 3; 1960, p. 8, Pl. 2, figs. 12, 13; 1964b, p. 75, Pl. 15, fig. 15.
Ordovician: Russia: Siberian Platform

Pentagonopentagonalis tscherkesovae Yeltyschewa and Stukalina, 1963, p. 40, Pl. 2, fig. 24; Fig. 9.
Upper Ordovician: Russia: Taymyr

Pentagonopentagonalis vitreus Stukalina, 1961 = *Medinecrinus vitreus*

PENTAGONOPTERNIX Moore and Jeffords, 1968. *P. insculptus* Moore and Jeffords, 1968. Moore and Jeffords, 1968, p. 55.
Pennsylvanian, Virgilian

Pentagonopternix insculptus Moore and Jeffords, 1968, p. 55, Pl. 9, fig. 10.
Pennsylvanian, Virgilian, Chaffin Limestone: United States: Texas

PENTAGONOSTAURUS Moore and Jeffords, 1968. *P. leptus* Moore and Jeffords, 1968. Moore and Jeffords, 1968, p. 54.
Mississippian, Chesterian

*Pentagonostaurus leptus Moore and Jeffords, 1968, p. 54, Pl. 9, figs. 4, 5.
   Mississippian, Chesterian, Paint Creek Formation: United States: Illinois

PENTAGONOSTIPES Moore and Jeffords, 1968. *P. petaloides Moore and Jeffords, 1968. Moore and Jeffords, 1968, p. 52.
   Devonian, Erian

*Pentagonostipes petaloides Moore and Jeffords, 1968, p. 52, Pl. 8, figs. 14, 15.
   Devonian, Erian, Wanakah Shale: United States: New York

PENTAGONOTRIGONALIS Yeltyschewa, 1956, nomen vetitum.
   Yeltyschewa, 1956, p. 42, fig. 1-3A; 1959, p. 233, fig. 1-3A; 1964b, p. 75, fig. 114-3A. Dubatolova, 1964, p. 9, fig. 1-3A.
   Theoretical genus, no known species.

PENTAGONOTETRAGONALIS Yeltyschewa, 1956, nomen vetitum.
   Yeltyschewa, 1956, p. 42, fig. 1-3B; 1959, p. 233, fig. 1-3B; 1964b, p. 75, fig. 114-3B. Dubatolova, 1964, p. 9, fig. 1-3B.
   Theoretical genus, no known species.

PENTAMEROSTELA Moore and Jeffords, 1968. *P. delicatula Moore and Jeffords, 1968. Moore and Jeffords, 1968, p. 56.
   Mississippian, Osagean

*Pentamerostela delicatula Moore and Jeffords, 1968, p. 56, Pl. 9, figs. 8, 9.
   Mississippian, Osagean, New Providence Shale: United States: Kentucky

Pentamerostela minuta Moore and Jeffords, 1968, p. 56, Pl. 9, figs. 6, 7.
   Mississippian, Osagean, New Providence Shale: United States: Kentucky

PENTARIDICA Moore and Jeffords, 1968. *P. rothi Moore and Jeffords, 1968. Moore and Jeffords, 1968, p. 54.
   Pennsylvanian, Desmoinesian-Virgilian

Pentaridica pentagonalis Moore and Jeffords, 1968, p. 55, Pl. 9, fig. 14.
   Pennsylvanian, Desmoinesian, Minturn Formation: United States: Colorado

*Pentaridica rothi Moore and Jeffords, 1968, p. 55, Pl. 9, figs. 12, 13.
   Pennsylvanian, Virgilian, Blach Ranch Limestone: United States: Texas

Pentaridica simplicis Moore and Jeffords, 1968, p. 55, Pl. 9, fig. 11.
   Pennsylvanian, Desmoinesian, Mingus Shale: United States: Texas

PETALERISMA Moore and Jeffords, 1968. *P. eriense Moore and Jeffords, 1968. Moore and Jeffords, 1968, p. 53.
   Devonian, Erian

*Petalerisma eriense Moore and Jeffords, 1968, p. 53, Pl. 8, figs. 12, 13.
   Devonian, Erian, Wanakah Shale: United States: New York

Phialocrinus Eichwald, 1856. Moore and Laudon, 1943b, p. 104.
   Ordovician

PLATYCION Moore and Jeffords, 1968. *P. mingusensis Moore and Jeffords, 1968. Moore and Jeffords, 1968, p. 46.
   Pennsylvanian, Desmoinesian

*Platycion mingusensis Moore and Jeffords, 1968, p. 46, Pl. 5, figs. 1, 2.
   Pennsylvanian, Desmoinesian, Mingus Shale: United States: Texas

PLATYCLONUS Moore and Jeffords, 1968. *P. dispar Moore and Jeffords, 1968. Moore and Jeffords, 1968, p. 44.
   Mississippian, Osagean

Platyclonus dispar Moore and Jeffords, 1968, p. 44, Pl. 4, fig. 2.
  Mississippian, Osagean, New Providence Shale: United States: Kentucky
Platycrinites? irroratus Moore and Jeffords, 1968, p. 41, Pl. 3, fig. 4.
  Mississippian, Osagean, Burlington Limestone: United States: Iowa
Platycrinites maiensis Sizova, 1960, p. 55, nomen nudum.
  Carboniferous, Visean-Namurian: Russia: Kazakhstan
Platycrinites permiensis (Yakovlev, *in* Yakovlev and Ivanov), 1956, nomen correctum.
  Platycrinus permiensis Yakovlev, *in* Yakovlev and Ivanov, 1956, p. 66, Pl. 10, figs. 14-20.
  Permian: Russia: Ural Mountains
Platycrinites cf. *saffordi* Hall, 1858. Yeltyschewa and Dubatolova, 1967, p. 28, Pl. 6, figs. 16-19 (misspelled as *P.* cf. *caffordi* on plate explanation).
  Lower Carboniferous: Russia: Upper Priamur
Platycrinites schmidtii Stuckenberg, 1875, nomen correctum.
  Platycrinus schmidtii Stuckenberg, 1875. Yakovlev, *in* Yakovlev and Ivanov, 1956, p. 76, Pl. 18, fig. 2.
  Permian: Russia: Timan Ridge
Platycrinites(?) *tuberculatus* Yakovlev, *in* Yakovlev and Ivanov, 1956, nomen correctum.
  Platycrinus(?) *tuberculatus* Yakovlev, *in* Yakovlev and Ivanov, 1956, p. 49, Pl. 12, figs. 8, 9.
  Carboniferous (C_3), Upper Coal succession: Russia: Kara-Chatir Range
Platycrinites spp.
  Ivanova, 1958, p. 134, Pl. 18, fig. 7.
    Carboniferous (C_2): Russia: Moscow Syncline
  Yeltyschewa and Dubatolova, 1967, p. 29, Pl. 6, figs. 12-15.
    Lower Carboniferous: Russia: Upper Priamur, Kazakhstan, Altay, Central Asia, Ural Mountains
  Elias, 1957, p. 386, Pl. 41, figs. 8-10 (*non* Pl. 4 as given in text, typographical error).
    Mississippian, Chesterian, Redoak Hollow Formation: United States: Oklahoma
  Ubaghs, 1953, p. 743, fig. 99.
    No age given
  Platycrinus sp. Sampelayo, 1946, p. 4, Pl. 1.
    Carboniferous: Spain
  Platycrinus sp. Plummer, 1950, p. 88, Pl. 19, fig. 14. Crinoid columnal. Heuer, 1958, p. 43, Pl. 3, fig. 14.
    Pennsylvanian, Strawn Group: United States: Texas
  Platycrinus sp. Termier and Termier, 1950, p. 86, Pl. 227, figs. 11-17 (*non* fig. 18 as given on p. 86).
    Carboniferous, Visean: Morocco
  Platycrinus sp. Termier and Termier, 1950, p. 87, Pl. 211, figs. 7-15 (*non* figs. 71-75 as given on p. 87).
    Carboniferous, Westphalian: Morocco
  Platycrinus sp. Termier and Termier, 1950, p. 87, Pl. 211, fig. 16.
    Carboniferous, Westphalian: Morocco
  Platycrinus sp. Termier and Termier, 1950, p. 87, Pl. 212, figs. 14, 15.
    Carboniferous, Strunian: Morocco
  Platycrinus sp. Termier and Termier, 1950, p. 87, Pl. 215, figs. 2-7.
    Carboniferous, Westphalian: Morocco
  Platycrinus sp. Termier and Termier, 1950, p. 206, Pl. 211, figs. 17, 18.

Carboniferous, Strunian: Morocco
Platycrinus sp. Termier and Termier, 1950, p. 206, Pl. 211, figs. 19-22.
Carboniferous, Strunian: Morocco
Platycrinus sp. Termier and Termier, 1950, p. 208, Pl. 212, figs. 16-19.
Carboniferous, Westphalian: Morocco
Platycrinus sp. Termier and Termier, 1950, p. 214, Pl. 215, figs. 10, 11.
Carboniferous, upper Visean: Morocco
Platycrinus? sp. Sieverts-Doreck, 1951a, p. 134, Pl. 9, fig. 10.
Lower Carboniferous, Erdbacker Limestone: Germany
Platycrinus sp. Sangil, 1955, p. 251, fig. 21.
No age given
Platycrinus sp. Yakovlev, *in* Yakovlev and Ivanov, 1956, p. 77, Pl. 13, fig. 4.
Permian: Russia: Timan Ridge
Platycrinus sp. Yakovlev and Ivanov, 1956, p. 33, fig. 8.
Carboniferous: Russia: Moscow Basin
Platycrinus sp. Yakovlev, 1961, p. 419, Pl. 2, fig. 10.
Lower Carboniferous: Russia: Donetz Basin
Platycrinus, Actinocrinus Hernández-Sampelayo, 1954, p. 25, Pl. 3, fig. 5.
Carboniferous: Spain

Platycrinus stellatus Eichwald, 1860 = *Pentagonocyclicus stellatus*

PLATYPARALLELUS Moore and Jeffords, 1968. *P. parilis* Moore and Jeffords, 1968. Moore and Jeffords, 1968, p. 43.
Mississippian, Osagean

*Platyparallelus parilis Moore and Jeffords, 1968, p. 43, Pl. 4, fig. 3.
Mississippian, Osagean, New Providence Shale: United States: Kentucky

PLATYPLATEIUM Moore and Jeffords, 1968. *P. texanum* Moore and Jeffords, 1968. Moore and Jeffords, 1968, p. 44.
Mississippian, Osagean-Pennsylvanian, Desmoinesian

Platyplateium providencense Moore and Jeffords, 1968, p. 45, Pl. 5, fig. 4.
Mississippian, Osagean, New Providence Shale: United States: Kentucky

*Platyplateium texanum Moore and Jeffords, 1968, p. 44, Pl. 4, figs. 4-6; Pl. 5, figs. 5-8.
Pennsylvanian, Desmoinesian, Millsap Lake Formation: United States: Texas

Platyplateium sp. Moore and Jeffords, 1968, p. 45, Pl. 5, fig. 9.
Pennsylvanian, Desmoinesian, Millsap Lake Formation: United States: Texas

PLATYSTELA Moore and Jeffords, 1968. *P. proiecta* Moore and Jeffords, 1968. Moore and Jeffords, 1968, p. 41.
Devonian, Erian

*Platystela proiecta Moore and Jeffords, 1968, p. 42, Pl. 3, figs. 8-10.
Devonian, Erian, Wanakah Shale: United States: New York

PLUMMERANTERIS Moore and Jeffords, 1968. *P. sansaba* Moore and Jeffords, 1968. Moore and Jeffords, 1968, p. 78.
Pennsylvanian, Atokan

*Plummeranteris sansaba Moore and Jeffords, 1968, p. 78, Pl. 24, figs. 10-14.
Pennsylvanian, Atokan, Marble Falls Limestone: United States: Texas

PLUSSACRINUS Yeltyschewa, 1957.   *P. flabellum Yeltyschewa, 1957.
    Yeltyschewa, 1957, p. 228.
    Middle Ordovician

Plussacrinus dentatus Yeltyschewa, 1957, p. 229, Pl. 3, fig. 3; Fig. 9.
    Middle Ordovician, Ekhinosteritoviy Limestone: Russia: Leningrad
        Province

*Plussacrinus flabellum Yeltyschewa, 1957, p. 228, Pl. 3, fig. 2; Fig. 8.
    Middle Ordovician, Ievskie layer: Russia: Estonia

PODOLIOCRINUS Yeltyschewa, 1957.   *P. nikiforovae Yeltyschewa, 1957.
    Yeltyschewa, 1957, p. 227.
    Silurian, Ludlovian

*Podoliocrinus nikiforovae Yeltyschewa, 1957, p. 227, Pl. 3, fig. 1;
    Fig. 7; 1964b, p. 80, Pl. 15, fig. 20. Schewtschenko, 1966,
    p. 149, Pl. 6, figs. 3-8; Fig. 19. Yeltyschewa, 1968, p. 40,
    Pl. 3, figs. 5-25.
    Silurian, Ludlovian, Borschovskian horizon, Shishkatskian horizon:
        Russia: Tien-Shan Mountains, Podolsk

Podolithus Sardeson, 1908. Moore and Laudon, 1943b, p. 104.
    Ordovician

Poteriocrinites cf. conicus Philip, 1836.
    Poteriocrinus cf. conicus Philip, 1836. Termier and Termier, 1950,
        p. 89, Pl. 218, figs. 37, 38.
    Carboniferous, Moscovian: Morocco

Poteriocrinites crassus Miller, 1821. Cain, 1968, p. 196, fig. 1-bii.
    Poteriocrinus crassus Miller, 1821. Termier and Termier, 1950, p. 89,
        Pl. 218, figs. 2-10; Pl. 226, fig. 9; Pl. 215, fig. 1; Pl. 223,
        fig. 1; Pl. 225, figs. 1-3, 6-9 (typographical error, given as
        Pl. 225, figs. 1-3, 6-7, on p. 89). Almela and Sanz, 1958, Pl. 18,
        fig. 1.
    Carboniferous, Visean: England; Morocco; Spain

Poteriocrinites minutus Hernández-Sampelayo, 1954, nomen correctum.
    Poteriocrinus minutus Hernández-Sampelayo, 1954, p. 25, Pl. 3, fig. 4.
    Carboniferous: Spain

Poteriocrinites spp.
    Poteriocrinus sp. Termier and Termier, 1950, p. 216, Pl. 216,
        figs. 5, 6.
    Carboniferous, Visean: Morocco
    Poteriocrinus sp. Termier and Termier, 1950, p. 220, Pl. 218,
        figs. 11, 12.
    Devonian, Famennian: Morocco
    Poteriocrinus sp. Termier and Termier, 1950, p. 220, Pl. 218,
        figs. 33-36.
    Carboniferous, Westphalian: Morocco
    Poteriocrinus sp. Hernández-Sampelayo, 1954, p. 25, Pl. 3, fig. 3.
    Carboniferous: Spain
    Poteriocrinus sp. Sangil, 1955, p. 251, fig. 2D.
    No age given

Poteriocrinus biplex Eichwald, 1860 = Pentagonocyclicus biplex

PREPTOPREMNUM Moore and Jeffords, 1968.   *P. rugosum Moore and Jeffords,
    1968. Moore and Jeffords, 1968, p. 80.
    Pennsylvanian, Desmoinesian-Virgilian

Preptopremnum laeve Moore and Jeffords, 1968, p. 81, Pl. 27, fig. 11.
    Pennsylvanian, Desmoinesian, Mingus Shale: United States: Texas

*Preptopremnum rugosum Moore and Jeffords, 1968, p. 81, Pl. 27, figs. 1-9;
    Pl. 28, fig. 1. Moore and others, 1968a, p. 19, Pl. 3, figs. 4-8;

Pl. 4, fig. 7. Jeffords and Miller, 1968, p. 12, Pl. 2, figs. 1-10; Figs. 1B, 2B, 3, 4G, H.
Pennsylvanian, Virgilian, Chaffin Limestone, Gunsight Limestone, Wayland Shale: United States: Texas

Proctothylacocrinus esseri Kesling, 1965. Moore and Jeffords, 1968, p. 35, Pl. 2, fig. 6.
Devonian, Erian, Silica Formation: United States: Ohio

Proctothylacocrinus longus Kier, 1952. Moore and Jeffords, 1968, p. 34, Pl. 2, figs. 4, 5.
Devonian, Erian, Wanakah Shale: United States: New York

RAMULICRINUS Stukalina, 1968. *Pentagonopentagonalis multipartitus Yeltyschewa, 1955. Stukalina, 1968c, p. 89.
Middle-Upper Ordovician

Ramulicrinus hrustalnjensis (Yeltyschewa and Stukalina), 1963. Stukalina, 1968c, p. 89.
Pentagonopentagonalis hrustalnjensis Yeltyschewa and Stukalina, 1963, p. 44, Pl. 1, fig. 8; Fig. 13. Stukalina, 1968c, p. 89.
Upper Ordovician: Russia: Taymyr

Ramulicrinus multipartitus (Yeltyschewa), 1955. Stukalina, 1968c, p. 89.
Pentagonopentagonalis multipartitus Yeltyschewa, 1955a, p. 44, Pl. 37, fig. 3; 1960, p. 12, Pl. 2, figs. 10, 11. Yeltyschewa and Stukalina, 1963, p. 37, Pl. 2, figs. 15-18; Fig. 7. Yeltyschewa, 1964b, p. 75, Pl. 15, fig. 18. Stukalina, 1968c, p. 89.
Middle-Upper Ordovician: Russia: Siberian Platform, Taymyr

Rhodocrinites canaliculatus Goldfuss, 1826 = *Myelodactylus canaliculatus*

Rhodocrinites skourensis Delpey, 1939, nomen correctum.
Rhodocrinus skourensis Delpey, 1939. Termier and Termier, 1950, p. 83, Pl. 208, figs. 5, 6.
Carboniferous, upper Visean: Morocco

Rhodocrinites cf. *verus* Miller, 1821. Termier and Termier, 1950, Pl. 83; Pl. 208, figs. 13-15 (incorrectly given as *Rhodocrinus* cf. *verus*).
Carboniferous, Moscovian-Westphalian: Morocco

Rhodocrinites spp.
Rhodocrinus sp. Termier and Termier, 1950, p. 200, Pl. 208, figs. 7-9.
Carboniferous, upper Visean: Morocco
Rhodocrinus sp. Termier and Termier, 1950, p. 200, Pl. 208, figs. 10-12.
Carboniferous, upper Visean: Morocco
Rhodocrinus sp. Termier and Termier, 1950, p. 200, Pl. 208, figs. 25-28.
Carboniferous, Westphalian: Morocco
Rhodocrinus? sp. Termier and Termier, 1950, p. 220, Pl. 218, figs. 13, 14.
Carboniferous, Strunian: Morocco
Rhodocrinus? sp. Termier and Termier, 1950, p. 222, Pl. 219, figs. 21-23.
Carboniferous, upper Visean: Morocco

RHYSOCAMAX Moore and Jeffords, 1968. *R. cristata Moore and Jeffords, 1968. Moore and Jeffords, 1968, p. 60.
Mississippian, Osagean

Rhysocamax cristata Moore and Jeffords, 1968, p. 60, Pl. 12, figs. 3, 4.
Mississippian, Osagean, Burlington Limestone: United States: Iowa

Rhysocamax grandis Moore and Jeffords, 1968, p. 60, Pl. 12, fig. 9.
Mississippian, Osagean, Burlington Limestone: United States: Iowa

Rhysocamax tuberculata Moore and Jeffords, 1968, p. 60, Pl. 12, fig. 8.
  Mississippian, Osagean, Burlington Limestone: United States: Iowa
Ristnacrinus angulatus Yeltyschewa, 1966, p. 55, Pl. 2, figs. 5-10.
  Middle Ordovician: Russia: Estonia
Ristnacrinus marinus Öpik, 1934. Yeltyschewa, 1966, p. 53, Pl. 2,
    figs. 15-22, 27.
  Middle Ordovician: Russia: Estonia, Baltic Sea region

SALAGASTIANA Rusconi, 1952. *S. simetrica* Rusconi, 1952. Rusconi, 1951,
    p. 255, nomen nudum; 1952, p. 38.
  Ordovician?
Salagastiana simetrica Rusconi, 1952. Rusconi, 1951, p. 257, nomen
    nudum; 1952, p. 38, figs. 10, 11. Sieverts-Doreck, 1957b, p. 152,
    figs. 1, 2.
  Ordovician or younger (Sieverts-Doreck, 1957), Salagastense horizon:
    Argentina
SCELIDIOPTERNIX Moore and Jeffords, 1968. *S. norops* Moore and Jeffords,
    1968. Moore and Jeffords, 1968, p. 74.
  Silurian, Niagaran
Scelidiopternix norops Moore and Jeffords, 1968, p. 74, Pl. 21,
    figs. 11, 12.
  Silurian, Niagaran, Waldron Shale: United States: Indiana
Schizocrinus kuckersiensis Yeltyschewa, 1966, p. 57, Pl. 1, figs. 13-17.
  Middle Ordovician: Russia: Estonia
Schizocrinus aff. *nodosus* Hall, 1847. Yeltyschewa, 1966, p. 56, Pl. 1,
    figs. 18, 19.
  Middle Ordovician: Russia: Estonia
Schizocrinus cf. *nodosus* Hall, 1847. Termier and Termier, 1950, p. 93,
    Pl. 231, fig. 11.
  Ordovician: Morocco
Schizocrinus sp. Sahni and Gupta, 1965b, p. 247.
  Silurian: India: Kashmir
Scillus Gregorio, 1930. Moore and Laudon, 1943b, p. 105.
  Permian
Scyphocrinites elegans Zenker, 1833.
  Scyphocrinus elegans Zenker, 1833. Termier and Termier, 1950, p. 85,
    Pl. 213, figs. 6-13. Sieverts-Doreck, 1964, p. 3, Pl. 2.
  Silurian, Gotlandian: Morocco
Scyphocrinites marianne Yakovlev, 1953.
  Scyphocrinus marianne Yakovlev, 1953, p. 30, Pl. 3, fig. 3; 1956a,
    p. 106, fig. 39, no. 3; 1958, p. 95, fig. 39, no. 3; 1964a, p. 101,
    fig. 41, no. 3.
  Silurian: Russia
Scyphocrinites spp. nomen correctum.
  Loboliths. Yakovlev, 1953, p. 27, Pl. 2, figs. 1-6; Fig. 4; 1956a,
    p. 110, figs. 41, nos. 1-4; 42, nos. 3, 4; 1958, p. 95, figs. 41,
    nos. 1-4; 42, nos. 3, 4; 1964a, p. 99, figs. 42, nos. 3, 4;
    44, nos. 1-4 (root bulbs).
  Silurian: Russia: Kazakhstan; Czechoslovakia
  Scyphocrinus sp. Termier and Termier, 1950, p. 204, Pl. 210,
    figs. 7-9.
  Silurian, Gotlandian: Morocco
  Scyphocrinus sp. Sangil, 1955, p. 251, fig. 2E.
  No age given

SIDERICRINUS Stukalina, 1968. *S. depressus Stukalina, 1968. Stukalina,
    1968c, p. 84.
  Ordovician, Caradocian

*Sidericrinus depressus Stukalina, 1968c, p. 84, figs. 2v, 2s.
  Upper Ordovician, Caradocian: Russia: Kazakhstan

SOKOLOVICRINUS Yeltyschewa, 1968. *S. dnestrowensis Yeltyschewa, 1968.
    Yeltyschewa, 1968, p. 42.
  Ordovician-Silurian

Sokolovicrinus bifidus Yeltyschewa, 1968, p. 43, Pl. 5, fig. 16.
  Silurian, Skalskian horizon: Russia: Podolsk

*Sokolovicrinus dnestrowensis Yeltyschewa, 1968, p. 42, Pl. 5,
    figs. 17-19.
  Silurian, Skalskian horizon: Russia: Podolsk

Sokolovicrinus estonus (Yeltyschewa). Yeltyschewa, 1968, p. 42.
  Lower Ordovician: Russia

Sokolovicrinus lesnikovae (Yeltyschewa), 1964. Yeltyschewa, 1968, p. 42.
  Pentagonocyclicus lesnikovae Yeltyschewa, 1964a, p. 74, Pl. 2,
    fig. 15. Yeltyschewa, 1968, p. 42.
  Lower Ordovician: Russia: Estonia

Sphenocrinus iruensis Yeltyschewa, 1964a, p. 64, Pl. 2, figs. 16-18.
  Lower Ordovician: Russia: Estonia

Sphenocrinus multisulcatus Yeltyschewa, 1964a, p. 63, Pl. 2, fig. 9.
  Lower Ordovician: Russia: Estonia

Sphenocrinus obtusus Eichwald, 1860. Yeltyschewa, 1964a, p. 61, Pl. 2,
    figs. 1-8; Pl. 4, fig. 25 (non Pl. 1, figs. 1-8 as given in text).
  Lower Ordovician: Russia: Estonia

Sphenocrinus quinquevalatus Yeltyschewa, 1964a, p. 63, Pl. 2, fig. 11.
  Lower Ordovician: Russia: Estonia

Sphenocrinus rarisulcatus Yeltyschewa, 1964a, p. 62, Pl. 2, figs. 12-14;
    Pl. 4, fig. 26.
  Lower Ordovician: Russia: Estonia

SQUAMEOCRINUS Stukalina, 1968. *Pentagonopentagonalis privus
    Yeltyschewa, 1964. Stukalina, 1968c, p. 83.
  Lower Ordovician

*Squameocrinus privus (Yeltyschewa), 1964. Stukalina, 1968c, p. 83.
  Pentagonopentagonalis privus Yeltyschewa, 1964a, p. 69, Pl. 3,
    figs. 10, 11.
  Lower Ordovician: Russia: Estonia

Stereobrachiocrinus pustullosus Mather, 1915. Moore and Laudon, 1944,
    P. 209, Pl. 79, fig. 12 (misspelled as Stereobrachicrinus
    pustullosus).
  Pennsylvanian, Morrowan: United States: Oklahoma

STIBEROSTAURUS Moore and Jeffords, 1968. *S. aestimatus Moore and
    Jeffords, 1968. Moore and Jeffords, 1968, p. 61.
  Mississippian, Osagean

*Stiberostaurus aestimatus Moore and Jeffords, 1968, p. 61, Pl. 12,
    figs. 5-7; Pl. 13, figs. 4, 5.
  Mississippian, Osagean, New Providence Shale: United States: Kentucky

Streptocrinus crotalurus Bather, 1893. Stukalina, 1967, p. 203,
    fig. 4, no. 3.
  Silurian: Gotland

Syndetocrinus(?) *natus* Stukalina, 1965b, p. 135, Pl. 1, figs. 1-6; figs. 1, 2.
    Upper Silurian, Karaespinskian horizon: Russia: Kazakhstan

Taxocrinus cf. *colletti* White, 1880. Termier and Termier, 1950, p. 92, Pl. 224, fig. 32 (*non* figs. 30, 31 as given on p. 92).
    Lower Devonian: Morocco

Taxocrinus sp. Sangil, 1955, p. 253, fig. 3B.
    No age given

Tebagacrinus sp. Termier and Termier, 1958, p. 53, Pl. 1, figs. d-g.
    Permian: Tunisia

TETRAGONOCRINUS Yeltyschewa, 1964. *Goniaster pygmaeus* Eichwald, 1860. Yeltyschewa, 1964a, p. 65.
    Ordovician

Tetragonocrinus pygmaeus (Eichwald), 1860. Yeltyschewa, 1964a, p. 65, Pl. 1, figs. 8-14; Pl. 4, figs. 18, 19.
    Goniaster pygmaeus Eichwald, 1860. Yeltyschewa, 1964a, p. 65.
    Lower Ordovician: Russia: Estonia

Tetragonocrinus quadratus Stukalina and Tuyutyan, 1967, p. 75, figs. 7-9, 12.
    Upper Ordovician, Anderkenskian horizon: Russia: Kazakhstan

TETRAGONOCYCLICUS Yeltyschewa, 1956, nomen vetitum.
    Yeltyschewa, 1956, p. 42, figs. 1-2E, 2e-g; 1959, p. 233, fig. 1-2E; 1964b, p. 75, fig. 114-2E. Dubatolova, 1964, p. 9, fig. 1-2E.
    Lower-Middle Devonian

Tetragonocyclicus deflexus Dubatolova, 1964, p. 67, Pl. 8, fig. 17; Pl. 9, figs. 1, 2.
    Lower Devonian: Russia: Kuznetz Basin

Tetragonocyclicus filicatus Dubatolova, 1964, p. 65, Pl. 8, figs. 7-9.
    Lower Devonian: Russia: Kuznetz Basin

Tetragonocyclicus infinitus Dubatolova, 1964, p. 64, Pl. 8, fig. 6.
    Lower Devonian: Russia: Kuznetz Basin

Tetragonocyclicus fimbriatus Dubatolova, 1964, p. 66, Pl. 8, figs. 13-16.
    Lower Devonian: Russia: Kuznetz Basin

Tetragonocyclicus fuscus Dubatolova, 1964, p. 67, Pl. 9, figs. 3, 4.
    Lower Devonian: Russia: Kuznetz Basin

Tetragonocyclicus ignotus Dubatolova, 1964, p. 69, Pl. 9, fig. 7.
    Devonian, Eifelian: Russia: Kuznetz Basin

Tetragonocyclicus indefinitus Dubatolova, 1967, p. 38, Pl. 6, figs. 16-18; Fig. 5.
    Devonian, Eifelian: Russia: Tas-Hayahtah Mountains

Tetragonocyclicus multiforabilis Dubatolova, 1964, p. 69, Pl. 9, figs. 8-10; Fig. 8.
    Devonian, Givetian: Russia: Kuznetz Basin

Tetragonocyclicus permirus Dubatolova, 1964, p. 64, Pl. 8, fig. 5.
    Lower Devonian: Russia: Kuznetz Basin

Tetragonocyclicus perplexus Dubatolova, 1964, p. 68, Pl. 9, figs. 5, 6.
    Lower Devonian: Russia: Kuznetz Basin

TETRAGONOELLIPTICUS Yeltyschewa, 1956, nomen vetitum.
    Yeltyschewa, 1956, p. 42, fig. 1-2F; 1959, p. 233, fig. 1-2F; 1964b p. 75, fig. 114-2F. Dubatolova, 1964, p. 9, fig. 1-2F.
    Theoretical genus, no known species.

TETRAGONOHEXAGONALIS Yeltyschewa, 1956, nomen vetitum.
Yeltyschewa, 1956, p. 42, fig. 1-2D; 1959, p. 233, fig. 1-2D; 1964b,
p. 75, fig. 114-2D. Dubatolova, 1964, p. 9, fig. 1-2D.
Theoretical genus, no known species.

TETRAGONOPENTAGONALIS Yeltyschewa, 1956, nomen vetitum.
Yeltyschewa, 1956, p. 42, fig. 1-2C; 1959, p. 233, fig. 1-2C; 1964b,
p. 75, fig. 114-2B. Dubatolova, 1964, p. 9, fig. 1-2B.
Theoretical genus, no known species.

TETRAGONOTETRAGONALIS Yeltyschewa, 1956, nomen vetitum.
Yeltyschewa, 1956, p. 42, figs. 1-2B, 2a-d; 1959, p. 233, fig. 1-2B;
1964b, p. 75, fig. 114-2B. Dubatolova, 1964, p. 9, fig. 1-2B.
Lower Silurian-Devonian

Tetragonotetragonalis altaicus Yeltyschewa, *in* Avrov and Stukalina, 196 ,
p. 27, fig. 1.
Lower Silurian: Russia: Altay Mountains

Tetragonotetragonalis gratus Dubatolova, 1967, p. 37, Pl. 6, figs. 11- ;
Fig. 4.
Devonian, Eifelian: Russia: Tas-Hayahtah Mountains

Tetragonotetragonalis nudus Dubatolova, 1964, p. 62, Pl. 8, figs. 3, 4;
Fig. 7.
Devonian, Eifelian: Russia: Kuznetz Basin

Tetragonotetragonalis nutabundus Dubatolova, 1964, p. 63, Pl. 8,
figs. 10-12.
Devonian, Eifelian: Russia: Kuznetz Basin

Tetragonotetragonalis quadrihamatus Yeltyschewa, 1963 = *Dvortsowaecrinus quadrihamatus*

TETRAGONOTRIGONALIS Yeltyschewa, 1956, nomen vetitum.
Yeltyschewa, 1956, p. 42, fig. 1-2A; 1959, p. 233, fig. 1-2A; 1964b,
p. 75, fig. 114-2A. Dubatolova, 1964, p. 9, fig. 2A.
Theoretical genus, no known species.

TRIGONOCYCLICUS Yeltyschewa, 1956, nomen vetitum.
Yeltyschewa, 1956, p. 42, fig. 1-1E; 1959, p. 233, fig. 1-1E; 1964b,
p. 75, fig. 114-1E. Dubatolova, 1964, p. 9, fig. 1-1E.
Middle Ordovician-Lower Devonian

Trigonocyclicus acceptus Dubatolova, 1964, p. 61, Pl. 8, fig. 2; 1967,
p. 36, Pl. 6, fig. 10; Fig. 3.
Lower Devonian: Russia: Kuznetz Basin, Tas-Hayahtah Mountains

Trigonocyclicus vajgatschensis Yeltyschewa and Stukalina, 1963, p. 56,
Pl. 2, figs. 19-21; Fig. 22. Yeltyschewa, 1966, p. 57, Pl. 3,
figs. 9-11.
Middle Ordovician: Russia: Estonia, Vaigach

TRIGONOELLIPTICUS Yeltyschewa, 1956, nomen vetitum.
Yeltyschewa, 1956, p. 42, fig. 1-1F; 1959, p. 233, fig. 1-1F; 1964b,
p. 75, fig. 114-1F. Dubatolova, 1964, p. 9, fig. 1-1F.
Theoretical genus, no known species.

TRIGONOHEXAGONALIS Yeltyschewa, 1956, nomen vetitum.
Yeltyschewa, 1956, p. 42, fig. 1-1D; 1959, p. 233, fig. 1-1D; 1964b,
p. 75, fig. 114-1D. Dubatolova, 1964, p. 9, fig. 1-1D.
Theoretical genus, no known species.

TRIGONOPENTAGONALIS Yeltyschewa, 1956, nomen vetitum.
Yeltyschewa, 1956, p. 42, fig. 1-1C; 1959, p. 233, fig. 1-1C; 1964b,
p. 75, fig. 114-1C. Dubatolova, 1964, p. 9, fig. 1-1C.
Theoretical genus, no known species.

TRIGONOTETRAGONALIS Yeltyschewa, 1956, nomen vetitum.
 Yeltyschewa, 1956, p. 42, fig. 1-1B; 1959, p. 233, fig. 1-1B; 1964b,
 p. 75, fig. 114-1B. Dubatolova, 1964, p. 9, fig. 1-1B.
 Theoretical genus, no known species.

TRIGONOTRIGONALIS Yeltyschewa, 1956, nomen vetitum.
 Yeltyschewa, 1956, p. 42, fig. 1-1A; 1959, p. 233, fig. 1-1A; 1964b,
 p. 75, fig. 114-1A. Dubatolova, 1964, p. 8, fig. 1-1A.
 Lower Carboniferous

Trigonotrigonalis asymmetricus Yeltyschewa, *in* Dubatolova and Shao,
 1959, p. 42, Pl. 1, fig. 1.
 Lower Carboniferous, Visean: China: Hunan Province; Russia:
 Kazakhstan

Vasocrinus alveatus Schewtschenko, 1966, p. 132, Pl. 4, figs. 8, 15;
 Fig. 3.
 Lower Devonian, Pandzhrutskian horizon: Russia: Tien-Shan Mountains

Vasocrinus tuberculifer Schewtschenko, 1966, p. 132, Pl. 4, figs. 3, 9;
 Fig. 2.
 Lower Devonian, Pandzhrutskian horizon: Russia: Tien-Shan Mountains

Vasocrinus yeltyschewae Schewtschenko, 1966, p. 131, Pl. 4, fig. 10;
 Fig. 1.
 Lower Devonian, Pandzhrutskian horizon: Russia: Tien-Shan Mountains

WANAKASTAURUS Moore and Jeffords, 1968. **W.* delicatus* Moore and Jeffords,
 1968. Moore and Jeffords, 1968, p. 53.
 Devonian, Erian

**Wanakastaurus delicatus* Moore and Jeffords, 1968, p. 53, Pl. 8,
 figs. 9, 10.
 Devonian, Erian, Wanakah Shale: United States: New York

Woodocrinus sp. Hernández-Sampelayo, 1954, p. 25, Pl. 3, fig. 1c.
 Carboniferous: Spain

Xenobasis Faber, 1929. Moore and Laudon, 1943b, p. 105.
 Ordovician

ZERAVSCHANOCRINUS Schewtschenko, 1966. **Z.* barbulatus* Schewtschenko,
 1966. Schewtschenko, 1966, p. 153.
 Lower Devonian

Zeravschanocrinus apiculatus Schewtschenko, 1966, p. 158, Pl. 4, fig. 12;
 Fig. 28.
 Lower Devonian, Pandzhrutskian horizon: Russia: Tien-Shan Mountains

Zeravschanocrinus arenosus Schewtschenko, 1966, p. 156, Pl. 4, fig. 11;
 Fig. 25.
 Lower Devonian, Pandzhrutskian horizon: Russia: Tien-Shan Mountains

**Zeravschanocrinus barbulatus* Schewtschenko, 1966, p. 154, Pl. 4,
 figs. 6, 7; Fig. 23.
 Lower Devonian, Pandzhrutskian horizon: Russia: Tien-Shan Mountains

Zeravschanocrinus binarius Schewtschenko, 1966, p. 156, Pl. 4,
 figs. 4, 5; Fig. 26.
 Lower Devonian, Pandzhrutskian horizon: Russia: Tien-Shan Mountains

Zeravschanocrinus incubus Schewtschenko, 1966, p. 157, Pl. 4, fig. 17;
 Fig. 27.
 Lower Devonian, Pandzhrutskian horizon: Russia: Tien-Shan Mountains

Zeravschanocrinus quinquelobus Schewtschenko, 1966, p. 155, Pl. 4,
 fig. 14; Fig. 24.
 Lower Devonian, Pandzhrutskian horizon: Russia: Tien-Shan Mountains

SECTION 2. UNIDENTIFIED COLUMNALS

Artejos de crinoides. Larrauri y Mercadillo, 1944, Pl. 7, fig. 3.
  Devonian: Spanish Sahara

Bras de Crinoide non pinnulé Flexibilia. Termier and Termier, 1950,
  p. 222, Pl. 219, figs. 29-31.
  Devonian, Eifelian: Morocco

Bras de Crinoide portant des parasites. Termier and Termier, 1950,
  p. 222, Pl. 219, fig. 33.
  Devonian, Emsian: Morocco

Columnaire de Camerata. Termier and Termier, 1950, p. 204, Pl. 210,
  fig. 18.
  Devonian, Eifelian: Morocco

Columnaires de Camerata. Termier and Termier, 1950, p. 236, Pl. 226,
  figs. 1-4.
  Silurian, upper Gotlandian: Morocco

Columnaire de Diplobathra. Termier and Termier, 1950, p. 204, Pl. 210,
  figs. 13, 14.
  Devonian, Siegenian: Morocco

Columnaire de Diplobathra. Termier and Termier, 1950, p. 204, Pl. 210,
  fig. 15.
  Devonian, Emsian: Morocco

Columnaire de Diplobathra. Termier and Termier, 1950, p. 204, Pl. 210,
  fig. 17.
  Lower Devonian: Morocco

Columnaire de Flexibilia indéterminé. Termier and Termier, 1950, p. 210,
  Pl. 213, figs. 34, 35.
  Middle Devonian: Morocco

Columnaire de Monobathra indéterminé. Termier and Termier, 1950, p. 206,
  Pl. 211, fig. 44.
  Upper Gotlandian or Lower Devonian: Morocco

Columnaire de Monobathra indéterminé. Termier and Termier, 1950, p. 240,
  Pl. 288, figs. 25, 26.
  Carboniferous, upper Visean: Morocco

Columnaire d'un genre indéterminé. Termier and Termier, 1950, p. 238,
  Pl. 227, fig. 18.
  Carboniferous, Visean: Morocco

Columnaire indéterminée. Termier and Termier, 1950, p. 204, Pl. 210,
  figs. 19, 20.
  Devonian, Emsian: Morocco

Columnaire indéterminée. Termier and Termier, 1950, p. 204, Pl. 210,
  fig. 21.
  Carboniferous, Westphalian: Morocco

Columnaire indéterminée. Termier and Termier, 1950, p. 206, Pl. 211,
  figs. 45-48.
  Devonian, Siegenian: Morocco

Columnaire indéterminée. Termier and Termier, 1950, p. 204, Pl. 210, fig. 16.
Devonian, Emsian: Morocco

Columnaire indéterminée. Termier and Termier, 1950, p. 236, Pl. 226, figs. 32, 33.
Middle Devonian: Morocco

Columnaire indéterminée. Termier and Termier, 1950, p. 240, Pl. 228, figs. 27, 28.
Carboniferous, upper Visean: Morocco

Columnal a. Strimple, 1963d, p. 126, Pl. 12, fig. 1.
Silurian, Niagaran, Henryhouse Formation: United States: Oklahoma

Columnal b. Strimple, 1963d, p. 126, Pl. 12, fig. 2.
Silurian, Niagaran, Henryhouse Formation: United States: Oklahoma

Columnal c. Strimple, 1963d, p. 127, Pl. 12, fig. 4.
Silurian, Niagaran, Henryhouse Formation: United States: Oklahoma

Columnal d. Strimple, 1963d, p. 127, Pl. 11, fig. 4.
Silurian, Niagaran, Henryhouse Formation: United States: Oklahoma

Columnal dispar (Moore), 1939. Moore and Laudon, 1944, p. 209, Pl. 79, fig. 9.
Pennsylvanian, Missourian, Coffeyville Formation: United States: Kansas

Columnal ellipticus Strimple, 1962a, p. 4, figs. 1-3.
Pennsylvanian, Desmoinesian, Pumpkin Creek Limestone: United States: Oklahoma

Columnal excentricus (Moore), 1939. Moore and Laudon, 1944, p. 209, Pl. 79, fig. 14. Strimple, 1962e, p. 56, Pl. 7, figs. 3, 4.
Pennsylvanian, Desmoinesian, Mineral Coal cap, Oologah Limestone: United States: Kansas, Oklahoma

Columnal granulosus (Moore), 1939. Moore and Laudon, 1944, p. 209, Pl. 79, fig. 13.
Pennsylvanian, Desmoinesian, Savanna Formation: United States: Oklahoma

Columnal intermedia Byrne and Seeberger, 1942, p. 222, Pl. 1, figs. 3, 4.
Lower Permian, Americus Limestone: United States: Kansas

Columnal maxima Byrne and Seeberger, 1942, p. 223, Pl. 1, figs. 5, 6.
Lower Permian, Hughes Creek Shale: United States: Kansas

Columnal minima Byrne and Seeberger, 1942, p. 222, Pl. 1, figs. 1, 2.
Lower Permian, Hughes Creek Shale: United States: Kansas

Columnal nodosus Strimple, 1962e, p. 55, Pl. 7, figs. 1, 2.
Pennsylvanian, Desmoinesian, Oologah Limestone: United States: Oklahoma

Columnal quadrangulatus Strimple, 1962a, p. 3, figs. 4-7.
Pennsylvanian, Desmoinesian, Pumpkin Creek Limestone: United States: Oklahoma

Columnal spicatus (Moore), 1939. Moore and Laudon, 1944, p. 209, Pl. 79, fig. 20.
Pennsylvanian, Missourian, Coffeyville Formation: United States: Kansas

Columnal termination. Strimple, 1963d, fig. 3c.
Silurian?: United States: Oklahoma

Columnal termination. Strimple, 1963d, p. 27, fig. 3b.
Silurian?: United States: Oklahoma

Columnal termination. Strimple, 1963d, p. 26, Pl. 11, fig. 1.
Silurian?: United States: Oklahoma

Crinoid columnals. Talent, 1963, p. 49, Pl. 19, figs. 9-12.
Devonian, Wentworth Group: Australia: Victoria

Crinoid columnals. Schewtschenko, 1966, Pls. 7, 8.
Upper Silurian-Lower Devonian: Russia: Tien-Shan Mountains

Crinoid plates. Easton, 1962, p. 37, Pl. 4, figs. 6, 7.
Mississippian or Pennsylvanian, Cameron Creek Formation: United States: Montana

Crinoid plates. Easton, 1962, p. 37, Pl. 4, fig. 8.
Mississippian or Pennsylvanian, Alaska Bench Limestone: United States: Montana

Crinoid plates. Easton, 1962, p. 37, fig. 10.
Mississippian, Heath Formation: United States: Montana

Crinoid plates. Easton, 1962, p. 37, fig. 11.
Mississippian, Otter Formation: United States: Montana

Crinoid roots, coiled stems, etc. Wright, 1958, p. 327, Pl. 80, figs. 18, 20; Pl. 81.
Carboniferous: Scotland; Ireland

Crinoid stem. Tesmer, 1964, p. 3, fig. G.
Upper Devonian, Awkwright Group: United States: New York

Crinoid stems. Sahni and Gupta, 1965b, p. 248, fig. 1.
Silurian: India: Kashmir

Crinoid stems. Newell, and others, 1953, p. 165, Pl. 35, fig. 12.
Lower Permian, Copacabana Group: Peru

Crinoid stems and preserved root system. Prokop, 1967, p. 367, Pls. 1, 2.
Lower Devonian, Upper Koněprusy Beds: Bohemia

Crinoidal remains. Newell, and others, 1953, Pl. 23, fig. 47.
Permian, Guadalupian, Lamar Limestone: United States: Texas

Crinoides. Hernández-Sampelayo, 1954, p. 25, Pl. 3, fig. 1.
Carboniferous: Spain

Crotalocrinitidae gen. et sp. indet. Bouška, 1950, p. 21, Pl. 4, figs. 5, 5a.
Lower Devonian, Koněprusy Limestone: Bohemia

Dendrocrinoide sp. Termier and Termier, 1950, p. 200, Pl. 208, figs. 16-21.
Carboniferous, Visean-Namurian transition: Morocco

Dendrocrinoidea indéterminé. Termier and Termier, 1950, p. 200, Pl. 208, figs. 23, 24.
Carboniferous, Visean-Namurian transition: Morocco

Dendrocrinoidea indéterminé. Termier and Termier, 1950, p. 214, Pl. 215, figs. 17, 18.
Carboniferous, upper Visean: Morocco

Dendrocrinoidea indéterminé. Termier and Termier, 1950, p. 216, Pl. 216, figs. 7, 8.
Carboniferous, Namurian: Morocco

Dendrocrinoidea indéterminé. Termier and Termier, 1950, p. 230, Pl. 223, figs. 2, 3.
Carboniferous, upper Visean: Morocco

Diplobathra indéterminés. Termier and Termier, 1950, p. 230, Pl. 223, figs. 4-7.
Carboniferous, upper Visean: Morocco

Diplobathra? indéterminé. Termier and Termier, 1950, p. 230, Pl. 223, figs. 8, 9.
Carboniferous, Namurian: Morocco

Diplobathra indéterminé. Termier and Termier, 1950, p. 232, Pl. 224, fig. 35.
Devonian, Siegenian: Morocco

Diplobathra indéterminé. Termier and Termier, 1950, p. 232, Pl. 224, figs. 46, 47.
Carboniferous, upper Visean: Morocco

Diplobathra indéterminé. Termier and Termier, 1950, p. 232, Pl. 224, figs. 50, 51.
Devonian, Coblentzian: Morocco

Family, genus, and species unknown, species E. Webster and Lane, 1967, p. 31, Pl. 2, fig. 11 (root holdfast).
Permian, Wolfcampian, Bird Spring Formation: United States: Nevada

Family Platycrinidae—stem segment. Arendt and Hecker, 1964, p. 98, Pl. 15, fig. 8.
Middle Carboniferous: Russia: Moscow Basin

Fragment de la surface d'une tige d'*Arthroacantha*? sp. Termier and Termier, 1950, p. 214, Pl. 215, fig. 29.
Devonian, Siegenian: Morocco

Fragment de tige de Dendrocrinoidea. Termier and Termier, 1950, p. 236, Pl. 226, figs. 5, 6 (misspelled as Deudrocrinoidea).
Silurian, upper Gotlandian: Morocco

Fragment de tige indéterminé. Termier and Termier, 1950, p. 236, Pl. 226, figs. 7, 8.
Carboniferous, Visean: Morocco

Fragments de Crinoides indéterminés. Termier and Termier, 1950, p. 236, Pl. 226, figs. 13-22.
Devonian, Siegenian: Morocco

Gen. et sp. indet. Sieverts-Doreck, 1942, p. 230, fig. 7.
Permian: Australia: Tasmania

Indéterminé. Termier and Termier, 1950, p. 200, Pl. 208, fig. 22.
Carboniferous, Visean-Namurian transition: Morocco

Indéterminé. Termier and Termier, 1950, p. 206, Pl. 211, figs. 35, 36.
Devonian, Eifelian: Morocco

Indéterminé. Termier and Termier, 1950, p. 214, Pl. 215, figs. 14-16.
Carboniferous, Westphalian: Morocco

Indéterminé. Termier and Termier, 1950, p. 214, Pl. 215, figs. 19, 20.
Carboniferous, upper Visean: Morocco

Indéterminé. Termier and Termier, 1950, p. 216, Pl. 216, figs. 17, 18.
Middle Devonian: Morocco

Indéterminé. Termier and Termier, 1950, p. 216, Pl. 216, figs. 19, 20.
Carboniferous, Namurian: Morocco

Indéterminé. Termier and Termier, 1950, p. 216, Pl. 216, figs. 21, 22.
Carboniferous, Namurian: Morocco

Indéterminés. Termier and Termier, 1950, p. 222, Pl. 219, figs. 17-20.
Carboniferous, upper Visean: Morocco

Indéterminés. Termier and Termier, 1950, p. 222, Pl. 219, figs. 27, 28.
Devonian, Eifelian: Morocco

Indéterminé. Termier and Termier, 1950, p. 230, Pl. 223, figs. 16-18.
Carboniferous, upper Visean: Morocco

Indéterminé. Termier and Termier, 1950, p. 230, Pl. 223, figs. 25, 26.
Carboniferous, Westphalian: Morocco

Indéterminé. Termier and Termier, 1950, p. 232, Pl. 224, figs. 1-8.
Devonian, Siegenian: Morocco

Indéterminé. Termier and Termier, 1950, p. 232, Pl. 224, figs. 9, 10.
Devonian, Siegenian: Morocco

Indéterminé. Termier and Termier, 1950, p. 232, Pl. 224, figs. 11-17.
Devonian, Siegenian: Morocco

Indéterminé. Termier and Termier, 1950, p. 232, Pl. 224, figs. 18, 19
(*non* figs. 8-19 as given on p. 232).
Devonian, Famennian: Morocco

Indéterminés. Termier and Termier, 1950, p. 232, Pl. 224, figs. 20-29.
Devonian, Siegenian: Morocco

Indéterminé. Termier and Termier, 1950, p. 232, Pl. 224, figs. 30, 31.
Devonian, Siegenian: Morocco

Indéterminé. Termier and Termier, 1950, p. 232, Pl. 224, figs. 36, 37.
Devonian, Coblentzian: Morocco

Indéterminé. Termier and Termier, 1950, p. 232, Pl. 224, figs. 38-45.
Carboniferous, upper Visean: Morocco

Indéterminé. Termier and Termier, 1950, p. 234, Pl. 225, figs. 4, 5.
Carboniferous, upper Visean: Morocco

Indéterminé. Termier and Termier, 1950, p. 234, Pl. 225, figs. 10, 11.
Carboniferous, upper Visean: Morocco

Indéterminé. Termier and Termier, 1950, p. 234, Pl. 225, figs. 21, 22.
Carboniferous, lower Westphalian: Morocco

Indéterminé. Termier and Termier, 1950, p. 234, Pl. 225, figs. 23-28.
Carboniferous, lower Westphalian: Morocco

Monobathra indéterminé. Termier and Termier, 1950, p. 214, Pl. 215,
figs. 8, 9.
Middle Devonian: Morocco

Moule interne du canal pentagonal central d'une tige de Crinoide. Termier
and Termier, 1950, p. 222, Pl. 219, fig. 26.
No age given: Morocco

Pédoncule de *Arthroacantha*? sp. Termier and Termier, 1950, p. 224,
Pl. 220, figs. 5-7.
Devonian, Emsian: Morocco

Pédoncule de *Botryocrinus*. Termier and Termier, 1950, p. 210, Pl. 213,
figs. 1-5.
Lower Devonian: Morocco

Pédoncule de Crinoide pourteur d'expansions calciformes. Termier and
Termier, 1950, p. 224, Pl. 220, figs. 2-4.
Devonian, Emsian: Morocco

Pédoncule de Dendrocrinoidea indéterminé. Termier and Termier, 1950,
p. 240, Pl. 228, figs. 15, 16.
Carboniferous, upper Visean: Morocco

Pédoncules de Dendrocrinoides indéterminés. Termier and Termier, 1950, p. 240, Pl. 228, figs. 20-24.
Carboniferous, upper Visean: Morocco

Pédoncule de Flexibilia. Termier and Termier, 1950, p. 222, Pl. 219, fig. 6.
Devonian, Famennian: Morocco

Pédoncule de Monobathra indéterminé. Termier and Termier, 1950, p. 236, Pl. 226, figs. 34, 35.
Middle Devonian: Morocco

Pédoncules d'un Camerata indéterminé. Termier and Termier, 1950, p. 236, Pl. 226, figs. 25-29.
Middle Devonian: Morocco

Pédoncules d'un Dendrocrinoide indéterminé. Termier and Termier, 1950, p. 234, Pl. 225, figs. 12-16.
Carboniferous, lower Westphalian: Morocco

Pédoncule enroulé en spire plane indéterminé. Termier and Termier, 1950, p. 214, Pl. 215, figs. 12, 13.
Ordovician, Caradocian: Morocco

Pédoncule indéterminé. Termier and Termier, 1950, p. 206, Pl. 211, fig. 29.
Devonian, Emsian: Morocco

Pédoncule indéterminé. Termier and Termier, 1950, p. 206, Pl. 211, figs. 30-32.
Carboniferous, Strunian: Morocco

Pédoncule indéterminé. Termier and Termier, 1950, p. 210, Pl. 213, figs. 14, 15.
Lower Devonian: Morocco

Pédoncule indéterminé. Termier and Termier, 1950, p. 210, Pl. 213, figs. 30, 31.
Carboniferous, upper Visean: Morocco

Pédoncule indéterminé. Termier and Termier, 1950, p. 210, Pl. 213, figs. 32, 33.
Middle Devonian: Morocco

Pédoncule indéterminé. Termier and Termier, 1950, p. 220, Pl. 218, figs. 15, 17 (*non* fig. 16 as given on p. 220).
Carboniferous, Strunian: Morocco

Pédoncule indéterminé. Termier and Termier, 1950, p. 220, Pl. 218, fig. 53.
Carboniferous, Visean: Morocco

Pédoncule indéterminé. Termier and Termier, 1950, p. 222, Pl. 219, fig. 1.
Ordovician, Caradocian: Morocco

Pédoncule indéterminé. Termier and Termier, 1950, p. 222, Pl. 219, fig. 7.
Carboniferous, upper Visean: Morocco

Pédoncule indéterminé. Termier and Termier, 1950, p. 222, Pl. 219, figs. 8, 9.
Devonian, Famennian: Morocco

Pédoncule indéterminé. Termier and Termier, 1950, p. 222, Pl. 219, fig. 10.
Lower Devonian: Morocco

Pédoncule indéterminé. Termier and Termier, 1950, p. 222, Pl. 219, figs. 11, 12.
Devonian, Eifelian: Morocco

Pédoncule indéterminé. Termier and Termier, 1950, p. 222, Pl. 219, figs. 13, 14.
Carboniferous, upper Visean: Morocco

Pédoncule indéterminé. Termier and Termier, 1950, p. 222, Pl. 219, figs. 15, 16.
Carboniferous, upper Visean: Morocco

Pédoncule indéterminé. Termier and Termier, 1950, p. 224, Pl. 220, fig. 8.
Age unknown: Morocco

Pédoncule indéterminé. Termier and Termier, 1950, p. 224, Pl. 220, fig. 9.
Age unknown: Morocco

Pédoncule indéterminé. Termier and Termier, 1950, p. 224, Pl. 220, figs. 10, 11.
Age unknown: Morocco

Pédoncule indéterminé. Termier and Termier, 1950, p. 224, Pl. 220, figs. 12, 13.
Age unknown: Morocco

Pédoncule indéterminé. Termier and Termier, 1950, p. 234, Pl. 225, fig. 17.
Carboniferous, lower Westphalian: Morocco

Pédoncule indéterminé. Termier and Termier, 1950, p. 236, Pl. 226, fig. 10.
Carboniferous, Strunian: Morocco

Pédoncule indéterminé. Termier and Termier, 1950, p. 236, Pl. 226, figs. 30, 31.
Carboniferous, upper Visean: Morocco

Pédoncule indéterminé. Termier and Termier, 1950, p. 240, Pl. 228, fig. 17.
Carboniferous, upper Visean: Morocco

Pédoncule indéterminé. Termier and Termier, 1950, p. 240, Pl. 228, figs. 18, 19.
Carboniferous, upper Visean: Morocco

Pédoncule indéterminé, parasite. Termier and Termier, 1950, p. 234, Pl. 225, figs. 19, 20.
Carboniferous, lower Westphalian: Morocco

Pédoncule indéterminé probablement parasite. Termier and Termier, 1950, p. 234, Pl. 225, fig. 18.
Carboniferous, lower Westphalian: Morocco

Pédoncule indéterminé portant des incrustations de racines. Termier and Termier, 1950, p. 224, Pl. 220, figs. 18-20.
Age unknown: Morocco

Pédoncule indéterminé portant des insertions de cirres. Termier and Termier, 1950, p. 224, Pl. 220, figs. 21-25.
Age unknown: Morocco

Pédoncule indéterminé servant de supports aux racines d'individus jeunes. Termier and Termier, 1950, p. 224, Pl. 220, figs. 26-33.
Age unknown: Morocco

Pédoncule ou fragments de bras. Termier and Termier, 1950, p. 224,
Pl. 220, figs. 14-17.
Age unknown: Morocco

Pédoncule portant les racines d'un individu jeune. Termier and Termier,
1950, p. 222, Pl. 219, fig. 34.
Devonian, Eifelian: Morocco

Slab of loose columnals. Yeltyschewa, 1966, p. 68, Pl. 1, fig. 20.
Middle Ordovician: Russia: Estonia

Stem holdfast. Lane, 1963c, p. 1003, Pl. 128, fig. 2.
Mississippian, Osagean: United States: Indiana

Subaxial section d'une tige déformée d'un Crinoide. Termier and Termier,
1950, p. 238, Pl. 227, fig. 27.
Carboniferous, upper Visean: Morocco

Surface articulaire indéterminée. Termier and Termier, 1950, p. 230,
Pl. 223, fig. 11.
Carboniferous, upper Visean: Morocco

Twin-holed crinoid stems. Nelson, 1965, p. 19, figs. 3, 4.
Lower Devonian: northern Canada

Tige de Crinoide enroulé. Termier and Termier, 1950, p. 222, Pl. 219,
fig. 5.
Devonian, Gedinnian: Morocco

Tiges de Flexibilia indéterminé. Termier and Termier, 1950, p. 214,
Pl. 215, figs. 23-28.
Devonian, Famennian: Morocco

Tige indéterminé. Termier and Termier, 1950, p. 230, Pl. 223, fig. 10.
Carboniferous, upper Visean: Morocco

Tige indéterminée. Termier and Termier, 1950, p. 230, Pl. 223,
figs. 12, 13.
Carboniferous, Westphalian: Morocco

Tige indéterminée. Termier and Termier, 1950, p. 230, Pl. 223,
figs. 14, 15.
Carboniferous, upper Visean: Morocco

Unidentified columnals. Yakovlev, *in* Yakovlev and Ivanov, 1956, Pl. 10,
figs. 28-30, 32.
Lower Permian: Russia: Ural Mountains

Unidentified columnals. Elias, 1957, p. 384, Pl. 41, figs. 16-20.
Mississippian, Chesterian, Redoak Hollow Formation: United States:
Oklahoma

Unidentified columnals. Yeltyschewa, 1959, p. 234.
Paleozoic and Mesozoic: Russia

Unidentified crinoid columnal. Breitbach, 1966, p. 121, fig. 6.
Ordovician, Lederschiefer: Germany

Unidentified crinoid columnals. Boucot and others, 1958, p. 860, Pl. 3,
figs. 4, 5.
Lower Devonian, Bernardston Formation: United States: Massachusetts

Unidentified crinoid stem fragment. Genser, 1965, p. 155, fig. 3.
Carboniferous, Tournaisian: Germany

Unidentified stem fragment. Ivanova, 1958, Pl. 11, fig. 5; Pl. 14,
fig. 6; Figs. 4zh, 56a.
Middle Carboniferous: Russia: Moscow Syncline

Unnamed columnals. Arendt, 1961, p. 103, fig. 2.
Middle Carboniferous, Myachkovskian horizon: Russia: Moscow Basin

Unnamed columnals. Kleinschmidt, 1966, p. 707, figs. 1-9.
    Paleozoic: Karnic Alps

Appendix. New Genera Introduced Since 1942

Aacocrinus Bowsher, 1955
Abactinocrinus Laudon and Severson, 1953
Abathocrinus Strimple, 1963
Abatocrinus Lane, 1963
Abyssocrinus Strimple, 1963
Acylocrinus Kirk, 1947
Agathocrinus Schewtschenko, 1967
Aglaocrinus Strimple, 1961 = *Parethelocrinus*
Agnostocrinus Webster and Lane, 1967
Allosocrinus Strimple, 1949
Ambicocrinus Kirk, 1945
Ameliacrinus Kesling, 1968
Amonohexacrinus Schewtschenko, 1967
φ*Ampholenium* Moore and Jeffords, 1968
φ*Amsdenanteris* Moore and Jeffords, 1968
Anobasicrinus Strimple, 1961
φ*Anthinocrinus* Yeltyschewa and Sizova, *in* Stukalina, 1961
Anthracocrinus Strimple and Watkins, 1955
Anulocrinus Ramsbottom, 1960
φ*Apertocrinus* Stukalina, 1968
Aphelecrinus Kirk, 1944
φ*Apiastrum* Moore and Jeffords, 1968
Arroyocrinus Lane and Webster, 1966
Aryballocrinus Breimer, 1962
φ*Asteromischus* Moore and Jeffords, 1968
Asuturaecrinus Yakovlev, 1956
Ataxiacrinus Strimple, 1961 = *Tarachiocrinus*
Athabascacrinus Laudon, Parks, and Spreng, 1952
Atractocrinus Kirk, 1948
Aulodesocrinus Wright, 1942
Aviadocrinus Almela and Revilla, 1950 = *Cupressocrinites*
φ*Avicantus* Moore and Jeffords, 1968
Azygocrinus Lane, 1963
Balacrinus Ramsbottom, 1961

φ*Baryschyr* Moore and Jeffords, 1968
Bathronocrinus Strimple, 1962
φ*Bazaricrinus* Stukalina, 1968
Becharocrinus Termier and Termier, 1955
φ*Blothronagma* Moore and Jeffords, 1968
Bollandocrinus Wright, 1951
Bridgerocrinus Laudon and Severson, 1953
Bronaughocrinus Strimple, 1951
φ*Bystrowicrinus* Yeltyschewa, *in* Yeltyschewa and Stukalina, 1963
Cadiscocrinus Kirk, 1945
Caelocrinus Xu, 1962
Caldenocrinus Wright, 1946
Cantharocrinus Breimer, 1962
φ*Catagraphiocrinus* Stukalina, 1968
φ*Catholicorhachis* Moore and Jeffords, 1968
Celonocrinus Lane and Webster, 1966
Chiropinna Moore, 1962
φ*Cionerisma* Moore and Jeffords, 1968
φ*Clematidiscus* Moore and Jeffords, 1968
Cornucrinus Regnell, 1948
Corythocrinus Kirk, 1946
Corythocrinus Strimple, 1961 = *Tholiacrinus*
φ*Crenatames* Moore and Jeffords, 1968
Cribanocrinus Kirk, 1944
Crinobrachiatus Moore, 1962
Cusacrinus Bowsher, 1955
φ*Cyclocaudex* Moore and Jeffords, 1968
φ*Cyclocaudiculus* Moore and Jeffords, 1968
φ*Cyclocharax* Moore and Jeffords, 1968
φ*Cyclocion* Moore and Jeffords, 1968
φ*Cyclocrista* Moore and Jeffords, 1968
φ*Cyclocyclicus* Yeltyschewa, 1955
φ*Cyclomischus* Moore and Jeffords, 1968
φ*Cyclomonile* Moore and Jeffords, 1968
φ*Cyclopagoda* Moore and Jeffords, 1968
φ*Cyclopentagonalis* Yeltyschewa, 1964
φ*Cycloscapus* Moore and Jeffords, 1968
φ*Cyclostelechus* Moore and Jeffords, 1968
φ*Cylindrocauliscus* Moore and Jeffords, 1968
Cymbiocrinus Kirk, 1944
φ*Cyphostelechus* Moore and Jeffords, 1968

APPENDIX - NEW GENERA SINCE 1942

Cyrtocrinus Kirk, 1943 = *Cytidocrinus*
Cytidocrinus Kirk, 1944

φ*Decacrinus* Yeltyschewa, 1957
Denariocrinus Schmidt, 1941
φ*Dentiferocrinus* Stukalina, 1968
Derbiocrinus Wright, 1951
φ*Desidiamphidia* Moore and Jeffords, 1968
Desmacriocrinus Strimple, 1966
Dialutocrinus Wright, 1955
φ*Dianthicoeloma* Moore and Jeffords, 1968
Diatorocrinus Wright, 1955
φ*Dierocalipter* Moore and Jeffords, 1968
Dieuryocrinus Wright, 1954
φ*Dilanteris* Moore and Jeffords, 1968
φ*Dwortsowaecrinus* Stukalina, *in* Stukalina and Tuyutyan, 1967

Ectocrinus Wright, 1955
Eireocrinus Wright, 1951
Elpidocrinus Strimple, 1963
φ*Elytroclon* Moore and Jeffords, 1968
Epihalysiocrinus Arendt, 1965
Epipetschoracrinus Yakovlev, 1956
φ*Euloncherostigma* Moore and Jeffords, 1968
Eumorphocrinus Wright, 1955
φ*Eurax* Moore and Jeffords, 1968
φ*Exaesiodiscus* Moore and Jeffords, 1968
φ*Exedrodiscus* Moore and Jeffords, 1968
Exocrinus Strimple, 1949

φ*Fabalium* Moore and Jeffords, 1968
φ*Facetocrinus* Stukalina, 1968
φ*Fascicrinus* Stukalina, 1968
Fifeocrinus Wright, 1951
φ*Floricyclus* Moore and Jeffords, 1968
φ*Floripila* Moore and Jeffords, 1968
φ*Flucticharax* Moore and Jeffords, 1968
Forthocrinus Wright, 1942

Gaulocrinus Kirk, 1945
Glaukosocrinus Strimple, 1951
Globocrinus Washburn, 1968 (homonyn of *Globocrinus* Weller and others, 1920)
φ*Goniocion* Moore and Jeffords, 1968
φ*Goniostathmus* Moore and Jeffords, 1968

Graffhamicrinus Strimple, 1961
ɸ*Graphosterigma* Moore and Jeffords, 1968
ɸ*Gregariocrinus* Stukalina, 1968
Grenprisia Moore, 1962
Griphocrinus Kirk, 1945
Grypocrinus Strimple, 1963

ɸ*Hattinanteris* Moore and Jeffords, 1968
Heliosocrinus Strimple, 1951
ɸ*Heterostaurus* Moore and Jeffords, 1968
ɸ*Heterostelechus* Moore and Jeffords, 1968
Holcocrinus Kirk, 1945
Holynocrinus Bouška, 1948
Hosieocrinus Wright, 1952
ɸ*Hyperexochus* Moore and Jeffords, 1968
Hypermorphocrinus Arendt, 1968

Iberocrinus Sieverts-Doreck, 1951
Idosocrinus Wright, 1954
ɸ*Ilematerisma* Moore and Jeffords, 1968
Isoallagecrinus Strimple, 1966
Ivanovicrinus Yakovlev, *in* Arendt and Hecker, 1964 = *Trautscholdicrinus*
Jaekelicrinus Yakovlev, 1949
Jimbacrinus Teichert, 1954
Kalpidocrinus Goldring, 1954
ɸ*Kasachstanocrinus* Sizova, *in* Schewtschenko, 1966
Kopficrinus Goldring, 1954
Kophinocrinus Goldring, 1954
ɸ*Kstutocrinus* Schewtschenko, 1966
ɸ*Kuzbassocrinus* Yeltyschewa, 1957 = *Melocrinites* Goldfuss, 1826
Lampadosocrinus Strimple and Koenig, 1956
ɸ*Lamprosterigma* Moore and Jeffords, 1968
ɸ*Laudonomphalus* Moore and Jeffords, 1968
ɸ*Leptocarphium* Moore and Jeffords, 1968
Liomolgocrinus Strimple, 1963
ɸ*Lomalegnum* Moore and Jeffords, 1968

Maligneocrinus Laudon, Parks, and Spreng, 1952
ɸ*Malovicrinus* Stukalina, 1968
Mantikosocrinus Strimple, 1951
ɸ*Medinecrinus* Stukalina, 1965
ɸ*Mediocrinus* Stukalina, 1965
Megaliocrinus Moore and Laudon, 1942
Metacatillocrinus Moore and Strimple, 1942

APPENDIX - NEW GENERA SINCE 1942

Metacromyocrinus Strimple, 1961
Metaindocrinus Strimple, 1966
Metallagecrinus Strimple, 1966
Metaperimestocrinus Strimple, 1961
Miracrinus Bowsher, 1953
Moapacrinus Lane and Webster, 1966
Monstrocrinus Schmidt, 1941
φ*Mooreanteris* Miller, *in* Moore and Jeffords, 1968
Mooreocrinus Wright and Strimple, 1945
φ*Musivocrinus* Termier and Termier, 1958

Nactocrinus Kirk, 1947
Neoarchaeocrinus Strimple and Watkins, 1955
Nevadacrinus Lane and Webster, 1966
φ*Nothrosterigma* Moore and Jeffords, 1968
Nunnacrinus Bowsher, 1955

φ*Obuticrinus* Yeltyschewa and Stukalina, 1963
Ochlerocrinus Strimple, 1963
Oenochoacrinus Breimer, 1962
Ollulocrinus Bouška, 1956 = *Pisocrinus* (*Parapisocrinus*)
Opsiocrinus Kier, 1952
Orocrinus Sieverts-Doreck, 1951

Paianocrinus Strimple, 1951
φ*Pandocrinus* Stukalina, 1965
Parabotryocrinus Yakovlev, 1941
Paracosmetocrinus Strimple, 1967
Paracromyocrinus Strimple, 1966
Parahexacrinus Schewtschenko, 1967
Paramelocrinus Ubaghs, 1958
Parapernerocrinus Yakovlev, 1949
Pararchaeocrinus Strimple and Watkins, 1955
Parastachyocrinus Wanner, 1949
Parazophocrinus Strimple, 1963
Parethelocrinus Strimple, 1961
φ*Particrinus* Stukalina, 1968
Pedinocrinus Wright, 1951
Peniculocrinus Moore, 1962
φ*Pennatocrinus* Stukalina, 1968
φ*Pentaculiscus* Moore and Jeffords, 1968
φ*Pentagonomischus* Moore and Jeffords, 1968
φ*Pentagonopentagonalis* Yeltyschewa, 1955

φ*Pentagonopternix* Moore and Jeffords, 1968
φ*Pentagonostaurus* Moore and Jeffords, 1968
φ*Pentagonostipes* Moore and Jeffords, 1968
φ*Pentamerostela* Moore and Jeffords, 1968
φ*Pentaridica* Moore and Jeffords, 1968
Pentececrinus Koenig and Niewoehner, 1959
Permiocrinus Wanner, 1949
Pernerocrinus Bouška, 1950
φ*Petalerisma* Moore and Jeffords, 1968
Pimlicocrinus Wright, 1943
Pisocrinus (Parapisocrinus) Mu, 1954
Pithocrinus Kirk, 1945
φ*Platycion* Moore and Jeffords, 1968
φ*Platyclonus* Moore and Jeffords, 1968
φ*Platyparallelus* Moore and Jeffords, 1968
φ*Platyplateium* Moore and Jeffords, 1968
φ*Platystela* Moore and Jeffords, 1968
Plemnocrinus Kirk, 1946
φ*Plummeranteris* Moore and Jeffords, 1968
Plummericrinus Moore and Laudon, 1943
φ*Plussacrinus* Yeltyschewa, 1957
φ*Podoliocrinus* Yeltyschewa, 1957
Polusocrinus Strimple, 1951
Polygonocrinus Strimple, 1961
Pottsicrinus Jillson, 1960
φ*Preptopremnum* Moore and Jeffords, 1968
Proctothylacocrinus Kier, 1952
Prohexacrinus Yakovlev, 1946
Pygmaeocrinus Bouška, 1946

Quantoxocrinus Webby, 1965
Quiniocrinus Schmidt, 1941

Ramseyocrinus Bates, 1968
φ*Ramulicrinus* Stukalina, 1968
Ramulocrinus Laudon, Parks, and Spreng, 1952
Rhabdocrinus Wright, 1944
φ*Rhysocamax* Moore and Jeffords, 1968

φ*Salagastiana* Rusconi, 1952
φ*Scelidiopternix* Moore and Jeffords, 1968
Schedexocrinus Strimple, 1961
Scotiacrinus Wright, 1945
φ*Sidericrinus* Stukalina, 1968

Sigambrocrinus Schmidt, 1941
Situlacrinus Breimer, 1962
φ*Sokolovicrinus* Yeltyschewa, 1968
φ*Squameocrinus* Stukalina, 1968
Stamnocrinus Breimer, 1962
Stegocrinus Sieverts-Doreck, 1962
Stenopecrinus Strimple, 1961
φ*Stiberostaurus* Moore and Jeffords, 1968
Stiptocrinus Kirk, 1946
Streblocrinus Koenig and Meyer, 1965
Sunwaptacrinus Laudon, Parks, and Spreng, 1952
Synarmocrinus Lane, 1964

Tarachiocrinus Strimple, 1962
Tebagacrinus Termier and Termier, 1958
Telikosocrinus Strimple, 1951
φ*Tetragonocrinus* Yeltyschewa, 1964
Tetrapleurocrinus Wanner, 1942
Theloreus Moore, 1962
Tholiacrinus Strimple, 1962 = *Endelocrinus*
Thyridocrinus Kirk, 1944
Trampidocrinus Lane and Webster, 1966
Trautscholdicrinus Yakovlev and Ivanov, *in* Yakovlev, 1939
Triacrinus Termier and Termier, 1950 = *Becharocrinus*
Triboloporus Kesling and Paul, 1968
Triceracrinus Bramlette, 1943
Trichinocrinus Moore and Laudon, 1943
Tropiocrinus Kirk, 1947
Tunguskocrinus Arendt, 1963
Tyrieocrinus Wright, 1945

Ureocrinus Wright and Strimple, 1945

φ*Wanakastaurus* Moore and Jeffords, 1968

φ*Zeravschanocrinus* Schewtschenko, 1966
Zeusocrinus Strimple, 1961
Zostocrinus Kirk, 1948
Zygotocrinus Kirk, 1943

WITHDRAWN